# Mathematical Engineering

**Series Editors**

Jörg Schröder, Institute of Mechanics, University of Duisburg-Essen, Essen, Germany

Bernhard Weigand, Institute of Aerospace Thermodynamics, University of Stuttgart, Stuttgart, Germany

Today, the development of high-tech systems is unthinkable without mathematical modeling and analysis of system behavior. As such, many fields in the modern engineering sciences (e.g. control engineering, communications engineering, mechanical engineering, and robotics) call for sophisticated mathematical methods in order to solve the tasks at hand.

The series Mathematical Engineering presents new or heretofore little-known methods to support engineers in finding suitable answers to their questions, presenting those methods in such manner as to make them ideally comprehensible and applicable in practice.

Therefore, the primary focus is—without neglecting mathematical accuracy—on comprehensibility and real-world applicability.

To submit a proposal or request further information, please use the PDF Proposal Form or contact directly: Dr. Thomas Ditzinger (thomas.ditzinger@springer.com)

More information about this series at http://www.springer.com/series/8445

Gianpietro Elvio Cossali · Simona Tonini

# Drop Heating and Evaporation: Analytical Solutions in Curvilinear Coordinate Systems

Gianpietro Elvio Cossali
Department of Engineering
and Applied Sciences
University of Bergamo
Dalmine, Bergamo, Italy

Simona Tonini
Department of Engineering
and Applied Sciences
University of Bergamo
Dalmine, Bergamo, Italy

ISSN 2192-4732　　　　　　　　ISSN 2192-4740　(electronic)
Mathematical Engineering
ISBN 978-3-030-49276-2　　　　ISBN 978-3-030-49274-8　(eBook)
https://doi.org/10.1007/978-3-030-49274-8

This Springer imprint is published by the registered company Springer Nature Switzerland AG
The registered company address is: Gewerbestrasse 11, 6330 Cham, Switzerland

# Preface

The evaporation of liquid drops in a gaseous atmosphere is an everyday phenomenon, which is of importance in many industrial and civil applications. Despite an apparent simplicity, the study of drop heating and evaporation involves disciplines like transport phenomena in continua, fluid mechanics, interfacial phenomena and two-phase flows. Such a complexity, together with the increased availability of efficient numerical simulation methods and computer power, is the main reason for the growth in the number of numerical studies on this phenomenon in the last decades. On the other hand, analytical approaches provide an understanding of the underlying physics that cannot be reached by even the most accurate numerical simulation, despite the simplifications usually necessary to find analytical solutions to the differential equations that describe the many processes involved. The analytical modelling of drop heating and evaporation can be empowered by the use of mathematical tools, which are not so commonly used in engineering, and among them, the use of curvilinear coordinates often allows to simplify problems of apparently geometric complexity.

Our past experience in dealing with spray flows and droplet dynamics and with modelling of drop heating and evaporation has convinced us that there still exists a gap between the available mathematical tools that applied mathematics provides and their use in engineering research, particularly in the field of drop heating and evaporation, and this is the main motivation of this book.

The book is divided into three parts, dealing, respectively, with mathematical tools, conservation and constitutive equations and application to modelling of evaporating drops. The idea is to take the reader from mathematics through physical description to engineering applications, trying to show the effective power of applied mathematics when properly adapted to the real world.

The first part of this book deals with fundamental mathematical tools like tensor calculus and differential geometry of surfaces, as well as with generalised series and Sturm-Liouville theory. Since the book is intended for research engineers, we decided to present these topics in a way that may be accessible rather than rigorous, taking care to avoid a strict formalisation of the subjects, trying to maintain a connection with the real world of engineering and applications. A chapter is entirely devoted to orthogonal curvilinear systems of coordinates and to an accurate description of sixteen different systems, with explicit forms of the main differential operators (gradient, divergence, curl, Laplacian) and transformation matrices.

The second part is devoted to the differential equations used to describe the mass, chemical species, momentum and energy conservation in continua and with their generalisation to general curvilinear coordinates. Constitutive equations are used to characterise the material properties, and they are presented with particular attention to the case of multicomponent mixtures.

The third part describes the use of the developed tools to modelling heating and evaporation of liquid drops. Most of the reported examples come from direct experience, but also classical approaches from the literature, like for example heat conduction in spheroids, are revisited in the light of the tools described in the previous parts.

Professor Bernhard Weigand has been an enthusiastic supporter in the writing of this book and we are indebted to him, for his help in reading and correcting the drafts of the book and for his invaluable advice and fruitful discussions.

Dalmine, Italy                                                              Gianpietro Elvio Cossali
                                                                                      Simona Tonini

# Contents

**Part I Mathematical Tools**

**1 Introduction to Tensor Analysis** ........................... 3
   1.1    Space and Transformation of Coordinates. ................. 3
   1.2    Scalars, Vectors and Tensors ......................... 5
   1.3    Covariant, Contravariant and Mixed Tensors .............. 7
   1.4    Algebraic Operations on Tensors ...................... 10
   1.5    The Metric Tensor ................................ 14
   1.6    Lowering and Raising Indices ........................ 16
   1.7    The Angles Between Two Vectors...................... 18
   1.8    Relative Tensors ................................. 19
   1.9    Geometrical Interpretation of Tensors. .................. 21
   1.10   The Volume Element ............................. 24
   1.11   Physical Components of Tensors ...................... 26
   1.12   Vector and Tensor Derivation ........................ 29
   1.13   The Intrinsic Derivative of Tensors .................... 35
   1.14   Differential Operators in General Curvilinear Coordinates ..... 36
   1.15   Orthogonal Curvilinear Coordinates .................... 39
   References ........................................... 40

**2 Elements of Differential Geometry of a Surface** ............... 41
   2.1    Parametric, Cartesian and Implicit Description of a Surface.... 41
   2.2    Surface Coordinates and Surface Tensors ................ 43
   2.3    Relation with the Ambient Space...................... 45
   2.4    The Normal Vector and the Curvature Tensor ............ 49
   2.5    Length and Area on a Surface ........................ 53
   2.6    The Fundamental Forms ............................ 57
   2.7    The Curvatures of a Surface ......................... 58
       2.7.1    Mean Curvature ............................ 61
       2.7.2    Gaussian Curvature ......................... 62

2.8    Alternative Relations to Calculate the Curvatures
       of a Surface .......................................    65
2.9    The Moving Surface ................................    67
       2.9.1   The Invariant Time Derivative ................    68
       2.9.2   The Normal Velocity of a Moving Surface ........    70
References ...............................................    71

**3    Separability of PDE** ......................................    73
3.1    Definition of Separability ..........................    73
3.2    The Stäckel Matrix ................................    74
3.3    The Conditions for Separability ......................    75
       3.3.1   Conditions for Simple Separability .............    76
       3.3.2   Conditions for R-Separability ................    79
3.4    Non-unicity of the Stäckel Matrix ...................    84
3.5    Finding a Stäckel Matrix ..........................    85
References ...............................................    88

**4    Orthogonal Curvilinear Coordinate Systems** ...................    89
4.1    Orthogonal Coordinate Systems in Euclidean 3D-space ......    89
       4.1.1   Transformation Matrices in Rotational Coordinate
               Systems ....................................    93
4.2    Gradient, Divergence, Curl, Laplacian in Orthogonal
       Curvilinear Coordinates ...........................    96
4.3    Laplace and Helmholtz Equations ....................    99
4.4    List of Curvilinear Coordinate Systems ...............   100
       4.4.1   Circular-Cylinder Coordinate System ............   101
       4.4.2   Elliptic-Cylinder Coordinate System ............   103
       4.4.3   Parabolic-Cylinder Coordinate System ...........   105
       4.4.4   Spherical Coordinate System ..................   107
       4.4.5   Prolate Spheroidal Coordinate System ...........   111
       4.4.6   Oblate Spheroidal Coordinate System ...........   115
       4.4.7   Parabolic Coordinate System ..................   119
       4.4.8   Conical Coordinate System ...................   121
       4.4.9   Ellipsoidal Coordinate System .................   124
       4.4.10  Paraboloidal Coordinate System ...............   127
       4.4.11  Tangent-Sphere Coordinate System .............   130
       4.4.12  Cardioid Coordinate System ..................   132
       4.4.13  Bispherical Coordinate System ................   134
       4.4.14  Toroidal Coordinate System ..................   137
       4.4.15  Inverse Prolate Spheroidal Coordinate System .......   139
       4.4.16  Inverse Oblate Spheroidal Coordinate System .......   144
References ...............................................   148

**5   Sturm–Liouville Problems** . . . . . . . . . . . . . . . . . . . . . . . . . . . . . . . . .   149
   5.1   Orthogonal Functions . . . . . . . . . . . . . . . . . . . . . . . . . . . . . . . .   149
      5.1.1   Linear Spaces, Hilbert Space and the Inner
           Product . . . . . . . . . . . . . . . . . . . . . . . . . . . . . . . . .   149
      5.1.2   Orthogonality of Functions . . . . . . . . . . . . . . . . . . . .   151
   5.2   Orthogonal Sequences . . . . . . . . . . . . . . . . . . . . . . . . . . . . . . .   154
   5.3   Generalised Fourier Series . . . . . . . . . . . . . . . . . . . . . . . . . . . .   160
   5.4   Elements of Sturm–Liouville Theory . . . . . . . . . . . . . . . . . . . .   168
      5.4.1   Eigenvalue Problems . . . . . . . . . . . . . . . . . . . . . . . . .   171
      5.4.2   The Sturm–Liouville System . . . . . . . . . . . . . . . . . . . .   174
      5.4.3   The Periodic Sturm–Liouville Problem . . . . . . . . . . . .   176
      5.4.4   The Singular Sturm–Liouville Problem . . . . . . . . . . . .   178
   References . . . . . . . . . . . . . . . . . . . . . . . . . . . . . . . . . . . . . . . . . . . .   180

**Part II   Mass, Momentum and Energy Conservation Equations
in Curvilinear Coordinates**

**6   Conservation Equations** . . . . . . . . . . . . . . . . . . . . . . . . . . . . . . . . .   185
   6.1   The Divergence Theorem . . . . . . . . . . . . . . . . . . . . . . . . . . . . .   185
   6.2   The Transport Theorem . . . . . . . . . . . . . . . . . . . . . . . . . . . . . . .   187
   6.3   The Conservation Equations . . . . . . . . . . . . . . . . . . . . . . . . . . .   189
      6.3.1   The Mass Conservation Equation . . . . . . . . . . . . . . . .   189
      6.3.2   The Species Conservation Equations . . . . . . . . . . . . . .   190
      6.3.3   The Momentum Conservation Equation . . . . . . . . . . . .   195
      6.3.4   The Energy Conservation Equation . . . . . . . . . . . . . . .   197
   6.4   The Balances at Interfaces . . . . . . . . . . . . . . . . . . . . . . . . . . . . .   199
      6.4.1   The Transport Theorem in the Presence
           of Interfaces . . . . . . . . . . . . . . . . . . . . . . . . . . . . . . .   200
      6.4.2   The Jump Mass Balance at Interface . . . . . . . . . . . . . .   201
      6.4.3   The Jump Species Balance at the Interface . . . . . . . . . .   202
      6.4.4   The Jump Momentum Balance at the Interface . . . . . . .   204
      6.4.5   The Jump Energy Balance at the Interface . . . . . . . . . .   205
   References . . . . . . . . . . . . . . . . . . . . . . . . . . . . . . . . . . . . . . . . . . . .   206

**7   Introduction to Constitutive Equations** . . . . . . . . . . . . . . . . . . . . .   207
   7.1   Generalised Maxwell–Stefan Equations . . . . . . . . . . . . . . . . . . .   207
   7.2   The Diffusional Forces . . . . . . . . . . . . . . . . . . . . . . . . . . . . . . .   209
   7.3   Components of the Diffusive Fluxes . . . . . . . . . . . . . . . . . . . . . .   211
   7.4   Dilute Gas Mixtures . . . . . . . . . . . . . . . . . . . . . . . . . . . . . . . . .   212
   7.5   Binary Mixtures and Fick's Law . . . . . . . . . . . . . . . . . . . . . . . .   213
   7.6   Fick's Law in Multi-component Mixtures . . . . . . . . . . . . . . . . . .   215
   7.7   Constitutive Equations for the Heat Flux . . . . . . . . . . . . . . . . . .   218
   7.8   Constitutive Equations for the Deviatoric Stress Tensor . . . . . . .   220
   References . . . . . . . . . . . . . . . . . . . . . . . . . . . . . . . . . . . . . . . . . . . .   222

**8    Conservation and Constitutive Equations in Curvilinear**
      **Coordinates** . . . . . . . . . . . . . . . . . . . . . . . . . . . . . . . . . . . . . . . . . . .   225
      8.1    The Conservation Equations  . . . . . . . . . . . . . . . . . . . . . . . . . . .   225
             8.1.1    The Mass and Species Conservation Equations  . . . . . .   226
             8.1.2    The Momentum Conservation Equation . . . . . . . . . . . .   228
             8.1.3    The Energy Conservation Equation . . . . . . . . . . . . . . .   230
      8.2    The Constitutive Equations . . . . . . . . . . . . . . . . . . . . . . . . . . . .   231
             8.2.1    Example: The Steady State Energy Equation . . . . . . . .   232

**9    Modelling Heat and Mass Transfer from an Evaporating**
      **Drop** . . . . . . . . . . . . . . . . . . . . . . . . . . . . . . . . . . . . . . . . . . . . . . . . .   235
      9.1    Mass Transfer . . . . . . . . . . . . . . . . . . . . . . . . . . . . . . . . . . . . . . .   235
      9.2    Energy Transfer . . . . . . . . . . . . . . . . . . . . . . . . . . . . . . . . . . . . . .   240
             9.2.1    The Heat Transfer inside a Single-Component
                      Drop . . . . . . . . . . . . . . . . . . . . . . . . . . . . . . . . . . . . . . .   242
             9.2.2    The Quasi-steady Single-Component Drop
                      Evaporation . . . . . . . . . . . . . . . . . . . . . . . . . . . . . . . . .   243
             9.2.3    Time-Dependent Integral Balances on a Single
                      Component Drop . . . . . . . . . . . . . . . . . . . . . . . . . . . . .   245
             9.2.4    The Effective Temperature . . . . . . . . . . . . . . . . . . . . . .   248
      References . . . . . . . . . . . . . . . . . . . . . . . . . . . . . . . . . . . . . . . . . . . . . .   249

**Part III    Analytical Modelling of Drop Heating and Evaporation**

**10   One-Dimensional Modelling of Drop Heating and Evaporation**
      **Under Steady Conditions** . . . . . . . . . . . . . . . . . . . . . . . . . . . . . . . . .   253
      10.1   Analytical Modelling of the Heat and Mass Transfer
             from a Single-Component Drop in a Gaseous Mixture . . . . . . .   253
             10.1.1   The Analytical Solution . . . . . . . . . . . . . . . . . . . . . . .   259
             10.1.2   The Species and Energy Fluxes. . . . . . . . . . . . . . . . . .   268
             10.1.3   The Effect of Temperature Dependence
                      of the Thermo-Physical Properties . . . . . . . . . . . . . . .   269
             10.1.4   The Effect of Drop Shape . . . . . . . . . . . . . . . . . . . . . .   273
      10.2   The Spherical Drop . . . . . . . . . . . . . . . . . . . . . . . . . . . . . . . . . .   274
             10.2.1   Effect of Convection . . . . . . . . . . . . . . . . . . . . . . . . . .   276
      10.3   The Spheroidal Drop . . . . . . . . . . . . . . . . . . . . . . . . . . . . . . . . .   278
             10.3.1   The Prolate Spheroidal Drop . . . . . . . . . . . . . . . . . . . .   279
             10.3.2   The Oblate Spheroidal Drop . . . . . . . . . . . . . . . . . . . .   281
             10.3.3   The Vapour Fluxes from Spheroidal Drops . . . . . . . . .   283
      10.4   The Triaxial Ellipsoidal Drop . . . . . . . . . . . . . . . . . . . . . . . . . .   287
             10.4.1   The Fluxes and Evaporation Rates . . . . . . . . . . . . . . .   291
      References . . . . . . . . . . . . . . . . . . . . . . . . . . . . . . . . . . . . . . . . . . . . . .   294

**11   Two- and Three-Dimensional Modelling of Heating
       and Evaporation Under Steady Conditions** ................... 297
       11.1   A Pair of Interacting Spherical Drops. .................. 297
              11.1.1   The Vapour and Heat Fluxes .................. 302
       11.2   The Inverse Spheroidal Drops ....................... 304
              11.2.1   Effect of Drop Surface Curvature ............... 312
       11.3   The Sessile Drop ............................... 317
              11.3.1   The Boundary Conditions .................. 320
              11.3.2   Explicit Solutions ....................... 321
       11.4   Drop Evaporation with Non-uniform Boundary Conditions .... 322
              11.4.1   The Vapour Fluxes ...................... 327
       References ......................................... 328

**12   Drop Evaporation Under Unsteady Conditions** ................. 329
       12.1   Oscillating Drops .............................. 329
              12.1.1   Evaporation of an Oscillating Spheroidal Drop ....... 333
       12.2   Unsteady Heat and Mass Transfer in a Shrinking Spherical
              Drop ..................................... 336
       12.3   Unsteady Heat and Mass Transfer Outside a Shrinking
              Drop ..................................... 341
              12.3.1   Initial and Boundary Conditions ............... 343
              12.3.2   Energy and Mass Balances over the Liquid Drop ..... 345
              12.3.3   A Comparison with a Quasi-steady Solution ....... 347
       12.4   Transient Heat Conduction in Ellipsoids of Revolution ....... 348
              12.4.1   The Energy Equation in Spheroidal Coordinates...... 349
              12.4.2   The Niven Solution for the Energy Equation ........ 351
              12.4.3   The Orthogonality of $T_{nk}$ ..................... 353
       References ......................................... 355

**13   Multi-component Drop Evaporation** ...................... 357
       13.1   A Multi-component Model for Spherical Drops ............ 357
       13.2   Multi-component Drop Evaporation Solving
              the Stefan–Maxwell Equations........................ 363
       13.3   Drop Evaporation from Multi-component Ellipsoidal
              Drops..................................... 366
       References ......................................... 369

**Appendix: Special Functions** ................................ 371
       A.1 The Gamma, Digamma and Beta Functions and Elliptic
            Integrals..................................... 371
       A.2 The Bessel Functions............................. 377
       A.3 The Legendre Functions .......................... 386
       A.4 Spheroidal Wave Functions........................ 394
       A.5 The Hypergeometric Function ...................... 396
       References ......................................... 398

# Part I
# Mathematical Tools

The first part of this book is devoted to the mathematical tools, which are necessary when dealing with analytical modelling of heating and evaporation of drops. Apart of ideal cases, drops are inherently non-spherical, since gravity and drop-gas interaction tend to deform the drop against the restoring action of surface tension. The solution of the partial differential conservation equations is facilitated choosing proper coordinate systems where the problem may simplify; for example, if the drop has an oblate shape the use of an oblate spheroidal coordinate system may reduce the dimensionality of the differential problem.

It is then important to reformulate the differential problem in different coordinate systems and this is why the first chapter is devoted to tensor calculus, the ideal tool to generalise physical statements (like conservation equations) to any coordinate system. In two-phase systems interfaces play a fundamental role since they connect two regions having different properties. Interfaces may be schematised as singular surfaces and their geometrical characteristics (like curvatures) may play a fundamental role when modelling the physics of evaporating drops; the second chapter is meant to introduce the reader to the fundamentals of differential geometry of surfaces, taking full advantage of the tensor calculus described in the first chapter. The third chapter treats the problem of the PDE separability, a powerful tool to search for analytical solutions and the fourth chapter analyses with a certain detail the so-called orthogonal curvilinear coordinates, reporting a description of sixteen different systems. The fifth and last chapter introduces the most important elements of functional analysis (like the concept of orthogonality of functions and that of generalised Fourier series) necessary to the discussion of Sturm-Liouville theory for second order ordinary differential equations.

The mathematical tools discussed in this part are meant to be instrumental to the analytical modelling, therefore the focus is more on their usability, examples are often used and sometimes demonstrations are skipped re-directing the reader to specialised books.

# Chapter 1
# Introduction to Tensor Analysis

Tensor theory provides a remarkably concise mathematical framework for the formalisation of problems in many branches of physics and engineering. The strength of tensor calculus lays on the fact that any tensor equation can be written in exactly the same way in any coordinate system, and this allows avoiding the choice of any particular coordinate system when formulating a physical problem. In this first chapter, the basic elements of tensor algebra and tensor calculus will be explored; as in the rest of the book, proofs will be given for many but not all statements, and for the missing ones the interested reader is re-directed to classical books on the subject.

## 1.1 Space and Transformation of Coordinates

The position of a point in a 3D space is defined by a set of three values $x^1, x^2, x^3$, which are called coordinates. The collection of all points, given possible restrictions on the values of $x^1, x^2, x^3$, is what we can call a *space*. Although tensor calculus was born to deal in a general way with spaces of any dimension and general characteristics, we will restrict the following discussion to 3D Euclidean spaces. However, surfaces embedded into this space are in general 2D non-Euclidean spaces and that is why we shall consider general coordinate transformations and become acquainted with the general rules of tensorial calculus.

Consider the following set of single valued, continuous and differentiable functions

$$u^j = u^j \left( x^1, x^2, x^3 \right) \quad (j = 1 \ldots 3) \tag{1.1}$$

where for convenience of notation the same symbol $u^j$ is used to indicate the variable and the function. These equations define a coordinate transformation, from the original coordinates system $x^1, x^2, x^3$ to the new coordinate system $u^1, u^2, u^3$. The Jacobian determinant (named after the German mathematician Carl Gustav Jacob Jacobi, 1804–1851) of this transformation is defined as

© Springer Nature Switzerland AG 2021
G. E. Cossali and S. Tonini, *Drop Heating and Evaporation: Analytical Solutions in Curvilinear Coordinate Systems*, Mathematical Engineering,
https://doi.org/10.1007/978-3-030-49274-8_1

$$J = \begin{vmatrix} \frac{\partial u^1}{\partial x^1} & \frac{\partial u^1}{\partial x^2} & \frac{\partial u^1}{\partial x^3} \\ \frac{\partial u^2}{\partial x^1} & \frac{\partial u^2}{\partial x^2} & \frac{\partial u^2}{\partial x^3} \\ \frac{\partial u^3}{\partial x^1} & \frac{\partial u^3}{\partial x^2} & \frac{\partial u^3}{\partial x^3} \end{vmatrix} = \frac{\partial \left( u^1, u^2, u^3 \right)}{\partial \left( x^1, x^2, x^3 \right)} \tag{1.2}$$

the notation on the right hand side of Eq. (1.2) is sometimes used since it may remind some useful properties of the Jacobian determinants, among which

$$\frac{\partial \left( u^1, u^2, u^3 \right)}{\partial \left( x^1, x^2, x^3 \right)} = \left[ \frac{\partial \left( x^1, x^2, x^3 \right)}{\partial \left( u^1, u^2, u^3 \right)} \right]^{-1} \tag{1.3a}$$

$$\frac{\partial \left( u^1, x^2, x^3 \right)}{\partial \left( x^1, x^2, x^3 \right)} = \left( \frac{\partial u^1}{\partial x^1} \right)_{x^2, x^3} \tag{1.3b}$$

$$\frac{\partial \left( u^1, u^2, u^3 \right)}{\partial \left( x^1, x^2, x^3 \right)} = \frac{\partial \left( u^1, u^2, u^3 \right)}{\partial \left( v^1, v^2, v^3 \right)} \frac{\partial \left( v^1, v^2, v^3 \right)}{\partial \left( x^1, x^2, x^3 \right)} \tag{1.3c}$$

If the Jacobian of a transformation does not vanish, Eq. (1.1) can be solved for $x^k$ to yield

$$x^k = x^k \left( u^1, u^2, u^3 \right); \quad k = 1, 2, 3 \tag{1.4}$$

which is the inverse coordinate transformation. The Jacobian of this coordinate transformation is

$$J' = \frac{\partial \left( x^1, x^2, x^3 \right)}{\partial \left( u^1, u^2, u^3 \right)} = \left[ \frac{\partial \left( u^1, u^2, u^3 \right)}{\partial \left( x^1, x^2, x^3 \right)} \right]^{-1} = J^{-1} \tag{1.5}$$

Let now suppose to perform a further coordinate transformation defined as

$$v^j = v^j \left( u^1, u^2, u^3 \right) \tag{1.6}$$

Substitution of Eq. (1.1) into (1.6) would yield the transformation from the system $x^1, x^2, x^3$ to the system $v^1, v^2, v^3$, and the Jacobian of this last transformation can be obtained multiplying the Jacobians of the two consecutive transformations, i.e.

$$\frac{\partial \left( v^1, v^2, v^3 \right)}{\partial \left( x^1, x^2, x^3 \right)} = \frac{\partial \left( v^1, v^2, v^3 \right)}{\partial \left( u^1, u^2, u^3 \right)} \frac{\partial \left( u^1, u^2, u^3 \right)}{\partial \left( x^1, x^2, x^3 \right)} \tag{1.7}$$

where the equality comes from the properties of the Jacobian determinants.

The use of indices for the coordinates and, as we will see, for vector and tensor components, implies often the summation over such indices, and equations may become quite bulky. To mitigate this problem we will adopt the so-called *summation convention*.

**Def.** *Summation convention.*

Consider a term in an equation that contains quantities with suffixes, either subscripts or superscripts, the following convention will be adopted unless otherwise specified: (i) if the suffix is repeated, summation with respect to that index, over the range 1 to N is understood; (ii) if the suffix is not repeated, it is assumed to take all the values from 1 to N; here N is the dimension of the space. In some cases, a capital letter may be used as a suffix, then those suffixes are excluded from the summation convention.

As an example of application, consider the so-called *Kronecker delta* (named after the German mathematician Leopold Kronecker, 1823–1891), a symbol $\delta_j^k$ defined as

$$\delta_j^k = \begin{cases} 1 & \text{if } k = j \\ 0 & \text{if } k \neq j \end{cases} \tag{1.8}$$

which will be used quite often in the following. It is obvious that $\frac{\partial x^k}{\partial x^j} = \delta_j^k$ but since

$$\delta_j^k = \frac{\partial x^k}{\partial x^j} = \sum_{l=1}^{3} \frac{\partial x^k}{\partial u^l} \frac{\partial u^l}{\partial x^j} = \frac{\partial x^k}{\partial u^l} \frac{\partial u^l}{\partial x^j} \tag{1.9}$$

the two matrices

$$a_l^k = \frac{\partial x^k}{\partial u^l}; \quad b_j^l = \frac{\partial u^l}{\partial x^j} \tag{1.10}$$

are one the inverse of the other, a results that will become useful later. In the last equality of Eq. (1.9) the summation convention has been used.

## 1.2  Scalars, Vectors and Tensors

Coordinate transformations may have effect on the objects used to describe physical quantities, like temperature, velocities, stresses, etc. When a transformation from a coordinate system $(x^j)$ to another $(u^j)$ is performed, we will indicate the object in the second coordinate system by the same letter used in the first one, adding a prime, i.e. if $A$ is the object in the original system, $A'$ is the same object in the new system. Consider first a quantity that is defined at any coordinate value by a single number, such as the temperature of a solid body, or the pressure in a fluid. It is understood that a coordinate transformation does not affect the value of such quantity. Quantities like these are called *scalar invariants* (or sometime simply *invariants* or *scalars*) and will be represented by a letter without suffixes, and the following trivial transformation rule holds

$$T' = T \tag{1.11}$$

Beside the scalar quantities, other more complex objects are used to represent physical quantities. For example: velocity $(V_j)$, heat flux $(q_j)$, force $(F_j)$ are represented by objects, called vectors, that can be defined by an ordered set of numbers, called *components*, i.e.

$$\mathbf{V} = [V_1, V_2, V_3] \tag{1.12a}$$

$$\mathbf{q} = [q_1, q_2, q_3] \tag{1.12b}$$

$$\mathbf{F} = [F_1, F_2, F_3] \tag{1.12c}$$

The most important property of these kind of objects is the *transformation rule* that relates the values of the components in a coordinate system to those in the other one. Details on this transformation rule will be discussed in the following subsection, but its most striking characteristics is the fact that such transformation is linear and homogeneous (sometimes called *affine*), i.e.

$$V'_j = a^k_j V_k \tag{1.13}$$

This characteristic is quite important since a vanishing vector (i.e. a vector that has all nil components) in a system will remain a vanishing vector also in the new system. A fundamental consequence is that a physical law that can be written as a vanishing vector will maintain its form in any coordinate system. For example, Newton's second law of dynamics states that the acceleration of an object caused by a net force is proportional to the net force and inversely proportional to its mass: i.e. $F_j = ma_j$. The same equation can be written as

$$R_j = F_j - ma_j = 0 \tag{1.14}$$

which states that the vector $R_j$ obtained summing (algebraically) the vector $F_j$ (net force) and the vector $-ma_j$ (which is obtained by multiplying the vector $a_j$ by the scalar $m$) vanishes. The affine transformation (1.13) assures that this happens in any coordinate system, thus Newton's second law has the same form in each system of coordinates, although the components of $F_j$ and $a_j$ may be different in different coordinate systems.

Physical objects with added complexity can be defined following this path. It has been shown that a vector in a space of dimension $N$ has $N$ components (i.e. it is defined by an ordered set of $N$ numbers and by the affine transformation rule); from a couple of such objects, say $A_k$ and $B_j$, we can define a new object defined by the $N^2$ numbers $A_k B_j$. Since the change of coordinate system yields, for each vector

$$A'_k = a^p_k A_p; \quad B'_j = a^q_j B_q \tag{1.15}$$

the new object consequently transforms as

$$A'_k B'_j = \left(a^p_k a^q_j\right) A_p B_q \tag{1.16}$$

An object that transforms following the rule (1.16) is called a *tensor*, and the components are expressed by a letter with two indices (say $T_{kj}$) , i.e.

$$\mathbf{T} = \begin{bmatrix} T_{11} & T_{12} & T_{13} \\ T_{21} & T_{22} & T_{23} \\ T_{31} & T_{32} & T_{33} \end{bmatrix} \tag{1.17}$$

and the transformation rule is:

$$T'_{kj} = \left( a^p_k a^q_j \right) T_{pq} \tag{1.18}$$

For convenience of notation symbols like $T_{kj}$ are often used to indicate both the object $\mathbf{T}$ (the tensor) and its components, like for functions where the same symbol is often used to indicate both the variable and the function.

The term *tensor*, in its contemporary meaning, was introduced by the German physicist Woldemar Voigt (1850–1919) in 1898, in the treatment of some physical properties of crystals [1]. A typical example is the stress tensor, which is used to define the state of stress at a point inside a material in a deformed state.

Strictly speaking, the structure that assigns to each point of the space a tensor should be called a *tensor field*, but commonly the word *tensor* is used also to indicate tensor fields. Tensors like the one above described are specifically defined as tensors *of order two*, since their components transform like the product of two vector components, but it is straightforward to define tensors of higher order. Moreover, one can interpret vectors as tensors of order one, and scalars as tensors of order zero. This observation leads to treat each of those objects, namely scalars, vectors and tensors, by a unique theoretical approach, the so-called *tensor calculus*, first introduced by Ricci (Gregorio Ricci-Curbasto, an Italian mathematician, 1853–1925) and Levi-Civita (Tullio Levi-Civita, an Italian mathematician, 1873–1941) [2].

As above stated for the case of vectors, the fundamental characteristic of tensors in physics is that, due to the affine transformation, a physical rule that can be stated as the vanishing of a tensor in a coordinate system, say $T_{kj} = 0$, it can be written in exactly the same way, $T'_{kj} = 0$, in any other coordinate system.

## 1.3  Covariant, Contravariant and Mixed Tensors

A simple geometrical definition of a vector is the following: consider two points A and B in a 3D Euclidean space, they define a displacement or vector $\mathbf{V} = \overrightarrow{AB}$. Vectors are entities that can be associated to physical quantities (like velocity, heat flux, etc.) and they have an absolute meaning, i.e. a meaning that does not depend on the coordinate system. However, in different coordinate systems, the numbers that describe a vector are different. It is easier, for the sake of obtaining the rule for transforming vectors among different coordinate systems, to start with infinitesimal displacements. If the two points $A$ and $B$ are infinitesimally close to each other, i.e. the coordinates of $A$ are $x^k$ and those of $B$ are $x^k + dx^k$, the infinitesimal displacement is defined by the three numbers $dx^k$, which are the components of such vector. Transforming the coordinate system, the points $A$ and $B$ remain the same, but the values of $dx^k$ change. If the transformation is defined by the functions

$$u^j = u^j \left( x^1, x^2, x^3 \right) \tag{1.19}$$

the displacement $\overrightarrow{AB}$ has the components:

$$du^j = \frac{\partial u^j}{\partial x^k} dx^k \tag{1.20}$$

where summation rule has been used. This rule of transformation is then used to define the objects called *contravariant* vectors.

**Def.** An ordered set of numbers $V^k$ associated to any point $A$ in space that, on a change of coordinates, transforms like

$$V'^j = \frac{\partial u^j}{\partial x^k} V^k \tag{1.21}$$

(where $\frac{\partial u^j}{\partial x^k}$ are evaluated at point $A$) is called a *contravariant* vector.

Beside these objects, there exists another class of objects that show similar, but not identical, properties. In a Cartesian coordinate system, the partial derivative of an invariant function of coordinates, $T$, i.e. $U_k = \frac{\partial T}{\partial x^k}$, is what we call a *gradient*, which is a vector as well.

However, let evaluate the same vector in a different coordinate system, defined by (1.19), like $U'_j = \frac{\partial T}{\partial u^j}$. From the chain rule of derivation one obtains

$$U'_j = \frac{\partial T}{\partial u^j} = \frac{\partial T}{\partial x^k} \frac{\partial x^k}{\partial u^j} = \frac{\partial x^k}{\partial u^j} U_k \tag{1.22}$$

and it is easy to see that this transformation rule is different from that of a contravariant vector (1.21). This new rule of transformation is then used to define a second type of vectors.

**Def.** An ordered set of numbers $U_k$ associated to any point $A$ in space that, on a change of coordinates, transforms like

$$U'_j = \frac{\partial x^k}{\partial u^j} U_k \tag{1.23}$$

(where $\frac{\partial x^k}{\partial u^j}$ are evaluated at point $A$) is called a *covariant* vector.

The generally accepted way to distinguish between the two types of vectors is the following: the contravariant type is indicated by a superscript ($V^k$), while the covariant one by a subscript ($U_k$).

As anticipated in paragraph 1.2, the rules of transformation given by Eqs. (1.21) and (1.23) allow defining more general objects called *tensors*. Starting with contravariant tensors we can give the following definition.

**Def**. An ordered set of numbers $T^{kj}$ associated to any point $A$ in space that, on a change of coordinates, transforms like

$$T'^{kj} = \frac{\partial u^k}{\partial x^p} \frac{\partial u^j}{\partial x^q} T^{pq} \tag{1.24}$$

(where $\frac{\partial u^j}{\partial x^k}$ are evaluated at point $A$) is called a *contravariant tensor*, of order two.

The definition of a contravariant tensor of higher order is straightforward, while contravariant vectors can be understood as contravariant tensors of order one. Similarly to a covariant vector, a *covariant tensor* can be defined as follows.

**Def**. An ordered set of numbers $T_{kj}$ associated to any point $A$ in space that, on a change of coordinates, transforms like

$$T'_{jk} = \frac{\partial x^p}{\partial u^j} \frac{\partial x^q}{\partial u^k} T_{pq} \tag{1.25}$$

(where $\frac{\partial x^k}{\partial u^j}$ are evaluated at point $A$) is called a *covariant tensor*, of order two.

In a similar way covariant tensors of higher order can be defined. The characteristic of a tensor to be covariant or contravariant is called *type*, or sometimes *character*, of the tensor.

As a consequence of the existence of contravariant and covariant vectors, the concept of mixed tensor is easy to define. For example, an object that transforms like the product $A_j B^k$ is called a *mixed* tensor. Generally, a mixed tensor of order two $T_j^k$ transforms like

$$T'^{k}_{j} = \frac{\partial x^p}{\partial u^j} \frac{\partial u^k}{\partial x^q} T_p^q \tag{1.26}$$

a mixed tensor of order three, $T_{jm}^k$, transforms like

$$T'^{k}_{jm} = \frac{\partial x^p}{\partial u^j} \frac{\partial u^k}{\partial x^q} \frac{\partial x^r}{\partial u^m} T_{pr}^q \tag{1.27}$$

and so on. The Kronecker delta $\delta_j^k$ is a mixed tensor of order two, in fact consider the transformation

$$\delta'^{p}_{q} = \frac{\partial u^p}{\partial x^j} \frac{\partial x^k}{\partial u^q} \delta_k^j \tag{1.28}$$

the product $\frac{\partial x^k}{\partial u^q} \delta_k^j$ (summation over the index $q$) is equal to $\frac{\partial x^j}{\partial u^q}$. In Sect. 1.2 (Eq. 1.10) it was shown that the matrices $a_q^j = \frac{\partial x^j}{\partial u^q}$ and $b_j^p = \frac{\partial u^p}{\partial x^j}$ are one the inverse of the other, then $\frac{\partial u^p}{\partial x^j} \frac{\partial x^j}{\partial u^q}$ is the identity matrix, that proves the statement.

In Sect. 1.2 it was shown that the Jacobian determinant of two consecutive transformations of coordinates can be obtained from the Jacobians of each single transformation (see Eq. (1.7)). In a similar way, consider two consecutive transformations
of coordinates: $u^j = u^j \left( x^1, x^2, x^3 \right)$ and $v^j = v^j \left( u^1, u^2, u^3 \right)$, and a second order
covariant tensor $T_{pq}$, but the same can be shown for any order and for any type
(covariant, contravariant, mixed) tensors. Since

$$T'_{mn} = \frac{\partial x^p}{\partial u^m} \frac{\partial x^q}{\partial u^n} T_{pq} \tag{1.29a}$$

$$T''_{jk} = \frac{\partial u^m}{\partial v^j} \frac{\partial u^n}{\partial v^k} T'_{mn} \tag{1.29b}$$

then

$$T''_{jk} = \frac{\partial u^m}{\partial v^j} \frac{\partial u^n}{\partial v^k} \frac{\partial x^p}{\partial u^m} \frac{\partial x^q}{\partial u^n} T_{pq} = \left( \frac{\partial u^m}{\partial v^j} \frac{\partial x^p}{\partial u^m} \right) \left( \frac{\partial u^n}{\partial v^k} \frac{\partial x^q}{\partial u^n} \right) T_{pq} = \frac{\partial x^p}{\partial v^j} \frac{\partial x^q}{\partial v^k} T_{pq} \tag{1.30}$$

which is equivalent to a direct transformation from $\left( x^1, x^2, x^3 \right)$ to $\left( v^1, v^2, v^3 \right)$.

## 1.4  Algebraic Operations on Tensors

In this section the term *tensor* will be used to indicate any order and type, including
vectors (order one) and scalars (order zero). The basic operations of sum and product
can be defined on tensors, with some important distinctions. Consider two tensors
of the same order and type, like $P_k^j$ and $T_k^j$, the object $C_k^j$, whose components are
defined as

$$C_k^j = P_k^j + T_k^j \tag{1.31}$$

is still a tensor of the same order and type. In fact the transformation rule from
$\left( x^1, x^2, x^3 \right)$ to $\left( u^1, u^2, u^3 \right)$, which is inherited from that of $P_k^j$ and $T_k^j$, is

$$C_j'^k = P_j'^k + T_j'^k = \frac{\partial x^p}{\partial u^j} \frac{\partial u^k}{\partial x^q} P_p^q + \frac{\partial x^p}{\partial u^j} \frac{\partial u^k}{\partial x^q} T_p^q = \frac{\partial x^p}{\partial u^j} \frac{\partial u^k}{\partial x^q} \left( P_p^q + T_p^q \right) = \frac{\partial x^p}{\partial u^j} \frac{\partial u^k}{\partial x^q} C_p^q \tag{1.32}$$

The sum of tensors is then allowed only for tensors of the same order and type.
Product among tensors are more varied. A first product is that among a scalar and
a tensor of any order and type. Consider a scalar $a$ and a tensor, say $T_{jm}^k$, the product
among them is still a tensor of the same order and type of $T_{jm}^k$, with components
equal to

$$P_{jm}^k = a T_{jm}^k \tag{1.33}$$

the proof is straightforward. This operation allows to define some symmetry properties of tensors. Consider a second order covariant tensor $T_{jk}$, then calculate the two tensors

$$T_{\underline{jk}} = \frac{1}{2}\left(T_{jk} + T_{kj}\right) \tag{1.34a}$$

$$T_{\underset{\smile}{jk}} = \frac{1}{2}\left(T_{jk} - T_{kj}\right) \tag{1.34b}$$

having the following properties

$$T_{\underline{jk}} = T_{\underline{kj}} \tag{1.35a}$$

$$T_{\underset{\smile}{jk}} = -T_{\underset{\smile}{kj}} \tag{1.35b}$$

the first one, $T_{\underline{jk}}$, is said to be a *symmetric tensor* while $T_{\underset{\smile}{jk}}$ is said to be *skew-symmetric* (or *anti-symmetric*). The symmetry property is preserved by coordinate transformations, in fact the tensorial equations

$$T_{\underline{jk}} - T_{\underline{kj}} = 0 \tag{1.36a}$$

$$T_{\underset{\smile}{jk}} + T_{\underset{\smile}{kj}} = 0 \tag{1.36b}$$

must hold, as already stated, in any coordinate system. It is remarkable that any second order tensor $T_{jk}$ can be always written as the sum of a symmetric and a skew-symmetric tensors, precisely

$$T_{jk} = T_{\underline{jk}} + T_{\underset{\smile}{jk}} \tag{1.37}$$

which again can be easily proven. The same results can be extended to second order contravariant tensors, $T^{jk}$, but they cannot, in general, be extended to second order mixed tensors. The symmetry property can be extended to higher order tensors, for example taking the fourth order mixed tensor $P^k_{jmn}$, we can say that it is symmetric (or skew-symmetric) with respect to the indices $m$ and $n$ if:

$$P^k_{jmn} = P^k_{jnm} \text{ symmetric} \tag{1.38a}$$

$$P^k_{jmn} = -P^k_{jnm} \text{ skew-symmetric} \tag{1.38b}$$

To generalise, given a tensor of any order and type, and considering two suffixes (both superscripts or both subscripts) we may say that the tensor is symmetric or skew-symmetric with respect to these suffixes, if after interchanging them the components remain unchanged (symmetric) or change sign (skew-symmetric).

We can now consider the most general multiplication rule between two tensors of any order and type. Consider two tensors like $A^{jk\cdots}_{mn\cdots}$ and $B^{rs\cdots}_{pq\cdots}$, the object having components obtained by the simple product of the components of each of the two tensors, i.e.

$$C_{mn...pq...}^{jk...rs...} = A_{mn...}^{jk...} B_{pq...}^{rs...} \tag{1.39}$$

is still a tensor, of order equal to the sum of the orders of the two tensors and type given as in equation (1.39). This kind of product is sometimes called *outer product*, to distinguish it from the *inner product*, explained below. That the new object is a tensor can be proven by showing that the usual transformation rule holds.

A second kind of product (the *inner product*) between two tensors can be defined through the process of *index contraction*.

The *index contraction* is a process consisting on equating a superscript and a subscript index and summing over the entire range of values (1 to $N$ in a $N$-dimensional space). As an example, consider the tensor of order three $T_{km}^{j}$ and apply the described process to the indices $j$ and $k$ (one superscript and one subscript), this yields

$$C_m = T_{jm}^{j} = T_{1m}^{1} + T_{2m}^{2} + T_{3m}^{3} \tag{1.40}$$

where the summation convention is used in the first equality and the last term shows the explicit form, for $N = 3$. The resulting object, $C_m$, is a vector. In fact, in a change of coordinates

$$T_{sp}^{\prime r} = \frac{\partial u^r}{\partial x^j} \frac{\partial x^k}{\partial u^s} \frac{\partial x^m}{\partial u^p} T_{km}^{j} \tag{1.41}$$

$$C_p^{\prime} = \frac{\partial x^m}{\partial u^p} C_m \tag{1.42}$$

and contracting $r$ and $s$ in Eq. (1.41)

$$T_{sp}^{\prime s} = \left( \frac{\partial u^s}{\partial x^j} \frac{\partial x^k}{\partial u^s} \right) \frac{\partial x^m}{\partial u^p} T_{km}^{j} = \delta_j^k \frac{\partial x^m}{\partial u^p} T_{km}^{j} = \frac{\partial x^m}{\partial u^p} \left( \delta_j^k T_{km}^{j} \right) = \frac{\partial x^m}{\partial u^p} T_{jm}^{j} \tag{1.43}$$

which shows that $T_{jm}^{j}$ transforms like $C_m$. For the general case, contraction diminishes the order of the tensor by two, while the tensor character is that of the remaining, non-contracted, indices.

The *inner product* can now be defined by applying the contraction process to an outer product. For example, the inner product of two vectors $A_m$ and $B^j$ is obtained taking the outer product $A_m B^j$ and then contracting the superscript $j$ with the subscript $m$

$$C = A_m B^m \tag{1.44}$$

the result is, in this case, a scalar, and the inner product is the usual *scalar product* of two vectors.

In general, taking two tensors of any order and character, like $A_{mn...}^{jk...}$ and $B_{pq...}^{rs...}$, the object having components obtained by the outer product of the components of each of the two vectors, followed by the contraction of one or more superscripts with the same number of subscripts, is the inner product of the two tensors; for example, contracting $j$ with $p$, $m$ with $r$ and $n$ with $s$

$$A^{jk\ldots}_{mn\ldots} B^{mn\ldots}_{jq\ldots} = D^{k\ldots}_{q\ldots} \tag{1.45}$$

(the contracted indices are evidenced). An important result is the following: given a symmetric tensor $A_{jk}$ and a skew-symmetric tensor $B^{mn}_{\vee}$ then the inner product $A_{jk} B^{jk}_{\vee}$ is nil, in fact

$$A_{jk} B^{jk}_{\vee} = A_{kj} B^{jk}_{\vee} = -A_{kj} B^{kj}_{\vee} = -A_{mn} B^{mn}_{\vee} = -A_{jk} B^{jk}_{\vee} = 0 \tag{1.46}$$

where, the third equality is obtained by simply using $m$ instead of $k$ and $n$ instead of $j$, and the fourth equality is obtained again by changing the name of the suffixes. This result can be extended to any order of tensors, for example

$$A^{mjkp} B_{qjks}_{\vee} = 0 \tag{1.47}$$

It is useful now to introduce the permutation symbol $e^{jkl}$, which is skew-symmetric with respect to any two indexes and

$$e^{jkl} = \begin{cases} +1 & \text{if } [j\ k\ l] \text{ is an even permutation of } [1\ 2\ 3] \\ -1 & \text{if } [j\ k\ l] \text{ is an odd permutation of } [1\ 2\ 3] \\ 0 & \text{if } [j\ k\ l] \text{ contains a repetition} \end{cases} \tag{1.48}$$

so that, for example, $e^{123} = e^{231} = e^{312} = 1$ while $e^{213} = e^{132} = e^{321} = -1$, and $e^{113} = e^{221} = e^{333} = 0$.

The *covariant* form $e_{jkl}$ is defined analogously

$$e_{jkl} = \begin{cases} +1 & \text{if } [j\ k\ l] \text{ is an even permutation of } [1\ 2\ 3] \\ -1 & \text{if } [j\ k\ l] \text{ is an odd permutation of } [1\ 2\ 3] \\ 0 & \text{if } [j\ k\ l] \text{ contains a repetition} \end{cases} \tag{1.49}$$

It should be noticed that neither $e^{jkl}$ nor $e_{jkl}$ are tensors, as this will be made clear in the next section. Among other uses, this symbol is useful to evaluate a determinant in a compact way. Consider a matrix $\mathbf{M}$ defined by the coefficients $M^k_j$, then the determinant det $[\mathbf{M}]$ can be calculated (see [3] for a clear demonstration) as

$$\det [\mathbf{M}] = \frac{1}{3!} e^{ijk} e_{rsp} M^r_i M^s_j M^p_k \tag{1.50}$$

and analogously for the matrix $\mathbf{P}$ defined by the coefficients $P_{jk}$ and the matrix $\mathbf{Q}$ defined by the coefficients $Q^{jk}$

$$\det [\mathbf{P}] = \frac{1}{3!} e^{ijk} e^{rsp} P_{ir} P_{js} P_{kp} \tag{1.51}$$

$$\det [\mathbf{Q}] = \frac{1}{3!} e_{ijk} e_{rsp} Q^{ir} Q^{js} Q^{kp} \tag{1.52}$$

As an example, take $M_j^k = \frac{\partial u^k}{\partial u^{j\prime}}$ and set $J = \det \left[ \frac{\partial u^k}{\partial u^{j\prime}} \right]$ then

$$J = \frac{1}{3!} e^{ijk} e_{rsp} \frac{\partial u^r}{\partial u^{i\prime}} \frac{\partial u^s}{\partial u^{j\prime}} \frac{\partial u^p}{\partial u^{k\prime}}. \tag{1.53}$$

## 1.5   The Metric Tensor

When dealing with curvilinear coordinate systems the concept of *metric tensors* becomes pivotal to any treatment of differential equations, and consequently to mathematical modelling of physical laws, since it defines the intrinsic structure of the geometrical space. The metric tensor can be introduced by a simple geometric reasoning. Here we are considering the usual 3D Euclidean space, but we will see that the basic concept can be extended to surfaces embedded into this space, which are in general non-Euclidean 2D spaces, and even to spaces of higher dimensionality and more complex structures than that of an Euclidean space, called Riemannian spaces (named after the German mathematician Bernhard Riemann, 1826–1866). We will not deal with higher dimension Riemannian spaces in this book, the interested reader is referred to classical books like [4, 5].

In this section and below we will use the symbols $x^j$ (or specifically $x, y, z$ if needed) when referring to Cartesian orthogonal coordinates. Consider two points in space infinitely close to each other and define $ds$ as the distance between these points. In 3D Euclidean space such distance can be calculated in a Cartesian coordinate system by the well known rule

$$ds^2 = dx^2 + dy^2 + dz^2 \tag{1.54}$$

If we change the coordinate system to the spherical one $(R, \theta, \varphi)$, defined by the equations

$$x = R \sin (\theta) \cos (\varphi) \tag{1.55a}$$
$$y = R \sin (\theta) \sin (\varphi) \tag{1.55b}$$
$$z = R \cos (\theta) \tag{1.55c}$$

the same distance can be calculated through the equation found substituting Eqs. (1.55a) into Eq. (1.54), i.e.

$$ds^2 = dR^2 + R^2 d\theta^2 + R^2 \sin^2 \theta d\varphi^2 \tag{1.56}$$

If we choose the coordinate system $(u, v, w)$ defined by

$$x = a_x u + b_x v + c_x w \tag{1.57a}$$
$$y = a_y u + b_y v + c_y w \tag{1.57b}$$
$$z = a_z w + b_z v + c_z w \tag{1.57c}$$

i.e. a non-orthogonal Cartesian system, the same distance can be calculated as

$$ds^2 = \left(a_x^2 + a_y^2 + a_z^2\right) du^2 + \left(b_x^2 + b_y^2 + b_z^2\right) dv^2 + \left(c_x^2 + c_y^2 + c_z^2\right) dw^2 +$$
$$+2\left(a_x b_x + a_y b_y + a_z b_z\right) du\, dv + 2\left(a_x c_x + a_y c_y + a_z c_z\right) du\, dw +$$
$$+2\left(b_x c_x + b_y c_y + b_z c_z\right) dw\, dv \tag{1.58}$$

These three ways to calculate $ds^2$ have in common the fact that the form on the right hand side is a quadratic one. For a general coordinate transformation $x^j\left(u^1, u^2, u^3\right)$ the element $dx^j$ is

$$dx^j = \frac{\partial x^j}{\partial u^k} du^k \tag{1.59}$$

and substitution into Eq. (1.54) yields the most general form to calculate the distance $ds$

$$ds^2 = g_{mn}\, du^m du^n \tag{1.60}$$

Since $ds^2$ is a scalar and $du^m$, $du^n$ are contravariant vectors, the coefficients $g_{mn}$ are the components of a covariant tensor, called the *metric tensor*. In a Cartesian coordinate system, where $ds^2 = dx^2 + dy^2 + dz^2$, the metric tensor is simply

$$g_{mn} = \delta_{mn} \tag{1.61}$$

where $\delta_{mn}$ is the Kronecker symbol. The metric tensor is assumed to be symmetric, and for the case we are dealing with this can be shown considering that

$$ds^2 = \delta_{jk} dx^j dx^k = \delta_{jk} \frac{\partial x^j}{\partial u^n} \frac{\partial x^k}{\partial u^m} du^m du^n \tag{1.62}$$

and then

$$g_{mn} = \delta_{jk} \frac{\partial x^j}{\partial u^n} \frac{\partial x^k}{\partial u^m} \tag{1.63}$$

The components of the metric tensor for the three coordinate systems above used are

$$g_{mn} = \begin{bmatrix} 1 & 0 & 0 \\ 0 & 1 & 0 \\ 0 & 0 & 1 \end{bmatrix} \text{ orthogonal Cartesian coordinates} \tag{1.64}$$

$$g_{mn} = \begin{bmatrix} 1 & 0 & 0 \\ 0 & R^2 & 0 \\ 0 & 0 & R^2 \sin^2(\theta) \end{bmatrix} \text{ spherical coordinates} \tag{1.65}$$

$$g_{mn} = \begin{bmatrix} a_x^2 + a_y^2 + a_z^2 & a_x b_x + a_y b_y + a_z b_z & a_x c_x + a_y c_y + a_z c_z \\ a_x b_x + a_y b_y + a_z b_z & b_x^2 + b_y^2 + b_z^2 & b_x c_x + b_y c_y + b_z c_z \\ a_x c_x + a_y c_y + a_z c_z & b_x c_x + b_y c_y + b_z c_z & c_x^2 + c_y^2 + c_z^2 \end{bmatrix} \text{ skew coordinates}$$

$$\tag{1.66}$$

Let now introduce the set of coefficients $\hat{g}^{jn}$, symmetric with respect to the suffixes, and defined by the following system of linear equations

$$\hat{g}^{jn} g_{nk} = \delta_k^j \tag{1.67}$$

In the framework of matrix algebra, $\left[\hat{g}^{jn}\right]$ is then the inverse matrix of $[g_{nk}]$. Taking $\text{cof}^{nk}$ as the cofactors of the element $g_{nk}$, the rule to evaluate the determinant of such matrix, $g = \det[g_{nk}]$, allows writing

$$g_{nk} \text{cof}^{pk} = g\delta_n^p \tag{1.68}$$

and multiplying both sides of Eq. (1.67) by $\text{cof}^{kp}$ and contracting on the index $k$ yields

$$\hat{g}^{jp} = \frac{\text{cof}^{jp}}{g} \tag{1.69}$$

This definition of $\hat{g}^{jp}$ holds in any coordinate system, then Eq. (1.67) holds in any coordinate system and the set of coefficients $\hat{g}^{jp}$ are then the component of a symmetric contravariant tensor of second order which is called *conjugate* of the metric tensor, and it will be hereinafter indicated by $g^{jp}$.

It is important to notice that $g^{jp}$ is fully determined by the metric tensor, through Eq. (1.67), and it can then be understood as the contravariant form of the metric tensor: $g^{jp}$ and $g_{jp}$ are essentially the same object under two different forms. This interpretation of covariant and contravariant tensors can be generalised by the so-called *lowering and raising index* rule.

## 1.6   Lowering and Raising Indices

Consider first the covariant vector $V_j$ and calculate the inner product $U^m = g^{mj} V_j$, the contravariant vector obtained in this way is, in the given coordinate system, completely defined by the components of the covariant vector $V_j$, and these can be obtained by the contravariant components $U^m$

$$U^m g_{mj} = g^{mk} V_k g_{mj} = V_j \tag{1.70}$$

where the last equality is obtained using Eq. (1.67). The two vectors represented by $U^m$ and $V_j$ are then the same object, and hereinafter we will use the same letter to indicate covariant, $V_j$, and contravariant, $V^j$, forms of the same vector. This process can be extended to tensors of any order and type.

The components of a tensor of second order or higher are defined by a letter with two or more indices, either superscripts or subscripts. The order of the indices is clearly important and in fully contravariant, say $T^{ijk}$, or covariant, say $U_{mnp}$, tensors the ordering is straightforward. The problem may rise in mixed tensors, like $P_k^j$, where it is not clear which is the first index, and this may be a problem when raising and lowering indices procedure is applied, for example: $P_k^j g_{jn}$ may be written either $P_{nk}$ or $P_{kn}$, and if $P_{nk}$ is not symmetric this introduces an ambiguity. A way to define the order of the indices in a mixed tensor is based on the use of slots for each index, for example: $P_{.k}^j$ indicates that $j$ is the first index (the slot is marked with a dot), while $Q_k^{.j}$ indicates that $k$ is the first index. Then $P_{.k}^j g_{jn} = P_{nk}$ while $Q_k^{.j} g_{jn} = Q_{kn}$. This method will be used whenever an ambiguity can raise.

Hereinafter, we can represent physical objects by tensors and we can choose to have them as contravariant, covariant or mixed as it is more suitable to our purposes.

Following up, we can now define the magnitude of a vector, either covariant or contravariant, by the following rule

$$|V|^2 = V_k V^k = g^{jk} V_k V_j = g_{jk} V^j V^k \tag{1.71}$$

since $|V|^2$, the square of the magnitude of vector $V_j$, is a scalar, the definition holds in any coordinate system. It is now worth to notice that although all coordinate systems are equivalent in modelling the physical world, the Cartesian orthogonal system has, in Euclidean spaces, a particular prominent position, which is related to the intrinsically simpler form that tensor equations may take. In particular it must be noticed that if $U^j$ is a contravariant vector in such coordinate system, the particular form of the metric tensor (see Eq. 1.64) yields

$$U_j = g_{jk} U^k = \delta_{jk} U^k = U^j \tag{1.72}$$

i.e., the contravariant components are exactly the same as the covariant ones. The same can be shown for any other tensor, even the metric tensor and its conjugate have the same components. This simplification can sometimes be used to prove the correctness of a tensor equation: writing the equation in orthogonal Cartesian coordinates, where it may be simpler to prove its validity, the tensorial character of the equation is enough to say that it holds in any other coordinate system. For example, we can use this result to introduce the general way to evaluate the angle between two vectors.

## 1.7 The Angles Between Two Vectors

Consider two vectors, say $A'_j$ and $B'_j$, in any coordinate system and transform them to the orthogonal Cartesian system, obtaining $A_j$ and $B_j$. In this system the cosine of the angle between the two vectors can be evaluated as the ratio between the dot product of the two vectors and the product of the magnitudes of each vector

$$\cos\theta = \frac{\mathbf{A}\cdot\mathbf{B}}{|\mathbf{A}|\,|\mathbf{B}|} \tag{1.73}$$

The RHS of this equation can be written using tensorial symbolism; to this end, let first introduce the unit vectors as the product between the vector components and the reciprocal of the magnitude (that is a scalar), i.e.

$$a_j = \frac{1}{|\mathbf{A}|}A_j; \ b_j = \frac{1}{|\mathbf{B}|}B_j; \tag{1.74}$$

and then consider that the inner product can be written as $\mathbf{A}\cdot\mathbf{B} = A_j B^j = g^{jk}A_j B_k$. Then Eq. (1.73) becomes

$$\cos\theta = g^{jk}a_j b_k = g_{jk}a^j b^k \tag{1.75}$$

Returning to the original system of coordinates, the definitions of unit vectors still hold, since $|A|$ and $|B|$ are scalars

$$a'_j = \frac{1}{|\mathbf{A}|}A'_j; \ b'_j = \frac{1}{|\mathbf{B}|}B'_j \tag{1.76}$$

and the scalar $g^{jk}a_j b_k$ transforms to $g'^{jk}a'_j b'_k$, then in any coordinate system the angle between two vectors can be defined by

$$\cos\theta = g^{jk}a_j b_k = g_{jk}a^j b^k = \frac{g^{jk}A_j B_k}{|\mathbf{A}|\,|\mathbf{B}|} = \frac{g_{jk}A^j B^k}{|\mathbf{A}|\,|\mathbf{B}|} = \frac{A_k B^k}{|\mathbf{A}|\,|\mathbf{B}|} \tag{1.77}$$

It is worth to notice, as a consequence, that two vectors are orthogonal when their inner product is nil

$$g_{jk}A^j B^k = A_k B^k = 0 \tag{1.78}$$

and they remain orthogonal in any coordinate system.

## 1.8  Relative Tensors

The determinant $g$ of the metric tensor, defined in Eq. (1.68), is a quantity of great importance in tensor calculus and it needs some further analysis. It is easy to show that $g$ is not a scalar invariant. In fact, consider two coordinate systems, say $u^j$ and $u^{j'}$, with the corresponding metric tensors $g_{mn}$ and $g'_{mn}$ respectively, and see how $g$ transforms. Since

$$g'_{mn} = g_{jk} \frac{\partial u^j}{\partial u^{m'}} \frac{\partial u^k}{\partial u^{n'}} \tag{1.79}$$

and defining $J = \det \left[ \frac{\partial u^k}{\partial u^{n'}} \right]$, then by the determinant product rule

$$g' = g J^2 \tag{1.80}$$

which proves the statement. However, due to its correlation with the Jacobian determinant, the quantity $g$ is called a *relative invariant of weight 2*. Relative tensors are often found in applications and it is worth to spend some time on them. Their definition is relatively simple. Consider, for example, an object defined by the coefficients $T_k^j$, this is said to be a relative tensor of weight $Q$ if it transforms like

$$T_k'^j = T_n^m \frac{\partial u^n}{\partial u^{k'}} \frac{\partial u^{j'}}{\partial u^m} J^Q \tag{1.81}$$

The usual tensors can be seen as relative tensor of weight zero, and to avoid confusion they are sometimes called *absolute tensors*. It is easy to show that the product of two relative tensors, of any order and type, of weight $Q_1$ and $Q_2$ is a relative tensor of weight $Q_1 + Q_2$.

An example of a relative tensor is the permutation symbol $e_{ijk}$. This symbol has the same definition in any coordinate system (see Eqs. 1.48, 1.49), assuming that it is a relative tensor of weight $Q$ and considering a change of coordinates

$$e'_{ijk} = e_{mnp} \frac{\partial u^m}{\partial u^{i'}} \frac{\partial u^n}{\partial u^{j'}} \frac{\partial u^p}{\partial u^{k'}} J^Q \tag{1.82}$$

where the weight $Q$ is still indeterminate. To notice that $J = \det \left[ \frac{\partial u^k}{\partial u^{n'}} \right]$ and $J' = \det \left[ \frac{\partial u^{k'}}{\partial u^n} \right] = J^{-1}$ since $\frac{\partial u^k}{\partial u^{n'}} \frac{\partial u^{n'}}{\partial u^j} = \delta_j^k$. Multiplying Eq. (1.82) on both sides by $e^{ijk}$ and contracting all indices we obtain, from Eq. (1.50)

$$e^{ijk} e'_{ijk} = e^{ijk} e_{mnp} \frac{\partial u^m}{\partial u^{i'}} \frac{\partial u^n}{\partial u^{j'}} \frac{\partial u^p}{\partial u^{k'}} J^Q = 3! J J^Q \tag{1.83}$$

and since $e^{ijk} e'_{ijk} = 6$, as it can be easily seen, then $Q = -1$ and $e_{mnp}$ is a covariant relative tensor of weight -1.

About the contravariant symbol, let us start again setting

$$e^{ijk\prime} = e^{mnp} \frac{\partial u^{i\prime}}{\partial u^m} \frac{\partial u^{j\prime}}{\partial u^n} \frac{\partial u^{k\prime}}{\partial u^p} J^M \tag{1.84}$$

where the weight $M$ is indeterminate, multiplying both sides of this equation by $e_{ijk}$ and contracting all indices

$$e^{ijk\prime} e_{ijk} = e_{ijk} e^{mnp} \frac{\partial u^{i\prime}}{\partial u^m} \frac{\partial u^{j\prime}}{\partial u^n} \frac{\partial u^{k\prime}}{\partial u^p} J^M = 3! J^{-1} J^M \tag{1.85}$$

which shows that $e^{mnp}$ is a contravariant relative tensor of weight 1.

Given a generic determinant of a second order contravariant (absolute) tensor $A^{ik}$, since

$$A = \det \left[ A^{ik} \right] = \frac{1}{3!} e_{ijk} e_{mnp} A^{im} A^{jn} A^{kp} \tag{1.86}$$

then $A$, which is the product of two relative tensors of weight $-1$ and three absolute tensors (weight $= 0$), must be a relative invariant of weight $-2$, while for a covariant absolute tensor $B_{ik}$

$$B = \det [B_{ik}] = \frac{1}{3!} e^{ijk} e^{mnp} B_{im} B_{jn} B_{kp} \tag{1.87}$$

$B$ is a relative invariant of weight 2.

It is possible to form absolute tensors from the relative tensors $e^{ijk}$ and $e_{mnp}$ in the following way. Let introduce the Levi-Civita tensor $\varepsilon^{jkl}$ (named after the Italian mathematician Tullio Levi-Civita, 1873–1941) as

$$\varepsilon^{jkl} = \frac{e^{jkl}}{\sqrt{g}}; \quad \varepsilon_{jkl} = \sqrt{g} e_{jkl} \tag{1.88}$$

it is straightforward to see that these objects transform as absolute tensors; moreover, since

$$\frac{1}{3!} \varepsilon^{jkl} \varepsilon_{jkl} = \frac{1}{3!} e^{jkl} e_{jkl} = 1 \tag{1.89}$$

and

$$\frac{1}{3!} \varepsilon^{mnp} \varepsilon^{jkl} g_{jm} g_{kn} g_{lp} = \frac{1}{3!} \frac{e^{mnp}}{\sqrt{g}} \frac{e^{jkl}}{\sqrt{g}} g_{jm} g_{kn} g_{lp} = 1 \tag{1.90}$$

where the last equality is obtained using Eq. (1.50) then

$$\varepsilon^{jkl} \varepsilon_{jkl} = \varepsilon^{mnp} \varepsilon^{jkl} g_{jm} g_{kn} g_{lp} \tag{1.91}$$

which shows that

$$\varepsilon_{jkl} = \varepsilon^{mnp} g_{jm} g_{kn} g_{lp} \tag{1.92}$$

are the covariant components of $\varepsilon^{mnp}$.

The Levi-Civita tensor is useful to express in invariant form the cross product of two vectors. We know that the vector $\mathbf{C} = \mathbf{A} \times \mathbf{B}$ can be written in Cartesian coordinates as

$$\mathbf{C} = (A_2 B_3 - A_3 B_2)\,\mathbf{i} + (A_3 B_1 - A_1 B_3)\,\mathbf{j} + (A_1 B_2 - A_2 B_1)\,\mathbf{k} \tag{1.93}$$

The generalisation in general curvilinear coordinates is then

$$C^j = \varepsilon^{jkl} A_k B_l. \tag{1.94}$$

## 1.9 Geometrical Interpretation of Tensors

The definition of tensors given in the previous sections is complete and enough to develop the necessary theory. However, tensors are geometrical objects, and we can take advantage of the self-imposed limitation to deal with a 3D Euclidean space for exploring some geometrical interpretations of those objects. The interested reader can also refer to [6]. Consider first an orthogonal Cartesian coordinate system in a 3D Euclidean space. A basis in this system is a triplet of *orthonormal* vectors $(\mathbf{i}, \mathbf{j}, \mathbf{k})$ (see Fig. 1.1), which means that $\mathbf{i}, \mathbf{j}$ and $\mathbf{k}$ are orthogonal and they have unitary length.

Any vector can then be represented as a linear combination of the three basis vectors, for example vector $\mathbf{r}$ (position vector, see Fig. 1.1) in this coordinate system can be written as

$$\mathbf{r} = x\mathbf{i} + y\mathbf{j} + z\mathbf{k} \tag{1.95}$$

where $(x, y, z)$ are the vector components. Below we will use more often the symbols $\mathbf{e}_1, \mathbf{e}_2, \mathbf{e}_3$ for $\mathbf{i}, \mathbf{j}, \mathbf{k}$ since it helps to compact formulae. The vector $\mathbf{e}_j$ is tangent to the

**Fig. 1.1** Vector representation in a Cartesian coordinate system

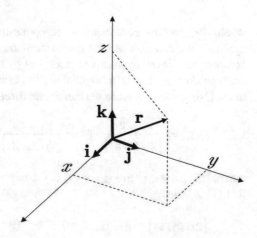

$x^j$-axis and it can be related to the position vector observing that, since the position vector $\mathbf{r}$ is function of $(x^j)$, then the vector displacement between two infinitesimally close points is

$$dr = \frac{\partial \mathbf{r}}{\partial x^k} dx^k = \mathbf{e}_k dx^k \tag{1.96}$$

and then

$$\frac{\partial \mathbf{r}}{\partial x^k} = \mathbf{e}_k \tag{1.97}$$

which shows that each $\mathbf{e}_k$ is tangent to the coordinate line $x^k$. The distance length $ds$ between these points can be calculated as $ds^2 = dr \cdot dr$, and using Eq. (1.96)

$$ds^2 = dx^j \mathbf{e}_j \cdot dx^k \mathbf{e}_k = dx^k dx^j \mathbf{e}_j \cdot \mathbf{e}_k = dx^j dx^j \tag{1.98}$$

and the last equality comes from the orthonormality of the basis $(\mathbf{e}_1, \mathbf{e}_2, \mathbf{e}_3)$. A further important property of this basis is that it is independent of the position.

Any vector $\mathbf{V}$ can then be represented in this basis as

$$\mathbf{V} = V^1 \mathbf{i} + V^2 \mathbf{j} + V^3 \mathbf{k} = V^j \mathbf{e}_j \tag{1.99}$$

where the summation convention is used in the last equality. Let now consider a general curvilinear system of coordinates $(u^1, u^2, u^3)$, we can define the basis using the same method as above

$$\frac{\partial \mathbf{r}}{\partial u^j} = \mathbf{g}_j \tag{1.100}$$

where the *covariant* vector basis $\mathbf{g}_j$ is again tangential to the corresponding curvilinear coordinate $u^j$ (see Fig. 1.2). It must be noticed that in general the vectors $\mathbf{g}_j$ are neither unitary nor orthogonal to each other. The same vector $\mathbf{V}$ can now be represented in this basis as

$$\mathbf{V} = V^{j'} \mathbf{g}_j \tag{1.101}$$

where $V^{j'}$ are the *contravariant* components of $\mathbf{V}$. Thanks to this geometrical approach we can now see the geometrical meaning of contravariant and covariant components. Taken two basis vectors, say $\mathbf{g}_1$ and $\mathbf{g}_2$, they generate a plane and the cross product $\mathbf{g}_1 \times \mathbf{g}_2$ is orthogonal to this plane, but in general it will not be parallel to $\mathbf{g}_3$. Using this procedure we can define three new vectors as

$$\mathbf{g}^1 = \frac{\mathbf{g}_2 \times \mathbf{g}_3}{[\mathbf{g}_1, \mathbf{g}_2, \mathbf{g}_3]}; \quad \mathbf{g}^2 = \frac{\mathbf{g}_3 \times \mathbf{g}_1}{[\mathbf{g}_1, \mathbf{g}_2, \mathbf{g}_3]}; \quad \mathbf{g}^3 = \frac{\mathbf{g}_1 \times \mathbf{g}_2}{[\mathbf{g}_1, \mathbf{g}_2, \mathbf{g}_3]} \tag{1.102}$$

where $[\mathbf{g}_1, \mathbf{g}_2, \mathbf{g}_3] = \mathbf{g}_1 \cdot (\mathbf{g}_2 \times \mathbf{g}_3)$ is the triple product of $\mathbf{g}_j$ and it is easy to show that [7]

$$[\mathbf{g}_1, \mathbf{g}_2, \mathbf{g}_3] = \mathbf{g}_1 \cdot (\mathbf{g}_2 \times \mathbf{g}_3) = \mathbf{g}_2 \cdot (\mathbf{g}_3 \times \mathbf{g}_1) = \mathbf{g}_3 \cdot (\mathbf{g}_1 \times \mathbf{g}_2) \tag{1.103}$$

**Fig. 1.2** Covariant vector basis, $\mathbf{g}_j$, *and contravariant vector basis, $\mathbf{g}^j$, in a general curvilinear system of coordinates*

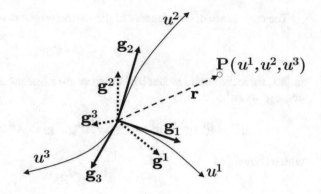

The vectors $\mathbf{g}^j$ are called the *contravariant* basis and the same vector $\mathbf{V}$ can be represented in this basis as

$$\mathbf{V} = V'_j \mathbf{g}^j \tag{1.104}$$

where $V'_j$ are the *covariant* components. Now it is quite clear that covariant and contravariant components of a vector are simply a different way to represent the same object, using two different basis. If we try to apply the same procedure to the Cartesian basis $\mathbf{e}_1, \mathbf{e}_2, \mathbf{e}_3$, we immediately discover that $\mathbf{e}^k = \mathbf{e}_k$ and as a consequence $V^k = V_k$, i.e. in Cartesian coordinates contravariant and covariant components of a vector are the same.

Consider now the dot product between a covariant basis vector and a contravariant one, from the definition (1.102) it is promptly obtained

$$\mathbf{g}^k \cdot \mathbf{g}_j = \delta^k_j \tag{1.105}$$

The distance between two neighbouring points (see Eq. 1.98) is now

$$ds^2 = \mathbf{g}_j du^j \cdot \mathbf{g}_k du^k = \mathbf{g}_j \cdot \mathbf{g}_k du^j du^k = g_{jk} du^j du^k \tag{1.106}$$

and

$$g_{jk} = \mathbf{g}_j \cdot \mathbf{g}_k \tag{1.107}$$

are the metric tensor components, and relation (1.63) can be obtained recalling that the position vector $\mathbf{r}$ can be seen as a function of $x^j$ or of $u^j$, from Eq. (1.100)

$$g_{jk} = \mathbf{g}_j \cdot \mathbf{g}_k = \frac{\partial \mathbf{r}}{\partial u^j} \cdot \frac{\partial \mathbf{r}}{\partial u^k} = \frac{\partial \mathbf{r}}{\partial x^p} \frac{\partial x^p}{\partial u^j} \cdot \frac{\partial \mathbf{r}}{\partial x^q} \frac{\partial x^q}{\partial u^k} = \frac{\partial x^p}{\partial u^j} \frac{\partial x^q}{\partial u^k} (\mathbf{e}_p \cdot \mathbf{e}_q) = \frac{\partial x^p}{\partial u^j} \frac{\partial x^q}{\partial u^k} \delta_{pq} \tag{1.108}$$

There are special coordinate systems where $\mathbf{g}_j \cdot \mathbf{g}_k = 0$ when $j \neq k$, i.e. the vector basis is orthogonal. These systems, called *orthogonal curvilinear coordinates*, have special properties that will be analysed in details in Chap. 4.

The contravariant components of the metric tensor can be evaluated as

$$g^{jk} = \mathbf{g}^j \cdot \mathbf{g}^k \tag{1.109}$$

in fact, since the contravariant basis vectors are a linear combination of the covariant ones: $\mathbf{g}^k = A^{kl}\mathbf{g}_l$, then

$$g^{jk} = \mathbf{g}^j \cdot \mathbf{g}^k = \mathbf{g}^j \cdot A^{kl}\mathbf{g}_l = A^{kl}\mathbf{g}^j \cdot \mathbf{g}_l = A^{kl}\delta_l^j = A^{kj} \tag{1.110}$$

which shows that

$$\mathbf{g}^k = g^{kj}\mathbf{g}_j \tag{1.111}$$

then $g^{jk} = \mathbf{g}^j \cdot \mathbf{g}^k = g^{jp}\mathbf{g}_p \cdot g^{kq}\mathbf{g}_q = g^{jp}g^{kq}g_{pq}$ which yields: $g^{kq}g_{pq} = \delta_p^k$ .

The components $g^{jk}$ and $g_{jk}$ can be used to raise or lower the indices also for the basis vectors

$$\mathbf{g}_p = \delta_p^k\mathbf{g}_k = g^{kq}g_{pq}\mathbf{g}_k = g_{pq}g^{kq}\mathbf{g}_k = g_{pq}\mathbf{g}^q \tag{1.112}$$

where the last equality comes from Eq. (1.111).

In the present context a tensor of order 2 can be defined extending the definition of vector

$$\mathbf{T} = T^{jk}\mathbf{g}_j\mathbf{g}_k \tag{1.113}$$

where $\mathbf{g}_j\mathbf{g}_k$ is the so-called dyadic or outer product and $T^{jk}$ are contravariant components. Similarly for the other type of components (contravariant, mixed), then the lowering and raising index procedure is promptly re-obtained

$$\mathbf{T} = T^{jk}\mathbf{g}_j\mathbf{g}_k = T^{jk}g_{jq}\mathbf{g}^q\mathbf{g}_k = T_q^k\mathbf{g}^q\mathbf{g}_k \Rightarrow T^{jk}g_{jq} = T_q^k \tag{1.114}$$

$$\mathbf{T} = T^{jk}\mathbf{g}_j\mathbf{g}_k = T^{jk}g_{jq}g_{kp}\mathbf{g}^q\mathbf{g}^p = T_{qp}\mathbf{g}^q\mathbf{g}^p \Rightarrow T^{jk}g_{jq}g_{kp} = T_{qp} \tag{1.115}$$

This way to describe a tensor (of any order) makes clear that the covariant, contravariant and mixed components are simply different ways to describe the same entity.

## 1.10  The Volume Element

The infinitesimal volume element in Cartesian coordinates is simply given by

$$dV = dx\,dy\,dz \tag{1.116}$$

and, since this is a geometrical object, it must be independent of the choice of the coordinate system, thus it must be an invariant. In a general coordinate system, the

**Fig. 1.3** Infinitesimal element along the coordinate lines ($d\mathbf{a}, d\mathbf{b}, d\mathbf{c}$)

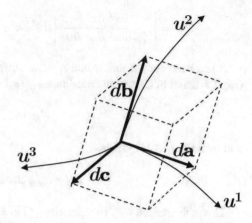

volume element can be calculated taking the volume of the parallelepiped formed by the infinitesimal elements of length along the coordinate lines (see Fig. 1.3).

It is well known [7] that the volume of such parallelepiped can be evaluated taking the absolute value of the triple product of the side vectors

$$dV = |d\mathbf{a} \times d\mathbf{b} \cdot d\mathbf{c}| \tag{1.117}$$

In the present case, the side vectors are infinitesimal and given by $\frac{\partial \mathbf{r}}{\partial u^j} du^j$ then

$$dV = \left| \frac{\partial \mathbf{r}}{\partial u^1} \times \frac{\partial \mathbf{r}}{\partial u^2} \cdot \frac{\partial \mathbf{r}}{\partial u^3} \right| du^1 du^2 du^3 = \left| \frac{\partial \mathbf{r}}{\partial x^j} \frac{\partial x^j}{\partial u^1} \times \frac{\partial \mathbf{r}}{\partial x^k} \frac{\partial x^k}{\partial u^2} \cdot \frac{\partial \mathbf{r}}{\partial x^m} \frac{\partial x^m}{\partial u^3} \right| du^1 du^2 du^3$$

$$= \left| \mathbf{e}_j \times \mathbf{e}_k \cdot \mathbf{e}_m \frac{\partial x^j}{\partial u^1} \frac{\partial x^k}{\partial u^2} \frac{\partial x^m}{\partial u^3} \right| du^1 du^2 du^3$$

$$\tag{1.118}$$

where it is easy to observe that

$$\mathbf{e}_j \times \mathbf{e}_k \cdot \mathbf{e}_m = e_{jkm} \tag{1.119}$$

since the triple product is always nil when it contains two identical vectors, while it is equal to $\pm 1$ depending on the order of the vectors. The Jacobian determinant of the coordinate transformation can be calculated as

$$J = e_{jkm} \frac{\partial x^j}{\partial u^1} \frac{\partial x^k}{\partial u^2} \frac{\partial x^m}{\partial u^3} \tag{1.120}$$

(it can be proven by the expansion of the triple summation) and then

$$dV = \left| e_{jkm} \frac{\partial x^j}{\partial u^1} \frac{\partial x^k}{\partial u^2} \frac{\partial x^m}{\partial u^3} \right| du^1 du^2 du^3 = |J| \, du^1 du^2 du^3 \tag{1.121}$$

It has been shown that $g$ is a relative invariant of weight 2, and then $g' = gJ^2$ (Eq. 1.80) and in Cartesian coordinates $g = 1$ then in a general coordinate system

$$\sqrt{g'} = |J| \tag{1.122}$$

and then Eq. (1.121) yields

$$dV = \sqrt{g'} du^1 du^2 du^3 \tag{1.123}$$

This is the reason why the quantity $\sqrt{g}$ is sometimes called *volume element*. The volume of a portion of space can then be calculated in general coordinate system as

$$V = \int_V dV = \int \int \int \sqrt{g'} du^1 du^2 du^3 \tag{1.124}$$

and the volume integral of a quantity $T$ is then

$$\int_V T dV = \int \int \int T \sqrt{g'} du^1 du^2 du^3. \tag{1.125}$$

## 1.11   Physical Components of Tensors

Consider a vector $\mathbf{V}$ in a 3D Cartesian system of coordinates. In such a system, covariant and contravariant components are the same: $V^k = V_k$, as shown above. The components $V^k = V_k$ in this coordinate system are sometimes called *physical components* of the vector. When the vector is transformed by a change of coordinates

$$V^{j\,'} = \frac{\partial u^j}{\partial x^k} V^k; \quad V_j' = \frac{\partial x^k}{\partial u^j} V_k \tag{1.126}$$

the contravariant and covariant components are, in general, no longer equal. Moreover, they may even have different physical dimensions. Consider for example the velocity vector $\mathbf{V}$ in a spherical coordinate system, which is defined by the equations (see Fig. 1.4)

$$x = R \sin(\theta) \cos(\varphi) \quad R = \sqrt{x^2 + y^2 + z^2}$$
$$y = R \sin(\theta) \sin(\varphi) \quad \theta = \arctan\left(\frac{\sqrt{x^2+y^2}}{z}\right) \tag{1.127}$$
$$z = R \cos(\theta) \quad\quad \varphi = \arctan\left(\frac{y}{x}\right)$$

**Fig. 1.4** Spherical coordinate systems

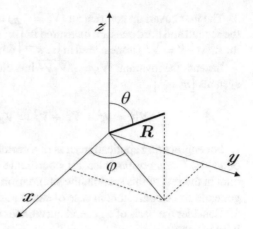

The components (covariant or contravariant) in Cartesian coordinates are $V_x$, $V_y$, $V_z$ and they are all measured in $\left[ m \ s^{-1} \right]$.

Let now evaluate the covariant components in the spherical system, using Eq. (1.126), where $(u^1, u^2, u^3) = (R, \theta, \varphi)$ and $(x^1, x^2, x^3) = (x, y, z)$. Since

$$\frac{\partial x^k}{\partial u^j} = \begin{bmatrix} \sin(\theta)\cos(\varphi) & R\cos(\theta)\cos(\varphi) & -R\sin(\theta)\sin(\varphi) \\ \sin(\theta)\sin(\varphi) & R\cos(\theta)\sin(\varphi) & R\sin(\theta)\cos(\varphi) \\ \cos(\theta) & -R\sin(\theta) & 0 \end{bmatrix} \tag{1.128}$$

$$\frac{\partial u^k}{\partial x^j} = \begin{bmatrix} \sin(\theta)\cos(\varphi) & \sin(\theta)\sin(\varphi) & \cos(\theta) \\ \frac{\cos(\varphi)\cos(\theta)}{R} & \frac{\sin(\varphi)\cos(\theta)}{R} & -\frac{\sin(\theta)}{R} \\ -\frac{\sin(\varphi)}{R\sin(\theta)} & \frac{\cos(\varphi)}{R\sin(\theta)} & 0 \end{bmatrix} \tag{1.129}$$

then

$$V_1{}' = V_R = \sin(\theta)\cos(\varphi)\,V_x + \sin(\theta)\sin(\varphi)\,V_y + \cos(\theta)\,V_z \tag{1.130a}$$

$$V_2{}' = V_\theta = R\cos(\theta)\cos(\varphi)\,V_x + R\cos(\theta)\sin(\varphi)\,V_y - R\sin(\theta)\,V_z \tag{1.130b}$$

$$V_3{}' = V_\varphi = -R\sin(\theta)\sin(\varphi)\,V_x + R\sin(\theta)\cos(\varphi)\,V_y \tag{1.130c}$$

and

$$V^{1}{}' = V^R = \sin(\theta)\cos(\varphi)\,V_x + \sin(\theta)\sin(\varphi)\,V_y + \cos(\theta)\,V_z \tag{1.131a}$$

$$V^{2}{}' = V^\theta = \frac{\cos(\theta)\cos(\varphi)}{R}V_x + \frac{\cos(\theta)\sin(\varphi)}{R}V_y - \frac{\sin(\theta)}{R}V_z \tag{1.131b}$$

$$V^{3}{}' = V^\varphi = -\frac{\sin(\varphi)}{R\sin(\theta)}V_x + \frac{\cos(\varphi)}{R\sin(\theta)}V_y \tag{1.131c}$$

The first covariant component ($V_1' = V_R$) has the right dimension $[m \ s^{-1}]$, while the second and third ones are measured in $[m^2 s^{-1}]$. For the contravariant component: the first ($V^{1'} = V^R$) is measured in $[m \ s^{-1}]$ while the other two are measured in $[s^{-1}]$.

Instead, the invariant $|\mathbf{V}| = \sqrt{V_j V^j}$ has clearly the same measuring units in both systems $[m \ s^{-1}]$

$$|\mathbf{V}|^2 = V_x^2 + V_y^2 + V_z^2 = V_R V^R + V_\theta V^\theta + V_\varphi V^\varphi \tag{1.132}$$

For engineering applications it is of a certain importance to introduce the concept of *physical components* also for coordinate systems different from the Cartesian one. In the rest of this section, the summation convention is suspended, since it may generate some confusion; in case of summation the explicit symbol $\sum$ will be used.

Consider the basis of a general curvilinear system $\mathbf{g}_j$ and normalise it by dividing it by its length

$$\tilde{\mathbf{g}}_j = \frac{1}{|\mathbf{g}_j|} \mathbf{g}_j \tag{1.133}$$

where

$$|\mathbf{g}_j| = \sqrt{\mathbf{g}_j \cdot \mathbf{g}_j} = \sqrt{g_{jj}} = h_j \tag{1.134}$$

with no summation on the indices. The length of the basis $\mathbf{g}_j$ is called *scale factor* and

$$\mathbf{g}_j = h_j \tilde{\mathbf{g}}_j \tag{1.135}$$

The physical components of a vector $\mathbf{U}$ can now be defined by the dot product

$$\mathbf{U} \cdot \tilde{\mathbf{g}}_j = \hat{U}_j \tag{1.136}$$

which is simply the projection of the vector along the coordinate direction. The relation between these components and the covariant and contravariant ones are obtained as

$$\hat{U}_j = \mathbf{U} \cdot \tilde{\mathbf{g}}_j = \sum_k U_k \mathbf{g}^k \cdot \tilde{\mathbf{g}}_j = \sum_k U_k \mathbf{g}^k \cdot \frac{1}{h_j} \mathbf{g}_j = U_j \frac{1}{h_j} \tag{1.137}$$

$$\hat{U}_j = \mathbf{U} \cdot \tilde{\mathbf{g}}_j = \sum_k U^k \mathbf{g}_k \cdot \tilde{\mathbf{g}}_j = \sum_k U^k \mathbf{g}_k \cdot \frac{1}{h_j} \mathbf{g}_j = \sum_k U^k \frac{g_{kj}}{h_j} \tag{1.138}$$

When using general curvilinear coordinates, the physical components of vectors and tensors "[...] represent quantities possessed of the natural physical dimensions of the field and capable of immediate physical interpretation" [8]. They have the same measuring units as in Cartesian coordinate systems, and they can be interpreted as the projection of vectors (and tensor) along the curvilinear coordinates.

For the case of orthogonal coordinates, which are the most used in applications, some further useful properties hold. For example the general formulae connecting physical, covariant and contravariant components for vectors and second order tensors become

$$\hat{U}_j = \sum_k U^k \frac{g_{kj}}{h_j} = U^j h_j = U_j \frac{1}{h_j} \tag{1.139}$$

$$\hat{T}_{jk} = T_{jk} \frac{1}{h_j h_j} = T^{kj} h_j h_k = T^k_j \frac{h_k}{h_j} \tag{1.140}$$

and similarly for higher order. In that case, the contravariant basis vectors are also orthogonal to each other, and they can be normalised obtaining

$$\tilde{\mathbf{g}}^j = \frac{1}{|\mathbf{g}^j|} \mathbf{g}^j \tag{1.141}$$

where

$$|\mathbf{g}^j| = \sqrt{\mathbf{g}^j \cdot \mathbf{g}^j} = \sqrt{g^{jj}} \tag{1.142}$$

and since

$$1 = \mathbf{g}^j \cdot \mathbf{g}_j = \sqrt{g^{jj}} \tilde{\mathbf{g}}^j h_j \tilde{\mathbf{g}}_j = \sqrt{g^{jj}} h_j \tag{1.143}$$

then $\sqrt{g^{jj}} = \frac{1}{h_j}$ and finally

$$\mathbf{g}^j = \frac{1}{h_j} \tilde{\mathbf{g}}^j \tag{1.144}$$

It is of practical importance to be able to transform physical components in a coordinate system to physical components in another coordinate system, particularly for orthogonal curvilinear system, and this point will be treated in details in Chap. 4.

## 1.12  Vector and Tensor Derivation

In this section, the differential properties of tensor fields will be explored by introducing the *covariant derivatives* of vectors and tensors, which generalises to any coordinate system the concept of partial derivatives of vector and tensor fields. The summation convention is restored hereinafter.

Consider a curve $C$ defined parametrically in a general coordinate system by

$$u^j = u^j(t) \tag{1.145}$$

when $t$ is changing, the point $P(u^j)$ moves along the curve. The basis vectors $(\mathbf{g}_1, \mathbf{g}_2, \mathbf{g}_3)$ at point $P$ also depend on position, then the derivatives of $\mathbf{g}_j$ with respect to $t$ can be calculated as

$$\frac{\partial \mathbf{g}_j}{\partial t} = \frac{\partial \mathbf{g}_j}{\partial u^k} \frac{\partial u^k}{\partial t} = \mathbf{g}_{j,k} \frac{\partial u^k}{\partial t} \tag{1.146}$$

where $\mathbf{g}_{j,k} = \frac{\partial \mathbf{g}_j}{\partial u^k}$ and the comma as a subscript is a commonly accepted way to indicate the usual partial derivative.

From the definition of the basis vectors

$$\mathbf{g}_j = \frac{\partial \mathbf{r}}{\partial u^j} = \frac{\partial \mathbf{r}}{\partial x^p} \frac{\partial x^p}{\partial u^j} = \mathbf{e}_p \frac{\partial x^p}{\partial u^j} \tag{1.147}$$

and in a similar way

$$\mathbf{e}_p = \frac{\partial \mathbf{r}}{\partial x^p} = \frac{\partial \mathbf{r}}{\partial u^n} \frac{\partial u^n}{\partial x^p} = \mathbf{g}_n \frac{\partial u^n}{\partial x^p} \tag{1.148}$$

then, recalling that the basis of the Cartesian coordinates $\mathbf{e}_p$ does not change with position

$$\mathbf{g}_{j,k} = \frac{\partial \mathbf{g}_j}{\partial u^k} = \frac{\partial \left(\mathbf{e}_p \frac{\partial x^p}{\partial u^j}\right)}{\partial u^k} = \mathbf{e}_p \frac{\partial^2 x^p}{\partial u^k \partial u^j} = \left(\frac{\partial u^n}{\partial x^p} \frac{\partial^2 x^p}{\partial u^k \partial u^j}\right) \mathbf{g}_n = \left\{ {n \atop k\ j} \right\} \mathbf{g}_n \tag{1.149}$$

where the second equality is obtained using Eq. (1.147), the fourth equality using Eq. (1.148) and $\left\{ {n \atop k\ j} \right\}$ is the *Christoffel symbol* of second kind (named after the German mathematician Elwin Bruno Christoffel, 1829–1900), and it is easy to see that it is symmetric with respect to the lower indices.

The Christoffel symbols can be more conveniently expressed through the derivative of the metric tensor $g_{jk}$. Consider the derivative of $g_{jk}$ with respect to $u^m$

$$g_{jk,m} = \frac{\partial g_{jk}}{\partial u^m} = \frac{\partial \left(\mathbf{g}_j \cdot \mathbf{g}_k\right)}{\partial u^m} = \mathbf{g}_{j,m} \cdot \mathbf{g}_k + \mathbf{g}_j \cdot \mathbf{g}_{k,m} \tag{1.150}$$

where Eq. (1.107) was used, and using Eq. (1.149) to eliminate $\mathbf{g}_{a,b}$

$$g_{jk,m} = \left\{ {n \atop m\ j} \right\} \mathbf{g}_n \cdot \mathbf{g}_k + \left\{ {n \atop k\ m} \right\} \mathbf{g}_j \cdot \mathbf{g}_n = \left\{ {n \atop m\ j} \right\} g_{nk} + \left\{ {n \atop k\ m} \right\} g_{jn} \tag{1.151}$$

By permutations of the indices the following two relations are found

$$g_{mk,j} = \left\{ {n \atop j\ m} \right\} g_{nk} + \left\{ {n \atop k\ j} \right\} g_{mn} \tag{1.152a}$$

$$g_{jm,k} = \left\{ {n \atop k\ j} \right\} g_{nm} + \left\{ {n \atop m\ k} \right\} g_{jn} \tag{1.152b}$$

and the following identity can be proven by substitution

$$\frac{1}{2}\left(g_{jk,m} + g_{mk,j} - g_{jm,k}\right) = \begin{Bmatrix} n \\ m \ j \end{Bmatrix} g_{nk} \tag{1.153}$$

Finally, multiplying Eq. (1.153) by $g^{kp}$ and contracting on $k$ gives

$$\begin{Bmatrix} p \\ m \ j \end{Bmatrix} = \frac{1}{2} g^{kp}\left(g_{jk,m} + g_{mk,j} - g_{jm,k}\right) \tag{1.154}$$

The Christoffel symbols are not tensors, and this can be easily seen by the fact that they are nil in a Cartesian coordinate system (in a Cartesian coordinate system the metric tensor is independent of coordinates and its derivatives are nil), while they are not nil in a general curvilinear system.

The Christoffel symbols of second kind are called like that because Christoffel symbols of *first kind* exists, which are connected to the Christoffel symbols of second kind by

$$[m \ n, r] = g_{rs}\begin{Bmatrix} s \\ m \ n \end{Bmatrix} = \frac{1}{2}\left(g_{nr,m} + g_{mr,n} - g_{nm,r}\right) \tag{1.155}$$

Let now take the vector **V** defined by its contravariant components

$$\mathbf{V} = V^j \mathbf{g}_j \tag{1.156}$$

and take the partial derivative $\frac{\partial \mathbf{V}}{\partial u^k} = \mathbf{V}_{,k}$. Using the rules for the partial derivative

$$\mathbf{V}_{,k} = \frac{\partial\left(V^j \mathbf{g}_j\right)}{\partial u^k} = V^j_{,k}\mathbf{g}_j + V^j \mathbf{g}_{j,k} = V^j_{,k}\mathbf{g}_j + V^j\begin{Bmatrix} n \\ k \ j \end{Bmatrix}\mathbf{g}_n = \left[V^n_{,k} + V^j\begin{Bmatrix} n \\ k \ j \end{Bmatrix}\right]\mathbf{g}_n \tag{1.157}$$

where Eq. (1.149) was used. The quantities in square brackets in the last term of Eq. (1.157)

$$V^n_{/k} = V^n_{,k} + V^j\begin{Bmatrix} n \\ k \ j \end{Bmatrix} = \mathbf{V}_{,k} \cdot \mathbf{g}^n \tag{1.158}$$

are the components of the derivative of the vector **V** and Eq. (1.158) has a fundamental property: it is the same in any coordinate system, which means that $V^n_{/k}$ are the components of a second order mixed tensor. The slash is used to indicate this derivative that, thanks to its invariant property, is called *covariant derivative*. The tensorial property of $V^n_{/k}$ can be proven by showing that it transforms following the rule for tensors, i.e. considering the change of coordinates $u^j = u^j\left(v^1, v^2, v^3\right)$, the following rule should hold

$$\bar{V}^n_{/k} = V^q_{/p}\frac{\partial v^n}{\partial u^q}\frac{\partial u^p}{\partial v^k} \tag{1.159}$$

where the bar is used to indicate the component of vectors and tensors in $\left(v^1, v^2, v^3\right)$. To show this, consider the same vector in Cartesian coordinates $\mathbf{V} = \check{V}^j \mathbf{e}_j$ (here the symbol $^\vee$ indicates the component of vectors and tensors in Cartesian coordinates) and take again the derivative with respect to the coordinate $x^k$, then

$$\mathbf{V}_{,k} = \frac{\partial \mathbf{V}}{\partial x^k} = \frac{\partial \left( \check{V}^j \mathbf{e}_j \right)}{\partial u^k} = \check{V}^j_{,k} \mathbf{e}_j \tag{1.160}$$

and $\mathbf{V}_{,k} \cdot \mathbf{e}^n = \check{V}^n_{,k} = \check{V}^n_{/k}$, consistently with the fact that the Christoffel symbols are nil in Cartesian coordinates. Let now try to transform the quantities $\check{V}^j_{/k} = \check{V}^j_{,k}$, from the Cartesian coordinates to the new coordinate system $(v^1, v^2, v^3)$, by the following way. First notice that, since from the transformation rule for contravariant vectors $\check{V}^j = \overline{V}^p \frac{\partial x^j}{\partial v^p}$, then

$$\check{V}^j_{/k} = \check{V}^j_{,k} = \frac{\partial \check{V}^j}{\partial x^k} = \frac{\partial}{\partial x^k}\left[ \overline{V}^p \frac{\partial x^j}{\partial v^p} \right] = \frac{\partial \overline{V}^p}{\partial x^k}\frac{\partial x^j}{\partial v^p} + \overline{V}^p \frac{\partial}{\partial x^k}\left[ \frac{\partial x^j}{\partial v^p} \right] =$$
$$= \frac{\partial \overline{V}^p}{\partial v^m}\frac{\partial v^m}{\partial x^k}\frac{\partial x^j}{\partial v^p} + \overline{V}^p \left[ \frac{\partial^2 x^j}{\partial v^p \partial v^m}\frac{\partial v^m}{\partial x^k} \right] \tag{1.161}$$

where the chain-rule $\frac{\partial}{\partial x^k}\left[ \frac{\partial x^j}{\partial v^p} \right] = \frac{\partial}{\partial v^m}\left[ \frac{\partial x^j}{\partial v^p} \right]\frac{\partial v^m}{\partial x^k}$ was used.

Let now multiply both sides by $\frac{\partial v^q}{\partial x^j}\frac{\partial x^k}{\partial v^s}$ and contract the indices $k$ and $j$

$$\check{V}^j_{/k}\frac{\partial v^q}{\partial x^j}\frac{\partial x^k}{\partial v^s} = \frac{\partial \overline{V}^p}{\partial v^m}\frac{\partial v^m}{\partial x^k}\frac{\partial x^j}{\partial v^p}\frac{\partial v^q}{\partial x^j}\frac{\partial x^k}{\partial v^s} + \overline{V}^p\left[ \frac{\partial^2 x^j}{\partial v^p \partial v^m}\frac{\partial v^m}{\partial x^k} \right]\frac{\partial v^q}{\partial x^j}\frac{\partial x^k}{\partial v^s} = \tag{1.162}$$
$$= \frac{\partial \overline{V}^p}{\partial v^m}\delta^m_s \delta^q_p + \overline{V}^p \frac{\partial^2 x^j}{\partial v^p \partial v^m}\delta^m_s \frac{\partial v^q}{\partial x^j} = \frac{\partial \overline{V}^q}{\partial v^s} + \overline{V}^p \frac{\partial^2 x^j}{\partial v^p \partial v^s}\frac{\partial v^q}{\partial x^j} = \frac{\partial \overline{V}^q}{\partial v^s} + \overline{V}^p \left\{ \begin{matrix} q \\ p\ s \end{matrix} \right\}$$

where the relations $\frac{\partial v^m}{\partial x^k}\frac{\partial x^k}{\partial v^s} = \delta^m_s$ and $\frac{\partial v^q}{\partial x^j}\frac{\partial x^j}{\partial v^p} = \delta^q_p$ have been used and the last equality comes from the definition (1.149) of the Christoffel symbols in $(v^j)$ coordinates. From Eq. (1.158), the last term is exactly $\overline{V}^q_{/s}$, then

$$\overline{V}^q_{/s} = \check{V}^j_{/k}\frac{\partial v^q}{\partial x^j}\frac{\partial x^k}{\partial v^s} \tag{1.163}$$

and repeating the procedure for the transformation from Cartesian coordinates to $u^j$ coordinates

$$V^q_{/s} = \check{V}^j_{/k}\frac{\partial u^q}{\partial x^j}\frac{\partial x^k}{\partial u^s} \tag{1.164}$$

It is now straightforward to prove Eq. (1.159), multiply both sides of Eq. (1.164) by $\frac{\partial x^n}{\partial u^q}\frac{\partial u^s}{\partial x^m}$ and contract the repeated indices, then remembering that $\frac{\partial x^n}{\partial u^q}\frac{\partial u^q}{\partial x^j} = \delta^n_j$; $\frac{\partial x^k}{\partial u^s}\frac{\partial u^s}{\partial x^m} = \delta^k_m$ the following is obtained

$$\check{V}^n_{/m} = V^q_{/s} \frac{\partial x^n}{\partial u^q} \frac{\partial u^s}{\partial x^m} \tag{1.165}$$

then multiply both sides of Eq. (1.165) by $\frac{\partial v^p}{\partial x^n} \frac{\partial x^m}{\partial v^r}$ and contract the indexes $m, n$ to obtain

$$\check{V}^n_{/m} \frac{\partial v^p}{\partial x^n} \frac{\partial x^m}{\partial v^r} = V^q_{/s} \frac{\partial x^n}{\partial u^q} \frac{\partial u^s}{\partial x^m} \frac{\partial v^p}{\partial x^n} \frac{\partial x^m}{\partial v^r} = V^q_{/s} \frac{\partial v^p}{\partial u^q} \frac{\partial u^s}{\partial v^r} \tag{1.166}$$

where the last equality is obtained observing that $\frac{\partial v^p}{\partial x^n} \frac{\partial x^n}{\partial u^q} = \frac{\partial v^p}{\partial u^q}$ and $\frac{\partial u^s}{\partial x^m} \frac{\partial x^m}{\partial v^r} = \frac{\partial u^s}{\partial v^r}$. Finally, from Eq. (1.163) $\check{V}^n_{/m} \frac{\partial v^p}{\partial x^n} \frac{\partial x^m}{\partial v^r} = \bar{V}^p_{/r}$ and Eq. (1.166) yields

$$\bar{V}^p_{/r} = V^q_{/s} \frac{\partial v^p}{\partial u^q} \frac{\partial u^s}{\partial v^r} \tag{1.167}$$

which is the transformation rule of a mixed second order tensor. Since the coordinates $v^j$ and $u^j$ are arbitrary, this proves the tensorial character of $V^q_{/s}$, as defined by Eq. (1.158). It must be noticed that neither $V^n_{,k}$ nor $V^j \left\{ {n \atop k\ j} \right\}$ are tensors (as it could be proven by transforming them into a new coordinate system), but their sum is a tensor.

Since $V^n_{/k}$ is a tensor, also $V^n_{/k} g_{np}$ is a tensor

$$V^n_{/k} g_{np} = V^n_{,k} g_{np} + V^j \left\{ {n \atop k\ j} \right\} g_{np} = V_{p,k} + V^s \left[ \left\{ {n \atop k\ s} \right\} g_{np} - g_{sp,k} \right] \tag{1.168}$$

$$= V_{p,k} + V^s \left[ \frac{1}{2} \left( g_{sp,k} + g_{kp,s} - g_{sk,p} \right) - g_{sp,k} \right] =$$

$$= V_{p,k} - V_n \left[ \frac{1}{2} g^{sn} \left( g_{sp,k} + g_{sk,p} - g_{kp,s} \right) \right] = V_{p,k} - V_n \left\{ {n \atop k\ p} \right\}$$

and it is the covariant derivative of a covariant vector, i.e.

$$V_{p/k} = V_{p,k} - V_n \left\{ {n \atop k\ p} \right\} \tag{1.169}$$

The covariant derivatives of tensors can be obtained in similar ways, for second order tensors

$$T^{rs}_{/n} = T^{rs}_{,n} + \left\{ {r \atop m\ n} \right\} T^{ms} + \left\{ {s \atop m\ n} \right\} T^{rm} \tag{1.170}$$

$$T^r_{.s/n} = T^r_{.s,n} + \left\{ {r \atop m\ n} \right\} T^m_{.s} - \left\{ {m \atop s\ n} \right\} T^r_{.m} \tag{1.171}$$

$$T_{rs/n} = T_{rs,n} - \left\{ {m \atop r\ n} \right\} T_{ms} - \left\{ {m \atop s\ n} \right\} T_{rm} \tag{1.172}$$

and when applied to the metric tensor yields an interesting result

$$g_{rs/n} = g_{rs,n} - \begin{Bmatrix} m \\ r\,n \end{Bmatrix} g_{ms} - \begin{Bmatrix} m \\ s\,n \end{Bmatrix} g_{rm} = 0 \tag{1.173}$$

as it can be easily proven by direct substitution of Eq. (1.154), and analogously $g^{rs}_{/n} = 0$.

The general rule for high order tensors is

$$T^{r_1..r_M}_{.s_1...s_R\ /n} = T^{r_1..r_M}_{.s_1...s_R\ ,n} + \sum_{\alpha=1}^{M} \begin{Bmatrix} r_\alpha \\ m\,n \end{Bmatrix} T^{r_1...m...r_M}_{.s_1...s_R} - \sum_{\alpha=1}^{R} \begin{Bmatrix} m \\ s_\alpha\,n \end{Bmatrix} T^{r_1...r_M}_{.s_1...m...sR} \tag{1.174}$$

and for a scalar (tensor of order zero) it is evident that

$$T_{/n} = \frac{\partial T}{\partial u^n} = T_{,n} \tag{1.175}$$

The covariant derivative is then the generalisation of the partial derivative in Cartesian coordinates to general curvilinear coordinates, and it follows the same rules. The covariant derivative is a linear operator (like the partial derivative), then the derivative of a linear combination of tensors is the linear combination of the derivatives of each tensor. The covariant derivative of product follows the same rule as for partial derivative, for example

$$\left( V_m T^{rs} \right)_{/n} = V_m T^{rs}_{/n} + V_{m/n} T^{rs} \tag{1.176}$$

or

$$V_{r/n} = \left( V^s g_{sr} \right)_{/n} = V^s g_{sr/n} + V^s_{/n} g_{sr} = V^s_{/n} g_{sr} \;. \tag{1.177}$$

The last equality is obtained using Eq. (1.173).

Derivatives of higher order can be obtained by applying repeatedly the above described operation. For example, the second covariant derivative of the vector $V^n$ is obtained applying the covariant derivative operation to the tensor $V^n_{/k}$ i.e.

$$V^n_{/k/p} = V^n_{/k,p} + V^j_{/k} \begin{Bmatrix} n \\ p\,j \end{Bmatrix} - V^n_{/j} \begin{Bmatrix} j \\ p\,k \end{Bmatrix} \tag{1.178}$$

and similarly for higher order tensors and derivatives. In the formula above we have indicated the order of derivation, leaving the slash symbol among the suffixes, and this needs some discussion. A tensorial equation, like (1.178), holds in any coordinate system. Since Cartesian coordinates exist in a 3D Euclidean space, we may write the same equation is such a system, where the Christoffel symbols vanish, and find out the following result

$$V^n_{/k/p} = \check{V}^n_{/k,p} = \check{V}^n_{,k,p} = \check{V}^n_{,p,k} = \check{V}^n_{/p,k} = \check{V}^n_{/p/k} \tag{1.179}$$

i.e. the second covariant derivatives commute, where the third equality comes from the well known Schwarz's theorem (named after the German mathematician Herman Schwarz, 1843–1921) about changing the order of partial derivatives. Then we may drop the double slash symbol when dealing with second derivatives.

However, the results just obtained are not generally valid, since they rely on the existence of a Cartesian coordinate system, which is true only in Euclidean spaces. When we will talk about surfaces from a general point of view, we will see that the commutativity property is lost, since general surfaces cannot be considered Euclidean 2D spaces, and we will see that this property is strictly connected with the concept of curvature.

## 1.13   The Intrinsic Derivative of Tensors

Consider a curve $C$ defined parametrically in a general coordinate system by

$$u^j = u^j(t) \tag{1.180}$$

where $t$ is the parameter. Taking two infinitely close points on such curve, calculate the difference between the values of a tensor in such points. Starting with a vector $\mathbf{V}$, the difference between the values of $\mathbf{V}$ between two infinitely close points along the curve, $\delta \mathbf{V}$ is

$$\delta \mathbf{V} = \frac{d\mathbf{V}}{dt}dt = \frac{d\left(V^j \mathbf{g}_j\right)}{dt}dt \tag{1.181}$$

Using the normal rules of derivation with respect to a parameter

$$\frac{d\left(V^j \mathbf{g}_j\right)}{dt} = \frac{dV^j}{dt}\mathbf{g}_j + V^j\frac{d\left(\mathbf{g}_j\right)}{dt} = \frac{\partial V^j}{\partial u^k}\frac{du^k}{dt}\mathbf{g}_j + V^j\mathbf{g}_{j,k}\frac{du^k}{dt} \tag{1.182}$$

$$= V^n_{,k}\frac{du^k}{dt}\mathbf{g}_n + V^j\begin{Bmatrix} n \\ k\ j \end{Bmatrix}\frac{du^k}{dt}\mathbf{g}_n = \left[V^n_{,k} + V^j\begin{Bmatrix} n \\ k\ j \end{Bmatrix}\right]\frac{du^k}{dt}\mathbf{g}_n = V^n_{/k}\frac{du^k}{dt}\mathbf{g}_n$$

and the quantity

$$\frac{\delta V^n}{\delta t} = V^n_{/k}\frac{du^k}{dt} \tag{1.183}$$

is called *intrinsic* (or sometimes *absolute*) derivative of the vector $\mathbf{V}$, along the curve $u^j = u^j(t)$ and it is a vector. Similar results can be obtained by tensor of any order; for example: the components of the *intrinsic* derivative of the third order mixed tensor $T^{rs}_m$ are

$$\frac{\delta T^{rs}_m}{\delta t} = T^{rs}_{m/k}\frac{du^k}{dt}. \tag{1.184}$$

## 1.14 Differential Operators in General Curvilinear Coordinates

Conservation and constitutive equations contain some specific differential operators like gradient, divergence, etc. It is then worth to find their covariant expressions to be able to write them in any coordinate system.

Let us start with the simplest one, the *gradient* of a scalar $S$. We have already seen this in Sect. 1.4 and found that it is a covariant vector. The general way to indicate its component is $S_{/k}$

$$S_{/k} = \frac{\partial S}{\partial u^k} \tag{1.185}$$

A widely used operator on a vector field is the *divergence*; the divergence of a contravariant vector in Cartesian coordinates is

$$\text{div } \mathbf{V} = \frac{\partial V^j}{\partial x^j} = V^j_{,j} \tag{1.186}$$

i.e. the derivative of the vector with respect to its coordinate followed by a contraction. Its generalisation is straightforward

$$\text{div } \mathbf{V} = V^j_{/j} = V^j_{,j} + V^s \begin{Bmatrix} j \\ j \ s \end{Bmatrix} \tag{1.187}$$

and a simple form can be found observing that

$$
\begin{Bmatrix} j \\ j \ s \end{Bmatrix} = \frac{1}{2} g^{kj} \left( g_{jk,s} + g_{sk,j} - g_{js,k} \right) = \frac{1}{2} \left( g^{kj} g_{jk,s} + g^{kj} g_{sk,j} - g^{kj} g_{js,k} \right)
$$
$$
= \frac{1}{2} \left( g^{kj} g_{jk,s} + g^{pm} g_{sp,m} - g^{pm} g_{ps,m} \right) = \frac{1}{2} g^{kj} g_{jk,s}
\tag{1.188}
$$

where the last but one equality is obtained by changing the names of the repeated indices and using the symmetry of $g_{sp}$. This last expression can be related to the partial derivatives of the determinant $g$.

Suspending the summation convention in the next paragraph, the expansion of the determinant $g$ (see Eq. 4.103) is

$$g = \sum_k g_{pk} \text{cof}^{pk} \tag{1.189}$$

where it must be noticed that the index $p$ (no summation on it!) can be chosen arbitrarily. Considering now $g$ as a function of $g_{mn}$ (which may be contained also in $\text{cof}^{pk}$), let calculate the derivative $\frac{\partial g}{\partial g_{mn}}$

$$\frac{\partial g}{\partial g_{mn}} = \sum_k \frac{\partial g_{pk} \mathrm{cof}^{pk}}{\partial g_{mn}} = \sum_k \frac{\partial g_{mk} \mathrm{cof}^{mk}}{\partial g_{mn}} \tag{1.190}$$

where in the last term the arbitrary index $p$ has been chosen equal to $m$ (please remember that there is no summation over repeated indexes different from $k$). Developing the derivatives

$$\frac{\partial g}{\partial g_{mn}} = \sum_k \frac{\partial g_{mk}}{\partial g_{mn}} \mathrm{cof}^{mk} + \sum_k g_{mk} \frac{\partial \mathrm{cof}^{mk}}{\partial g_{mn}} \tag{1.191}$$

and observing that by its own definition $\mathrm{cof}^{mk}$ cannot be function of $g_{mn}$ and that $\frac{\partial g_{mk}}{\partial g_{mn}}$ is nil unless $k = n$, the equation yields

$$\frac{\partial g}{\partial g_{mn}} = \mathrm{cof}^{mn} = g g^{mn} \tag{1.192}$$

The last equality results from the definition of $g^{mn}$, Eq. (1.69). This result could have been obtained from Jacobi's rule for the derivative of a determinant, see [9] for a classical reference. From now on we resume the summation convention.

Considering now $g$ as a function of coordinates

$$\frac{\partial g}{\partial u^j} = \frac{\partial g}{\partial g_{mn}} g_{mn,j} = g g^{mn} g_{mn,j} . \tag{1.193}$$

Then the contracted Christoffel symbol of Eq. (1.188) becomes

$$\left\{ \begin{matrix} j \\ j \ s \end{matrix} \right\} = \frac{1}{2} g^{kj} g_{jk,s} = \frac{1}{2} \frac{1}{g} \frac{\partial g}{\partial u^s} = \frac{\partial \ln \sqrt{g}}{\partial u^s} = \frac{1}{\sqrt{g}} \frac{\partial \sqrt{g}}{\partial u^s} \tag{1.194}$$

and the covariant divergence becomes

$$V^j_{/j} = \frac{\partial V^s}{\partial u^s} + V^s \frac{1}{\sqrt{g}} \frac{\partial \sqrt{g}}{\partial u^s} = \frac{1}{\sqrt{g}} \frac{\partial \left( \sqrt{g} V^s \right)}{\partial u^s} \tag{1.195}$$

which is a compact way to express it.

The *Laplacian* of a scalar (named after the French mathematician Pier Simon Laplace, 1749–1827) is defined in Cartesian coordinates as $\nabla^2 S = \sum_j \frac{\partial^2 S}{\partial x^{j2}}$. It can be more generally defined as the divergence of the gradient of the scalar $S$, and this interpretation allows the extension to general curvilinear systems. The gradient of a scalar is a covariant vector $S_{/j} = \frac{\partial S}{\partial u^j} = S_{,j}$ and the corresponding contravariant vector is $g^{jk} S_{/k}$; taking its divergence

$$\nabla^2 S = \left( g^{jk} S_{/k} \right)_{/j} = \frac{1}{\sqrt{g}} \frac{\partial}{\partial u^j} \left( \sqrt{g} g^{jk} \frac{\partial S}{\partial u^k} \right) \tag{1.196}$$

where the equivalence $S_{/k} = S_{,k} = \frac{\partial S}{\partial u^k}$ is used.

The *curl* of a vector $\mathbf{V}$ is a vector $\boldsymbol{\omega}$ and its definition in this coordinates is

$$\boldsymbol{\omega} = \text{curl } \mathbf{V} = \left( \frac{\partial V_z}{\partial y} - \frac{\partial V_y}{\partial z} \right) \mathbf{i} + \left( \frac{\partial V_x}{\partial z} - \frac{\partial V_y}{\partial x} \right) \mathbf{j} + \left( \frac{\partial V_y}{\partial x} - \frac{\partial V_x}{\partial y} \right) \mathbf{k} \quad (1.197)$$

and using the Levi-Civita tensors $\varepsilon^{jkl}$ in Cartesian coordinate system, the components of the curl can be written as

$$\omega^j = \varepsilon^{jkm} V_{m,k} \quad (1.198)$$

The generalisation in curvilinear coordinates is then

$$\omega^j = \varepsilon^{jkm} V_{m/k} \quad (1.199)$$

and the ordinary properties of this operator can easily be obtained. For example, it is known that in Cartesian coordinates the divergence of a curl is always nil; in curvilinear coordinates, if $\omega^j$ are the contravariant components of the curl, then

$$\text{div curl } \mathbf{V} = \omega^j_{/j} = \left( \varepsilon^{jkm} V_{m/k} \right)_{/j} = \varepsilon^{jkm} V_{m/kj} + \left( \varepsilon^{jkm} \right)_{/j} V_{m/k} \quad (1.200)$$

The term $\varepsilon^{jkm} V_{m/kj}$ is nil since the second covariant derivative, in a 3D Euclidean space, commutes and then $V_{m/kj}$ is symmetric with respect to the indices $k$ and $j$ while $\varepsilon^{jkm}$ is skew-symmmetric with respect to the same indices. The last term is also nil, in fact, from Eq. (1.88)

$$\varepsilon^{jkm}_{/j} = \left( g^{-1/2} e^{jkl} \right)_{/j} = g^{-1/2} e^{jkm}_{/j} - \frac{1}{2} e^{jkm} g^{-3/2} \frac{\partial g}{\partial u^j} \quad (1.201)$$

and, noticing that $e^{jkm}_{,j} = 0$, the covariant derivative $\left( e^{jkm} \right)_{/j}$ is

$$\left( e^{jkm} \right)_{/j} = e^{pkm} \left\{ \begin{matrix} j \\ j \ p \end{matrix} \right\} + e^{jpm} \left\{ \begin{matrix} k \\ j \ p \end{matrix} \right\} + e^{jkp} \left\{ \begin{matrix} m \\ j \ p \end{matrix} \right\} = e^{pkm} \left\{ \begin{matrix} j \\ j \ p \end{matrix} \right\} \quad (1.202)$$

where the symmetry of the Christoffel symbols with respect to the lower indexes and the skew symmetry of $e^{pkm}$ were used. From Eq. (1.194) $\left\{ \begin{matrix} j \\ j \ p \end{matrix} \right\} = \frac{1}{2} \frac{1}{g} \frac{\partial g}{\partial u^p}$, then

$$\varepsilon^{jkm}_{/j} = g^{-1/2} e^{pkm} \left\{ \begin{matrix} j \\ j \ p \end{matrix} \right\} - \frac{1}{2} e^{jkm} g^{-3/2} \frac{\partial g}{\partial u^j} = g^{-1/2} e^{pkm} \frac{1}{2} \frac{1}{g} \frac{\partial g}{\partial u^p} - \frac{1}{2} e^{pkm} g^{-3/2} \frac{\partial g}{\partial u^p} = 0 \quad (1.203)$$

It should be noticed that in general, also for $p \neq j, \varepsilon^{jkm}_{/p} = 0$, a proof can be found in [3].

It is interesting to notice that in a non Euclidean space, where second covariant derivatives do not commute: div curl $\mathbf{V} \neq 0$.

## 1.15 Orthogonal Curvilinear Coordinates

Among the infinity of general curvilinear coordinate systems, some of them are preferred to describe a physical problem since they have special properties. These are the orthogonal curvilinear coordinates, which are defined to be generally curvilinear, but such that at any intersection the tangents to the coordinate curves are orthogonal to each other. For example, in a plane, polar coordinate curves are circles centered at the origin and straight lines exiting from the origin, whenever they intersect each other they are orthogonal. A general definition of orthogonal coordinates can be given using the covariant basis $\mathbf{g}_j$ (see Sect. 1.10), which are vector tangent to the coordinate curves. A coordinate system is orthogonal if

$$\mathbf{g}_j \cdot \mathbf{g}_k = 0 \tag{1.204}$$

everywhere, which is equivalent to say that the off-diagonal components of the metric tensor vanish: $g_{jk} = 0$ if $j \neq k$. In such coordinate systems the distance between two infinitely close points $ds$ is given by

$$ds^2 = h_1^2 \left(du^1\right)^2 + h_2^2 \left(du^2\right)^2 + h_3^2 \left(du^3\right)^2 \tag{1.205}$$

where $h_j = \left|g_{jj}\right|^{1/2}$ (no summation over the index $j$) are the so-called *scale factors* (see Sect. 1.12). In these systems the physical components of a vector are

$$\left(\hat{V}_1, \hat{V}_2, \hat{V}_3\right) = \left(\frac{V_1}{h_1}, \frac{V_2}{h_2}, \frac{V_3}{h_3}\right) = \left(h_1 V^1, h_2 V^2, h_3 V^3\right) \tag{1.206}$$

while for second order tensors (no summation over repeated indexes)

$$\hat{T}_{jk} = \frac{1}{h_j h_k} T_{jk} \tag{1.207}$$

$$\hat{T}^{jk*} = h_j h_k T^{jk} \tag{1.208}$$

$$\hat{T}_k^j = \frac{h_j}{h_k} T_k^j \tag{1.209}$$

and similarly for higher order tensors. The simplification induced by this particular choice of coordinates reflects itself also on the form of the differential operators like divergence, Laplacian, curl. These and other details regarding orthogonal coordinates will be treated in Chap. 4.

# References

1. Voigt, W.:Die fundamentalen physikalischen Eigenschaften der Krystalle in elementarer Darstellung (in German). Von Veit, Leipzig (1898)
2. Ricci, G., Levi-Civita, T.: Méthodes de calcul différentiel absolu et leurs applications (in French). Math. Annalen. **54**(1–2), 125–201 (1900)
3. Grinfeld, P.: Introduction to tensor analysis and the calculus of moving surfaces. Springer, New York (2013)
4. Eisenhart, L.P.: Riemannian geometry. Princeton Univ Press, Princeton, NJ (1949)
5. Cartan, E.: Geometry of Riemannian Spaces. Math Science Press, Berkeley (1983)
6. Nguyen-Schäfer, H., Schmidt, J.P.: Tensor analysis and elementary differential geometry for physicists and engineers. Springer, Berlin (2014)
7. Lass, H.: Vector and Tensor Analysis. McGraw-Hill Book Company, Inc., New York (1950)
8. Truesdell, C.: The physical components of vectors and tensors. Z. Angew. Math. Mech. (J. Appl. Math. Mech.). **33**(10–11), 345–356 (1953)
9. Magnus, J.R., Neudecker, H.: Matrix differential calculus with applications in statistics and econometrics. Wiley, Hoboken (1999)

# Chapter 2
# Elements of Differential Geometry of a Surface

When dealing with multiphase systems, like evaporating liquid drops or vapour bubbles in a liquid, etc., interfaces are present and they can be schematised as surfaces separating two different phases. The properties of interfaces often depend on the geometric properties of such surfaces, then it is of a certain interest to explore the geometrical properties of these objects.

In this chapter the basics of differential geometry of surfaces will be given, also making profit of the mathematical tools described in Chap. 1. The concept of curvature will be introduced using the formalism of differential geometry and of tensor calculus and interpreted from a geometrical point of view. The treatment of moving surfaces follows that of [1], to which the interested reader is invited to refer for a more complete treatment of the subject.

## 2.1 Parametric, Cartesian and Implicit Description of a Surface

A surface embedded in a 3D Euclidean space can be conveniently defined in a parametric way, that is

$$x = x(u, v); \quad y = y(u, v); \quad z = z(u, v) \tag{2.1}$$

where $u$ and $v$ are two parameters. To write this definition in a general way, the parameters will be written as $v^1$ and $v^2$ or $v^\alpha$ with the convention that Greek indices assume values $1, 2$ while as usual Latin indices range from 1 to 3; thus equations (2.1) can be written in a compact form as

$$x^j = x^j(v^\alpha) \tag{2.2}$$

© Springer Nature Switzerland AG 2021
G. E. Cossali and S. Tonini, *Drop Heating and Evaporation: Analytical Solutions in Curvilinear Coordinate Systems*, Mathematical Engineering,
https://doi.org/10.1007/978-3-030-49274-8_2

It is often convenient to write this parametric form using a general coordinate system $u^j$, then the surface may be described by the equations

$$u^j = u^j (v^\alpha) \tag{2.3}$$

As an example, the parametric equations of a sphere with radius $R_0$ in Cartesian coordinates is, writing $u$ instead of $\cos(\theta)$

$$x = R_0\sqrt{1 - u^2} \cos(v) \tag{2.4a}$$
$$y = R_0\sqrt{1 - u^2} \sin(v) \tag{2.4b}$$
$$z = R_0 u \tag{2.4c}$$

while in spherical coordinates ($\eta = \cos(\theta)$) it is

$$R = R_0 \tag{2.5a}$$
$$\eta = u \tag{2.5b}$$
$$\varphi = v \tag{2.5c}$$

The implicit form in Cartesian coordinates of a surface is an equation relating the Cartesian coordinates, which can be obtained eliminating $v^\alpha$ from Eq. (2.2) to obtain

$$F\left(x^j\right) = 0 \tag{2.6}$$

For example, for the sphere the implicit form is:

$$x^2 + y^2 + z^2 - R_0^2 = 0 \tag{2.7}$$

If the same is done starting with general coordinates we obtain the implicit form

$$G\left(u^j\right) = 0 \tag{2.8}$$

Again for the sphere this is simply

$$R - R_0 = 0 \tag{2.9}$$

Each of these ways to define a surface are equivalent, but sometimes one form is more convenient than another. Below we will make large use of the parametric forms given by Eqs. (2.2) and (2.3).

## 2.2  Surface Coordinates and Surface Tensors

Starting from the general parametric form (2.3), the couple of variables $v^\alpha$ are called *surface coordinates,* and Greek indices (spanning over 1, 2) will be used, while the coordinates of the 3D space $u^j$ will be called *ambient coordinates* and the Latin indices span from 1 to 3. A surface can be considered a 2D space, embedded in a 3D space, i.e. a subspace of the 3D space. Below we will continue to limit ourself to the case of a 3D Euclidean space, but a general treatment can be done for more general spaces (Riemannian spaces) and even for higher dimensions, the interested reader can refer to [2, 3].

The treatment used in Chap. 1 can be extended to these 2D spaces, with one warning: surfaces are not, in general, Euclidean spaces and this is what makes the treatment of these lower dimensional spaces sometimes more complex than that of the ambient 3D Euclidean space.

We may start again from the geometrical object $\mathbf{r}$, the position vector in 3D Euclidean space, which exists, as all geometrical objects, independently of the choice of the coordinate system. The basis of the coordinate system on the surfaces can again be defined as

$$\mathbf{s}_\alpha = \frac{\partial \mathbf{r}}{\partial v^\alpha} \tag{2.10}$$

and from this definition, the vectors $\mathbf{s}_\alpha$ are tangent to the surface. Vectors that can be built by linear combination of these bases are then tangent to the surface, and the space spanned by those vectors is called *tangential plane.* Any vector (tensor) defined with respect to these bases is then tangent to the surface and we will call them *surface vectors* (tensors). As for the 3D case, a surface tensor is such if it transforms in a linear and homogeneous (affine) way when *surface* coordinates are changed. Considering a new coordinate system on the surface, say $v^{\alpha'}$, then a surface vector $\mathbf{V} = V^\alpha \mathbf{s}_\alpha$ transforms to the new system like

$$V'^\alpha = V^\beta \frac{\partial v^{\alpha'}}{\partial v^\beta} \tag{2.11}$$

and so on for tensors of higher order. For example the surface tensor $T^{\alpha\gamma}_{\beta\omega}$ transforms like

$$T'^{\alpha\gamma}_{\beta\omega} = T^{\delta\mu}_{\varepsilon\tau} \frac{\partial v^{\alpha'}}{\partial v^\delta} \frac{\partial v^{\gamma'}}{\partial v^\mu} \frac{\partial v^\varepsilon}{\partial v^{\beta'}} \frac{\partial v^\tau}{\partial v^{\omega'}} \tag{2.12}$$

Having defined the surface bases, the surface metric tensor can be defined as in Sect. 1.10, Eq. (1.107)

$$s_{\alpha\beta} = \mathbf{s}_\alpha \cdot \mathbf{s}_\beta \tag{2.13}$$

and its contravariant form is found again from the definition

$$s^{\alpha\beta} s_{\beta\gamma} = \delta^\alpha_\gamma . \tag{2.14}$$

The components $s^{\alpha\beta}$, $s_{\beta\gamma}$ can be used for raising and lowering indices. For example:

$$\mathbf{s}^\alpha = s^{\alpha\beta}\mathbf{s}_\beta \tag{2.15}$$

and

$$\mathbf{s}^\alpha \cdot \mathbf{s}_\gamma = s^{\alpha\beta}\mathbf{s}_\beta \cdot \mathbf{s}_\gamma = s^{\alpha\beta}s_{\beta\gamma} = \delta^\alpha_\gamma \tag{2.16}$$

Again the distance $ds$ between two infinitely close points on the surface can be measured as

$$ds^2 = s_{\alpha\beta}dv^\alpha dv^\beta \tag{2.17}$$

The Levi-Civita tensors $\varepsilon^{\alpha\beta}$, $\varepsilon_{\alpha\beta}$ can again be obtained from the permutation symbols $e^{\alpha\beta}$ and $e_{\alpha\beta}$

$$e^{\alpha\beta} = \begin{bmatrix} 0 & 1 \\ -1 & 0 \end{bmatrix}; \quad e_{\alpha\beta} = \begin{bmatrix} 0 & 1 \\ -1 & 0 \end{bmatrix} \tag{2.18}$$

by the equations:

$$\varepsilon^{\alpha\beta} = \frac{e^{\alpha\beta}}{\sqrt{\breve{\sigma}}}; \quad \varepsilon_{\alpha\beta} = e_{\alpha\beta}\sqrt{\breve{\sigma}} \tag{2.19}$$

where $\breve{\sigma} = \det\left[s_{\alpha\beta}\right]$.

The whole apparatus of covariant derivatives applies also to the surface tensors.

Defining the 2D Christoffel symbols of first kind from the metric tensor as (see also Eq. 1.154)

$$\begin{Bmatrix} \gamma \\ \alpha\,\beta \end{Bmatrix} = \frac{1}{2}s^{\omega\gamma}\left(s_{\beta\omega,\alpha} + s_{\alpha\omega,\beta} - s_{\beta\alpha,\omega}\right) \tag{2.20}$$

then the covariant derivative of contravariant and covariant surface vectors are

$$V^\alpha_{/\beta} = V^\alpha_{,\beta} + \begin{Bmatrix} \alpha \\ \gamma\,\beta \end{Bmatrix}V^\gamma \tag{2.21}$$

$$V_{\alpha/\beta} = V_{\alpha,\beta} - \begin{Bmatrix} \gamma \\ \alpha\,\beta \end{Bmatrix}V_\gamma \tag{2.22}$$

and in general the same rule as reported in Sect. 1.13 can be used to evaluate the derivatives of higher order surface tensors of any kind. A particular result is (see also Sect. 1.13, Eq. 1.173)

$$s_{\alpha\beta/\gamma} = 0 \tag{2.23}$$

There are, however, important differences with the case of the 3D Euclidean space, which can all be related to the following statement: *in the surface space a system of coordinates where the basis is independent of the position may not, in general, exist.*

We know that the Cartesian coordinates have this property, and we have used this when dealing with 3D Euclidean space. For example, we have shown that in that case, when Cartesian coordinates are used, all Christoffel symbols vanish. This property was used, for example, to show that second covariant derivatives do commute, simply observing that second covariant derivatives of a tensor are equal to second partial derivatives in Cartesian coordinates, and, since they commute in such a system, they must commute in any system.

For a general surface, this procedure cannot be applied and, in general, second covariant derivatives do not commute. This precise fact is related to the concept of curvature of the surface, concept that it does not exist in 3D Euclidean spaces, but it exists in general Riemannian spaces [2, 4].

## 2.3 Relation with the Ambient Space

From a geometrical point of view, it is somehow advantageous to consider the surface as a 2D space embedded into a 3D Euclidean space, and from the physical point of view it is necessary, since 2D spaces are used often to represent interfaces among two portions of a 3D space, and the interface physical behaviour cannot be separated from that of the remaining space. However, it is worth to notice that for a complete geometric treatment of a non-Euclidean $N$-dimensional space there is no necessity, in general, to consider embedding spaces of higher dimensions.

The first link between the surface space and the 3D-Euclidean space can be created by observing that the surface basis $\mathbf{s}_\alpha$ can be related to a basis of the 3D space $\mathbf{g}_j$ in the following simple manner

$$\mathbf{s}_\alpha = \frac{\partial \mathbf{r}}{\partial v^\alpha} = \frac{\partial \mathbf{r}}{\partial u^j} \frac{\partial u^j}{\partial v^\alpha} = \frac{\partial u^j}{\partial v^\alpha} \mathbf{g}_j \ . \tag{2.24}$$

The set of coefficients $\frac{\partial u^j}{\partial v^\alpha} = U_\alpha^j$ is also called *shift tensor* [1], and from the definition of the contravariant bases of the 3D space

$$\mathbf{s}_\alpha \cdot \mathbf{g}^j = U_\alpha^j \tag{2.25}$$

That the symbol $U_\alpha^j$ is a tensor can be seen observing how it transforms when coordinates are changed. Supposing to change *ambient* coordinates (from $u^j$ to $u^{j'}$) and *surface* coordinates (from $v^\alpha$ to $v^{\alpha'}$) then

$$U'^j_\alpha = \frac{\partial u^{j'}}{\partial v^{\alpha'}} = \frac{\partial u^{j'}}{\partial u^k} \frac{\partial u^k}{\partial v^\beta} \frac{\partial v^\beta}{\partial v^{\alpha'}} = U_\beta^k \frac{\partial u^{j'}}{\partial u^k} \frac{\partial v^\beta}{\partial v^{\alpha'}} \tag{2.26}$$

which proves the statement. This fact allows using the lowering and rising index procedure to obtain

$$U_j^\alpha = s^{\alpha\beta} g_{jk} U_\beta^k \tag{2.27}$$

It must be noticed that when dealing with *ambient* indices (Latin letters), the *ambient metric tensor* must be used, while the *surface metric tensor* is used for the *surface* indices (Greek letters). Let now calculate the following inner product, $\bar{U}_j^\alpha = \mathbf{s}^\alpha \cdot \mathbf{g}_j$, and using the surface and ambient metric tensors we can show that

$$\bar{U}_j^\alpha = \mathbf{s}^\alpha \cdot \mathbf{g}_j = s^{\alpha\beta}\mathbf{s}_\beta \cdot \mathbf{g}^k g_{jk} = s^{\alpha\beta}U_\beta^k g_{jk} = U_j^\alpha \tag{2.28}$$

The shift tensor can be used to relate the components of *surface tensors* in *surface* coordinates to those in *ambient* coordinates. To avoid confusion, in this section, and whenever necessary, we will indicate surface tensors with a *tilde*, for example $\tilde{\mathbf{V}}$ is a surface vector and its contravariant components in *surface* coordinates are $\tilde{V}^\alpha$ and in *ambient* coordinates are $\tilde{V}^j$, with the only exception of the surface bases, $\mathbf{s}_\alpha$, $\mathbf{s}^\alpha$. An ambient tensor will be indicated by letters without tilde. A surface vector (tensor) can be seen as a particular case of an ambient vector, i.e. a vector (tensor) with components only in the tangent plane at each point.

Given a surface vector $\tilde{\mathbf{V}}$, the contravariant *surface* components, $\tilde{V}^\alpha$, and the contravariant *ambient* components, $\hat{V}^j$, can be related to each other by

$$\tilde{V}^j \mathbf{g}_j = \tilde{\mathbf{V}} = \tilde{V}^\alpha \mathbf{s}_\alpha = \tilde{V}^\alpha U_\alpha^j \mathbf{g}_j \tag{2.29}$$

then

$$\tilde{V}^j = \tilde{V}^\alpha U_\alpha^j \tag{2.30}$$

and also

$$\tilde{V}^\alpha = \tilde{\mathbf{V}} \cdot \mathbf{s}^\alpha = \tilde{V}^j \mathbf{g}_j \cdot \mathbf{s}^\alpha = \tilde{V}^j U_j^\alpha \tag{2.31}$$

This holds also for covariant components, using the lowering and raising index procedure, the following results can be obtained

$$\tilde{V}_k = g_{kj}\tilde{V}^j = \tilde{V}^\alpha U_\alpha^j g_{kj} = \tilde{V}_\beta s^{\beta\alpha} U_\alpha^j g_{kj} = \tilde{V}_\beta U_k^\beta \tag{2.32}$$

$$\tilde{V}_\alpha = s_{\alpha\beta}\tilde{V}^\beta = s_{\alpha\beta}\tilde{V}^j U_j^\beta = s_{\alpha\beta}\tilde{V}_k g^{kj} U_j^\beta = \tilde{V}_k U_\alpha^k \tag{2.33}$$

The metric tensor of the 2D space can be related to that of the 3D space by (see Eq. 2.24)

$$s_{\alpha\beta} = \mathbf{s}_\alpha \cdot \mathbf{s}_\beta = U_\alpha^j U_\beta^k \mathbf{g}_j \cdot \mathbf{g}_k = U_\alpha^j U_\beta^k g_{jk} \tag{2.34}$$

and also, by some manipulations

$$s^{\alpha\beta} = U_m^\alpha U_n^\beta g^{mn} \tag{2.35}$$

As an example consider a sphere of radius $R_0$, in Cartesian ambient coordinates it can be defined as

$$x = R_0\sqrt{1-\eta^2}\cos(\varphi) \tag{2.36a}$$

$$y = R_0\sqrt{1-\eta^2}\sin(\varphi) \tag{2.36b}$$

$$z = R_0\eta \tag{2.36c}$$

here the surface coordinates are $(v^1, v^2) = (\eta, \varphi)$. The shift tensor is given by

$$U^j_\alpha = \frac{\partial x^j}{\partial v^\alpha} = \begin{bmatrix} -R_0\frac{\eta\cos(\varphi)}{\sqrt{1-\eta^2}} & -R_0\sqrt{1-\eta^2}\sin(\varphi) \\ -R_0\frac{\eta\sin(\varphi)}{\sqrt{1-\eta^2}} & R_0\sqrt{1-\eta^2}\cos(\varphi) \\ R_0 & 0 \end{bmatrix} \tag{2.37}$$

and the surface metric tensor can be obtained from Eq. (2.34), recalling that in Cartesian coordinates $g_{jk} = \delta_{jk}$, then

$$S_{\alpha\beta} = U^j_\alpha U^k_\beta g_{jk} = \sum_j U^j_\alpha U^j_\beta \tag{2.38}$$

the last term can be interpreted as the matrix product between the matrix $U^j_\alpha$ and its transpose.

For the present case

$$S_{\alpha\beta} = \begin{bmatrix} R_0^2\frac{1}{1-\eta^2} & 0 \\ 0 & R_0^2\left(1-\eta^2\right) \end{bmatrix} \tag{2.39}$$

Let now take as *ambient* coordinates the spherical ones:

$$x = R\sqrt{1-\eta^2}\cos(\varphi) \tag{2.40a}$$

$$y = R\sqrt{1-\eta^2}\sin(\varphi) \tag{2.40b}$$

$$z = R\eta \tag{2.40c}$$

where now $(R, \eta, \varphi) = (u^1, u^2, u^3)$, then the surface can be defined by the equations

$$u^1 = R_0 \tag{2.41a}$$

$$u^2 = \eta \tag{2.41b}$$

$$u^3 = \varphi \tag{2.41c}$$

where again $(\eta, \varphi) = (v^1, v^2)$. The shift tensor is now

$$U^j_\alpha = \frac{\partial u^j}{\partial v^\alpha} = \begin{bmatrix} \frac{\partial u^1}{\partial v^1} & \frac{\partial u^1}{\partial v^2} \\ \frac{\partial u^2}{\partial v^1} & \frac{\partial u^2}{\partial v^2} \\ \frac{\partial u^3}{\partial v^1} & \frac{\partial u^3}{\partial v^2} \end{bmatrix} = \begin{bmatrix} 0 & 0 \\ 1 & 0 \\ 0 & 1 \end{bmatrix} \tag{2.42}$$

and the metric tensor of the ambient coordinates is (see Chap. 4)

$$
g_{mn} = \begin{bmatrix} 1 & 0 & 0 \\ 0 & R^2\frac{1}{1-\eta^2} & 0 \\ 0 & 0 & R^2\left(1-\eta^2\right) \end{bmatrix} \tag{2.43}
$$

Applying again Eq. (2.34), noticing that the ambient metric tensor must be evaluated on the surface ($R = R_0$), yields

$$
\begin{aligned}
s_{\alpha\beta} &= U_\alpha^1 U_\beta^1 g_{11} + U_\alpha^2 U_\beta^2 g_{22} + U_\alpha^3 U_\beta^3 g_{33} \\
&= \delta_\alpha^1 \delta_\beta^1 g_{22} + \delta_\alpha^2 \delta_\beta^2 g_{33} = \begin{bmatrix} g_{22} & 0 \\ 0 & g_{33} \end{bmatrix} = \begin{bmatrix} R_0^2\frac{1}{1-\eta^2} & 0 \\ 0 & R_0^2\left(1-\eta^2\right) \end{bmatrix}
\end{aligned} \tag{2.44}
$$

as already obtained. In the last case, the surface coordinates are *inherited* from the ambient coordinates and the surface metric $s_{\alpha\beta}$ is then *inherited* from the ambient metric.

The following relation also holds $U_k^\alpha U_\beta^k = \delta_\beta^\alpha$, in fact from Eq. (2.16) and Eq. (2.34)

$$
\delta_\beta^\alpha = s^{\alpha\gamma} s_{\gamma\beta} = s^{\alpha\gamma} \mathbf{s}_\gamma \cdot \mathbf{s}_\beta = s^{\alpha\gamma} U_\gamma^j U_\beta^k \mathbf{g}_j \cdot \mathbf{g}_k = s^{\alpha\gamma} U_\gamma^j U_\beta^k g_{jk} = U_k^\alpha U_\beta^k \tag{2.45}
$$

Consider now an ambient vector $\mathbf{V}$ and calculate the derivative with respect to *surface* coordinates

$$
\begin{aligned}
\mathbf{V}_{,\alpha} &= \frac{\partial \mathbf{V}}{\partial v^\alpha} = \frac{\partial \left(V^j \mathbf{g}_j\right)}{\partial v^\alpha} = \frac{\partial \left(V^j\right)}{\partial v^\alpha}\mathbf{g}_j + V^j \frac{\partial \left(\mathbf{g}_j\right)}{\partial v^\alpha} = V_{,\alpha}^j \mathbf{g}_j + V^j \frac{\partial \mathbf{g}_j}{\partial u^k}\frac{\partial u^k}{\partial v^\alpha} \\
&= V_{,\alpha}^m \mathbf{g}_m + V^j \begin{Bmatrix} m \\ j\ k \end{Bmatrix} \mathbf{g}_m U_\alpha^k = \left[V_{,\alpha}^m + V^j \begin{Bmatrix} m \\ j\ k \end{Bmatrix} U_\alpha^k\right] \mathbf{g}_m
\end{aligned} \tag{2.46}
$$

where the equation $\frac{\partial \mathbf{g}_j}{\partial u^k} = \begin{Bmatrix} m \\ j\ k \end{Bmatrix}\mathbf{g}_m$ (see Sect. 1.13) has been used.

The *surface* covariant derivative of the contravariant components of the *ambient* vector can then be defined as

$$
V_{/\alpha}^j = V_{,\alpha}^j + V^n \begin{Bmatrix} j \\ n\ k \end{Bmatrix} U_\alpha^k \tag{2.47}
$$

and noticing that $V_{,\alpha}^m = \frac{\partial V^m}{\partial v^\alpha} = \frac{\partial V^m}{\partial u^k}\frac{\partial u^k}{\partial v^\alpha} = V_{,k}^m U_\alpha^k$ the following compact result holds

$$
V_{/\alpha}^j = \left[V_{,k}^j + V^n \begin{Bmatrix} j \\ n\ k \end{Bmatrix}\right] U_\alpha^k = V_{/k}^j U_\alpha^k \tag{2.48}
$$

that allows to find the *surface* derivative of an *ambient* vector from the *ambient* derivatives, quite consistently with the meaning given to the shift tensor (see above).

The relation can be extended to any tensor and two particular results are

$$g_{jk/\alpha} = g_{jk/m}U_\alpha^m = 0 \tag{2.49}$$

$$\mathbf{g}_{j/\alpha} = 0 \tag{2.50}$$

The first one comes from Eq. (1.173) (Chap. 1) while for the second one, from the definition of covariant derivative

$$\mathbf{g}_{j/\alpha} = \mathbf{g}_{j,\alpha} - \mathbf{g}_n \begin{Bmatrix} n \\ j\ k \end{Bmatrix} U_\alpha^k \tag{2.51}$$

and since

$$\mathbf{g}_{j,\alpha} = \mathbf{g}_{j,k}\frac{\partial u^k}{\partial v^\alpha} = \mathbf{g}_n \begin{Bmatrix} n \\ j\ k \end{Bmatrix} U_\alpha^k \tag{2.52}$$

the identity is proven.

For a tensor $\mathbf{T}$ with both kinds of indexes (surface and ambient) the rule can be understood from the following example:

$$T^i_{\beta/\gamma} = T^i_{\beta,\gamma} + \begin{Bmatrix} i \\ n\ k \end{Bmatrix} T^n_\beta U^k_\gamma - T^i_\omega \begin{Bmatrix} \omega \\ \gamma\ \beta \end{Bmatrix}. \tag{2.53}$$

## 2.4 The Normal Vector and the Curvature Tensor

As already pointed out, the concept of vectors and tensors, metric tensors, covariant derivatives, etc., on a 2D surface can be developed without even relating to the ambient coordinates, exactly as the same concepts for the 3D space were developed in Chap. 1. However, to appreciate the concept of surface curvature, the fact that we may consider the 2D space embedded in a 3D Euclidean space is of great help in interpreting it, since intuitively the surface curvature can be related to the deviation of the surface from the tangent plane. All those properties of a surface that can be derived without reference to the ambient space are called *intrinsic* (we will see that Gaussian curvature, named after the celebrated German mathematician Carl Friedrich Gauss, 1777–1855, is an intrinsic property of a surface), while those related to the ambient space are called *extrinsic* (and the mean curvature is one of them).

The normal vector $\check{\mathbf{N}}$ to a surface can be defined only relying on the 3D space, in fact by definition it is orthogonal to the tangential plane, i.e.:

$$\check{\mathbf{N}} \cdot \mathbf{s}_\alpha = 0 \tag{2.54}$$

moreover its length is unitary, $\left|\check{\mathbf{N}}\right| = 1$. The contravariant components of the normal vector in ambient coordinates are $N^j$ while all components on the surface are clearly nil, and this leads to the following relation with the shift tensor:

$$0 = \check{\mathbf{N}} \cdot \mathbf{s}_\alpha = \check{N}^j \mathbf{g}_j \cdot U^k_\alpha \mathbf{g}_k = \check{N}^j g_{jk} U^k_\alpha = \check{N}_k U^k_\alpha \qquad (2.55)$$

and

$$0 = \check{N}_k U^k_\alpha s^{\alpha\beta} = \check{N}^j g_{jk} U^k_\alpha s^{\alpha\beta} = \check{N}^j U^\beta_j \qquad (2.56)$$

There is another way to define the normal vector. Consider first the implicit definition of a surface in Cartesian coordinates

$$F\left(x^j\right) = 0 \qquad (2.57)$$

from ordinary calculus we know that the gradient of $F$, i.e. the vector with covariant components $F_{,k} = \frac{\partial F}{\partial x^k}$, is orthogonal to that surface. In fact, since the function $F$ is constant over the surface, then the derivatives $\frac{\partial F}{\partial v^\alpha}$, that are covariant components of a surface vector, $F_{,\alpha}$, are all nil, and

$$0 = \frac{\partial F}{\partial v^\alpha} = F_{,\alpha} = \frac{\partial F}{\partial x^k} \frac{\partial x^k}{\partial v^\alpha} = F_{,k} X^k_\alpha \qquad (2.58)$$

where $X^k_\alpha$ is the shift tensor in Cartesian coordinates (see the general definition 2.24), which shows that $F_{,\alpha}$ are the surface components of the gradient and they are all nil. The gradient is then normal to the surface and the normal vector can then be defined as

$$\check{N}_k = \frac{F_{,k}}{|F_{,k}|}; \quad \check{N}^j = g^{jk} \check{N}_k \qquad (2.59)$$

Starting, instead, from the implicit definition of a surface in a general coordinate system

$$G\left(u^j\right) = 0 \qquad (2.60)$$

again the gradient of $G$ is orthogonal to the surface, as it can be shown exactly as above, and

$$\check{N}_k = \frac{G_{,j}}{|G_{,j}|} \qquad (2.61)$$

We can observe that the set of versors $\left(\mathbf{s}_1, \mathbf{s}_2, \check{\mathbf{N}}\right)$ form a basis for the 3D ambient space at any point of the surface, then any ambient vector $\mathbf{V}$ at any point of the surface can be defined by its components with respect to this basis (see Fig. 2.1).

Given the contravariant components $V^j$ of a vector $\mathbf{V}$ in ambient coordinates, those with respect to the new basis can be evaluated in the following way. Let first evaluate the normal component of the vector $\mathbf{V}$, that can be found as:

$$V_n = \mathbf{V} \cdot \check{\mathbf{N}} = V^j \check{N}_j = V_j \check{N}^j \qquad (2.62)$$

**Fig. 2.1** An ambient vector **V** and its components tangent, **W**, and normal, $\mathbf{V}^{(N)}$, to the surface.

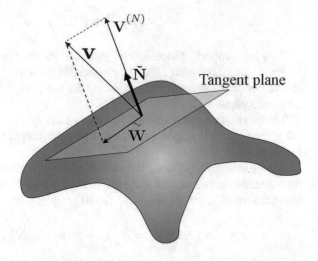

from its definition the vector $\mathbf{V}^{(N)} = V_n \check{\mathbf{N}} = V_j \check{N}^j \check{\mathbf{N}}$ has nil components along the tangential plane, and the vector $\mathbf{V}^{(T)} = \mathbf{V} - V_n \check{\mathbf{N}}$ has nil components normal to the surface, since

$$\mathbf{V}^{(T)} \cdot \check{\mathbf{N}} = \left(\mathbf{V} - V_n \check{\mathbf{N}}\right) \cdot \check{\mathbf{N}} = \mathbf{V} \cdot \check{\mathbf{N}} - V_n \check{\mathbf{N}} \cdot \check{\mathbf{N}} = 0 \qquad (2.63)$$

then $\mathbf{V}^{(T)} = \widetilde{\mathbf{W}}$ is a surface vector, $\mathbf{V} = \widetilde{\mathbf{W}} + \mathbf{V}^{(N)}$ and

$$\mathbf{V} \cdot \mathbf{s}^\alpha = \left(\widetilde{\mathbf{W}} + \mathbf{V}^{(N)}\right) \cdot \mathbf{s}^\alpha = \widetilde{\mathbf{W}} \cdot \mathbf{s}^\alpha = \widetilde{W}^{\,\alpha} \qquad (2.64)$$

The components of $\mathbf{V}^{(N)}$ in ambient coordinates are

$$V^{(N)k} = \mathbf{V}^{(N)} \cdot \mathbf{g}^k = V^j \check{N}_j \check{\mathbf{N}} \cdot \mathbf{g}^k = V^j \check{N}_j \check{N}^k \qquad (2.65)$$

and the components of $\widetilde{\mathbf{W}}$ in ambient coordinates are (using Eq. 2.64)

$$\widetilde{W}^k = U_\alpha^k \widetilde{\mathbf{W}} = U_\alpha^k \mathbf{V} \cdot \mathbf{s}^\alpha = U_\alpha^k V^j \mathbf{g}_j \cdot \mathbf{s}^\alpha = V^j U_\alpha^k U_j^\alpha \qquad (2.66)$$

so the components of $\mathbf{V}$ in ambient coordinates are

$$V^k = V^{(N)k} + \widetilde{W}^k = V^j N_j N^k + V^j U_\alpha^k U_j^\alpha = V^j \left(\check{N}_j \check{N}^k + U_\alpha^k U_j^\alpha\right) \qquad (2.67)$$

and the arbitrariness of $\mathbf{V}$ yields

$$\left(\check{N}_j \check{N}^k + U_\alpha^k U_j^\alpha\right) = \delta_j^k \tag{2.68}$$

This is an important identity that will be often used in the following.

From Eqs. (2.65) and (2.66), we can see that the tensor $\check{N}_j \check{N}^k$ acts as the *projector* of a vector on the normal direction, while the tensor $U_\alpha^k U_j^\alpha$ projects the vector on the tangential plane.

The fact that we can consider a surface as a 2D space embedded into a (Euclidean) 3D space helps also to give a first, intuitive definition of the concept of *curvature* of a surface.

For a flat surface (a plane) the normal vector is constant along the surface, while for a generic curved surface it is not. The derivative of the normal vector along the non-flat surface is not nil and, from (2.48)

$$\check{\mathbf{N}}_{,\alpha} = \left(\check{N}^j \mathbf{g}_j\right)_{,\alpha} = \check{N}_{,\alpha}^j \mathbf{g}_j + \check{N}^j \mathbf{g}_{j,\alpha} = \check{N}_{,\alpha}^j \mathbf{g}_j + \check{N}^j \mathbf{g}_{j,k} U_\alpha^k \tag{2.69}$$

$$= \left[\check{N}_{,\alpha}^n + \check{N}^j \left\{ \begin{matrix} n \\ k\ j \end{matrix} \right\} U_\alpha^k \right] \mathbf{g}_n = \check{N}_{/k}^n U_\alpha^k \mathbf{g}_n = \check{N}_{/\alpha}^n \mathbf{g}_n$$

From the covariant derivatives of the shift tensor a new tensor can be defined as

$$B_{\alpha\beta} = \check{N}_j U_{\alpha/\beta}^j \tag{2.70}$$

which is called *extrinsic curvature tensor*. The reason for such a name will become clearer in Sect. 2.8, but at present we can give a hint about it. Since Eq. (2.55) states that $\check{N}_j U_\alpha^j = 0$, taking the covariant derivative yields: $\check{N}_j U_{\alpha/\beta}^j + \check{N}_{j/\beta} U_\alpha^j = 0$ and then

$$B_{\alpha\beta} = \check{N}_j U_{\alpha/\beta}^j = -\check{N}_{j/\beta} U_\alpha^j = -g_{jk} \check{N}_{/\beta}^k U_\alpha^j \tag{2.71}$$

showing that the *extrinsic curvature tensor* $B_{\alpha\beta}$ is generally not nil when the derivative of the normal vector along the surface $\check{N}_{/\beta}^k$ is not nil, i.e. when the surface is not flat.

The mixed component of the extrinsic curvature tensor becomes

$$B_\beta^\gamma = s^{\gamma\alpha} B_{\alpha\beta} = -s^{\gamma\alpha} g_{jk} \check{N}_{/\beta}^k U_\alpha^j = -\check{N}_{/\beta}^k U_k^\gamma \tag{2.72}$$

where the last equality is obtained by the lowering and raising index procedure. Multiplying both sides by $U_\gamma^m$ and contracting on $\gamma$ yields: $U_\gamma^m B_\beta^\gamma = -\check{N}_{/\beta}^k U_k^\gamma U_\gamma^m$ and using Eq. (2.68) to eliminate $U_k^\gamma U_\gamma^m$ yields

$$U_\gamma^m B_\beta^\gamma = -\check{N}_{/\beta}^k \left[\delta_k^m - \check{N}^m \check{N}_k\right] = -\check{N}_{/\beta}^m - \check{N}_k \check{N}_{/\beta}^k N^m = -\check{N}_{/\beta}^m \tag{2.73}$$

where the last equality is obtained observing that $\check{N}_k \check{N}_{/\beta}^k = 0$ since $\left(\check{N}^k \check{N}_k\right)_{/\beta} = 0$. Equation (2.73) is called *Weingarten* equation (named after the German mathematician Julius Weingarten, 1836–1910), and shows the direct relation between the tensor $B_\beta^\gamma$ and the derivatives of the normal vector along the surface, which, as already said, are nil for a flat surface and generally not nil for a curved surface.

## 2.5 Length and Area on a Surface

A curve on a surface can be defined in a parametric way by the functions

$$v^\alpha = v^\alpha(t) \tag{2.74}$$

where $t$ is a parameter. A point $P$ on this curve, corresponding to a value $t$ of the parameter, can also be determined by its position vector in the Euclidean 3D space: $\mathbf{r}(t)$. Taking two infinitely close points on this curve, say $P$ and $Q$, their positions are defined by the position vectors $\mathbf{r}(t)$ and $\mathbf{r}(t+dt)$ and their distance can be calculated as

$$ds = |\mathbf{r}(t+dt) - \mathbf{r}(t)| = \sqrt{d\mathbf{r}{\cdot}d\mathbf{r}} \tag{2.75}$$

and since

$$d\mathbf{r} = \frac{d\mathbf{r}}{dt}dt = \frac{\partial \mathbf{r}}{\partial v^\alpha}\frac{\partial v^\alpha}{\partial t}dt = \mathbf{s}_\alpha\frac{\partial v^\alpha}{\partial t}dt \tag{2.76}$$

then

$$ds = \sqrt{\frac{\partial v^\alpha}{\partial t}\frac{\partial v^\beta}{\partial t}\mathbf{s}_\alpha{\cdot}\mathbf{s}_\beta}dt = \sqrt{\frac{\partial v^\alpha}{\partial t}\frac{\partial v^\beta}{\partial t}s_{\alpha\beta}}dt \tag{2.77}$$

Given now two points $A$ and $B$ on the curve, defined respectively by the two values $t_1$ and $t_2$ of the parameter $t$, the distance between the two points, measured along the curve, is

$$s = \int_{t_1}^{t_2} \sqrt{\frac{\partial v^\alpha}{\partial t}\frac{\partial v^\beta}{\partial t}s_{\alpha\beta}}dt \tag{2.78}$$

As an example, take again a sphere of radius $R_0$, defined by the equations

$$x = R_0\sqrt{1 - \eta^2}\cos(\varphi) \tag{2.79a}$$
$$y = R_0\sqrt{1 - \eta^2}\sin(\varphi) \tag{2.79b}$$
$$z = R_0\eta \tag{2.79c}$$

where the surface coordinates are $(v^1, v^2) = (\eta, \varphi)$ (see Fig. 2.2) and define a curve on it by the equations

**Fig. 2.2** Curve over a
spherical surface

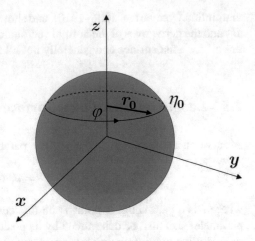

$$\eta = \eta_0 \tag{2.80a}$$

$$\varphi = t \tag{2.80b}$$

that for $t_1 = 0$ and $t_2 = 2\pi$ defines a circumference of radius $r_0 = \sqrt{R_0^2 \left(1 - \eta_0^2\right)}$
(see Fig. 2.2).

Lets calculate the length of the curve over the sphere by Eq. (2.78); we have
already seen that

$$s_{\alpha\beta} = \begin{bmatrix} R_0^2 \frac{1}{1-\eta^2} & 0 \\ 0 & R_0^2 \left(1 - \eta^2\right) \end{bmatrix} \tag{2.81}$$

and

$$\frac{\partial \eta}{\partial t} = 0; \quad \frac{\partial \varphi}{\partial t} = 1 \tag{2.82}$$

Then

$$\frac{\partial v^\alpha}{\partial t} \frac{\partial v^\beta}{\partial t} s_{\alpha\beta} = \frac{\partial v^1}{\partial t} \frac{\partial v^1}{\partial t} s_{11} + \frac{\partial v^2}{\partial t} \frac{\partial v^2}{\partial t} s_{22} = \tag{2.83}$$

$$= \left(\frac{\partial \eta}{\partial t}\right)^2 R_0^2 \frac{1}{1-\eta^2} + \left(\frac{\partial \varphi}{\partial t}\right)^2 R_0^2 \left(1 - \eta^2\right) = R_0^2 \left(1 - \eta^2\right)$$

Finally one obtains

$$\int_0^{2\pi} \sqrt{\frac{\partial v^\alpha}{\partial t} \frac{\partial v^\beta}{\partial t} s_{\alpha\beta}} dt = \int_0^{2\pi} \sqrt{R_0^2 \left(1 - \eta_0^2\right)} dt = 2\pi \sqrt{R_0^2 \left(1 - \eta_0^2\right)} = 2\pi r_0 \tag{2.84}$$

as expected for this simple case.

**Fig. 2.3** Infinitesimal
element of surface

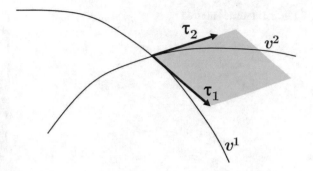

Consider now an infinitesimal element of the surface at any given point (see
Fig. 2.3), which is defined by the two infinitesimal vectors belonging to the tangential
plane and tangent to the coordinate lines.

As for the volume element (see Sect. 1.11), the area element is given by $\sqrt{\breve{\sigma}}$ where
$\breve{\sigma} = \det |s_{mn}|$ and the infinitesimal area on the surface is then

$$dA = \sqrt{\breve{\sigma}}dv^1 dv^2 \tag{2.85}$$

The area of a given portion of a surface can then be calculated as

$$A = \int_A dA = \int \int \sqrt{\breve{\sigma}}dv^1 dv^2 \tag{2.86}$$

and the surface integral of a quantity $T$ is then

$$\int_A T dA = \int \int T\sqrt{\breve{\sigma}}dv^1 dv^2 \tag{2.87}$$

As an example, let us calculate the surface of a prolate ellipsoid, defined by the
parametric equations

$$x = a_r\sqrt{1 - \eta^2}\cos(\varphi) \tag{2.88a}$$
$$y = a_r\sqrt{1 - \eta^2}\sin(\varphi) \tag{2.88b}$$
$$z = a_z\eta \tag{2.88c}$$

where the surface coordinates are $(v^1, v^2) = (\eta, \varphi)$ and $a_r, a_z$ are the semi-axes (see
Fig. 2.4) and $\alpha_z > a_r$.

In prolate spheroidal coordinates (see Chap. 4 for all details about this coordinate
system), which are defined by their relation with Cartesian coordinates

**Fig. 2.4** Prolate ellipsoid

$$x = a\sqrt{\zeta^2 - 1}\sqrt{1 - \eta^2}\cos(\varphi) \tag{2.89a}$$

$$y = a\sqrt{\zeta^2 - 1}\sqrt{1 - \eta^2}\sin(\varphi) \tag{2.89b}$$

$$z = a\zeta\eta \tag{2.89c}$$

where $\left(u^1, u^2, u^3\right) = (\zeta, \eta, \varphi)$, the surface can be described by the simple equation:

$$\zeta = \zeta_0 = \frac{\varepsilon}{\left(\varepsilon^2 - 1\right)^{1/2}} \tag{2.90}$$

where $\varepsilon = \frac{a_z}{a_r} > 1$ and $a = a_r\left(\varepsilon^2 - 1\right)^{1/2} = a_z\frac{\left(\varepsilon^2 - 1\right)^{1/2}}{\varepsilon}$. The ambient metric tensor in this coordinate system can be calculated by Eq. (1.63) yielding (see also Chap. 4)

$$g_{jk} = \begin{bmatrix} a^2\frac{\zeta^2 - \eta^2}{\zeta^2 + 1} & 0 & 0 \\ 0 & a^2\frac{\zeta^2 - \eta^2}{1 - \eta^2} & 0 \\ 0 & 0 & a^2\left(\zeta^2 - 1\right)\left(1 - \eta^2\right) \end{bmatrix} \tag{2.91}$$

and since the shift tensor in this case is simply

$$U_\alpha^j = \frac{\partial u^j}{\partial v^\alpha} = \begin{bmatrix} \frac{\partial u^1}{\partial v^1} & \frac{\partial u^1}{\partial v^2} \\ \frac{\partial u^2}{\partial v^1} & \frac{\partial u^2}{\partial v^2} \\ \frac{\partial u^3}{\partial v^1} & \frac{\partial u^3}{\partial v^2} \end{bmatrix} = \begin{bmatrix} 0 & 0 \\ 1 & 0 \\ 0 & 1 \end{bmatrix} \tag{2.92}$$

the surface metric tensor (*inherited* for the ambient one) is

$$s_{\alpha\beta} = U_\alpha^1 U_\beta^1 g_{11} + U_\alpha^2 U_\beta^2 g_{22} + U_\alpha^3 U_\beta^3 g_{33} = \delta_\alpha^1\delta_\beta^1 g_{22} + \delta_\alpha^2\delta_\beta^2 g_{33} \tag{2.93}$$

$$= \begin{bmatrix} g_{22} & 0 \\ 0 & g_{33} \end{bmatrix} = \begin{bmatrix} a^2\frac{\zeta_0^2 - \eta^2}{1 - \eta^2} & 0 \\ 0 & a^2\left(\zeta_0^2 - 1\right)\left(1 - \eta^2\right) \end{bmatrix}$$

and $\check{\sigma} = \det\left[s_{\alpha\beta}\right] = a^4\left(\zeta_0^2 - \eta^2\right)\left(\zeta_0^2 - 1\right)$. We can now calculate the surface of this ellipsoid by Eq. (2.86)

$$A = \int_A \sqrt{\check{\sigma}}d\eta d\varphi = a^2 \int \left[\int_{-1}^{1}\sqrt{\left(\zeta_0^2 - \eta^2\right)\left(\zeta_0^2 - 1\right)}d\eta\right]d\varphi \qquad (2.94)$$

$$= 2\pi a^2 \sqrt{\left(\zeta_0^2 - 1\right)} \int_{-1}^{1}\sqrt{\left(\zeta_0^2 - \eta^2\right)}d\eta = 2\pi a^2 \sqrt{\left(\zeta_0^2 - 1\right)}\left[\sqrt{\left(\zeta_0^2 - 1\right)} + \zeta_0^2 arccsc\left(\zeta_0\right)\right]$$

$$= 2\pi a_r^2 \left[1 + \frac{\varepsilon^2}{\left(\varepsilon^2 - 1\right)^{1/2}} arcsin\left(\frac{\left(\varepsilon^2 - 1\right)^{1/2}}{\varepsilon}\right)\right].$$

## 2.6 The Fundamental Forms

A classical approach to the differential geometry of surfaces starts by defining two quadratic forms, called *first* and *second fundamental forms*. Out of these quadratic forms it is possible to obtain the curvature characteristics of the surface. We are going to see that those quadratic forms can actually be expressed as two surface tensors, and that the curvature characteristics, among which the principal curvatures are the most fundamental ones, can be obtained out of those tensors.

The first fundamental form is the quadratic expression

$$I = Edv^1dv^1 + 2Fdv^1dv^2 + Gdv^2dv^2 \qquad (2.95)$$

where $I$ is the length element $ds^2 = s_{\alpha\beta}dv^\alpha dv^\beta$, then

$$\begin{bmatrix} E & F \\ F & G \end{bmatrix} = \begin{bmatrix} s_{11} & s_{12} \\ s_{21} & s_{22} \end{bmatrix} \qquad (2.96)$$

The second fundamental form is defined as

$$II = Ldv^1dv^1 + 2Mdv^1dv^2 + Ndv^2dv^2 \qquad (2.97)$$

where the coefficients $L$, $M$ and $N$ are related to the Hessian (named after the German mathematician Otto Hesse 1811–1874) of the position vector

$$\begin{bmatrix} L & M \\ M & N \end{bmatrix} = \begin{bmatrix} H_{11} & H_{12} \\ H_{21} & H_{22} \end{bmatrix} \qquad (2.98)$$

where [1]

$$H_{\alpha\beta} = \mathbf{r}_{/\alpha\beta} \cdot \check{\mathbf{N}} = \mathbf{s}_{\alpha/\beta} \cdot \check{\mathbf{N}} \qquad (2.99)$$

The term $\mathbf{s}_{\alpha/\beta} = \frac{\partial \mathbf{s}_\alpha}{\partial v^\beta} - \{{}^{\omega}_{\alpha\,\beta}\}\mathbf{s}_\omega$ can be developed considering that

$$\frac{\partial \mathbf{s}_\alpha}{\partial v^\beta} = \frac{\partial U^m_\alpha \mathbf{g}_m}{\partial v^\beta} = \mathbf{g}_m U^m_{\alpha,\beta} + U^m_\alpha \frac{\partial \mathbf{g}_m}{\partial u^k}\frac{\partial u^k}{\partial v^\beta} = \mathbf{g}_m U^m_{\alpha,\beta} + U^m_\alpha U^k_\beta \left\{{}^{\;p}_{m\;k}\right\}\mathbf{g}_p = \left(U^m_{\alpha,\beta} + U^j_\alpha U^k_\beta \left\{{}^{\;m}_{j\;k}\right\}\right)\mathbf{g}_m \tag{2.100}$$

where the definition $U^k_\beta = \frac{\partial u^k}{\partial v^\beta}$ and Eq. (1.149) have been used, then

$$\mathbf{s}_{\alpha/\beta} = \left(U^m_{\alpha,\beta} + U^j_\alpha \left\{{}^{\;m}_{k\;j}\right\}U^k_\beta\right)\mathbf{g}_m - \left\{{}^{\;\omega}_{\alpha\,\beta}\right\}U^m_\omega \mathbf{g}_m = U^m_{\alpha/\beta}\mathbf{g}_m \tag{2.101}$$

The coefficients of the second fundamental form $H_{\alpha\beta}$ then become

$$H_{\alpha\beta} = \mathbf{s}_{\alpha/\beta} \cdot \check{\mathbf{N}} = U^m_{\alpha/\beta}\mathbf{g}_m \cdot \check{\mathbf{N}} = U^m_{\alpha/\beta}N_m = B_{\alpha\beta} \tag{2.102}$$

which is the *extrinsic curvature tensor* $B_{\alpha\beta}$.

The two fundamental forms are used to find the so-called principal curvatures as the two eigensolutions $\kappa_{1,2}$ of the equation

$$\det\left[H_{\alpha\beta} - \kappa s_{\alpha\beta}\right] = 0. \tag{2.103}$$

## 2.7   The Curvatures of a Surface

The concept of curvature for a plane curve can be understood in a simple geometric way. Consider the curve of Fig. 2.5, that can be described in a parametric way by

$$x = x(t); \quad y = y(t) \tag{2.104}$$

At a given point $P$ of the curve, the tangent vector $\check{\mathbf{T}}$ is defined by the components

**Fig. 2.5** Plane curve with obsculating circle

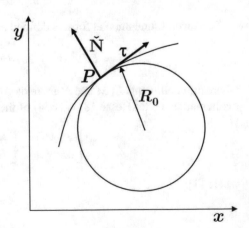

$$\check{T}_x = \frac{dx}{dt}; \quad \check{T}_y = \frac{dy}{dt} \tag{2.105}$$

and the corresponding unit vector $\tau$ as

$$\tau_\alpha = \frac{\check{T}_\alpha}{\left|\check{T}\right|} = \frac{\check{T}_\alpha}{\sqrt{\check{T}_x^2 + \check{T}_y^2}} \tag{2.106}$$

The curvature of this curve is defined as

$$\kappa = \left|\frac{d\tau}{ds}\right| \tag{2.107}$$

where $s$ is the arc length coordinate, which can be related to the generic parameter $t$ by the equation

$$ds = \sqrt{dx^2 + dy^2} = \left|\check{T}\right| dt \tag{2.108}$$

The simple reason why $\kappa$ is called *curvature* can be understood by applying the same definition to a circumference, given by

$$x = R_0 \cos t; \quad y = R_0 \sin t \tag{2.109}$$

in fact in this case

$$\check{T}_x = -R_0 \sin t; \quad \check{T}_y = R_0 \cos t \tag{2.110}$$

then $\left|\check{T}\right| = R_0$ and

$$\tau_x = -\sin t; \quad \tau_y = \cos t \tag{2.111}$$

Since now $ds = R_0 dt$, then

$$\kappa = \left|\frac{d\tau}{ds}\right| = \left|\frac{d\tau}{dt}\frac{dt}{ds}\right| = \frac{1}{R_0}\left|\frac{d\tau}{dt}\right| = \frac{1}{R_0} \tag{2.112}$$

For the generic plane curve the curvature is then the reciprocal of the radius of the osculating circle (see Fig. 2.5). Interestingly, defining the normal to the curve at the given point as $\check{\mathbf{N}} = \left(\check{N}_x, \check{N}_y\right) = \left(\tau_y, -\tau_x\right)$, the curvature may be defined as

$$\kappa = \left|\frac{d\check{\mathbf{N}}}{ds}\right| \tag{2.113}$$

and even as

$$\kappa = \frac{d\check{\mathbf{N}}}{ds} \cdot \tau \tag{2.114}$$

**Fig. 2.6** Curve defined by
the intersection of a plane
with a surface

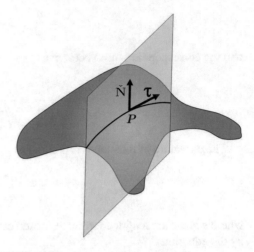

In fact: $\frac{d\check{\mathbf{N}}}{ds}$ has only a component parallel to $\boldsymbol{\tau}$ since: $0 = \frac{d\check{\mathbf{N}}\cdot\check{\mathbf{N}}}{ds} = 2\frac{d\check{\mathbf{N}}}{ds} \cdot \check{\mathbf{N}}$, then
its length is equal to the length of the tangential component, i.e. $\frac{d\check{\mathbf{N}}}{ds} \cdot \boldsymbol{\tau} = \left|\frac{d\check{\mathbf{N}}}{ds}\right|$.

Consider now a smooth surface, choosing any point $P$ we can find the normal
vector $\check{\mathbf{N}}$ at that point, and using the surface coordinates, an infinity of tangent vectors,
belonging to the tangent plane at that point, with unitary length can be defined as

$$\boldsymbol{\tau} = \tau^{\alpha}\mathbf{s}_{\alpha} \tag{2.115}$$

where $\tau^{\alpha}\tau_{\alpha} = 1$. The vectors $\boldsymbol{\tau}$ and $\check{\mathbf{N}}$ define a unique plane and the intersection
between this plane and the surface yields a plane curve (see Fig. 2.6), which can be
parametrically described as

$$v^{\alpha} = v^{\alpha}(t) \tag{2.116}$$

where $v^{\alpha}$ are the surface coordinates.

The components of the unitary tangent vector along the curve is then

$$\tau^{\alpha} = \frac{dv^{\alpha}(t)}{ds} \tag{2.117}$$

where $s$ is the arc length. The curvature of that curve at point $P$ is the surface curvature
in the direction of $\boldsymbol{\tau}$. The value of the curvature can be calculated as above, taking
the derivative of $\boldsymbol{\tau}$ with respect to the arc length, or equivalently the derivative of $\mathbf{N}$,
along the curve, and since this derivative lays in the plane, then, apart for a sign, the
curvature is

$$\kappa = \frac{d\check{\mathbf{N}}}{ds} \cdot \boldsymbol{\tau} \tag{2.118}$$

The intrinsic derivative of the vector $\check{\mathbf{N}}$ can be calculated as

$$\frac{d\check{\mathbf{N}}}{ds} = \frac{d\left(\check{N}^j\mathbf{g}_j\right)}{ds} = \frac{d\check{N}^j}{ds}\mathbf{g}_j + \check{N}^j\frac{d\mathbf{g}_j}{ds} = \left(\check{N}^j_{/\beta}\mathbf{g}_j + \check{N}^j\mathbf{g}_{j/\beta}\right)\frac{dv^\beta}{ds} \quad (2.119)$$

using Eq. (2.73)

$$\frac{d\check{\mathbf{N}}}{ds} = \left(-U^j_\gamma B^\gamma_\beta\mathbf{g}_j + \check{N}^j\mathbf{g}_{j/\beta}\right)\frac{dv^\beta}{ds} = -U^j_\gamma B^\gamma_\beta\frac{dv^\beta}{ds}\mathbf{g}_j = -B^\gamma_\beta\frac{dv^\beta}{ds}\mathbf{s}_\gamma = -B^\gamma_\beta\tau^\beta\mathbf{s}_\gamma \quad (2.120)$$

where the second equality comes from $\mathbf{g}_{j/\beta} = 0$ (Eq. 2.50), the third equality comes from the relation $\mathbf{s}_\gamma = U^j_\gamma\mathbf{g}_j$ (Eq. 2.24) and the last equality comes from Eq. (2.117).

Since the derivative $\frac{d\check{\mathbf{N}}}{ds}$ is taken along the curve with direction $\boldsymbol{\tau} = \tau^\alpha\mathbf{s}_\alpha$, then, apart for a sign

$$\kappa = \frac{d\check{\mathbf{N}}}{ds}\cdot\boldsymbol{\tau} = -B^\gamma_\beta\tau^\beta\tau^\alpha\mathbf{s}_\gamma\cdot\mathbf{s}_\alpha = -B^\gamma_\beta\tau^\beta\tau^\alpha s_{\gamma\alpha} = -B^\gamma_\beta\tau^\beta\tau_\gamma \quad (2.121)$$

This is the curvature along the direction of $\boldsymbol{\tau}$ and it is a quadratic form. From a matrix algebra point of view, $B^\gamma_\beta$ are the elements of a 2x2 symmetric matrix $\left[B^\gamma_\beta\right]$ and $[\tau^\alpha]$ a unit vector, then $\kappa = -\left[\tau^\beta\right]\left[B^\gamma_\beta\right][\tau_\alpha]$ is a quadratic form and it is known that the minimum and maximum values of $\kappa$ are given by the two eigenvalues of the matrix $\left[B^\beta_\alpha\right]$, i.e. by the solutions to the equation

$$\det\left[B^\beta_\alpha - \kappa\delta^\beta_\alpha\right] = 0 \quad (2.122)$$

that is identical to Eq. (2.103).

### 2.7.1 Mean Curvature

The mean curvature takes its name from the fact that it is usually defined as the arithmetic mean of the minimum and maximum curvatures, however often this name is given to twice that value, i.e.

$$C_m = \kappa_1 + \kappa_2 \quad (2.123)$$

and this definition is used hereinafter.

Since we have shown that $\kappa_1$ and $\kappa_2$ are eigenvalues of $B^\alpha_\beta$, their sum is the trace of $B^\alpha_\beta$ i.e. $C_m = B^\alpha_\alpha$, which is an invariant, as in fact it cannot depend on the choice of the coordinates. Another definition of the mean curvature is related again to the surface normal, precisely the mean curvature equals, apart from the sign, the divergence of the normal vector

$$C_m = -\nabla_j \check{N}^j = -\check{N}^j_{/j} \tag{2.124}$$

and this can be proven as follows. From Eq. (2.69): $N^j_{/\alpha} = N^j_{/k} U^k_\alpha$, multiplying both sides by $U^\alpha_m$ and contracting over the index $\alpha$

$$\check{N}^j_{/\alpha} U^\alpha_m = \check{N}^j_{/k} U^k_\alpha U^\alpha_m = \check{N}^j_{/k} \left( \delta^k_m - \check{N}^k \check{N}_m \right) = \check{N}^j_{/m} - \check{N}^j_{/k} \check{N}^k \check{N}_m \tag{2.125}$$

where the second equality comes from Eq. (2.68). Equation (2.125) can be rewritten as

$$\check{N}^j_{/m} = \check{N}^j_{/\alpha} U^\alpha_m + \check{N}^j_{/k} \check{N}^k \check{N}_m \tag{2.126}$$

Multiplying both sides by $\delta^m_j$ and contracting yields

$$\check{N}^j_{/j} = \delta^m_j \check{N}^j_{/m} = \check{N}^j_{/\alpha} U^\alpha_j + \check{N}_j \check{N}^j_{/k} \check{N}^k = \check{N}^j_{/\alpha} U^\alpha_j \tag{2.127}$$

where the last equality comes from the fact that $\left( \check{N}^j \check{N}_j \right)_{/k} = 0$ and then $\check{N}_j \check{N}^j_{/k} = 0$. On the other hand, using Eq. (2.72)

$$C_m = B^\alpha_\alpha = -\check{N}^k_{/\alpha} U^\alpha_k = -\check{N}^k_{/k} \tag{2.128}$$

that proves the statement.


## 2.7.2 Gaussian Curvature

The Gaussian curvature is often introduced as the ratio of the determinants of the matrices $[H_{\alpha\beta}]$ and $[s_{\alpha\beta}]$, i.e. the matrices of the coefficients of the two fundamental forms

$$K_G = \frac{\det [H_{\alpha\beta}]}{\det [s_{\alpha\beta}]} \tag{2.129}$$

but also as the product of the two principal curvatures. These two statements are equivalent and it can be shown recalling that $B_{\alpha\beta} = H_{\alpha\beta}$ (see Sect. 2.7), then, since $B^\gamma_\alpha = s^{\gamma\beta} B_{\alpha\beta}$

$$\det [B^\gamma_\alpha] = \det [s^{\gamma\beta}] \det [B_{\alpha\beta}] = \frac{\det [B_{\alpha\beta}]}{\det [s_{\alpha\gamma}]} = \frac{\det [H_{\alpha\beta}]}{\det [s_{\alpha\gamma}]} = K_G \tag{2.130}$$

but the determinant of a matrix can always be written as the product of its eigenvalues $\kappa_1$ and $\kappa_2$, then

$$K_G = \det\left[B_\alpha^\gamma\right] = \kappa_1\kappa_2 \tag{2.131}$$

There is a way to show that the Gaussian curvature is an intrinsic property of the 2D space and this is found by the following tensorial reasoning. We have seen that in an Euclidean space the second covariant derivative commutes, but this is not in general true in non-Euclidean spaces, like a curved surface. In fact in this case

$$T_{/\beta/\alpha}^\gamma - T_{/\alpha/\beta}^\gamma = T^\varepsilon R_{\cdot\varepsilon\alpha\beta}^\gamma \tag{2.132}$$

where

$$R_{\cdot\varepsilon\alpha\beta}^\gamma = \left[\left\{\begin{matrix}\gamma\\\beta\,\varepsilon\end{matrix}\right\}_{,\alpha} - \left\{\begin{matrix}\gamma\\\alpha\varepsilon\end{matrix}\right\}_{,\beta} + \left\{\begin{matrix}\delta\\\beta\,\varepsilon\end{matrix}\right\}\left\{\begin{matrix}\gamma\\\alpha\,\delta\end{matrix}\right\} - \left\{\begin{matrix}\delta\\\alpha\varepsilon\end{matrix}\right\}\left\{\begin{matrix}\gamma\\\beta\,\delta\end{matrix}\right\}\right] \tag{2.133}$$

is the so-called Riemann-Christoffel tensor, and $R_{\omega\varepsilon\alpha\beta} = s_{\omega\gamma}R_{\cdot\varepsilon\alpha\beta}^\gamma$ is the fully covariant form. The proof is the following. The first covariant derivative of the vector $T^\gamma$ is a second order tensor

$$A_\alpha^\gamma = T_{/\alpha}^\gamma = T_{,\alpha}^\gamma + T^\omega\left\{\begin{matrix}\gamma\\\alpha\omega\end{matrix}\right\} \tag{2.134}$$

Let now calculate the first covariant derivatives of $A_\alpha^\gamma$ and $A_\beta^\gamma$

$$A_{\beta/\alpha}^\gamma = A_{\beta,\alpha}^\gamma - A_\omega^\gamma\left\{\begin{matrix}\omega\\\beta\,\alpha\end{matrix}\right\} + A_\beta^\omega\left\{\begin{matrix}\gamma\\\alpha\,\omega\end{matrix}\right\} = T_{/\beta/\alpha}^\gamma \tag{2.135a}$$

$$A_{\alpha/\beta}^\gamma = A_{\alpha,\beta}^\gamma - A_\omega^\gamma\left\{\begin{matrix}\omega\\\beta\,\alpha\end{matrix}\right\} + A_\alpha^\omega\left\{\begin{matrix}\gamma\\\beta\,\omega\end{matrix}\right\} = T_{/\alpha/\beta}^\gamma \tag{2.135b}$$

which are the second covariant derivatives of the vector $T^\gamma$. Their difference can be calculated as

$$T_{/\beta/\alpha}^\gamma - T_{/\alpha/\beta}^\gamma = A_{\beta,\alpha}^\gamma - A_{\alpha,\beta}^\gamma + A_\beta^\omega\left\{\begin{matrix}\gamma\\\alpha\,\omega\end{matrix}\right\} - A_\alpha^\omega\left\{\begin{matrix}\gamma\\\beta\,\omega\end{matrix}\right\} \tag{2.136}$$

$$= \left[T_{,\beta}^\gamma + T^\omega\left\{\begin{matrix}\gamma\\\beta\,\omega\end{matrix}\right\}\right]_{,\alpha} - \left[T_{,\alpha}^\gamma + T^\omega\left\{\begin{matrix}\gamma\\\alpha\omega\end{matrix}\right\}\right]_{,\beta}$$

$$+ \left[T_{,\beta}^\omega + T^\varepsilon\left\{\begin{matrix}\omega\\\beta\,\varepsilon\end{matrix}\right\}\right]\left\{\begin{matrix}\gamma\\\alpha\,\omega\end{matrix}\right\} - \left[T_{,\alpha}^\omega + T^\varepsilon\left\{\begin{matrix}\omega\\\alpha\varepsilon\end{matrix}\right\}\right]\left\{\begin{matrix}\gamma\\\beta\,\omega\end{matrix}\right\}$$

$$= T_{,\alpha}^\omega\left\{\begin{matrix}\gamma\\\beta\,\omega\end{matrix}\right\} + T^\omega\left\{\begin{matrix}\gamma\\\beta\,\omega\end{matrix}\right\}_{,\alpha} - T_{,\beta}^\omega\left\{\begin{matrix}\gamma\\\alpha\omega\end{matrix}\right\} - T^\omega\left\{\begin{matrix}\gamma\\\alpha\omega\end{matrix}\right\}_{,\beta}$$

$$+ T_{,\beta}^\omega\left\{\begin{matrix}\gamma\\\alpha\,\omega\end{matrix}\right\} + T^\varepsilon\left\{\begin{matrix}\omega\\\beta\,\varepsilon\end{matrix}\right\}\left\{\begin{matrix}\gamma\\\alpha\,\omega\end{matrix}\right\} - T_{,\alpha}^\omega\left\{\begin{matrix}\gamma\\\beta\,\omega\end{matrix}\right\} - T^\varepsilon\left\{\begin{matrix}\omega\\\alpha\varepsilon\end{matrix}\right\}\left\{\begin{matrix}\gamma\\\beta\,\omega\end{matrix}\right\}$$

$$= T^\varepsilon \left[ \left\{ \begin{matrix} \gamma \\ \beta\,\varepsilon \end{matrix} \right\}_{,\alpha} - \left\{ \begin{matrix} \gamma \\ \alpha\varepsilon \end{matrix} \right\}_{,\beta} + \left\{ \begin{matrix} \omega \\ \beta\,\varepsilon \end{matrix} \right\} \left\{ \begin{matrix} \gamma \\ \alpha\,\omega \end{matrix} \right\} - \left\{ \begin{matrix} \omega \\ \alpha\varepsilon \end{matrix} \right\} \left\{ \begin{matrix} \gamma \\ \beta\,\omega \end{matrix} \right\} \right]$$

$$= T^\varepsilon R^\gamma_{\cdot\varepsilon\alpha\beta}$$

In an Euclidean space (i.e. for 2D spaces: a plane), where Cartesian coordinates exist, this tensor vanishes, as it can be easily shown by observing that in Cartesian coordinates all the Christoffel symbols vanish, then $R^\gamma_{\cdot\varepsilon\alpha\beta} = 0$ and since this is a tensorial equation, it must hold for all coordinate systems. On a curved surface this is not in general true, then a non-nil Riemann-Christoffel tensor must be related to the fact that the surface is curved.

We can observe from the definition (2.133) that $R_{\omega\varepsilon\alpha\beta}$ is skew-symmetric with respect to $\alpha, \beta$, from the definition of the Christoffel symbols, it is also skew-symmetric with respect to $\omega, \varepsilon$, and it is symmetric with respect to the switching of the first pair of symbols $(\omega, \varepsilon)$ with the second one $(\alpha, \beta)$, i.e.

$$R_{\omega\varepsilon\alpha\beta} = -R_{\varepsilon\omega\alpha\beta}; \quad R_{\omega\varepsilon\alpha\beta} = -R_{\omega\varepsilon\beta\alpha}; \quad R_{\omega\varepsilon\alpha\beta} = R_{\alpha\beta\omega\varepsilon} \tag{2.137}$$

For a surface, where greek symbols span over 1 and 2, these properties have an important consequence: the tensor is fully defined by a unique number, in fact, setting a value for $R_{1212}$ all the other 16 components are either nil or they have just this value, maybe with a different sign and they can then be connected to this one by the simple relation

$$R_{\omega\varepsilon\alpha\beta} = e_{\omega\varepsilon}e_{\alpha\beta}R_{1212} = \varepsilon_{\omega\varepsilon}\varepsilon_{\alpha\beta}\frac{R_{1212}}{\breve{\sigma}} \tag{2.138}$$

then, $\frac{R_{1212}}{\breve{\sigma}}$ is an invariant. The Riemann-Christoffel curvature tensor shares a relationship with the extrinsic curvature tensor that can be stated as (see [1] for a demonstration)

$$B_{\omega\alpha}B_{\beta\gamma} - B_{\omega\beta}B_{\alpha\gamma} = R_{\omega\gamma\alpha\beta} \tag{2.139}$$

$$B^\delta_\alpha B_{\beta\gamma} - B^\delta_\beta B_{\alpha\gamma} = R^\delta_{\cdot\gamma\alpha\beta} \tag{2.140}$$

Then

$$R_{1212} = B_{11}B_{22} - B_{12}B_{12} = \det\left[B_{\alpha\beta}\right] \tag{2.141}$$

and since, from Eq. (2.130), $\det\left[B_{\alpha\beta}\right] = K_G\breve{\sigma}$ then $K_G = \frac{R_{1212}}{\breve{\sigma}}$ and

$$R_{\omega\varepsilon\alpha\beta} = \varepsilon_{\omega\varepsilon}\varepsilon_{\alpha\beta}K_G \tag{2.142}$$

or, since $\varepsilon^{\omega\varepsilon}\varepsilon_{\omega\varepsilon} = 2$,

$$K_G = \frac{1}{4}\varepsilon^{\omega\varepsilon}\varepsilon^{\alpha\beta}R_{\omega\varepsilon\alpha\beta} \tag{2.143}$$

This last equation shows two important properties of the Gaussian curvature. First, it is clearly an invariant (it is obtained as inner product of tensors), then it is independent of the coordinate system, as it is expected. Second, the Gaussian curvature can be calculated from the intrinsic property of the space, precisely form the surface metric tensor, and it is then an intrinsic property of the surface.

## 2.8 Alternative Relations to Calculate the Curvatures of a Surface

The previous sections have shown the definitions of the surface curvatures, and it was made clear that Gaussian and mean curvatures can be calculated form the principal curvatures. The reverse is also true, since

$$C_m = \kappa_1 + \kappa_2 \tag{2.144}$$

$$K_G = \kappa_1 \kappa_2 \tag{2.145}$$

then

$$\kappa_{1,2} = \frac{C_m \pm \sqrt{C_m^2 - 4K_G}}{2} \tag{2.146}$$

From the practical point of view, it is often convenient to use specific formulae, that can be rigorously derived from what said in the previous sections, to calculate the mean and the Gaussian curvatures from the equations that describe the surface, for some special cases.

If the surface is described by the explicit form in Cartesian coordinates, also called Monge's form (named after the French mathematician Gaspard Monge, 1746–1818), then the Gaussian curvature can be calculated as

$$K_G = \frac{z_{xx} z_{yy} - z_{xy}^2}{\left(1 + z_x^2 + z_y^2\right)^2} \tag{2.147}$$

while the mean curvature can be calculated as

$$C_m = \frac{z_{xx}\left(1 + z_y^2\right) - 2z_{xy}z_x z_y + z_{yy}\left(1 + z_x^2\right)}{\left(1 + z_x^2 + z_y^2\right)^{3/2}} \tag{2.148}$$

where $z_x$, $z_y$ are the first derivatives with respect to $x$ and $y$, and $z_{xx}$, $z_{yy}$ and $z_{xy}$ are the second derivatives.

For a rotational symmetric surface, described by the parametric equations

$$x = \Phi(u)\cos(v) \tag{2.149a}$$
$$y = \Phi(u)\sin(v) \tag{2.149b}$$
$$z = \Psi(u) \tag{2.149c}$$

the Gaussian curvature can be calculated as

$$K_G = \frac{(\Psi_{uu}\Phi_u - \Phi_{uu}\Psi_u)\,\Psi_u}{\left(\Phi_u^2 + \Psi_u^2\right)^2 \Phi} \tag{2.150}$$

and the mean curvature as

$$C_m = \frac{(\Psi_{uu}\Phi_u - \Phi_{uu}\Psi_u)\,\Phi + \left(\Phi_u^2 + \Psi_u^2\right)\Psi_u}{\left(\Phi_u^2 + \Psi_u^2\right)^{3/2} |\Phi|} \tag{2.151}$$

When the surface is described in the implicit way by the equation

$$G\left(u^1, u^2, u^3\right) = G(u, v, w) = 0 \tag{2.152}$$

the Gaussian and mean curvatures can be calculated (see [5]) from the matrix

$$\mathbf{H}(G) = \left[G_{jk}\right] = \begin{bmatrix} G_{uu} & G_{uv} & G_{uw} \\ G_{vu} & G_{vv} & G_{vw} \\ G_{wu} & G_{wv} & G_{ww} \end{bmatrix} \tag{2.153}$$

and the vector $\nabla G = [G_u, G_v, G_w]$ as

$$K_G = \frac{\nabla G \cdot \mathbf{H}^* \cdot \nabla G^T}{|\nabla G|^4} \tag{2.154}$$

$$C_m = \frac{\nabla G \cdot \mathbf{H}^* \cdot \nabla G^T - |\nabla G|^2\, Tr(\mathbf{H})}{|\nabla G|^3} \tag{2.155}$$

where

$$\mathbf{H}^* = \mathbf{H}^{-1} \det \mathbf{H} \tag{2.156}$$
$$Tr(\mathbf{H}) = G_{uu} + G_{vv} + G_{ww} \tag{2.157}$$
$$|\nabla G|^2 = G_u^2 + G_v^2 + G_w^2 \tag{2.158}$$

The following explicit relationship between $K_G$ and the metric tensor $s_{\alpha\beta}$, valid for orthogonal curvilinear coordinates, can be derived from Eq. (2.143) [6]

$$K_G = -\frac{1}{2\sqrt{\bar{\sigma}}} \left[ \frac{\partial}{\partial v^1} \left( \frac{1}{\sqrt{\bar{\sigma}}} \frac{\partial s_{22}}{\partial v^1} \right) + \frac{\partial}{\partial v^2} \left( \frac{1}{\sqrt{\bar{\sigma}}} \frac{\partial s_{11}}{\partial v^2} \right) \right]. \tag{2.159}$$

## 2.9 The Moving Surface

When dealing with two-phase problems, like drop heating and evaporation, interfaces play a prominent role, since they are separating different phases. Such surfaces are usually without strong constraints, then they are free to move and deform with time. This particular problem, which can be found also in many other engineering fields, needs to be treated separately since it has its own peculiarities.

A moving surface is defined by a set of equations, in a 3D Euclidean space and Cartesian ambient coordinate system $x^j$, as

$$x^j = x^j (v^\alpha, t) \tag{2.160}$$

where $v^\alpha$ is a 2D surface coordinate system and $t$ is a scalar parameter (time).

After defining a new ambient coordinate system as

$$w^j = [v^1, v^2, t] \tag{2.161}$$

equations (2.160) can be interpreted as the definition of the new curvilinear coordinate system $w^j$

$$x^j = x^j (w^k) \tag{2.162}$$

The basis of $w^j$ are

$$\mathbf{g}_\alpha = \frac{\partial \mathbf{r}}{\partial v^\alpha}; \; \mathbf{g}_3 = \frac{\partial \mathbf{r}}{\partial t} \tag{2.163}$$

where $\mathbf{r}$ is the position vector and $\frac{\partial \mathbf{r}}{\partial t} = \mathbf{g}_3 = \mathbf{V}$ can be defined as the velocity of a particle on the moving surface with fixed surface coordinates. The system $w^j$ is non-orthogonal, in general, and then the vector $\mathbf{V}$ has normal and tangential components.

In the Cartesian coordinate system: $\mathbf{V} = \check{V}^j \mathbf{e}_j$, where the $\vee$-shaped symbol (inverted caret) is used to indicate Cartesian components of vectors and tensors, while in the system $w^j$: $\mathbf{V} = V^j \mathbf{g}_j$, and $V^j$ are the components of the vector.

From the definition of $\mathbf{V}$

$$\mathbf{V} = \frac{\partial \mathbf{r}}{\partial t} = \frac{\partial \mathbf{r}}{\partial x^j} \frac{\partial x^j}{\partial t} = \mathbf{e}_j \frac{\partial x^j}{\partial t} \tag{2.164a}$$

$$\mathbf{V} = \frac{\partial \mathbf{r}}{\partial t} = \frac{\partial \mathbf{r}}{\partial w^j} \frac{\partial w^j}{\partial t} = \mathbf{g}_j \frac{\partial w^j}{\partial t} \tag{2.164b}$$

and then

$$\check{V}^j = \frac{\partial x^j}{\partial t}; \; V^j = \frac{\partial w^j}{\partial t} = [0, 0, 1] \tag{2.165}$$

where the last equality is obtained from Eq. (2.161). The shift tensor, in Cartesian ambient coordinates and in the system $w^j$, is

$$\check{X}^k_\alpha = \frac{\partial x^k}{\partial v^\alpha} \quad \text{Cartesian components} \tag{2.166}$$

$$U^k_\alpha = \frac{\partial w^k}{\partial v^\alpha} \quad \text{components in } w^j;$$

$$U^k_\alpha = \begin{bmatrix} \frac{\partial w^1}{\partial v^1} & \frac{\partial w^1}{\partial v^2} \\ \frac{\partial w^2}{\partial v^1} & \frac{\partial w^2}{\partial v^2} \\ \frac{\partial w^3}{\partial v^1} & \frac{\partial w^3}{\partial v^2} \end{bmatrix} = \begin{bmatrix} 1 & 0 \\ 0 & 1 \\ 0 & 0 \end{bmatrix} = \delta^k_\alpha \tag{2.167}$$

Let us now consider the projection tensor $U^k_\alpha U^\alpha_j$ (see Eq. 2.66) that can be used to project a vector (or tensor) on the surface. The projection of the velocity vector on the surface, namely $\mathbf{W} = \check{W}^k \mathbf{e}_k = W^k \mathbf{g}_k$ , which is a surface vector, can be obtained as

$$\check{W}^k = \check{V}^j \check{X}^k_\alpha \check{X}^\alpha_j \tag{2.168}$$

$$W^k = W^j U^k_\alpha U^\alpha_j \tag{2.169}$$

It must be noticed that these are the *ambient* components of the *surface* vector $\mathbf{W}$; the *surface* components (which will be indicated by a tilde) of the *surface* vector $\mathbf{W}$ are instead (see Eq. 2.31)

$$\tilde{W}^\beta = \check{W}^k X^\beta_k = W^k \tilde{U}^\beta_k \tag{2.170}$$

to notice that Greek letters are used as indices of the surface components of $\mathbf{W}$ and they span between 1 and 2.

Using Eqs. (2.168), (2.170) we obtain

$$\tilde{W}^\beta = \check{W}^k \check{X}^\beta_k = \check{V}^j \check{X}^k_\alpha \check{X}^\alpha_j \check{X}^\beta_k = \check{V}^j \check{X}^\alpha_j \delta^\beta_\alpha = \check{V}^j \check{X}^\beta_j \tag{2.171}$$

$$\tilde{W}^\beta = W^k U^\beta_k = V^j U^k_\alpha U^\alpha_j U^\beta_k = V^j U^\alpha_j \delta^\beta_\alpha = V^j U^\beta_j \tag{2.172}$$

where the identity $\delta^\beta_\alpha = \check{X}^k_\alpha \check{X}^\beta_k = U^k_\alpha U^\beta_k$ has been used.

From the second of Eq. (2.165) the surface component of the surface velocity vector $\mathbf{W}$ is given by

$$\tilde{W}^\beta = V^j U^\beta_j = U^\beta_3 = U^\beta_t. \tag{2.173}$$

### 2.9.1  The Invariant Time Derivative

Consider a scalar field $G$ and take the gradient $\mathbf{T} = \check{T}_k \mathbf{e}^k = T_k \mathbf{g}^k$

$$\check{T}_k = \frac{\partial G}{\partial x^k}; \ T_k = \frac{\partial G}{\partial w^k} = \left[\frac{\partial G}{\partial v^1}, \frac{\partial G}{\partial v^2}, \frac{\partial G}{\partial t}\right] \tag{2.174}$$

Let now calculate the projection of this vector on the normal to the surface, say **P**; this can be obtained through the projection tensor $\check{N}^j \check{N}_k$ (see Eq. 2.65)

$$P_j = T_k \check{N}^k \check{N}_j \tag{2.175}$$

where $\check{N}^k$ are the components of the normal vector $\check{\mathbf{N}}$ in the curvilinear coordinate system and, since $\check{N}^k \check{N}_j = \delta_j^k - U_\alpha^k U_j^\alpha$, then

$$P_j = T_j - T_k U_\alpha^k U_j^\alpha = T_j - T_\alpha U_j^\alpha \tag{2.176}$$

where the last equality is obtained from Eq. (2.167). Consider now the time component of this vector in the system $w^j$ (i.e. $j = 3$), from Eq. (2.173) one obtains

$$P_t = T_t - T_\alpha U_t^\alpha = T_t - \tilde{W}^\alpha T_\alpha \tag{2.177}$$

which is the so-called *invariant time derivative* [1]. The reason for this name is the following.

Consider a general time dependent change of surface coordinates

$$v^\alpha = v^\alpha \left(v^{\beta'}, t'\right) \tag{2.178}$$
$$t = t' \tag{2.179}$$

and a corresponding change of ambient coordinates, from $w^j$ to $\left[w^{j'}\right] = \left[v^{1'}, v^{2'}, t'\right]$. The ambient vector $P_j$ components transform as usual

$$P_j' = P_k \frac{\partial w^k}{\partial w^{j'}} = T_k \frac{\partial w^k}{\partial w^{j'}} - T_\alpha U_k^\alpha \frac{\partial w^k}{\partial w^{j'}} \tag{2.180}$$

and for the time component (i.e. $j = 3$)

$$P_t' = T_k \frac{\partial w^k}{\partial t'} - T_\alpha U_k^\alpha \frac{\partial w^k}{\partial t'} \tag{2.181}$$
$$= \left(T_t \frac{\partial t}{\partial t'} + T_\beta \frac{\partial w^\beta}{\partial t'}\right) - T_\alpha \left(U_t^\alpha \frac{\partial t}{\partial t'} + U_\beta^\alpha \frac{\partial w^\beta}{\partial t'}\right)$$
$$= T_t - T_\alpha U_t^\alpha + \left(T_\beta - T_\alpha U_\beta^\alpha\right) \frac{\partial w^\beta}{\partial t'} = T_t - T_\alpha U_k^\alpha = P_t$$

since $U_\beta^\alpha = \delta_\beta^\alpha$, then $P_t$ is an invariant, under transformations that do not change the time parameter.

## 2.9.2   The Normal Velocity of a Moving Surface

The normal velocity vector can be obtained subtracting from the velocity vector its tangential projection

$$\mathbf{C} = \mathbf{V} - \mathbf{W} \tag{2.182}$$

Consider now a new ambient coordinate system defined as: $\left[w^{j'}\right] = \left[v^{1'}, v^{2'}, t'\right]$, with the characteristics of being orthogonal. The time derivative of the position vector $\mathbf{r}$ is then $\frac{\partial \mathbf{r}}{\partial t'} = \mathbf{C} = \mathbf{g}'_3$ and from Eqs. (2.182), (2.168), (2.170)

$$C^j = V^j - V^m U^j_\alpha U^\alpha_m \tag{2.183}$$

$$\check{C}^j = \check{V}^j - \check{V}^m \check{X}^j_\alpha \check{X}^\alpha_m \tag{2.184}$$

$$C^{j'} = V^{j'} \tag{2.185}$$

in this last system $V^{j'} = C^{j'}$, since $\hat{W}^{k'} = 0$

The selected change of coordinates is equivalent to set

$$v^\alpha = v^\alpha \left(v^{\beta'}, t'\right) \tag{2.186}$$

$$t = t' \tag{2.187}$$

with the further condition of orthogonality

$$\tilde{g}'_{mn} = 0 \text{ if } m \neq n \tag{2.188}$$

and in particular

$$\tilde{g}'_{mt} = 0 \text{ if } m \neq n \tag{2.189}$$

i.e.

$$\delta_{kj} \frac{\partial x^j}{w^{m'}} \frac{\partial x^k}{w^{n'}} = 0 \ \text{ if } m \neq n \tag{2.190}$$

From this it can be observed that the normal velocity vector has a special meaning, since its definition is independent of the choice of the surface coordinate system. In fact the magnitude of the vector $\mathbf{C}$ is:

$$C = C \cdot \check{\mathbf{N}} = V \cdot \check{\mathbf{N}} - W \cdot \check{\mathbf{N}} = V \cdot \check{N} \tag{2.191}$$

and it is an invariant, then $\mathbf{C} = C \, \check{\mathbf{N}}$, and all the information regarding the motion of the surface are contained in the scalar quantity $C$ (see [1] for a more comprehensive analysis).

# References

1. Grinfeld, P.: Introduction to tensor analysis and the calculus of moving surfaces. Springer, New York (2013)
2. Eisenhart, L.P.: Riemannian Geometry. Princeton Univ Press, Princeton, NJ (1949)
3. Petersen, P.: Riemannian Geometry. Springer, Berlin (2006)
4. Cartan, E.: Geometry of Riemannian Spaces. Math Science Press, Berkeley (1983)
5. Goldman, R.: Curvature formulas for implicit curves and surfaces. Comput. Aided Geom. Des. **22**, 632–658 (2005)
6. Synge, J.L., Schild, A.: Tensor calculus, vol. 5. Courier Corporation, Chelmsford (1978)

# Chapter 3
# Separability of PDE

The analytical modelling of heat and mass transfer phenomena relies on the analytical solutions to partial differential equations, which are used to describe the conservation of mass, chemical species, momentum and energy and their transfer mechanisms, as it will be shown in Part II of this book. Analytical methods for this kind of problems are widely used (see [1]) and among the several available techniques to solve Partial Differential Equations (PDE), separation of variable is generally the most valuable one since it may yield solutions in a form that is easily implementable for routine calculations. Separability of a PDE depends on the chosen coordinate system and this chapter is devoted to analyse conditions and methods for PDE separation. A deep analysis of the involved methods can be found in [2] and the interested reader is invited to refer to it for a complete treatment. In this chapter the summation convention will not be used.

## 3.1 Definition of Separability

Separation of variables is a powerful analytical tool that allows to reduce the original partial differential equation to a set of ordinary differential equations. More generally, the original PDE can be separated into equations of lower dimensionality, but here we will confine ourself to the first definition. As already stated, we will assume that the space is 3D and Euclidean and in this section we will further assume that the curvilinear coordinate system $\left(u^1, u^2, u^3\right)$ is orthogonal. The definitions given below can be easily extended to spaces of higher dimensionality. We assume that the function $\Psi\left(u^1, u^2, u^3\right)$ is the solution of the given PDE. Separability of PDEs can be of two kinds: simple separability and R-separability.

© Springer Nature Switzerland AG 2021
G. E. Cossali and S. Tonini, *Drop Heating and Evaporation: Analytical Solutions in Curvilinear Coordinate Systems*, Mathematical Engineering,
https://doi.org/10.1007/978-3-030-49274-8_3

**Definition** (*Simple separability*) When the assumption

$$\Psi = U^1 \left(u^1\right) U^2 \left(u^2\right) U^3 \left(u^3\right) \tag{3.1}$$

where $U^j \left(u^j\right)$ are functions of one variable only, allows the separation of the PDE into three ordinary differential equations (ODEs), the PDE is said to be *simply separable*.

**Definition** (*R-separability*) When the assumption

$$\Psi = \frac{U^1 \left(u^1\right) U^2 \left(u^2\right) U^3 \left(u^3\right)}{\tilde{R} \left(u^1, u^2, u^3\right)} \tag{3.2}$$

where $\tilde{R} \left(u^1, u^2, u^3\right)$ is a non-constant function, allows the separation of the PDE into three ODEs, the equation is said to be *R-separable*. The case $\tilde{R} = const.$ yields the simple separable case.

The definitions adopted here are those of [3–5], although other definitions were proposed (see for example [6, 7]), but they will not be considered here.

## 3.2 The Stäckel Matrix

The approach to separability reported in this section follows the one of [3] and [4], and a central role in this approach is played by the Stäckel matrix (named after the German mathematician Paul Stäckel, 1862–1919).

A Stäckel matrix is a $n \times n$ matrix, which for $n = 3$ is defined as

$$\mathbf{\Phi} = \begin{bmatrix} \Phi_{11} \left(u^1\right) & \Phi_{12} \left(u^1\right) & \Phi_{13} \left(u^1\right) \\ \Phi_{21} \left(u^2\right) & \Phi_{22} \left(u^2\right) & \Phi_{23} \left(u^2\right) \\ \Phi_{31} \left(u^3\right) & \Phi_{32} \left(u^3\right) & \Phi_{33} \left(u^3\right) \end{bmatrix} \tag{3.3}$$

where each of the $\Phi_{nk} \left(u^n\right)$ is a function of $u^n$ only, i.e. the functions on the $n$th row depend only on the $n$th coordinate. The determinant of this matrix: $S = \det \mathbf{\Phi}$ is termed Stäckel determinant. This matrix has useful properties, related to its cofactors.

Consider the cofactors of the elements of the first column

$$M_1 = (-1)^{1+1} \operatorname{minor}(\Phi_{11}) = \left[\Phi_{22} \left(u^2\right) \Phi_{33} \left(u^3\right) - \Phi_{32} \left(u^3\right) \Phi_{23} \left(u^2\right)\right] \tag{3.4}$$

$$M_2 = (-1)^{2+1} \operatorname{minor}(\Phi_{21}) = -\left[\Phi_{12} \left(u^1\right) \Phi_{33} \left(u^3\right) - \Phi_{32} \left(u^3\right) \Phi_{13} \left(u^1\right)\right] \tag{3.5}$$

$$M_3 = (-1)^{3+1} \operatorname{minor}(\Phi_{31}) = \left[\Phi_{12} \left(u^1\right) \Phi_{23} \left(u^2\right) - \Phi_{22} \left(u^2\right) \Phi_{13} \left(u^1\right)\right] \tag{3.6}$$

the functional form of each cofactor is

$$M_1 = M_1 \left(u^2, u^3\right); \quad M_2 = M_2 \left(u^1, u^3\right); \quad M_3 = M_3 \left(u^1, u^2\right) \tag{3.7}$$

i.e. the cofactor $M_n$ is not a function of $u^n$ ($\frac{\partial M_n}{\partial u^n} = 0$). Moreover, these cofactors satisfy the identities

$$\sum_{n=1}^{3} \Phi_{n1} M_n = S \qquad (3.8a)$$

$$\sum_{n=1}^{3} \Phi_{n2} M_n = 0 \qquad (3.8b)$$

$$\sum_{n=1}^{3} \Phi_{n3} M_n = 0 \qquad (3.8c)$$

the first one is the usual rule to compute a determinant, the second and third ones express the determinant of the matrix when the first column is substituted by the second and third ones, respectively, and in such case the determinant is nil.

In this chapter, to increase the equations' readability, the dependence of a function on its variable/s is not always explicitly shown, as in Eqs. (3.8a–3.8c).

## 3.3 The Conditions for Separability

We will consider here the condition for separability of Helmholtz equation (named after the German physicist Hermann Ludwig Ferdinand von Helmholtz, 1821–1894) and the Laplace equation. The Laplacian in curvilinear orthogonal coordinates has the general form (see Eq. 1.196)

$$\nabla^2 \Psi = \frac{1}{g^{1/2}} \sum_i \frac{\partial}{\partial u^i} \left[ \frac{g^{1/2}}{h_i^2} \frac{\partial \Psi}{\partial u^i} \right] \qquad (3.9)$$

where $h_j$ are the scale factors and $g^{1/2} = h_1 h_2 h_3$, and the general form of the Helmholtz equation is

$$\frac{1}{g^{1/2}} \sum_k \frac{\partial}{\partial u^k} \left[ W^{(k)} \frac{\partial \Psi}{\partial u^k} \right] + K^2 \Psi = 0 \qquad (3.10)$$

where $W^{(k)} = \frac{g^{1/2}}{h_k^2}$ depends only on the coordinate system. Separability of this equation means that the solution can be written as

$$\Psi = \frac{U^1\left(u^1\right) U^2\left(u^2\right) U^3\left(u^3\right)}{\tilde{R}\left(u^1, u^2, u^3\right)} \qquad (3.11)$$

where $R$ can also be a constant. Substitution into Eq. (3.10) yields

$$\frac{1}{g^{1/2}} \sum_{k=1}^{3} \frac{\tilde{R}}{U^k} \frac{\partial}{\partial u^k} \left[ W^{(k)} \frac{\partial U^k \tilde{R}^{-1}}{\partial u^k} \right] + K^2 = 0 \qquad (3.12)$$

The conditions for separability can be better understood by considering first the simple separability case.

### 3.3.1  Conditions for Simple Separability

In this case the function $\tilde{R}$ is a constant and it can be taken to be equal to one. Equation (3.12) then becomes

$$\sum_{k=1}^{3} \frac{1}{U^k} \frac{\partial}{\partial u^k} \left[ W^{(k)} \frac{dU^k}{du^k} \right] + K^2 g^{1/2} = 0 \qquad (3.13)$$

Separation of this equation is obtained when the LHS can be written as a sum like

$$\sum_{k=1}^{3} r_k \left\{ \frac{1}{p_k U^k} \frac{d}{du^k} \left( p_k \frac{dU^k}{du^k} \right) + q_k \right\} = 0 \qquad (3.14)$$

where the functions $p_k$ and $q_k$ depend only on the variable $u^k$. In fact, Eq. (3.14) is satisfied in general only when each term in the curl bracket is zero, thus decomposing the PDE into three ordinary differential equations.

A necessary condition for reaching this results is that the functions $W^{(k)} = \frac{g^{1/2}}{h_k^2}$ can be expressed as products like

$$W^{(k)} = f_k \left( u^k \right) F_k \left( u^{j \neq k} \right) \qquad (3.15)$$

where the functions $f_k$ depend only on $u^k$ while the functions $F_k$ do not depend on $u^k$. If this is satisfied then Eq. (3.13) can be written as

$$\sum_{k=1}^{3} W^{(k)} \left\{ \frac{1}{U^k f_k} \frac{d}{du^k} \left[ f_k \frac{dU^k}{du^k} \right] + q_k \right\} - \sum_{k=1}^{3} q_k W^{(k)} + K^2 g^{1/2} = 0 \qquad (3.16)$$

where $q_k \left( u^k \right)$ are arbitrary functions of one variable. Equation (3.16) can be reduced to the wanted form (3.14) by choosing $p_k = f_k \left( u^k \right)$ and defining the functions $q_k$ as

$$q_k \left( u^k \right) = \alpha_1 \Phi_{k1} \left( u^k \right) + \alpha_2 \Phi_{k2} \left( u^k \right) + \alpha_3 \Phi_{k3} \left( u^k \right) \qquad (3.17)$$

where the functions $\Phi_{kn} \left( u^k \right)$ satisfy the system

$$\begin{cases} \sum_{k=1}^{3} W^{(k)} \Phi_{k1}\left(u^k\right) = g^{1/2} \\ \sum_{k=1}^{3} W^{(k)} \Phi_{k2}\left(u^k\right) = 0 \\ \sum_{k=1}^{3} W^{(k)} \Phi_{k3}\left(u^k\right) = 0 \end{cases} \tag{3.18}$$

In fact, if the determinant of the Stäckel matrix (3.3) is not nil, i.e. $S \neq 0$ the functions $W^{(k)}$ have the form

$$W^{(k)} = \frac{g^{1/2}}{S} M_k \tag{3.19}$$

then, from the identities (3.8a–3.8c)

$$\sum_{k=1}^{3} q_k W^{(k)} = \left[ \alpha_1 \sum_{k=1}^{3} \Phi_{k1} W^{(k)} + \alpha_2 \sum_{k=1}^{3} \Phi_{k2} W^{(k)} + \alpha_3 \sum_{k=1}^{3} \Phi_{k3} W^{(k)} \right] = \alpha_1 g^{1/2} \tag{3.20}$$

and choosing $\alpha_1 = K^2$

$$-\sum_{k=1}^{3} q_k W^{(k)} + K^2 g^{1/2} = 0 \tag{3.21}$$

transforming Eq. (3.16) into the set of ordinary differential equations

$$\frac{1}{f_k} \frac{d}{du^k} \left[ f_k \frac{dU^k}{du^k} \right] + \sum_{n=1}^{3} \alpha_n \Phi_{kn}\left(u^k\right) U^k = 0 \tag{3.22}$$

The nine functions $\Phi_{kn}$ are then the elements of a Stäckel matrix, that are linked to the coordinate system through the three Eq. (3.19), which can also be written as

$$h_k^2 = \frac{S}{M_k} \tag{3.23}$$

In order to find the nine elements of the Stäckel matrix out of three conditions, the following ansatz, called Robertson condition (named after the American mathematician and physicist Howard Percy Robertson, 1903–1961), was proposed [8]

$$S = \frac{g^{1/2}}{f_1 f_2 f_3} \tag{3.24}$$

It is clear from the above analysis that the definition of the Stäckel matrix is non-unique and this will be discussed below. The special case of separability of the Laplace equation can be obtained by setting $K = 0$.

**Example: Spherical Coordinate System**
The spherical coordinate system $\left(u^1, u^2, u^3\right) = (r, \theta, \varphi)$ is defined by the equations (see Fig. 1.4)

$$x = r \sin(\theta) \cos(\varphi) \tag{3.25a}$$
$$y = r \sin(\theta) \sin(\varphi) \tag{3.25b}$$
$$z = r \cos(\theta) \tag{3.25c}$$

In this example the symbol $r$ is used for the first spherical coordinate instead of the usual symbol $R$, to avoid confusion with the function $\tilde{R}\left(u^1, u^2, u^3\right)$. The corresponding scale factors are

$$h_1 = h_r = 1 \tag{3.26a}$$
$$h_2 = h_\theta = r \tag{3.26b}$$
$$h_3 = h_\varphi = r \sin(\theta) \tag{3.26c}$$

then

$$\left(W^{(1)}, W^{(2)}, W^{(3)}\right) = \left(W^{(r)}, W^{(\theta)}, W^{(\varphi)}\right) = \left(r^2 \sin(\theta), \sin(\theta), \frac{1}{\sin(\theta)}\right) \tag{3.27}$$

These functions can be written as given by Eq. (3.15) choosing

$$f_1 = f_r = r^2; \qquad f_2 = f_\theta = \sin(\theta); \qquad f_3 = f_\varphi = 1 \tag{3.28}$$

then

$$W^{(r)} = r^2 \sin(\theta) = f_r \sin(\theta) \tag{3.29a}$$
$$W^{(\theta)} = \sin(\theta) = f_\theta \tag{3.29b}$$
$$W^{(\varphi)} = \frac{1}{\sin(\theta)} = f_\varphi \frac{1}{\sin(\theta)} \tag{3.29c}$$

A Stäckel matrix can then be written as

$$\Phi = \begin{bmatrix} 1 & -1/r^2 & 0 \\ 0 & 1 & -1/\sin^2(\theta) \\ 0 & 0 & 1 \end{bmatrix} \tag{3.30}$$

In fact

$$S = \det \Phi = 1 \tag{3.31}$$
$$M_1 = 1; \quad M_2 = \frac{1}{r^2}; \quad M_3 = \frac{1}{r^2 \sin^2(\theta)} \tag{3.32}$$

and the separability conditions (3.23) and Robertson condition (3.24) are verified. Choosing the functions $q_k$ as in Eq. (3.17) yields

$$q_1\left(u^1\right) = q_r\left(r\right) = \alpha_1 - \frac{\alpha_2}{r^2} \tag{3.33a}$$

$$q_2\left(u^2\right) = q_\theta\left(\theta\right) = \alpha_2 - \frac{\alpha_3}{\sin^2\left(\theta\right)} \tag{3.33b}$$

$$q_3\left(u^3\right) = q_\varphi\left(\varphi\right) = \alpha_3 \tag{3.33c}$$

and

$$-\sum_{k=1}^{3} q_k W^{(k)} + K^2 g^{1/2} = \left(\frac{\alpha_2}{r^2} - \alpha_1\right) W^{(r)} - \left(\alpha_2 - \frac{\alpha_3}{\sin^2\left(\theta\right)}\right) W^{(\theta)} - \alpha_3 W^{(\varphi)} + K^2 r^2 \sin\left(\theta\right) =$$

$$= \left(\frac{\alpha_2}{r^2} - \alpha_1\right) r^2 \sin\left(\theta\right) - \left(\alpha_2 - \frac{\alpha_3}{\sin^2\left(\theta\right)}\right) \sin\left(\theta\right) - \frac{\alpha_3}{\sin\left(\theta\right)} + K^2 r^2 \sin\theta =$$

$$= -\alpha_1 r^2 \sin\left(\theta\right) + K^2 r^2 \sin\left(\theta\right) \tag{3.34}$$

which is nil choosing $\alpha_1 = K^2$.

The Helmholtz equation reduces to the three ODEs equations

$$\frac{1}{r^2}\frac{d}{dr}\left(r^2\frac{dU^r}{dr}\right) + \left(K^2 - \frac{\alpha_2}{r^2}\right) U^r = 0 \tag{3.35a}$$

$$\frac{1}{\sin\left(\theta\right)}\frac{d}{d\theta}\left(\sin\left(\theta\right)\frac{dU^\theta}{d\theta}\right) + \left(\alpha_2 - \frac{\alpha_3}{\sin^2\left(\theta\right)}\right) U^\theta = 0 \tag{3.35b}$$

$$\frac{d^2 U^\varphi}{d\varphi^2} + \alpha_3 U^\varphi = 0 \tag{3.35c}$$

### 3.3.2  Conditions for R-Separability

In this section we will consider the case when the function $\tilde{R}\left(u^1, u^2, u^3\right)$ is not a constant. It should be noticed that this function only depends on the coordinate system.

Since

$$\frac{\partial U^k \tilde{R}^{-1}}{\partial u^k} = \tilde{R}^{-2}\left(\tilde{R}\frac{dU^k}{du^k} - U^k\frac{\partial\tilde{R}}{\partial u^k}\right) \tag{3.36}$$

Equation (3.12) takes the form

$$\sum_{k=1}^{3}\frac{\tilde{R}}{U^k}\frac{\partial}{\partial u^k}\left[W^{(k)}\tilde{R}^{-2}\left(\tilde{R}\frac{dU^k}{du^k} - U^k\frac{\partial\tilde{R}}{\partial u^k}\right)\right] + K^2 g^{1/2} = 0 \tag{3.37}$$

Now, if possible, select the function $\tilde{R}$ to satisfy the equations

$$W^{(k)}\tilde{R}^{-2} = f_k\left(u^k\right) F_k\left(u^{n\neq k}\right) \tag{3.38}$$

where again the functions $f_k$ depend only on $u^k$ and the functions $F_k$ do not depend on $u^k$. Inserting into the above equation yields

$$\sum_{k=1}^{3} \frac{W^{(k)} \tilde{R}^{-1}}{U^k f_k} \frac{\partial}{\partial u^k} \left[ f_k \left( \tilde{R} \frac{dU^k}{du^k} - U^k \frac{\partial \tilde{R}}{\partial u^k} \right) \right] + K^2 g^{1/2} = 0 \qquad (3.39)$$

and with the further transformation

$$\frac{\partial}{\partial u^k} \left[ f_k \left( \tilde{R} \frac{dU^k}{du^k} - U^k \frac{\partial \tilde{R}}{\partial u^k} \right) \right] = \tilde{R} \frac{d}{du^k} \left( f_k \frac{dU^k}{du^k} \right) - U^k \frac{\partial}{\partial u^k} \left( f_k \frac{\partial \tilde{R}}{\partial u^k} \right) \quad (3.40)$$

Equation (3.37) becomes

$$\sum_{k=1}^{3} \frac{W^{(k)}}{U^k f_k} \frac{d}{du^k} \left( f_k \frac{dU^k}{du^k} \right) - \sum_{k=1}^{3} \frac{W^{(k)} \tilde{R}^{-1}}{f_k} \frac{\partial}{\partial u^k} \left( f_k \frac{\partial \tilde{R}}{\partial u^k} \right) + K^2 g^{1/2} = 0 \quad (3.41)$$

where now the second term depends only on the coordinate system and can be written as

$$\tilde{R}^{-1} \sum_{k} \frac{W^{(k)}}{f_k} \frac{\partial}{\partial u^k} \left( f_k \frac{\partial \tilde{R}}{\partial u^k} \right) = -\frac{\alpha_1^2 g^{1/2}}{\tilde{Q}\left( u^1, u^2, u^3 \right)} \qquad (3.42)$$

where $\alpha_1$ is a constant and $\tilde{Q}$ is a suitable function. The LHS of this equation can be further transformed using Eq. (3.38)

$$\tilde{R}^{-1} \sum_{k} \frac{W^{(k)}}{f_k} \frac{\partial}{\partial u^k} \left( f_k \frac{\partial \tilde{R}}{\partial u^k} \right) = \tilde{R} \sum_{k} \frac{\tilde{R}^{-2} W^{(k)}}{f_k} \frac{\partial}{\partial u^k} \left( f_k \frac{\partial \tilde{R}}{\partial u^k} \right) = \qquad (3.43)$$

$$= \tilde{R} \sum_{k} F_k \left( u^{n \neq k} \right) \frac{\partial}{\partial u^k} \left( f_k \frac{\partial \tilde{R}}{\partial u^k} \right) =$$

$$= \tilde{R} \sum_{k} \frac{\partial}{\partial u^k} \left( W^{(k)} \tilde{R}^{-2} \frac{\partial \tilde{R}}{\partial u^k} \right) =$$

$$= -\tilde{R} \nabla^2 \tilde{R}^{-1}$$

and Eq. (3.42) becomes

$$\tilde{R} \nabla^2 \tilde{R}^{-1} = \frac{\alpha_1^2 g^{1/2}}{\tilde{Q}\left( u^1, u^2, u^3 \right)} \qquad (3.44)$$

which is a quite useful form to find the function $\tilde{Q}$.

The Helmholtz equation can now be written as

$$\sum_{k=1}^{3} W^{(k)} \left[ \frac{1}{U^k f_k} \frac{d}{du^k} \left( f_k \frac{dU^k}{du^k} \right) + q_k \right] - \sum_{k} q_k W^{(k)} + \frac{\alpha_1^2 g^{1/2}}{\tilde{Q}} + K^2 g^{1/2} = 0$$

(3.45)

Choosing now the functions $\Phi_{kn}$ to satisfy the system

$$\begin{cases} \sum_{k=1}^{3} W^{(k)} \Phi_{k1} \left( u^k \right) = g^{1/2} \left( 1 + \frac{1}{\tilde{Q}} \right) \\ \sum_{k=1}^{3} W^{(k)} \Phi_{k2} \left( u^k \right) = 0 \\ \sum_{k=1}^{3} W^{(k)} \Phi_{k3} \left( u^k \right) = 0 \end{cases}$$

(3.46)

the functions $W^{(k)}$ are such that

$$W^{(k)} = \frac{M_k}{S} g^{1/2} \left( 1 + \frac{1}{\tilde{Q}} \right)$$

(3.47)

and

$$\sum_{k=1}^{3} q_k W^{(k)} = \left[ \alpha_1 \sum_{k=1}^{3} \Phi_{k1} W^{(k)} + \alpha_2 \sum_{k=1}^{3} \Phi_{k2} W^{(k)} + \alpha_3 \sum_{k=1}^{3} \Phi_{k3} W^{(k)} \right] =$$

$$= \alpha_1 \sum_{k=1}^{3} \Phi_{k1} \frac{M_k}{S} g^{1/2} \left( 1 + \frac{1}{\tilde{Q}} \right) = \alpha_1 g^{1/2} \left( 1 + \frac{1}{\tilde{Q}} \right)$$

(3.48)

Then, choosing $\alpha_1 = K^2$, the sum of the last three terms in Eq. (3.45) vanishes and the Helmholtz equation reduces to the three ODEs

$$\frac{1}{f_k} \frac{d}{du^k} \left( f_k \frac{dU^k}{du^k} \right) + q_k U^k = 0$$

(3.49)

For the case of the Laplace equation ($K^2 = 0$) the constant $\alpha_1$ cannot be generally taken to be zero, as for the case of simple separability, unless $\nabla^2 \tilde{R}^{-1} = 0$ and Eq. (3.44) is satisfied choosing $\alpha_1 = 0$ (in Chap. 4 we will see examples of both cases). When $K^2 = 0$ the functions $\Phi_{kn}$ can be taken to satisfy the system

$$\begin{cases} \sum_{k=1}^{3} W^{(k)} \Phi_{k1} \left( u^k \right) = \frac{g^{1/2}}{\tilde{Q}} \\ \sum_{k=1}^{3} W^{(k)} \Phi_{k2} \left( u^k \right) = 0 \\ \sum_{k=1}^{3} W^{(k)} \Phi_{k3} \left( u^k \right) = 0 \end{cases}$$

(3.50)

and the conditions for separability become

$$W^{(k)} = \frac{M_k}{S} \frac{g^{1/2}}{\tilde{Q}}$$

(3.51)

and the Robertson condition in this case is generalised as

$$S = \frac{g^{1/2}}{f_1 f_2 f_3 \tilde{Q} \tilde{R}^2} \tag{3.52}$$

It is clear that when $\tilde{R} = 1$ (simple separation), Eq. (3.42) can be satisfied choosing $\alpha_1 = 0$ and $\tilde{Q} = 1$ and Eq. (3.24) is obtained. Moon and Spencer [2] state that no coordinate system is known where the Helmholtz equation is R-separable, while 11 coordinate systems exist where the Laplace equation is R-separable [2] and in Chap. 4 six of them (the most useful for the applications reported in this book) will be described in detail.

**Example: R-Separability in Bispherical Coordinates**
Bispherical coordinates $(u^1, u^2, u^3) = (\eta, \theta, \varphi)$ represent one of those coordinate systems that allows R-separability of the Laplace equation. The coordinates are defined by the equations

$$x = \frac{a}{\Theta} \sin(\theta) \cos(\varphi) \tag{3.53a}$$

$$y = \frac{a}{\Theta} \sin(\theta) \sin(\varphi) \tag{3.53b}$$

$$z = \frac{a}{\Theta} \sinh(\eta) \tag{3.53c}$$

with $\Theta = \cosh(\eta) - \cos(\theta)$ and $-\infty < \eta < +\infty$; $0 \le \theta < \pi$; $0 \le \varphi < 2\pi$ (see Chap. 4 for further details). The scale factors are

$$h_1 = h_\eta = \frac{a}{\Theta}; \quad h_2 = h_\theta = \frac{a}{\Theta}; \quad h_3 = h_\varphi = \frac{a}{\Theta} \sin(\theta); \quad g^{1/2} = \frac{a^3}{\Theta^3} \sin(\theta) \tag{3.54}$$

with this we obtain

$$\left(W^{(1)}, W^{(2)}, W^{(3)}\right) = \left(W^{(\eta)}, W^{(\theta)}, W^{(\varphi)}\right) = \left(\frac{a}{\Theta} \sin(\theta), \frac{a}{\Theta} \sin(\theta), \frac{a}{\Theta} \frac{1}{\sin(\theta)}\right) \tag{3.55}$$

These functions can be written as in Eq. (3.38), choosing

$$\tilde{R} = \Theta^{-1/2}$$
$$f_1 = f_\eta = 1; \quad f_2 = f_\theta = \sin(\theta); \quad f_3 = f_\varphi = a \tag{3.56}$$

then

$$W^{(\eta)} \tilde{R}^{-2} = a \sin(\theta) = f_\eta a \sin(\theta) \tag{3.57a}$$

$$W^{(\theta)} \tilde{R}^{-2} = a \sin(\theta) = f_\theta a \tag{3.57b}$$

$$W^{(\varphi)} \tilde{R}^{-2} = \frac{a}{\sin(\theta)} = f_\varphi \frac{1}{\sin(\eta)} \tag{3.57c}$$

Equation (3.42) can now be written as

$$\Theta^{-1/2}\nabla^2\Theta^{1/2} = \frac{\alpha_1^2}{\tilde{Q}} \tag{3.58}$$

and it is satisfied choosing

$$\tilde{Q} = \frac{a^2}{\Theta^2}; \quad \alpha_1 = -\frac{1}{4} \tag{3.59}$$

A Stäckel matrix can then be written as

$$\Phi = \begin{bmatrix} 1 & -1 & 0 \\ 0 & 1 & -1/\sin^2(\theta) \\ 0 & 0 & 1 \end{bmatrix} \tag{3.60}$$

In fact

$$S = \det \Phi = 1 \tag{3.61}$$

$$M_1 = 1; \quad M_2 = 1; \quad M_3 = \frac{1}{\sin^2(\theta)} \tag{3.62}$$

and the separability conditions (3.51) and the Robertson condition (3.52) are verified. Choosing the functions $q_k$ as in Eq. (3.17) yields

$$q_\eta = -\frac{1}{4} - \alpha_2 \tag{3.63a}$$

$$q_\theta = \alpha_2 - \frac{\alpha_3}{\sin^2(\theta)} \tag{3.63b}$$

$$q_\varphi = \alpha_3 \tag{3.63c}$$

and the Laplace equation reduces to the three ODEs

$$\frac{d^2U^\eta}{d\eta^2} - \left(\frac{1}{4} + \alpha_2\right)U^\eta = 0 \tag{3.64a}$$

$$\frac{1}{\sin(\theta)}\frac{d}{d\theta}\left(\sin(\theta)\frac{dU^\theta}{d\theta}\right) + \left(\alpha_2 - \frac{\alpha_3}{\sin^2(\theta)}\right)U^\theta = 0 \tag{3.64b}$$

$$\frac{d^2U^\varphi}{d^2\varphi} + \alpha_3 U^\varphi = 0 \tag{3.64c}$$

Then, the Laplace equation can be *R-separated* in this coordinate system.

## 3.4  Non-unicity of the Stäckel Matrix

The Stäckel matrix $\Phi$ must satisfy Eqs. (3.51) and (3.52), then supposing to have found one such matrix, there are an infinite number of other matrices that satisfy the same problem. Consider the following transformations of the Stäckel matrix:

(a) The multiplication of the second column of the given $\Phi$ by any real number and the third column by its reciprocal yields a new matrix that has the same cofactors and from Eq. (3.8a–3.8c) the same determinant $S = \det \Phi$, then Eqs. (3.51) and (3.52) are still satisfied.

(b) Exchanging the second and third column of the original $\Phi$ and multiplying by $-1$ one of them yields a matrix with the same cofactors and again with the same determinant (exchange of column and multiplying by $-1$ a column change the sign twice).

(c) Adding a multiple of one of the last two column to one of the other columns again maintains the same determinant and it does not change the cofactors.

All the above mentioned transformations of a given Stäckel matrix yield a new Stäckel matrix that solves the same problem.

**Example: The Stäckel Matrix of the Spherical Coordinate System**
Let us consider again the Stäckel matrix of the spherical coordinate system (in this example the symbol $r$ is used for the first spherical coordinate instead of the usual symbol $R$, to avoid confusion with the function $\tilde{R}\left(u^1, u^2, u^3\right)$)

$$\Phi = \begin{bmatrix} 1 & -1/r^2 & 0 \\ 0 & 1 & -1/\sin^2(\theta) \\ 0 & 0 & 1 \end{bmatrix} \tag{3.65}$$

and apply the above mentioned rules.

(a) Multiply the second column by the real number $\lambda$ and divide the third column by $\lambda$

$$\Phi = \begin{bmatrix} 1 & -\lambda/r^2 & 0 \\ 0 & \lambda & -1/\left[\lambda \sin^2(\theta)\right] \\ 0 & 0 & 1/\lambda \end{bmatrix} \tag{3.66}$$

clearly $S = \det \Phi = 1$ then

$$M_1 = 1; \quad M_2 = \frac{1}{r^2}; \quad M_3 = \frac{1}{r^2 \sin^2(\theta)} \tag{3.67}$$

as for the original matrix.

(b) Exchanging the second and third column of the original $\Phi$ and multiplying by $-1$ one of them yields

$$\Phi = \begin{bmatrix} 1 & 0 & -1/r^2 \\ 0 & 1/\sin^2(\theta) & 1 \\ 0 & -1 & 0 \end{bmatrix} \tag{3.68}$$

which again has the same determinant and the same cofactors $M_n$ as the original one.

(c) Adding a multiple of the last column to the second column yields

$$\Phi = \begin{bmatrix} 1 & -1/r^2 & 0 \\ 0 & 1 - \lambda/\sin^2(\theta) & -1/\sin^2(\theta) \\ 0 & \lambda & 1 \end{bmatrix} \tag{3.69}$$

and again

$$S = \det \Phi = 1 \tag{3.70}$$

$$M_1 = 1; \quad M_2 = \frac{1}{r^2}; \quad M_3 = \frac{1}{r^2 \sin^2(\theta)} \tag{3.71}$$

This gives a certain degree of freedom in defining the elements of a Stäckel matrix and it explains why different Stäckel matrices for the same coordinate system can be found in different books.

## 3.5 Finding a Stäckel Matrix

The procedure to find a Stäckel matrix for a given coordinate system is not straight-forward and some trial and error is needed. We will try to find a Stäckel matrix for one of the coordinate systems, choosing the spherical coordinates as an example.

The first step is the inspection of the functions $W^{(k)}$ to find the seven functions $f_k$, $F_k$ and $\tilde{R}$. From

$$\left(W^{(1)}, W^{(2)}, W^{(3)}\right) = \left(W^{(r)}, W^{(\eta)}, W^{(\varphi)}\right) = \left(r^2 \sin(\theta), \sin(\theta), \frac{1}{\sin(\theta)}\right) \tag{3.72}$$

a simple choice is (remembering that $f_k$ is a function of $u^k$ only, while $F_k$ is not a function of $u^k$)

$$\tilde{R} = 1 \tag{3.73}$$

$$f_r = r^2; \quad F_r = \sin(\theta) \tag{3.74}$$

$$f_\theta = \sin(\theta); \quad F_\theta = 1 \tag{3.75}$$

$$f_\varphi = 1; \quad F_\varphi = \frac{1}{\sin(\theta)} \tag{3.76}$$

and it is easy to see that there are no other choices available, then simple separability can be considered. The second step is to find the cofactors, and this is the point where the Robertson ansatz can be used. First, the determinant $S = \det \Phi$ can be calculated from Eq. (3.24) and substituting into Eq. (3.19) yields the three cofactors $M_k$

$$M_k = \frac{W^{(k)}}{f_1 f_2 f_3} \tag{3.77}$$

For the spherical coordinates case

$$S = \frac{g^{1/2}}{f_1 f_2 f_3} = 1 \tag{3.78}$$

then

$$M_1 = 1 \tag{3.79a}$$

$$M_2 = \frac{1}{r^2} \tag{3.79b}$$

$$M_3 = \frac{1}{r^2 \sin^2 (\theta)} \tag{3.79c}$$

Since the cofactors are related to the elements of the Stäckel matrix by the set of equations

$$M_1 = \Phi_{22} \Phi_{33} - \Phi_{32} \Phi_{23} \tag{3.80a}$$
$$M_2 = -\Phi_{12} \Phi_{33} + \Phi_{32} \Phi_{13} \tag{3.80b}$$
$$M_3 = \Phi_{12} \Phi_{23} - \Phi_{22} \Phi_{13} \tag{3.80c}$$

and

$$S = \Phi_{11} M_1 + \Phi_{21} M_2 + \Phi_{31} M_3 \tag{3.81}$$

the system of Eqs. (3.80), (3.81) is under-determined and the solution must be found by trial and error.

For example, for spherical coordinates, setting: $\Phi_{11} = 1$, Eq. (3.81) yields

$$0 = \Phi_{21} (\theta) + \Phi_{31} (\varphi) \frac{1}{\sin^2 (\theta)} \tag{3.82}$$

then two possibilities are left

$$\text{(a) } \Phi_{21} (\theta) = 0; \ \Phi_{31} (\varphi) = 0 \tag{3.83}$$

$$\text{(b) } \Phi_{21} (\theta) = \frac{1}{\sin^2 (\theta)}; \ \Phi_{31} (\varphi) = -1 \tag{3.84}$$

Equation (3.80) becomes

$$1 = \Phi_{22}(\theta)\,\Phi_{33}(\varphi) - \Phi_{32}(\varphi)\,\Phi_{23}(\theta) \tag{3.85a}$$

$$\frac{1}{r^2} = -\Phi_{12}(r)\,\Phi_{33}(\varphi) + \Phi_{32}(\varphi)\,\Phi_{13}(r) \tag{3.85b}$$

$$\frac{1}{r^2 \sin^2(\theta)} = \Phi_{12}(r)\,\Phi_{23}(\theta) - \Phi_{22}(\theta)\,\Phi_{13}(r) \tag{3.85c}$$

The simplest guess for the first equation is

$$\Phi_{22}(\theta) = 1; \quad \Phi_{33}(\varphi) = 1 \tag{3.86}$$

and as a consequence either $\Phi_{32}(\theta)$ or $\Phi_{23}(\varphi)$ or both must be nil.

The first choice, substituted into the two remaining equations, leads to

$$\Phi_{12}(r) = -\frac{1}{r^2} \tag{3.87a}$$

$$\frac{1}{r^2 \sin^2(\theta)} = -\frac{1}{r^2}\Phi_{23}(\theta) - \Phi_{13}(r) \tag{3.87b}$$

which forces

$$\Phi_{23}(\theta) = -\frac{1}{\sin^2(\theta)}; \quad \Phi_{13}(r) = 0 \tag{3.88}$$

The final matrix can then have two forms, following the two choices for $\Phi_{21}(\theta)$ and $\Phi_{31}(\varphi)$

$$\Phi_a = \begin{bmatrix} 1 & -1/r^2 & 0 \\ 0 & 1 & -1/\sin^2(\theta) \\ 0 & 0 & 1 \end{bmatrix} \tag{3.89}$$

$$\Phi_b = \begin{bmatrix} 1 & -1/r^2 & 0 \\ 1/\sin^2(\theta) & 1 & -1/\sin^2(\theta) \\ -1 & 0 & 1 \end{bmatrix} \tag{3.90}$$

Both matrices have the same cofactors

$$M_1 = 1; \quad M_2 = \frac{1}{r^2}; \quad M_3 = \frac{1}{r^2 \sin^2(\theta)} \tag{3.91}$$

and the same determinant: $S = 1$, and they are both a Stäckel matrix for the coordinate system. In fact, the matrix $\Phi_b$ can be transformed into the matrix $\Phi_a$ by the transformation (c) mentioned in Sect. 3.4, i.e. adding the third column to the first one. The functions $q_k$ are, for the $\Phi_a$

$$q_1^{(a)}(r) = \alpha_1^{(a)} - \alpha_2^{(a)}/r^2 \tag{3.92a}$$

$$q_2^{(a)}(\theta) = \alpha_2^{(a)} - \alpha_3^{(a)}/\sin^2(\theta) \tag{3.92b}$$

$$q_3^{(a)}(\varphi) = \alpha_3^{(a)} \tag{3.92c}$$

and for the $\Phi_b$

$$q_1^{(b)}(r) = \alpha_1^{(b)} - \alpha_2^{(b)}/r^2 \tag{3.93a}$$

$$q_2^{(b)}(\theta) = \alpha_2^{(b)} - \left(\alpha_3^{(b)} - \alpha_1^{(b)}\right)/\sin^2(\theta) \tag{3.93b}$$

$$q_3^{(b)}(\varphi) = \left(\alpha_3^{(b)} - \alpha_1^{(b)}\right) \tag{3.93c}$$

the two sets are equivalent by choosing the arbitrary constants as: $\left(\alpha_1^{(a)}, \alpha_2^{(a)}, \alpha_3^{(a)}\right) = \left(\alpha_1^{(b)}, \alpha_2^{(b)}, \alpha_3^{(b)} - \alpha_1^{(b)}\right)$. The final sets of ODE are identical for the two cases.

# References

1. Weigand, B.: Analytical Methods for Heat Transfer and Fluid Flow Problems. Springer, Berlin (2004)
2. Moon, P., Spencer, D.E.: Field Theory Handbook, 2nd edn. Springer, Berlin (1971)
3. Morse, P.M., Feshback, H.: Methods of Theoretical Physics. McGraw Hill, New York (1953)
4. Moon, P., Spencer, D.E.: Separability conditions for the Laplace and Helmholtz equations. J. Franklin Inst. **253**(6), 585–600 (1952)
5. Moon, P., Spencer, D.E.: Separability in a class of coordinate systems. J. Franklin Inst. **254**(3), 227–242 (1952)
6. Levison, N., Bogert, B., Redheffer, R.M.: Separation of Laplace's equation. Quart. Appl. Math. **7**(3), 241–262 (1949)
7. Moon, P., Spencer, D.E.: Recent investigation of the separation of Laplace's equation. Proc. Amer. Math. Soc. **4**(2), 302–307 (1953)
8. Robertson, H.P.: Bemerkung über separierbare systeme in der Wellenmechanik (in German). Math. Ann. **98**, 749–752 (1927)

# Chapter 4
# Orthogonal Curvilinear Coordinate Systems

Orthogonal curvilinear coordinates occupy a special place among general coordinate systems, due to their special properties. There exists a number of such coordinate systems where the Laplace or Helmholtz equations may be separable, thus yielding a powerful tool to solve them. Operations like gradients, divergence, Laplacian take on much simpler forms in orthogonal coordinates.

In this chapter many of the results found for general curvilinear coordinates will be recast to simpler forms of more practical use. The last section reports a detailed list of the most useful orthogonal coordinate systems and for each of them the explicit form of the most important differential operators is reported. In this chapter the summation convention will not be used.

## 4.1 Orthogonal Coordinate Systems in Euclidean 3D-space

Orthogonal curvilinear coordinates are characterised by the fact that coordinate lines always cross each other forming a right angle, which implies that the covariant coordinate basis vectors $\mathbf{g}_j$ are mutually orthogonal. The first consequence is that the metric tensor is diagonal, i.e. $g_{mn} = 0$ if $m \neq n$, in fact

$$g_{mn} = \mathbf{g}_m \cdot \mathbf{g}_n = \begin{cases} 0 & \text{if } m \neq n \\ h_n^2 & \text{if } m = n \end{cases} \tag{4.101}$$

where $h_n$ are the scale factors (see Chap. 1) and $h_n = |\mathbf{g}_n|$. From the definition of the contravariant basis (see Eq. 1.102 in Chap. 1), each vector $\mathbf{g}^j$ is in this case parallel to the corresponding covariant one ($\mathbf{g}_j$), then

$$g^{mn} = \mathbf{g}^m \cdot \mathbf{g}^n = \begin{cases} 0 & \text{if } m \neq n \\ 1/h_n^2 & \text{if } m = n \end{cases} \tag{4.102}$$

© Springer Nature Switzerland AG 2021
G. E. Cossali and S. Tonini, *Drop Heating and Evaporation: Analytical Solutions in Curvilinear Coordinate Systems*, Mathematical Engineering,
https://doi.org/10.1007/978-3-030-49274-8_4

The determinant of the metric tensor assumes then the following simple form

$$g = (h_1 h_2 h_3)^2 \tag{4.103}$$

In these coordinate systems, the physical components of vectors and tensors (see Chap. 1) are related to covariant and contravariant components by simple relations (in the following part of this section a *hat* will be used to indicate the physical components in any coordinate system)

$$\hat{V}_j = V^j h_j = V_j \frac{1}{h_j} \tag{4.104}$$

$$\hat{T}_{jk} = T^{jk} h_j h_k = T_k^j \frac{h_j}{h_k} = T_{jk} \frac{1}{h_j h_k} \tag{4.105}$$

We have seen that physical components are quite important in engineering applications since, differently from covariant and contravariant components, they have the same measuring units as in Cartesian coordinate systems, and they can have a more immediate physical interpretation.

It is of practical importance to be able to transform *physical* components from one orthogonal system to another one and this can be done introducing the transformation matrix for physical components. To this end, let transform a vector $\mathbf{Z}$ in the Cartesian coordinate system, where the physical components are $\hat{Z}_k = Z^k = Z_k$, to an orthogonal curvilinear system $(u^1, u^2, u^3)$, where the physical, $\hat{Z}'_k$, the contravariant, $Z'^k$, and the covariant, $Z'_k$, components are generally different. The transformation rules (1.126) and the definitions (1.11), (1.137) yield

$$\hat{Z}'_j = h_j Z^{j\,\prime} = h_j \sum_{k=1}^{3} \frac{\partial u^j}{\partial x^k} Z^k = \sum_{k=1}^{3} h_j \frac{\partial u^j}{\partial x^k} \hat{Z}_k \tag{4.106a}$$

$$\hat{Z}'_j = \frac{1}{h_j} Z'_j = \frac{1}{h_j} \sum_{k=1}^{3} \frac{\partial x^k}{\partial u^j} Z_k = \sum_{k=1}^{3} \frac{1}{h_j} \frac{\partial x^k}{\partial u^j} \hat{Z}_k \tag{4.106b}$$

The equivalence of Eqs. (4.106a) and (4.106b) implies that $h_j \frac{\partial u^j}{\partial x^k}$ and $\frac{1}{h_j} \frac{\partial x^k}{\partial u^j}$ are identical, but this can also be proven as follows. From the transformation rules for covariant and contravariant vectors

$$Z'_p = \sum_k \frac{\partial x^k}{\partial u^p} Z_k \tag{4.107}$$

$$Z'^j = \sum_k \frac{\partial u^j}{\partial x^k} Z^k \tag{4.108}$$

we can obtain

$$\sum_k \frac{\partial x^k}{\partial u^j} Z_k = Z'_j = \sum_s Z'^s g_{sj} = \sum_s \sum_k g_{sj} \frac{\partial u^s}{\partial x^k} Z^k \tag{4.109}$$

and since in Cartesian coordinates $Z_k = Z^k$ then

$$\frac{\partial x^k}{\partial u^j} = \sum_s g_{sj} \frac{\partial u^s}{\partial x^k} \tag{4.110}$$

For orthogonal coordinates, where $g_{js} = \delta_{js} h_j^2$

$$\frac{\partial x^k}{\partial u^j} = \sum_s \delta_{sj} h_s^2 \frac{\partial u^s}{\partial x^k} = h_j^2 \frac{\partial u^j}{\partial x^k} \tag{4.111}$$

which proves the statement.

The transformation matrix from the *physical* components in Cartesian coordinates to a general orthogonal curvilinear system, $\mathbf{M}^{x \to u}$, is then

$$\mathbf{M}^{x \to u} = \begin{bmatrix} h_1 \frac{\partial u^1}{\partial x^1} & h_1 \frac{\partial u^1}{\partial x^2} & h_1 \frac{\partial u^1}{\partial x^3} \\ h_2 \frac{\partial u^2}{\partial x^1} & h_2 \frac{\partial u^2}{\partial x^2} & h_2 \frac{\partial u^2}{\partial x^3} \\ h_3 \frac{\partial u^3}{\partial x^1} & h_3 \frac{\partial u^3}{\partial x^2} & h_3 \frac{\partial u^3}{\partial x^3} \end{bmatrix} = \begin{bmatrix} \frac{1}{h_1} \frac{\partial x^1}{\partial u^1} & \frac{1}{h_1} \frac{\partial x^2}{\partial u^1} & \frac{1}{h_1} \frac{\partial x^3}{\partial u^1} \\ \frac{1}{h_2} \frac{\partial x^1}{\partial u^2} & \frac{1}{h_2} \frac{\partial x^2}{\partial u^2} & \frac{1}{h_2} \frac{\partial x^3}{\partial u^2} \\ \frac{1}{h_3} \frac{\partial x^1}{\partial u^3} & \frac{1}{h_3} \frac{\partial x^2}{\partial u^3} & \frac{1}{h_3} \frac{\partial x^3}{\partial u^3} \end{bmatrix} \tag{4.112}$$

and the single element is: $M_{jk}^{x \to u} = h_j \frac{\partial u^j}{\partial x^k} = \frac{1}{h_j} \frac{\partial x^k}{\partial u^j}$. The inverse transformation matrix $\mathbf{M}^{u \to x}$ is defined by the equation:

$$\hat{Z}_k = \sum M_{kj}^{u \to x} \hat{Z}'_j \tag{4.113}$$

where

$$\mathbf{M}^{u \to x} = \left\{ \mathbf{M}^{x \to u} \right\}^{-1} \tag{4.114}$$

i.e. the transformation matrix from orthogonal curvilinear coordinates to Cartesian coordinates, $\mathbf{M}^{u \to x}$, is the inverse of the matrix $\mathbf{M}^{x \to u}$. The matrix $\mathbf{M}^{u \to x}$ can also be derived observing that, from the transformation rule (1.126)

$$\hat{Z}_k = Z^k = \sum_{j=1}^3 \frac{\partial x^k}{\partial u^j} Z'^j = \sum_{j=1}^3 \frac{1}{h_j} \frac{\partial x^k}{\partial u^j} \hat{Z}'_j \tag{4.115}$$

and the explicit form of $\mathbf{M}^{u \to x}$ is then

$$\mathbf{M}^{u \to x} = \begin{bmatrix} \frac{1}{h_1} \frac{\partial x^1}{\partial u^1} & \frac{1}{h_2} \frac{\partial x^1}{\partial u^2} & \frac{1}{h_3} \frac{\partial x^1}{\partial u^3} \\ \frac{1}{h_1} \frac{\partial x^2}{\partial u^1} & \frac{1}{h_2} \frac{\partial x^2}{\partial u^2} & \frac{1}{h_3} \frac{\partial x^2}{\partial u^3} \\ \frac{1}{h_1} \frac{\partial x^3}{\partial u^1} & \frac{1}{h_2} \frac{\partial x^3}{\partial u^2} & \frac{1}{h_3} \frac{\partial x^3}{\partial u^3} \end{bmatrix} \tag{4.116}$$

Comparing the matrix (4.116) with the matrix (4.112) we can notice that

$$\mathbf{M}^{u \to x} = \left\{ \mathbf{M}^{x \to u} \right\}^T \tag{4.117}$$

i.e. $\mathbf{M}^{u \to x}$ is also the transpose of $\mathbf{M}^{x \to u}$, then the matrices $\mathbf{M}^{u \to x}$ and $\mathbf{M}^{x \to u}$ are *orthogonal matrices*, and it follows that their determinants are either $+1$ or $-1$. This result can be directly derived from Eq. (4.116) (or from Eq. 4.112) observing that $\mathbf{M}^{u \to x}$ is obtained from the transformation matrix $\frac{\partial x^k}{\partial u^j}$ by multiplying each column by $\frac{1}{h_j}$. In such case

$$\det \mathbf{M}^{u \to x} = \frac{1}{h_1 h_2 h_3} \det \left\{ \frac{\partial x^k}{\partial u^j} \right\} = \frac{J}{\sqrt{g}} \tag{4.118}$$

where $J$ is the Jacobian determinant, and from Eq. (1.122) it stems that $\det \mathbf{M}^{u \to x} = \pm 1$.

It is also of some use sometimes to be able to transform physical components from an orthogonal curvilinear system to another one, and a general rule can be easily derived. Consider the two orthogonal coordinate systems defined by the two sets

$$u^j = u^j \left( x^1, x^2, x^3 \right) \tag{4.119}$$
$$v^j = v^j \left( x^1, x^2, x^3 \right) \tag{4.120}$$

Given a vector $\hat{Z}^k$ in the Cartesian system, the physical components in the coordinate system $\left( u^1, u^2, u^3 \right)$, i.e. $\hat{Z}'_j$, and those the coordinate system $\left( v^1, v^2, v^3 \right)$, i.e. $\hat{Z}''_j$ are

$$\hat{Z}'_j = \sum_{k=1}^{3} M_{jk}^{x \to u} \hat{Z}^k ; \quad \hat{Z}''_j = \sum_{k=1}^{3} M_{jk}^{x \to v} \hat{Z}^k \tag{4.121}$$

with

$$M_{jk}^{x \to u} = h_j \frac{\partial u^j}{\partial x^k} ; \quad M_{jk}^{x \to v} = h_j \frac{\partial v^j}{\partial x^k} \tag{4.122}$$

Using Eqs. (4.114) and (4.113) yields

$$\hat{Z}'_j = \sum_{k=1}^{3} M_{jk}^{x \to u} \hat{Z}^k = \sum_{k=1}^{3} \sum_{m=1}^{3} M_{jk}^{x \to u} M_{km}^{v \to x} \hat{Z}''_m \tag{4.123}$$

i.e.

$$\mathbf{M}^{v \to u} = \mathbf{M}^{x \to u} \mathbf{M}^{v \to x} \tag{4.124}$$

showing that the transformation matrix from one curvilinear system to the other one may be calculated from the transformation matrices from Cartesian to curvilinear systems. In Sect. 4.4 the matrix for the transformation of *physical* components from orthogonal curvilinear to Cartesian coordinate systems will be explicitly given for many orthogonal curvilinear coordinate systems.

### 4.1.1 Transformation Matrices in Rotational Coordinate Systems

The transformation matrix for physical components $\mathbf{M}^{u \to x}$ has a relatively simple form for all the rotational coordinate systems. In fact, a general rotational system of orthogonal coordinates can be described by the following equations

$$x = P\left(u^1, u^2\right) \cos(\varphi) \tag{4.125a}$$
$$y = P\left(u^1, u^2\right) \sin(\varphi) \tag{4.125b}$$
$$z = Q\left(u^1, u^2\right) \tag{4.125c}$$

where for simplicity $u^3 = \varphi$. In the present section the following notation will also be used $\frac{\partial P}{\partial u^k} = P_k$, $\frac{\partial Q}{\partial u^k} = Q_k$; remembering Eq. (1.63), the scale factors are then

$$h_1 = \sqrt{g_{11}} = \left(P_1^2 + Q_1^2\right)^{1/2} \tag{4.126a}$$
$$h_2 = \sqrt{g_{22}} = \left(P_2^2 + Q_2^2\right)^{1/2} \tag{4.126b}$$
$$h_3 = \sqrt{g_{33}} = P \tag{4.126c}$$

The orthogonality of the system implies that $g_{kj} = 0$ when $k \neq j$ then

$$g_{12} = P_1 P_2 + Q_1 Q_2 = 0 \tag{4.127a}$$
$$g_{13} = -P_1 P \cos(\varphi) \sin(\varphi) + P_1 P \cos(\varphi) \sin(\varphi) = 0 \tag{4.127b}$$
$$g_{23} = -P_2 P \cos(\varphi) \sin(\varphi) + P_2 P \cos(\varphi) \sin(\varphi) = 0 \tag{4.127c}$$

where (4.127b) and (4.127c) are identities whereas (4.127a) is a condition on the functions $P\left(u^1, u^2\right)$ and $Q\left(u^1, u^2\right)$. The transformation matrix for physical components can then be written in a general form as

$$\mathbf{M}^{u \to x} = \begin{bmatrix} A \cos(\varphi) & B \cos(\varphi) & -\sin\varphi \\ A \sin(\varphi) & B \sin(\varphi) & \cos(\varphi) \\ C & D & 0 \end{bmatrix} \tag{4.128}$$

where

$$A = \frac{P_1}{\left(P_1^2 + Q_1^2\right)^{1/2}}; \quad B = \frac{P_2}{\left(P_2^2 + Q_2^2\right)^{1/2}} \tag{4.129}$$

$$C = \frac{Q_1}{\left(P_1^2 + Q_1^2\right)^{1/2}}; \quad D = \frac{Q_2}{\left(P_2^2 + Q_2^2\right)^{1/2}} \tag{4.130}$$

i.e. it is fully defined by the four functions $A\left(u^1, u^2\right)$, $B\left(u^1, u^2\right)$, $C\left(u^1, u^2\right)$ and $D\left(u^1, u^2\right)$. However, these functions are not independent of each other.

Consider the two combinations $C - B$ and $C + B$

$$C - B = \frac{Q_1 \left(P_2^2 + Q_2^2\right)^{1/2} - P_2 \left(P_1^2 + Q_1^2\right)^{1/2}}{\left(P_1^2 + Q_1^2\right)^{1/2} \left(P_2^2 + Q_2^2\right)^{1/2}} \tag{4.131}$$

$$C + B = \frac{Q_1 \left(P_2^2 + Q_2^2\right)^{1/2} + P_2 \left(P_1^2 + Q_1^2\right)^{1/2}}{\left(P_1^2 + Q_1^2\right)^{1/2} \left(P_2^2 + Q_2^2\right)^{1/2}} \tag{4.132}$$

The orthogonality condition (4.127a) implies that

$$P_1 = -Q_1 \frac{Q_2}{P_2} \Rightarrow P_2 \left(P_1^2 + Q_1^2\right)^{1/2} = \pm Q_1 \left(P_2^2 + Q_2^2\right)^{1/2} \tag{4.133}$$

and it is easy to see that $C - B = 0$ when the positive sign holds in (4.133) while $C - B = \frac{2Q_1}{\left(P_1^2 + Q_1^2\right)^{1/2}}$ when the negative sign holds, i.e.

$$C - B = \begin{cases} 0 & + \\ \frac{2Q_1}{\left(P_1^2 + Q_1^2\right)^{1/2}} & - \end{cases} \tag{4.134}$$

Analogously, $C + B = 0$ when the negative sign holds in (4.133) while $C + B = \frac{2Q_1}{\left(P_1^2 + Q_1^2\right)^{1/2}}$ when the positive sign holds, i.e.

$$C + B = \begin{cases} 2\frac{Q_1}{\left(P_1^2 + Q_1^2\right)^{1/2}} & + \\ 0 & - \end{cases} \tag{4.135}$$

Re-writing the orthogonality condition (4.127a) as

$$P_2 = -Q_1 \frac{Q_2}{P_1} \Rightarrow P_1 \left(P_2^2 + Q_2^2\right)^{1/2} = \pm Q_2 \left(Q_1^2 + P_1^2\right)^{1/2} \tag{4.136}$$

and following the same procedure, it can be easily shown that $D - A = 0$ when the positive sign holds in (4.136) while $D - A = \frac{2Q_2}{\left(P_2^2 + Q_2^2\right)^{1/2}}$ when the negative sign holds, and the opposite is true for the combination $D + A$; summarising

$$D - A = \begin{cases} 0 & + \\ \dfrac{2Q_2}{\left(P_2^2 + Q_2^2\right)^{1/2}} & - \end{cases} \tag{4.137}$$

$$D + A = \begin{cases} \dfrac{2Q_2}{\left(P_2^2 + Q_2^2\right)^{1/2}} & + \\ 0 & - \end{cases} \tag{4.138}$$

Consider now the condition (4.133) with the positive sign and multiply both sides by $P_1$

$$P_1 P_2 \left(P_1^2 + Q_1^2\right)^{1/2} = +P_1 Q_1 \left(P_2^2 + Q_2^2\right)^{1/2} \tag{4.139}$$

and using (4.127a) to eliminate $P_1 P_2$ on the LHS yields

$$- Q_2 \left(P_1^2 + Q_1^2\right)^{1/2} = P_1 \left(P_2^2 + Q_2^2\right)^{1/2} \tag{4.140}$$

which is exactly the condition (4.136) with the negative sign; i.e. the condition (4.133) with the positive sign implies the condition (4.136) with the negative sign and vice versa.

When the positive sign in (4.133) holds, and then the negative sign in (4.136) holds too, we have

$$C = B \text{ and } D = -A \tag{4.141}$$

while when the negative sign holds in Eq. (4.133), and then the positive sign holds in Eq. (4.136) we have

$$C = -B \text{ and } D = A \tag{4.142}$$

thus for all the cases

$$|C| = |B| \text{ and } |D| = |A| \tag{4.143}$$

and the most general form for $\mathbf{M}^{u \to x}$ is then

$$\mathbf{M}^{u \to x} = \begin{bmatrix} A \cos(\varphi) & B \cos(\varphi) & -\sin\varphi \\ A \sin(\varphi) & B \sin(\varphi) & \cos(\varphi) \\ B & -A & 0 \end{bmatrix} \tag{4.144}$$

or

$$\mathbf{M}^{u \to x} = \begin{bmatrix} A \cos(\varphi) & B \cos(\varphi) & -\sin\varphi \\ A \sin(\varphi) & B \sin(\varphi) & \cos(\varphi) \\ -B & A & 0 \end{bmatrix} \tag{4.145}$$

## 4.2 Gradient, Divergence, Curl, Laplacian in Orthogonal Curvilinear Coordinates

Modelling of physical phenomena often relies on the use of differential operators, like gradient, divergence, curl, Laplacian. In Chap. 1 these operators have been casted in covariant form by quite general formulae, which is now the time to simplify for the case of orthogonal coordinates and, for a more practical use, in term of physical components.

The gradient of a scalar is a vector and its components in Cartesian coordinates are simply given by the corresponding partial derivatives

$$\nabla_k T = \frac{\partial T}{\partial x^k} \tag{4.146}$$

and these components can be interpreted in this coordinate system as covariant, contravariant or physical.

In orthogonal curvilinear coordinates, the partial derivatives of a scalar are the *covariant* component of a vector (the gradient)

$$G_k = \frac{\partial T}{\partial u^k} \tag{4.147}$$

and the *physical* components can be calculated as

$$\nabla_k T = \frac{1}{h_k} G_k = \frac{1}{h_k} \frac{\partial T}{\partial u^k} \tag{4.148}$$

where $h_k$ is the scale factor of coordinate $u^k$ and we will use the symbol $\nabla_k T$ to indicate the physical components of the gradient in any coordinate system. For example, the gradient of the scalar $T$ in Cartesian and in spherical coordinates

$$x = R \sin(\theta) \cos(\varphi) \tag{4.149a}$$
$$y = R \sin(\theta) \sin(\varphi) \tag{4.149b}$$
$$z = R \cos(\theta) \tag{4.149c}$$

in term of physical components is

$$\nabla_j T = \left( \frac{\partial T}{\partial x}, \frac{\partial T}{\partial y}, \frac{\partial T}{\partial z} \right) \text{ Cartesian coordinates} \tag{4.150}$$

$$\nabla_j T = \left( \frac{\partial T}{\partial R}, \frac{1}{R} \frac{\partial T}{\partial \theta}, \frac{1}{R \sin(\theta)} \frac{\partial T}{\partial \varphi} \right) \text{ spherical coordinates} \tag{4.151}$$

The divergence of a vector was defined in Chap. 1 for general coordinate system as the covariant derivative of a the vector followed by the index contraction, i.e.

$$\mathrm{div}\mathbf{V} = \sum_k V^k_{/k} \tag{4.152}$$

and the following compact formula was given

$$\sum_k V^k_{/k} = \frac{1}{\sqrt{g}} \sum_k \frac{\partial \left(\sqrt{g}V^k\right)}{\partial u^k} \tag{4.153}$$

Let now write it by using physical components. Equations (4.103) and (4.104) yields

$$\sum_k V^k_{/k} = \sum_k \frac{1}{h_1 h_2 h_3} \frac{\partial \left(W^{(k)} h_k \hat{V}^k\right)}{\partial u^k} \tag{4.154}$$

where

$$W^{(k)} = \frac{g^{1/2}}{h_k^2} \tag{4.155}$$

As an example, in spherical coordinates (see Sect. 4.4) defined as:

$$x = R\sqrt{1-\eta^2}\cos(\varphi); \; y = R\sqrt{1-\eta^2}\sin(\varphi); \; z = R\eta$$

The functions $W^k$ are

$$W^{(1)} = R^2; \quad W^{(2)} = 1-\eta^2; \quad W^{(3)} = \frac{1}{1-\eta^2} \tag{4.156}$$

and then

$$\begin{aligned}
\mathrm{div}\mathbf{V} &= \frac{1}{h_1 h_2 h_3} \sum_k \frac{\partial \left(W^{(k)} h_k \hat{V}^k\right)}{\partial u^k} \\
&= \frac{\partial \hat{V}^R}{\partial R} + \frac{2}{R}\hat{V}^R + \frac{\sqrt{1-\eta^2}}{R}\frac{\partial \hat{V}^\eta}{\partial \eta} - \frac{\eta}{R\sqrt{1-\eta^2}}\hat{V}^\eta + \frac{1}{R\sqrt{1-\eta^2}}\frac{\partial \hat{V}^\varphi}{\partial \varphi}
\end{aligned} \tag{4.157}$$

and as a second example, in bispherical coordinates, defined as (see Sect. 4.4)

$$x = a\frac{\sin(\psi)}{\Theta}\cos(\varphi); \quad y = a\frac{\sin(\psi)}{\Theta}\sin(\varphi); \quad z = a\frac{\sinh(\xi)}{\Theta} \tag{4.158}$$

where $\Theta = \cosh(\xi) - \cos(\psi)$, the function $W^{(k)}$ are

$$W^{(1)} = a\frac{\sin(\psi)}{\Theta}; \quad W^{(2)} = a\frac{\sin(\psi)}{\Theta}; \quad W^{(3)} = \frac{a}{\Theta\sin(\psi)} \tag{4.159}$$

and then

$$\text{div}\mathbf{V} = \frac{1}{h_1 h_2 h_3} \sum_k \frac{\partial \left( W^{(k)} h_k \hat{V}^k \right)}{\partial u^k} = \frac{\Phi}{a} \frac{\partial \hat{V}^\xi}{\partial \xi} - \frac{2 \sinh (\xi)}{a} \hat{V}^\xi + \Phi \frac{\partial \hat{V}^\psi}{\partial \psi}$$

$$+ \frac{\cosh (\xi) \cos (\psi) - 1 - \sin^2 (\psi)}{a \sin (\psi)} \hat{V}^\psi + \frac{\Phi}{a \sin (\psi)} \frac{\partial \hat{V}^\varphi}{\partial \varphi}$$

$$(4.160)$$

The curl of a vector is the vector (see Chap. 1)

$$\text{curl}\mathbf{V} = \omega^j = \sum_{k,m} \varepsilon^{jkm} V_{m/k} \qquad (4.161)$$

and in explicit form, due to the symmetry of the Christoffel symbols

$$\omega^1 = \frac{1}{g^{1/2}} \left( V_{3/2} - V_{2/3} \right) = \frac{1}{g^{1/2}} \left( V_{3,2} - V_{2,3} \right) \qquad (4.162a)$$

$$\omega^2 = \frac{1}{g^{1/2}} \left( V_{1/3} - V_{3/1} \right) = \frac{1}{g^{1/2}} \left( V_{1,3} - V_{3,1} \right) \qquad (4.162b)$$

$$\omega^3 = \frac{1}{g^{1/2}} \left( V_{2/1} - V_{1/2} \right) = \frac{1}{g^{1/2}} \left( V_{2,1} - V_{1,2} \right) \qquad (4.162c)$$

Using physical components: $\hat{\omega}_1 = \omega^j h_j$ and $\hat{V}_j = V_j \frac{1}{h_j}$, then

$$\hat{\omega}^1 = \frac{h_1}{g^{1/2}} \left( \frac{\partial \hat{V}_3 h_3}{\partial u^2} - \frac{\partial \hat{V}_2 h_2}{\partial u^3} \right) \qquad (4.163a)$$

$$\hat{\omega}^2 = \frac{h_2}{g^{1/2}} \left( \frac{\partial \hat{V}_1 h_1}{\partial u^3} - \frac{\partial \hat{V}_3 h_3}{\partial u^1} \right) \qquad (4.163b)$$

$$\hat{\omega}^3 = \frac{h_3}{g^{1/2}} \left( \frac{\partial \hat{V}_2 h_2}{\partial u^1} - \frac{\partial \hat{V}_1 h_1}{\partial u^2} \right) \qquad (4.163c)$$

The Laplacian of a scalar has been defined as (see Chap. 1)

$$\nabla^2 S = \sum_{kj} \left( g^{jk} S_{/k} \right)_{/j} = \frac{1}{\sqrt{g}} \sum_{kj} \frac{\partial}{\partial u^j} \left( \sqrt{g} g^{jk} \frac{\partial S}{\partial u^k} \right) \qquad (4.164)$$

and for the case of orthogonal coordinate it simplifies to

$$\nabla^2 S = \frac{1}{h_1 h_2 h_3} \sum_k \frac{\partial}{\partial u^k} \left( W^{(k)} \frac{\partial S}{\partial u^k} \right) \qquad (4.165)$$

## 4.3 Laplace and Helmholtz Equations

Laplace equation and Helmholtz equation are ubiquitous in physics and engineering, and thus it is worth to show their explicit form in orthogonal coordinates. The Helmholtz equation can be generally written as

$$\nabla^2 S + kS = 0 \tag{4.166}$$

where $k$ is a constant. It arises often from time dependent problems to which the separation of variable method is applied. As an example, consider the energy conservation equation in a solid or in a fluid at rest

$$\frac{\partial T}{\partial t} = \alpha \nabla^2 T \tag{4.167}$$

where $T$ is the temperature and $\alpha$ is the thermal diffusivity, known as *homogeneous heat equation* or sometimes as *Fourier equation* (named after the French mathematician and physicist Joseph Fourier, 1768–1830). The separation of variable approach assumes that

$$T\left(t, u^j\right) = T(t) U\left(u^j\right) \tag{4.168}$$

and substitution into Eq. (4.167) yields

$$\frac{1}{T(t)} \frac{dT(t)}{dt} = \alpha \frac{1}{U\left(u^j\right)} \nabla^2 U\left(u^j\right) \tag{4.169}$$

that can be separated into two differential equations

$$\frac{dT(t)}{dt} = k\alpha T(t) \tag{4.170}$$

$$\nabla^2 U\left(u^j\right) = kU\left(u^j\right) \tag{4.171}$$

where the second one is the Helmholtz equation.

Another typical case is that of the wave equation

$$\frac{\partial^2 P}{\partial t^2} = c^2 \nabla^2 P \tag{4.172}$$

where $c$ is the wave speed and again a separation of variables $P = T(t) U\left(u^j\right)$ yields the Helmholtz equation.

The Laplace equation, which can be seen as a special case of the Helmholtz equation, setting $k = 0$

$$\nabla^2 T = 0 \tag{4.173}$$

arises in many fields of science, often when stationary phenomena are studied. As an example the heat Eq. (4.167) for a stationary problem yields the Laplace equation (4.173). The solutions of Laplace's equation are called *harmonic functions*, and the general theory of solutions to Laplace's equation is known as *potential theory* (see [1] for more details).

Helmholtz and Laplace equations can be written in any orthogonal coordinate system as

$$\frac{1}{h_1 h_2 h_3} \sum_k \frac{\partial}{\partial u^k} \left( W^{(k)} \frac{\partial S}{\partial u^k} \right) + kS = 0 \text{ Helmholtz equation} \qquad (4.174)$$

$$\sum_k \frac{\partial}{\partial u^k} \left( W^{(k)} \frac{\partial S}{\partial u^k} \right) = 0 \text{ Laplace equation} \qquad (4.175)$$

## 4.4   List of Curvilinear Coordinate Systems

This section reports a collection of orthogonal curvilinear coordinates, comprising the *eleven* systems cited in the previous section where both Laplace and Helmholtz equations are separable, and other systems that may be of use in applications. Among the orthogonal curvilinear systems the rotational symmetric ones, see Table 4.1, are likely the most interesting when dealing with evaporation of lumps of fluids, since some coordinate surfaces may be taken to represent the interface between a finite quantity of liquid and the surrounding gas.
The selected list is not exhaustive and a complete collection can be found in [2].

Each coordinate system is described as follows. The defining equations

$$x^j = x^j \left( u^j \right) \qquad (4.176)$$

**Table 4.1**  *Eleven* and rotational coordinate systems

| Eleven | Rotational |
|---|---|
| Circular-cylinder | Tangent-sphere |
| Elliptic-cylinder | Cardioid |
| Parabolic-cylinder | Bispherical |
| Spherical | Toroidal |
| Spheroidal prolate | Inverse prolate spheroidal |
| Spheroidal oblate | Inverse oblate spheroidal |
| Parabolic | |
| Conical | |
| Ellipsoidal | |
| Paraboloidal | |

are reported, together with the domains of each $u^j$ and a 3D schematic of some iso-coordinate surface (i.e. $u^j = const$) to help in visualising the coordinates. For rotational systems, a 2D map, obtained by setting the azimuthal coordinate constant, is also reported, and for cylindrical coordinates 2D maps are also reported.

For each system, the Stäckel matrix is given, together with the functions $q_k$ and $f_k$ (see Chap. 3) necessary for separating Laplace and Helmholtz equations, when applicable.

Since it is of practical use, the matrix to transform the physical components of a vector in curvilinear coordinates to Cartesian coordinates $\mathbf{M}^{u \to x}$ is explicitly given (see Chap. 1).

The forms of the differential operators: gradient, divergence, curl and Laplacian, for each curvilinear system are explicitly reported in terms of physical components. Some of the coordinate systems can be defined using a different set of variables, and in such a case the information above mentioned are reported for both forms.

It must be mentioned that this tabulation is deliberately redundant, since for practical applications redundancy is an added value.

### 4.4.1 Circular-Cylinder Coordinate System

Circular-cylinder coordinates can be defined as

$$x = r \cos (\theta); \quad y = r \sin (\theta); \quad z = z \tag{4.177}$$

The coordinate space $(r, \theta, z)$ is limited as: $0 \le r < +\infty$, $0 \le \theta < 2\pi$, $-\infty < z < +\infty$.

The surface $r = r_0$ is a circular cylinder; the surface $\theta = \theta_0$ is a plane passing through the $z$-axis; the surface $z = z_0$ is a half plane orthogonal to the $z$-axis (see Fig. 4.1). The scale factors are

$$h_r = 1; \quad h_\theta = r; \quad h_z = 1; \quad g^{1/2} = r \tag{4.178}$$

and the functions $W^{(k)}$ are

$$W^{(r)} = r; \quad W^{(\theta)} = \frac{1}{r}; \quad W^{(z)} = r \tag{4.179}$$

The gradient of a scalar quantity $T$ is a vector and its physical components are

$$\nabla T = \text{grad } T = \left[ \frac{\partial T}{\partial r}, \frac{1}{r} \frac{\partial T}{\partial \theta}, \frac{\partial T}{\partial z} \right] \tag{4.180}$$

The divergence of a vector $\mathbf{V} = [\hat{V}_r, \hat{V}_\theta, \hat{V}_z]$ is the scalar

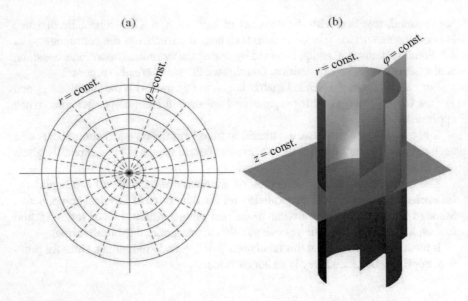

**Fig. 4.1** Circular-cylinder coordinate system: **a** coordinates over a plane $z = $ const.; **b** 3-D representation of the iso-surfaces

$$\mathbf{\nabla} \cdot \mathbf{V} = \text{div}\mathbf{V} = \frac{\partial \hat{V}_r}{\partial r} + \frac{1}{r}\hat{V}_r + \frac{1}{r}\frac{\partial \hat{V}_\theta}{\partial \theta} + \frac{\partial \hat{V}_z}{\partial z} \tag{4.181}$$

The curl of a vector $\mathbf{V} = [\hat{V}_r, \hat{V}_\theta, \hat{V}_z]$ is the vector $\boldsymbol{\omega} = \text{curl}\mathbf{V}$

$$\begin{bmatrix} \hat{\omega}_r \\ \hat{\omega}_\theta \\ \hat{\omega}_z \end{bmatrix} = \begin{bmatrix} \frac{1}{r}\frac{\partial \hat{V}_z}{\partial \theta} - \frac{\partial \hat{V}_\theta}{\partial z} \\ \frac{\partial \hat{V}_r}{\partial z} - \frac{\partial \hat{V}_z}{\partial r} \\ \frac{\partial \hat{V}_\theta}{\partial r} + \frac{1}{r}\hat{V}_\theta - \frac{1}{r}\frac{\partial \hat{V}_r}{\partial \theta} \end{bmatrix} \tag{4.182}$$

The Laplacian of the scalar $T$ is the scalar

$$\nabla^2 T = \frac{\partial^2 T}{\partial r^2} + \frac{1}{r}\frac{\partial T}{\partial r} + \frac{1}{r^2}\frac{\partial^2 T}{\partial \theta^2} + \frac{\partial^2 T}{\partial z^2} \tag{4.183}$$

The Laplace and Helmholtz equations are *simple-separable* in this system, the corresponding Stäckel matrix is

$$\mathbf{\Phi} = \begin{bmatrix} 0 & -\frac{1}{r^2} & -1 \\ 0 & 1 & 0 \\ 1 & 0 & 1 \end{bmatrix} \tag{4.184}$$

the functions $f_k$ are

$$f_1 = f_r(r) = r; \quad f_2 = f_\theta(\theta) = 1; \quad f_3 = f_z(z) = 1 \tag{4.185}$$

the functions $q_k$ are

$$q_1 = q_r(r) = -\frac{1}{r^2}\alpha_2 - \alpha_3; \quad q_2 = q_\theta(\theta) = \alpha_2; \quad q_3 = q_z(z) = \alpha_1 + \alpha_3 \tag{4.186}$$

for the Laplace equation $\alpha_1 = 0$ and for the Helmholtz equation $\alpha_1 = K^2$.

The transformation matrix for physical components from circular-cylinder to Cartesian coordinates, $\mathbf{M}^{u \to x}$ is

$$\mathbf{M}^{u \to x} = \begin{bmatrix} \cos(\theta) & -\sin(\theta) & 0 \\ \sin(\theta) & \cos(\theta) & 0 \\ 0 & 0 & 1 \end{bmatrix} \tag{4.187}$$

### 4.4.2 Elliptic-Cylinder Coordinate System

Elliptic-cylinder coordinates can be defined as

$$x = a\cosh(\xi)\cos(\psi); \quad y = a\sinh(\xi)\sin(\psi); \quad z = z \tag{4.188}$$

The coordinate space $(\xi, \psi, z)$ is limited as: $0 \le \xi < +\infty$, $0 \le \psi < 2\pi$, $-\infty < z < +\infty$.

The surface $\xi = \xi_0$ is an elliptic cylinder; the surface $\psi = \psi_0$ is an hyperbolic cylinder; the surface $z = z_0$ is a plane orthogonal to the $z$-axis (see Fig. 4.2). The scale factors are

$$h_\xi = a\sqrt{\cosh^2(\xi) - \cos^2(\psi)} = h_\psi; \quad h_z = 1; \quad g^{1/2} = a^2\left[\cosh^2(\xi) - \cos^2(\psi)\right] \tag{4.189}$$

and the functions $W^{(k)}$ are

$$W^{(\xi)} = 1 = W^{(\psi)}; \quad W^{(z)} = a^2\left[\cosh^2(\xi) - \cos^2(\psi)\right] \tag{4.190}$$

The gradient of a scalar quantity $T$ is a vector and its physical components are

$$\nabla T = \text{grad } T = \left[\frac{1}{a\sqrt{\cosh^2(\xi) - \cos^2(\psi)}}\frac{\partial T}{\partial \xi}, \frac{1}{a\sqrt{\cosh^2(\xi) - \cos^2(\psi)}}\frac{\partial T}{\partial \psi}, \frac{\partial T}{\partial z}\right] \tag{4.191}$$

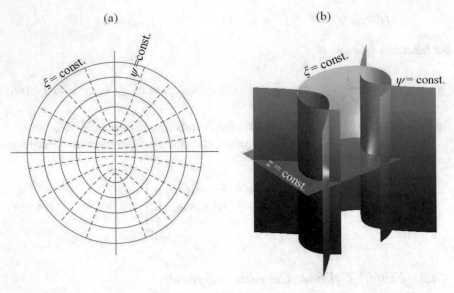

**Fig. 4.2** Elliptic-cylinder coordinate system: **a** coordinates over a plane $z$ =const.; **b** 3-D representation of the iso-surfaces

The divergence of a vector $\mathbf{V} = [\hat{V}_\xi, \hat{V}_\psi, \hat{V}_z]$ is the scalar

$$\nabla \cdot \mathbf{V} = \mathrm{div} \mathbf{V} = \frac{1}{a\sqrt{\cosh^2(\xi) - \cos^2(\psi)}} \left[ \frac{\partial \hat{V}_\xi}{\partial \xi} + \frac{\sinh(\xi)\cosh(\xi)}{\cosh^2(\xi) - \cos^2(\psi)} \hat{V}_\xi \right] +$$

$$+ \frac{1}{a\sqrt{\cosh^2(\xi) - \cos^2(\psi)}} \left[ \frac{\partial \hat{V}_\psi}{\partial \xi} + \frac{\sin(\psi)\cos(\psi)}{\cosh^2(\xi) - \cos^2(\psi)} \hat{V}_\psi \right] + \frac{\partial \hat{V}_z}{\partial z}$$

$$(4.192)$$

The curl of a vector $\mathbf{V} = [\hat{V}_\xi, \hat{V}_\psi, \hat{V}_z]$ is the vector $\boldsymbol{\omega} = \mathrm{curl} \mathbf{V}$

$$\begin{bmatrix} \hat{\omega}_\xi \\ \hat{\omega}_\psi \\ \hat{\omega}_z \end{bmatrix} = \begin{bmatrix} \frac{1}{a[\cosh^2(\xi)-\cos^2(\psi)]^{1/2}} \frac{\partial \hat{V}_z}{\partial \psi} - \frac{\partial \hat{V}_\psi}{\partial z} \\ \frac{\partial \hat{V}_\xi}{\partial z} - \frac{1}{a[\cosh^2(\xi)-\cos^2(\psi)]^{1/2}} \frac{\partial \hat{V}_z}{\partial \xi} \\ \frac{1}{a[\cosh^2(\xi)-\cos^2(\psi)]^{1/2}} \left[ \frac{\partial \hat{V}_\psi}{\partial \xi} + \frac{\sinh(\xi)\cosh(\xi)}{\cosh^2(\xi)-\cos^2(\psi)} \hat{V}_\psi - \frac{\partial \hat{V}_\xi}{\partial \psi} - \frac{\sin(\psi)\cos(\psi)}{\cosh^2(\xi)-\cos^2(\psi)} \hat{V}_\xi \right] \end{bmatrix}$$

$$(4.193)$$

The Laplacian of the scalar $T$ is the scalar

$$\nabla^2 T = \frac{1}{a^2 \left[ \cosh^2(\xi) - \cos^2(\psi) \right]} \left( \frac{\partial^2 T}{\partial \xi^2} + \frac{\partial^2 T}{\partial \psi^2} \right) + \frac{\partial^2 T}{\partial z^2} \qquad (4.194)$$

The Laplace and Helmholtz equations are *simple-separable* in this system, the corresponding Stäckel matrix is

$$\Phi = \begin{bmatrix} 0 & -1 & -a^2\cosh^2(\xi) \\ 0 & 1 & a^2\cos^2(\psi) \\ 1 & 0 & 1 \end{bmatrix} \tag{4.195}$$

the functions $f_k$ are

$$f_1 = f_\xi(\xi) = 1; \quad f_2 = f_\psi(\psi) = 1; \quad f_3 = f_z(z) = 1 \tag{4.196}$$

and the functions $q_k$ are

$$q_1 = q_\xi(\xi) = -\alpha_2 - a^2\cosh^2(\xi)\,\alpha_3; \quad q_2 = q_\psi(\psi) = \alpha_2 + a^2\cos^2(\psi)\,\alpha_3; \quad q_3 = q_z(z) = \alpha_1 + \alpha_3 \tag{4.197}$$

for the Laplace equation $\alpha_1 = 0$ and for the Helmholtz equation $\alpha_1 = K^2$.

The transformation matrix for physical components from elliptic-cylinder to Cartesian coordinates, $\mathbf{M}^{u \to x}$ is

$$\mathbf{M}^{u \to x} = \begin{bmatrix} \dfrac{\sinh(\xi)\cos(\psi)}{\left[\cosh^2(\xi)-\cos^2(\psi)\right]^{1/2}} & -\dfrac{\cosh(\xi)\sin(\psi)}{\left[\cosh^2(\xi)-\cos^2(\psi)\right]^{1/2}} & 0 \\ \dfrac{\cosh(\xi)\sin(\psi)}{\left[\cosh^2(\xi)-\cos^2(\psi)\right]^{1/2}} & \dfrac{\sinh(\xi)\cos(\psi)}{\left[\cosh^2(\xi)-\cos^2(\psi)\right]^{1/2}} & 0 \\ 0 & 0 & 1 \end{bmatrix} \tag{4.198}$$

### 4.4.3 Parabolic-Cylinder Coordinate System

Parabolic cylinder coordinates can be defined as

$$x = \frac{1}{2}\left(\mu^2 - \nu^2\right); \quad y = \mu\nu; \quad z = z \tag{4.199}$$

The coordinate space $(\mu, \nu, z)$ is limited as: $0 \le \mu < +\infty$, $-\infty < \nu < \infty$, $-\infty < z < +\infty$ (Fig. 4.3).

The surfaces $\mu = \mu_0$ and $\nu = \nu_0$ are parabolic cylinders; the surface $z = z_0$ is a plane orthogonal to the $z$-axis (see Fig. 4.3). The scale factors are

$$h_\mu = \sqrt{\mu^2 + \nu^2} = h_\nu; \quad h_z = 1; \quad g^{1/2} = \mu^2 + \nu^2 \tag{4.200}$$

and the functions $W^{(k)}$ are

$$W^{(\mu)} = 1; \quad W^{(\nu)} = 1; \quad W^{(z)} = \mu^2 + \nu^2 \tag{4.201}$$

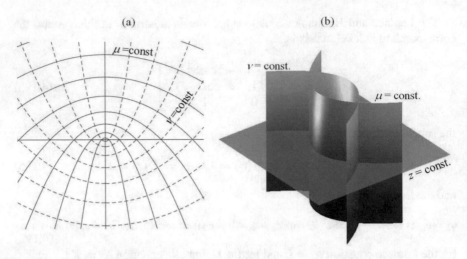

**Fig. 4.3** Parabolic-cylinder coordinate system: **a** coordinates over a plane $z =$const.; **b** 3-D representation of the iso-surfaces

The gradient of a scalar quantity $T$ is a vector and its physical components are

$$\nabla T = \text{grad } T = \left[ \frac{1}{\sqrt{\mu^2 + \nu^2}} \frac{\partial T}{\partial \mu}, \frac{1}{\sqrt{\mu^2 + \nu^2}} \frac{\partial T}{\partial \nu}, \frac{\partial T}{\partial z} \right] \tag{4.202}$$

The divergence of a vector $\mathbf{V} = [\hat{V}_\mu, \hat{V}_\nu, \hat{V}_z]$ is the scalar

$$\nabla \cdot \mathbf{V} = \text{div}\mathbf{V} = \frac{1}{\left(\mu^2 + \nu^2\right)^{\frac{1}{2}}} \left( \frac{\partial \hat{V}_\mu}{\partial \mu} + \frac{\mu}{\mu^2 + \nu^2} \hat{V}_\mu + \frac{\partial \hat{V}_\nu}{\partial \nu} + \frac{\nu}{\mu^2 + \nu^2} \hat{V}_\nu \right) + \frac{\partial \hat{V}_z}{\partial z} \tag{4.203}$$

The curl of a vector $\mathbf{V} = [\hat{V}_\mu, \hat{V}_\nu, \hat{V}_z]$ is the vector $\boldsymbol{\omega} = \text{curl}\mathbf{V}$

$$\begin{bmatrix} \hat{\omega}_\mu \\ \hat{\omega}_\nu \\ \hat{\omega}_z \end{bmatrix} = \begin{bmatrix} \frac{1}{(\mu^2+\nu^2)^{1/2}} \frac{\partial \hat{V}_z}{\partial \nu} - \frac{\partial \hat{V}_\nu}{\partial z} \\ \frac{\partial \hat{V}_\mu}{\partial z} - \frac{1}{(\mu^2+\nu^2)^{1/2}} \frac{\partial \hat{V}_z}{\partial \mu} \\ \frac{1}{(\mu^2+\nu^2)^{1/2}} \left[ \frac{\partial \hat{V}_\nu}{\partial \mu} + \frac{\mu}{\mu^2+\nu^2} \hat{V}_\nu - \frac{\partial \hat{V}_\mu}{\partial \nu} - \frac{\nu}{\mu^2+\nu^2} \hat{V}_\mu \right] \end{bmatrix} \tag{4.204}$$

The Laplacian of the scalar $T$ is the scalar

$$\nabla^2 T = \frac{1}{\mu^2 + \nu^2} \left( \frac{\partial^2 T}{\partial \mu^2} + \frac{\partial^2 T}{\partial \nu^2} \right) + \frac{\partial^2 T}{\partial z^2} \tag{4.205}$$

The Laplace and Helmholtz equations are *simple-separable* in this system, the corresponding Stäckel matrix is

$$\mathbf{\Phi} = \begin{bmatrix} 0 & -1 & -\mu^2 \\ 0 & 1 & -\nu^2 \\ 1 & 0 & 1 \end{bmatrix} \tag{4.206}$$

the functions $f_k$ are

$$f_1 = f_\mu(\mu) = 1; \quad f_2 = f_\nu(\nu) = 1; \quad f_3 = f_z(z) = 1 \tag{4.207}$$

and the functions $q_k$ are

$$q_1 = q_\mu(\mu) = -\alpha_2 - \mu^2\alpha_3; \quad q_2 = q_\nu(\nu) = \alpha_2 - \nu^2\alpha_3; \quad q_3 = q_z(z) = \alpha_1 + \alpha_3 \tag{4.208}$$

for the Laplace equation $\alpha_1 = 0$ and for the Helmholtz equation $\alpha_1 = K^2$.

The transformation matrix for physical components from parabolic-cylinder to Cartesian coordinates, $\mathbf{M}^{u \to x}$ is

$$\mathbf{M}^{u \to x} = \begin{bmatrix} \dfrac{\mu}{(\mu^2+\nu^2)^{1/2}} & -\dfrac{\nu}{(\mu^2+\nu^2)^{1/2}} & 0 \\ \dfrac{\nu}{(\mu^2+\nu^2)^{1/2}} & \dfrac{\mu}{(\mu^2+\nu^2)^{1/2}} & 0 \\ 0 & 0 & 1 \end{bmatrix} \tag{4.209}$$

### 4.4.4  Spherical Coordinate System

Spherical coordinates can be defined as

$$x = R\sqrt{1 - \eta^2}\cos(\varphi); \quad y = R\sqrt{1 - \eta^2}\sin(\varphi); \quad z = R\eta \tag{4.210}$$

The coordinate space $(R, \eta, \varphi)$ is limited as: $0 \le R < +\infty$, $-1 \le \eta \le 1$, $0 \le \varphi < 2\pi$.

The surface $R = R_0$ is a sphere, the surface $\eta = \eta_0$ is a circular cone around the $z$-axis. The surface $\varphi = \varphi_0$ is a half plane passing through the $z$-axis (see Fig. 4.4). The scale factors are

$$h_R = 1; \quad h_\eta = \frac{R}{\sqrt{1 - \eta^2}}; \quad h_\varphi = R\sqrt{1 - \eta^2}; \quad g^{1/2} = R^2 \tag{4.211}$$

and the functions $W^{(k)}$ are

$$W^{(R)} = R^2; \quad W^{(\eta)} = 1 - \eta^2; \quad W^{(\varphi)} = \frac{1}{1 - \eta^2} \tag{4.212}$$

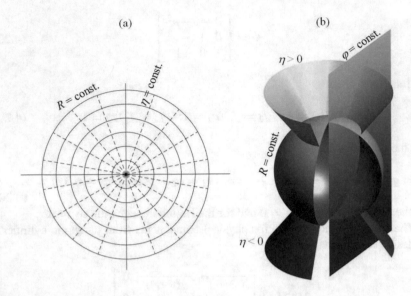

**Fig. 4.4** Spherical coordinate system: **a** coordinates over a plane passing through the $z$-axis; **b** 3-D representation of the iso-surfaces

The gradient of a scalar quantity $T$ is a vector and its physical components are

$$\nabla T = \text{grad } T = \left[ \frac{\partial T}{\partial R}, \frac{\sqrt{1-\eta^2}}{R} \frac{\partial T}{\partial \eta}, \frac{1}{R\sqrt{1-\eta^2}} \frac{\partial T}{\partial \varphi} \right] \tag{4.213}$$

The divergence of a vector $\mathbf{V} = [\hat{V}_R, \hat{V}_\eta, \hat{V}_\varphi]$ is the scalar

$$\nabla \cdot \mathbf{V} = \text{div} \mathbf{V} = \frac{\partial \hat{V}_R}{\partial R} + \frac{2}{R} \hat{V}_R + \frac{\sqrt{1-\eta^2}}{R} \frac{\partial \hat{V}_\eta}{\partial \eta} - \frac{\eta}{R\sqrt{1-\eta^2}} \hat{V}_\eta + \frac{1}{R\sqrt{1-\eta^2}} \frac{\partial \hat{V}_\varphi}{\partial \varphi} \tag{4.214}$$

The curl of a vector $\mathbf{V} = [\hat{V}_R, \hat{V}_\eta, \hat{V}_\varphi]$ is the vector $\omega = \text{curl} \mathbf{V}$

$$\begin{bmatrix} \hat{\omega}_R \\ \hat{\omega}_\eta \\ \hat{\omega}_\varphi \end{bmatrix} = \begin{bmatrix} \dfrac{(1-\eta^2)^{\frac{1}{2}}}{R} \dfrac{\partial \hat{V}_\varphi}{\partial \eta} - \dfrac{\eta}{R(1-\eta^2)^{\frac{1}{2}}} \hat{V}_\varphi - \dfrac{1}{R(1-\eta^2)^{\frac{1}{2}}} \dfrac{\partial \hat{V}_\eta}{\partial \varphi} \\ \dfrac{1}{R(1-\eta^2)^{\frac{1}{2}}} \dfrac{\partial \hat{V}_R}{\partial \varphi} - \dfrac{\partial \hat{V}_\varphi}{\partial R} - \dfrac{1}{R} \hat{V}_\varphi \\ \dfrac{\partial \hat{V}_\eta}{\partial R} + \dfrac{1}{R} \hat{V}_\eta - \dfrac{(1-\eta^2)^{\frac{1}{2}}}{R} \dfrac{\partial \hat{V}_R}{\partial \eta} \end{bmatrix} \tag{4.215}$$

The Laplacian of the scalar $T$ is the scalar

$$\nabla^2 T = \frac{\partial^2 T}{\partial R^2} + \frac{2}{R}\frac{\partial T}{\partial R} + \frac{1-\eta^2}{R^2}\frac{\partial^2 T}{\partial \eta^2} - \frac{2\eta}{R^2}\frac{\partial T}{\partial \eta} + \frac{1}{R^2\left(1-\eta^2\right)}\frac{\partial^2 T}{\partial \varphi^2} \qquad (4.216)$$

The Laplace and Helmholtz equations are *simple-separable* in this system, the corresponding Stäckel matrix is

$$\Phi = \begin{bmatrix} 1 & -\frac{1}{R^2} & 0 \\ 0 & \frac{1}{1-\eta^2} & -\frac{1}{\left(1-\eta^2\right)^2} \\ 0 & 0 & 1 \end{bmatrix} \qquad (4.217)$$

the functions $f_k$ are

$$f_1 = f_R\,(R) = R^2; \quad f_2 = f_\eta\,(\eta) = 1-\eta^2; \quad f_3 = f_\varphi\,(\varphi) = 1 \qquad (4.218)$$

the functions $q_k$ are

$$q_1 = q_R\,(R) = \alpha_1 - \frac{1}{R^2}\alpha_2; \quad q_2 = q_\eta\,(\eta) = \frac{\alpha_2}{1-\eta^2} - \frac{\alpha_3}{\left(1-\eta^2\right)^2}; \quad q_3 = q_\varphi\,(\varphi) = \alpha_3 \qquad (4.219)$$

for the Laplace equation $\alpha_1 = 0$ and for the Helmholtz equation $\alpha_1 = K^2$.

The transformation matrix for physical components from spherical to Cartesian coordinates, $\mathbf{M}^{u \to x}$ is

$$\mathbf{M}^{u \to x} = \begin{bmatrix} \sqrt{1-\eta^2}\cos(\varphi) & -\eta\cos(\varphi) & -\sin(\varphi) \\ \sqrt{1-\eta^2}\sin(\varphi) & -\eta\sin(\varphi) & \cos(\varphi) \\ \eta & \sqrt{1-\eta^2} & 0 \end{bmatrix} \qquad (4.220)$$

The spherical coordinate system is more often defined using the polar (zenith) angle $\theta$

$$x = R\sin(\theta)\cos(\varphi); \quad y = R\sin(\theta)\sin(\varphi); \quad z = R\cos(\theta) \qquad (4.221)$$

that can be related to the definition (4.210) by: $\eta = \cos(\theta)$.

The coordinate space $(R, \theta, \varphi)$ is limited as: $0 \le R < +\infty, 0 \le \theta \le \pi, 0 \le \varphi < 2\pi$.

The scale factors are

$$h_r = 1; \quad h_\theta = R; \quad h_\varphi = R\sin(\theta); \quad g^{1/2} = R^2\sin(\theta) \qquad (4.222)$$

and the functions $W^{(k)}$ are

$$W^{(R)} = R^2\sin(\theta); \quad W^{(\theta)} = \sin(\theta); \quad W^{(\varphi)} = \frac{1}{\sin(\theta)} \qquad (4.223)$$

The gradient of a scalar quantity $T$ is a vector and its physical components are

$$\nabla T = \text{grad } T = \left[ \frac{\partial T}{\partial R}, \ \frac{1}{R} \frac{\partial T}{\partial \theta}, \ \frac{1}{R \sin(\theta)} \frac{\partial T}{\partial \varphi} \right] \tag{4.224}$$

The divergence of a vector $\mathbf{V} = [\hat{V}_R, \hat{V}_\theta, \hat{V}_\varphi]$ is the scalar

$$\nabla \cdot \mathbf{V} = \text{div}\mathbf{V} = \frac{\partial \hat{V}_R}{\partial R} + \frac{2}{R} \hat{V}_R + \frac{1}{R} \frac{\partial \hat{V}_\theta}{\partial \theta} + \frac{\cot(\theta)}{R} \hat{V}_\theta + \frac{1}{R \sin(\theta)} \frac{\partial \hat{V}_\varphi}{\partial \varphi} \tag{4.225}$$

The curl of a vector $\mathbf{V} = [\hat{V}_R, \hat{V}_\theta, \hat{V}_\varphi]$ is the vector $\omega = \text{curl}\mathbf{V}$

$$\begin{bmatrix} \hat{\omega}_R \\ \hat{\omega}_\theta \\ \hat{\omega}_\varphi \end{bmatrix} = \begin{bmatrix} \frac{1}{R} \frac{\partial \hat{V}_\varphi}{\partial \theta} + \frac{\cot(\theta)}{R} \hat{V}_\varphi - \frac{1}{R \sin(\theta)} \frac{\partial \hat{V}_\theta}{\partial \varphi} \\ \frac{1}{R \sin(\theta)} \frac{\partial \hat{V}_R}{\partial \varphi} - \frac{\partial \hat{V}_\varphi}{\partial R} - \frac{1}{R} \hat{V}_\varphi \\ \frac{\partial \hat{V}_\theta}{\partial R} + \frac{1}{R} \hat{V}_\theta - \frac{1}{R} \frac{\partial \hat{V}_R}{\partial \theta} \end{bmatrix} \tag{4.226}$$

The Laplacian of the scalar $T$ is the scalar

$$\nabla^2 T = \frac{\partial^2 T}{\partial R^2} + \frac{2}{R} \frac{\partial T}{\partial R} + \frac{1}{R^2} \frac{\partial^2 T}{\partial \theta^2} + \frac{\cot(\theta)}{R^2} \frac{\partial T}{\partial \theta} + \frac{1}{R^2 \sin^2(\theta)} \frac{\partial^2 T}{\partial \varphi^2} \tag{4.227}$$

The Laplace and Helmholtz equations are *simple-separable* in this system, the corresponding Stäckel matrix is

$$\mathbf{\Phi} = \begin{bmatrix} 1 & -\frac{1}{R^2} & 0 \\ 0 & 1 & -\frac{1}{\sin^2(\theta)} \\ 0 & 0 & 1 \end{bmatrix} \tag{4.228}$$

the functions $f_k$ are

$$f_1 = f_R(R) = R^2; \quad f_2 = f_\theta(\theta) = \sin(\theta); \quad f_3 = f_\varphi(\varphi) = 1 \tag{4.229}$$

the functions $q_k$ are

$$q_1 = q_R(R) = \alpha_1 - \frac{1}{R^2} \alpha_2; \quad q_2 = q_\theta(\theta) = \alpha_2 - \frac{\alpha_3}{\sin^2(\theta)}; \quad q_3 = q_\varphi(\varphi) = \alpha_3 \tag{4.230}$$

for the Laplace equation $\alpha_1 = 0$ and for the Helmholtz equation $\alpha_1 = K^2$.

The transformation matrix for physical components from spherical to Cartesian coordinates, $\mathbf{M}^{u \to x}$ is

$$\mathbf{M}^{u \to x} = \begin{bmatrix} \sin(\theta)\cos(\varphi) & \cos(\theta)\cos(\varphi) & -\sin(\varphi) \\ \sin(\theta)\sin(\varphi) & \cos(\theta)\sin(\varphi) & \cos(\varphi) \\ \cos(\theta) & -\sin(\theta) & 0 \end{bmatrix} \qquad (4.231)$$

### 4.4.5   Prolate Spheroidal Coordinate System

Prolate spheroidal coordinates can be defined as

$$x = a\sqrt{\zeta^2 - 1}\sqrt{1 - \eta^2}\cos(\varphi); \quad y = a\sqrt{\zeta^2 - 1}\sqrt{1 - \eta^2}\sin(\varphi); \quad z = a\zeta\eta \tag{4.232}$$

The coordinate space $(\zeta, \eta, \varphi)$ is limited as: $1 \le \zeta < +\infty, -1 \le \eta \le 1, 0 \le \varphi < 2\pi$.

The surface $\zeta = \zeta_0$ is a prolate spheroid; the surface $\eta = \eta_0$ is a two-sheets hyperboloid of revolution around the $z$-axis; the surface $\varphi = \varphi_0$ is a half plane passing through the $z$-axis (see Fig. 4.5). The scale factors are

$$h_\zeta = a\frac{\sqrt{\zeta^2 - \eta^2}}{\sqrt{\zeta^2 - 1}}; \quad h_\eta = a\frac{\sqrt{\zeta^2 - \eta^2}}{\sqrt{1 - \eta^2}}; \quad h_\varphi = a\sqrt{\zeta^2 - 1}\sqrt{1 - \eta^2}; \quad g^{1/2} = a^3\left(\zeta^2 - \eta^2\right)$$

$$(4.233)$$

and the functions $W^{(k)}$ are

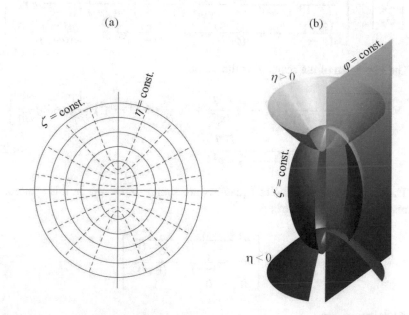

(a)                                              (b)

**Fig. 4.5** Prolate spheroidal coordinate system: **a** coordinates over a plane passing through the $z$-axis; **b** 3-D representation of the iso-surfaces

$$W^{(\zeta)} = a\left(\zeta^2 - 1\right); \quad W^{(\eta)} = a\left(1 - \eta^2\right); \quad W^{(\varphi)} = a\frac{\zeta^2 - \eta^2}{\left(\zeta^2 - 1\right)\left(1 - \eta^2\right)} \quad (4.234)$$

The gradient of a scalar quantity $T$ is a vector and its physical components are

$$\nabla T = \text{grad } T = \frac{1}{a}\left[\frac{\sqrt{\zeta^2 - 1}}{\sqrt{\zeta^2 - \eta^2}}\frac{\partial T}{\partial \zeta}, \frac{\sqrt{1 - \eta^2}}{\sqrt{\zeta^2 - \eta^2}}\frac{\partial T}{\partial \eta}, \frac{1}{\sqrt{\zeta^2 - 1}\sqrt{1 - \eta^2}}\frac{\partial T}{\partial \varphi}\right]$$
$$(4.235)$$

The divergence of a vector $\mathbf{V} = [\hat{V}_\zeta, \hat{V}_\eta, \hat{V}_\varphi]$ is the scalar

$$\nabla \cdot \mathbf{V} = \text{div}\mathbf{V} = \frac{\sqrt{\zeta^2 - 1}}{a\sqrt{\zeta^2 - \eta^2}}\frac{\partial \hat{V}_\zeta}{\partial \zeta} - \frac{\zeta\left(1 - 2\zeta^2 + \eta^2\right)}{a\sqrt{\zeta^2 - 1}\left(\zeta^2 - \eta^2\right)^{\frac{3}{2}}}\hat{V}_\zeta +$$

$$+\frac{\sqrt{1 - \eta^2}}{a\sqrt{\zeta^2 - \eta^2}}\frac{\partial \hat{V}_\eta}{\partial \eta} - \frac{\eta\left(1 + \zeta^2 - 2\eta^2\right)}{a\sqrt{1 - \eta^2}\left(\zeta^2 - \eta^2\right)^{\frac{3}{2}}}\hat{V}_\eta + \frac{1}{a\sqrt{\zeta^2 - 1}\sqrt{1 - \eta^2}}\frac{\partial \hat{V}_\varphi}{\partial \varphi}$$
$$(4.236)$$

The curl of a vector $\mathbf{V} = [\hat{V}_\zeta, \hat{V}_\eta, \hat{V}_\varphi]$ is the vector $\boldsymbol{\omega} = \text{curl}\mathbf{V}$

$$\begin{bmatrix} \hat{\omega}_\zeta \\ \hat{\omega}_\eta \\ \hat{\omega}_\varphi \end{bmatrix} = \begin{bmatrix} \frac{(1-\eta^2)^{1/2}}{a(\zeta^2-\eta^2)^{1/2}}\frac{\partial \hat{V}_\varphi}{\partial \eta} - \frac{\eta}{a(\zeta^2-\eta^2)^{1/2}(1-\eta^2)^{1/2}}\hat{V}_\varphi - \frac{1}{a(\zeta^2-1)^{1/2}(1-\eta^2)^{1/2}}\frac{\partial \hat{V}_\eta}{\partial \varphi} \\ \frac{1}{a(\zeta^2-1)^{1/2}(1-\eta^2)^{1/2}}\frac{\partial \hat{V}_\zeta}{\partial \varphi} - \frac{(\zeta^2-1)^{1/2}}{a(\zeta^2-\eta^2)^{1/2}}\frac{\partial \hat{V}_\varphi}{\partial \zeta} - \frac{\zeta}{a(\zeta^2-1)^{1/2}(\zeta^2-\eta^2)^{1/2}}\hat{V}_\varphi \\ \frac{(\zeta^2-1)^{1/2}}{a(\zeta^2-\eta^2)^{1/2}}\frac{\partial \hat{V}_\eta}{\partial \zeta} + \frac{\zeta(\zeta^2-1)^{1/2}}{a(\zeta^2-\eta^2)^{3/2}}\hat{V}_\eta - \frac{(1-\eta^2)^{1/2}}{a(\zeta^2-\eta^2)^{1/2}}\frac{\partial \hat{V}_\zeta}{\partial \eta} + \frac{\eta(1-\eta^2)^{1/2}}{a(\zeta^2-\eta^2)^{3/2}}\hat{V}_\zeta \end{bmatrix}$$
$$(4.237)$$

The Laplacian of the scalar $T$ is the scalar

$$\nabla^2 T = \frac{1}{a^2\left(\zeta^2 - \eta^2\right)}\left[\left(\zeta^2 - 1\right)\frac{\partial^2 T}{\partial \zeta^2} + 2\zeta\frac{\partial T}{\partial \zeta} + \left(1 - \eta^2\right)\frac{\partial^2 T}{\partial \eta^2} - 2\eta\frac{\partial T}{\partial \eta}\right]$$

$$+\frac{1}{a^2\left(\zeta^2 - 1\right)\left(1 - \eta^2\right)}\frac{\partial^2 T}{\partial \varphi^2} \quad (4.238)$$

The Laplace and Helmholtz equations are *simple-separable* in this system, the corresponding Stäckel matrix is

$$\boldsymbol{\Phi} = \begin{bmatrix} a^2 & -\frac{1}{a(\zeta^2-1)} & -\frac{1}{(\zeta^2-1)^2} \\ a^2 & \frac{1}{a(1-\eta^2)} & -\frac{1}{(1-\eta^2)^2} \\ 0 & 0 & 1 \end{bmatrix} \quad (4.239)$$

the functions $f_k$ are

$$f_1 = f_\zeta\left(\zeta\right) = a\left(\zeta^2 - 1\right); \quad f_2 = f_\eta\left(\eta\right) = a\left(1 - \eta^2\right); \quad f_3 = f_\varphi\left(\varphi\right) = 1 \quad (4.240)$$

the functions $q_k$ are

$$q_1 = q_\zeta(\zeta) = a^2\alpha_1 - \frac{\alpha_2}{a(\zeta^2-1)} - \frac{\alpha_3}{(\zeta^2-1)^2} \tag{4.241a}$$

$$q_2 = q_\eta(\eta) = a^2\alpha_1 + \frac{\alpha_2}{a(1-\eta^2)} - \frac{\alpha_3}{(1-\eta^2)^2} \tag{4.241b}$$

$$q_3 = q_\varphi(\varphi) = \alpha_3 \tag{4.241c}$$

for the Laplace equation $\alpha_1 = 0$ and for the Helmholtz equation $\alpha_1 = K^2$.

The transformation matrix for physical components from prolate spheroidal to Cartesian coordinates, $\mathbf{M}^{u \to x}$ is

$$\mathbf{M}^{u \to x} = \begin{bmatrix} \zeta\frac{(1-\eta^2)^{1/2}}{(\zeta^2-\eta^2)^{1/2}}\cos(\varphi) & -\eta\frac{(\zeta^2-1)^{1/2}}{(\zeta^2-\eta^2)^{1/2}}\cos(\varphi) & -\sin(\varphi) \\ \zeta\frac{(1-\eta^2)^{1/2}}{(\zeta^2-\eta^2)^{1/2}}\sin(\varphi) & -\eta\frac{(\zeta^2-1)^{1/2}}{(\zeta^2-\eta^2)^{1/2}}\sin(\varphi) & \cos(\varphi) \\ \eta\frac{(\zeta^2-1)^{1/2}}{(\zeta^2-\eta^2)^{1/2}} & \zeta\frac{(1-\eta^2)^{1/2}}{(\zeta^2-\eta^2)^{1/2}} & 0 \end{bmatrix} \tag{4.242}$$

The prolate spheroidal coordinate system can also be defined using the coordinates $(\xi, \psi, \varphi)$ as

$$x = a\sinh(\xi)\sin(\psi)\cos(\varphi); \quad y = a\sinh(\xi)\sin(\psi)\sin(\varphi); \quad z = a\cosh(\xi)\cos(\psi) \tag{4.243}$$

The coordinate space $(\xi, \psi, \varphi)$ is limited as: $0 \le \xi < +\infty, 0 \le \psi \le \pi, 0 \le \varphi < 2\pi$ and the relation with the system (4.232) is given by: $\zeta = \cosh(\xi)$ and $\eta = \cos(\psi)$.

The scale factors are

$$h_\xi = a\sqrt{\sinh^2(\xi) + \sin^2(\psi)} = h_\psi; \quad h_\varphi = a\sinh(\xi)\sin(\psi) \tag{4.244}$$

$$g^{1/2} = a^3\left[\sinh^2(\xi) + \sin^2(\psi)\right]\sinh(\xi)\sin(\psi) \tag{4.245}$$

and the functions $W^{(k)}$ are

$$W^{(\xi)} = a\sinh(\xi)\sin(\psi) = W^{(\psi)}; \quad W^{(\varphi)} = a\frac{\sinh^2(\xi) + \sin^2(\psi)}{\sin(\psi)\sinh(\xi)} \tag{4.246}$$

The gradient of a scalar quantity $T$ is a vector

$$\nabla T = \text{grad } T = \frac{1}{a\sqrt{\sinh^2(\xi) + \sin^2(\psi)}}\left[\frac{\partial T}{\partial \xi}, \frac{\partial T}{\partial \psi}, \frac{\sqrt{\sinh^2(\xi) + \sin^2(\psi)}}{\sinh(\xi)\sin(\psi)}\frac{\partial T}{\partial \varphi}\right] \tag{4.247}$$

The divergence of a vector $\mathbf{V} = [\hat{V}_\xi, \hat{V}_\psi, \hat{V}_\varphi]$ is the scalar

$$\boldsymbol{\nabla} \cdot \mathbf{V} = \operatorname{div}\mathbf{V} = \frac{1}{a\sqrt{\sinh^2(\xi) + \sin^2(\psi)}} \left[ \frac{\partial \hat{V}_\xi}{\partial \xi} + \frac{2\sinh^2(\xi) + \sin^2(\psi)}{\sinh^2(\xi) + \sin^2(\psi)} \coth(\xi)\, \hat{V}_\xi \right] +$$

$$+ \frac{1}{a\sqrt{\sinh^2(\xi) + \sin^2(\psi)}} \left[ \frac{\partial \hat{V}_\psi}{\partial \psi} + \frac{\sinh^2(\xi) + 2\sin^2(\psi)}{\sinh^2(\xi) + \sin^2(\psi)} \cot(\psi)\, \hat{V}_\psi \right] +$$

$$+ \frac{1}{a\sinh(\xi)\sin(\psi)} \frac{\partial \hat{V}_\varphi}{\partial \varphi} \tag{4.248}$$

The curl of a vector $\mathbf{V} = [\hat{V}_\xi, \hat{V}_\psi, \hat{V}_\varphi]$ is the vector $\boldsymbol{\omega} = \operatorname{curl}\mathbf{V}$

$$\begin{bmatrix} \hat{\omega}_\xi \\ \hat{\omega}_\psi \\ \hat{\omega}_\varphi \end{bmatrix} = \begin{bmatrix} \frac{1}{a[\sinh^2(\xi)+\sin^2(\psi)]^{1/2}} \left[ \frac{\partial \hat{V}_\varphi}{\partial \psi} + \cot(\psi)\,\hat{V}_\varphi \right] - \frac{1}{a\sinh(\xi)\sin(\psi)} \frac{\partial \hat{V}_\psi}{\partial \varphi} \\ \frac{1}{a\sinh(\xi)\sin(\psi)} \frac{\partial \hat{V}_\xi}{\partial \varphi} - \frac{1}{a[\sinh^2(\xi)+\sin^2(\psi)]^{1/2}} \left[ \frac{\partial \hat{V}_\varphi}{\partial \xi} + \coth(\xi)\,\hat{V}_\varphi \right] \\ \frac{1}{a[\sinh^2(\xi)+\sin^2(\psi)]^{1/2}} \left[ \frac{\partial \hat{V}_\psi}{\partial \xi} + \frac{\sinh(\xi)\cosh(\xi)}{\sinh^2(\xi)+\sin^2(\psi)}\,\hat{V}_\psi - \frac{\partial \hat{V}_\xi}{\partial \psi} - \frac{\sin(\psi)\cos(\psi)}{\sinh^2(\xi)+\sin^2(\psi)}\,\hat{V}_\xi \right] \end{bmatrix} \tag{4.249}$$

The Laplacian of the scalar $T$ is the scalar

$$\nabla^2 T = \frac{1}{a^2\left[\sinh^2(\xi) + \sin^2(\psi)\right]} \left[ \frac{\partial^2 T}{\partial \xi^2} + \coth(\xi)\frac{\partial T}{\partial \xi} + \frac{\partial^2 T}{\partial \psi^2} + \cot(\psi)\frac{\partial T}{\partial \psi} \right]$$

$$+ \frac{1}{a^2\sinh^2(\xi)\sin^2(\psi)} \frac{\partial^2 T}{\partial \varphi^2} \tag{4.250}$$

The Laplace and Helmholtz equations are *simple-separable* in this system, the corresponding Stäckel matrix is

$$\boldsymbol{\Phi} = \begin{bmatrix} a^2\sinh^2(\xi) & -1 & -\frac{1}{\sinh^2(\xi)} \\ a^2\sin^2(\psi) & 1 & -\frac{1}{\sin^2(\psi)} \\ 0 & 0 & 1 \end{bmatrix} \tag{4.251}$$

the functions $f_k$ are

$$f_1 = f_\xi(\xi) = \sinh(\xi); \qquad f_2 = f_\psi(\psi) = \sin(\psi); \qquad f_3 = f_\varphi(\varphi) = a \tag{4.252}$$

the functions $q_k$ are

$$q_1 = q_\xi(\xi) = a^2 \sinh^2(\xi)\,\alpha_1 - \alpha_2 - \frac{\alpha_3}{\sinh^2(\xi)} \qquad (4.253a)$$

$$q_2 = q_\psi(\psi) = a^2 \sin^2(\psi)\,\alpha_1 + \alpha_2 - \frac{\alpha_3}{\sin^2(\psi)} \qquad (4.253b)$$

$$q_3 = q_\varphi(\varphi) = \alpha_3 \qquad (4.253c)$$

for the Laplace equation $\alpha_1 = 0$ and for the Helmholtz equation $\alpha_1 = K^2$.

The transformation matrix for physical components from prolate spheroidal to Cartesian coordinates, $\mathbf{M}^{u \to x}$ is

$$\mathbf{M}^{u \to x} = \begin{bmatrix} \frac{\cosh(\xi)\sin(\psi)}{\left[\sinh^2(\xi)+\sin^2(\psi)\right]^{1/2}}\cos(\varphi) & \frac{\sinh(\xi)\cos(\psi)}{\left[\sinh^2(\xi)+\sin^2(\psi)\right]^{1/2}}\cos(\varphi) & -\sin(\varphi) \\ \frac{\cosh(\xi)\sin(\psi)}{\left[\sinh^2(\xi)+\sin^2(\psi)\right]^{1/2}}\sin(\varphi) & \frac{\sinh(\xi)\cos(\psi)}{\left[\sinh^2(\xi)+\sin^2(\psi)\right]^{1/2}}\sin(\varphi) & \cos(\varphi) \\ \frac{\sinh(\xi)\cos(\psi)}{\left[\sinh^2(\xi)+\sin^2(\psi)\right]^{1/2}} & -\frac{\cosh(\xi)\sin(\psi)}{\left[\sinh^2(\xi)+\sin^2(\psi)\right]^{1/2}} & 0 \end{bmatrix}$$

$$(4.254)$$

### 4.4.6  Oblate Spheroidal Coordinate System

Oblate spheroidal coordinates can be defined as

$$x = a\sqrt{\zeta^2+1}\sqrt{1-\eta^2}\cos(\varphi); \quad y = a\sqrt{\zeta^2+1}\sqrt{1-\eta^2}\sin(\varphi); \quad z = a\zeta\eta \qquad (4.255)$$

The coordinate space $(\zeta, \eta, \varphi)$ is limited as: $0 \le \zeta < +\infty$, $-1 \le \eta \le 1$, $0 \le \varphi < 2\pi$.

The surface $\zeta = \zeta_0$ is an oblate spheroid; the surface $\eta = \eta_0$ is a one-sheet hyperboloid of revolution around the $z$-axis; the surface $\varphi = \varphi_0$ is a half plane passing through the $z$-axis (see Fig. 4.6). The scale factors are

$$h_\zeta = a\frac{\sqrt{\zeta^2+\eta^2}}{\sqrt{\zeta^2+1}}; \quad h_\eta = a\frac{\sqrt{\zeta^2+\eta^2}}{\sqrt{1-\eta^2}}; \quad h_\varphi = a\sqrt{\zeta^2+1}\sqrt{1-\eta^2}; \quad g^{1/2} = a^3\left(\zeta^2+\eta^2\right) \qquad (4.256)$$

and the functions $W^{(k)}$ are

$$W^{(\zeta)} = a\left(\zeta^2+1\right); \quad W^{(\eta)} = a\left(1-\eta^2\right); \quad W^{(\varphi)} = a\frac{\zeta^2+\eta^2}{\left(\zeta^2+1\right)\left(1-\eta^2\right)} \qquad (4.257)$$

The gradient of a scalar quantity $T$ is a vector and its physical components are

$$\nabla T = \text{grad}\,T = \frac{1}{a}\left[\frac{\sqrt{\zeta^2+1}}{\sqrt{\zeta^2+\eta^2}}\frac{\partial T}{\partial \zeta}, \frac{\sqrt{1-\eta^2}}{\sqrt{\zeta^2+\eta^2}}\frac{\partial T}{\partial \eta}, \frac{1}{\sqrt{\zeta^2+1}\sqrt{1-\eta^2}}\frac{\partial T}{\partial \varphi}\right] \qquad (4.258)$$

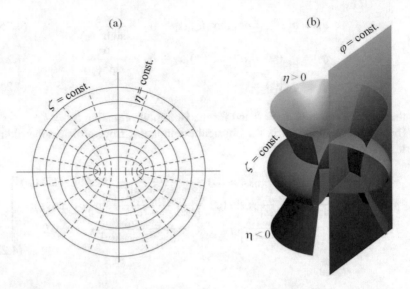

**Fig. 4.6** Oblate spheroidal coordinate system: **a** coordinates over a plane passing through the z-axis; **b** 3-D representation of the iso-surfaces

The divergence of a vector $\mathbf{V} = [\hat{V}_\zeta, \hat{V}_\eta, \hat{V}_\varphi]$ is the scalar

$$\nabla \cdot \mathbf{V} = \mathrm{div}\mathbf{V} = \frac{\sqrt{\zeta^2 + 1}}{a\sqrt{\zeta^2 + \eta^2}}\frac{\partial \hat{V}_\zeta}{\partial \zeta} + \frac{\zeta\left(1 + 2\zeta^2 + \eta^2\right)}{a\sqrt{\zeta^2 + 1}\left(\zeta^2 + \eta^2\right)^{3/2}}\hat{V}_\zeta +$$

$$+ \frac{\sqrt{1 - \eta^2}}{a\sqrt{\zeta^2 + \eta^2}}\frac{\partial \hat{V}_\eta}{\partial \eta} + \frac{\eta\left(1 - \zeta^2 - 2\eta^2\right)}{a\sqrt{1 - \eta^2}\left(\zeta^2 + \eta^2\right)^{3/2}}\hat{V}_\eta + \frac{1}{a\sqrt{\zeta^2 + 1}\sqrt{1 - \eta^2}}\frac{\partial \hat{V}_\varphi}{\partial \varphi}$$

$$\tag{4.259}$$

The curl of a vector $\mathbf{V} = [\hat{V}_\zeta, \hat{V}_\eta, \hat{V}_\varphi]$ is the vector $\boldsymbol{\omega} = \mathrm{curl}\mathbf{V}$

$$\begin{bmatrix} \hat{\omega}_\zeta \\ \hat{\omega}_\eta \\ \hat{\omega}_\varphi \end{bmatrix} = \begin{bmatrix} \frac{(1-\eta^2)^{1/2}}{a(\zeta^2+\eta^2)^{1/2}}\frac{\partial \hat{V}_\varphi}{\partial \eta} - \frac{\eta}{a(\zeta^2+\eta^2)^{1/2}(1-\eta^2)^{1/2}}\hat{V}_\varphi - \frac{1}{a(\zeta^2+1)^{1/2}(1-\eta^2)^{1/2}}\frac{\partial \hat{V}_\eta}{\partial \varphi} \\ \frac{1}{a(\zeta^2+1)^{1/2}(1-\eta^2)^{1/2}}\frac{\partial \hat{V}_\zeta}{\partial \varphi} - \frac{(\zeta^2+1)^{1/2}}{a(\zeta^2+\eta^2)^{1/2}}\frac{\partial \hat{V}_\varphi}{\partial \zeta} - \frac{\zeta}{a(\zeta^2+1)^{1/2}(\zeta^2+\eta^2)^{1/2}}\hat{V}_\varphi \\ \frac{(\zeta^2+1)^{1/2}}{a(\zeta^2+\eta^2)^{1/2}}\frac{\partial \hat{V}_\eta}{\partial \zeta} + \frac{\zeta(\zeta^2+1)^{1/2}}{a(\zeta^2+\eta^2)^{3/2}}\hat{V}_\eta - \frac{(1-\eta^2)^{1/2}}{a(\zeta^2+\eta^2)^{1/2}}\frac{\partial \hat{V}_\zeta}{\partial \eta} - \frac{\eta(1-\eta^2)^{1/2}}{a(\zeta^2+\eta^2)^{3/2}}\hat{V}_\zeta \end{bmatrix}$$

$$\tag{4.260}$$

The Laplacian of the scalar $T$ is the scalar

$$\nabla^2 T = \frac{1}{a^2\left(\zeta^2 + \eta^2\right)}\left[\left(\zeta^2 + 1\right)\frac{\partial^2 T}{\partial \zeta^2} + 2\zeta\frac{\partial T}{\partial \zeta} + \left(1 - \eta^2\right)\frac{\partial^2 T}{\partial \eta^2} - 2\eta\frac{\partial T}{\partial \eta}\right] +$$

$$+ \frac{1}{\left(\zeta^2 + 1\right)\left(1 - \eta^2\right)}\frac{\partial^2 T}{\partial \varphi^2}$$

$$\tag{4.261}$$

The Laplace and Helmholtz equations are *simple-separable* in this system, the corresponding Stäckel matrix is

$$
\mathbf{\Phi} = \begin{bmatrix} a^2 & -\dfrac{1}{a(\zeta^2+1)} & \dfrac{1}{(\zeta^2+1)^2} \\[2mm] -a^2 & \dfrac{1}{a(1-\eta^2)} & -\dfrac{1}{(1-\eta^2)^2} \\[2mm] 0 & 0 & 1 \end{bmatrix} \tag{4.262}
$$

the functions $f_k$ are

$$
f_1 = f_\zeta (\zeta) = a\left(\zeta^2 + 1\right); \quad f_2 = f_\eta (\eta) = a\left(1 - \eta^2\right); \quad f_3 = f_\varphi (\varphi) = 1 \tag{4.263}
$$

the functions $q_k$ are

$$
q_1 = q_\zeta (\zeta) = a^2\alpha_1 - \frac{\alpha_2}{a\left(\zeta^2 + 1\right)} + \frac{\alpha_3}{\left(\zeta^2 + 1\right)^2} \tag{4.264a}
$$

$$
q_2 = q_\eta (\eta) = -a^2\alpha_1 + \frac{\alpha_2}{a\left(1 - \eta^2\right)} - \frac{\alpha_3}{\left(1 - \eta^2\right)^2} \tag{4.264b}
$$

$$
q_3 = q_\varphi (\varphi) = \alpha_3 \tag{4.264c}
$$

for the Laplace equation $\alpha_1 = 0$ and for the Helmholtz equation $\alpha_1 = K^2$.

The transformation matrix for physical components from oblate spheroidal to Cartesian coordinates, $\mathbf{M}^{u \to x}$ is

$$
\mathbf{M}^{u \to x} = \begin{bmatrix} \zeta\dfrac{(1-\eta^2)^{1/2}}{(\zeta^2+\eta^2)^{1/2}} \cos(\varphi) & -\eta\dfrac{(\zeta^2+1)^{1/2}}{(\zeta^2+\eta^2)^{1/2}} \cos(\varphi) & -\sin(\varphi) \\[3mm] \zeta\dfrac{(1-\eta^2)^{1/2}}{(\zeta^2+\eta^2)^{1/2}} \sin(\varphi) & -\eta\dfrac{(\zeta^2+1)^{1/2}}{(\zeta^2+\eta^2)^{1/2}} \sin(\varphi) & \cos(\varphi) \\[3mm] \eta\dfrac{(\zeta^2+1)^{1/2}}{(\zeta^2+\eta^2)^{1/2}} & \zeta\dfrac{(1-\eta^2)^{1/2}}{(\zeta^2+\eta^2)^{1/2}} & 0 \end{bmatrix} \tag{4.265}
$$

The oblate spheroidal coordinate system can also be defined using the coordinates $(\xi, \psi, \varphi)$ as

$$
x = a \cosh(\xi) \sin(\psi) \cos(\varphi); \quad y = a \cosh(\xi) \sin(\psi) \sin(\varphi); \quad z = a \sinh(\xi) \cos(\psi) \tag{4.266}
$$

The coordinate space $(\xi, \psi, \varphi)$ is limited as: $0 \leq \xi < +\infty, 0 \leq \psi \leq \pi, 0 \leq \varphi < 2\pi$ and the relation with the system (4.255) is given by: $\zeta = \sinh(\xi)$ and $\eta = \cos(\psi)$.

The scale factors are

$$h_\xi = a\sqrt{\cosh^2{(\xi)} - \sin^2{(\psi)}} = h_\psi; \quad h_\varphi = a\cosh{(\xi)}\sin{(\psi)}; \quad (4.267)$$

$$g^{1/2} = a^3 \left[\cosh^2{(\xi)} - \sin^2{(\psi)}\right]\cosh{(\xi)}\sin{(\psi)} \tag{4.268}$$

and the functions $W^{(k)}$ are

$$W^{(\xi)} = a\cosh{(\xi)}\sin{(\psi)} = W^{(\psi)}; \quad W^{(\varphi)} = a\frac{\cosh^2{(\xi)} - \sin^2{(\psi)}}{\cosh{(\xi)}\sin{(\psi)}} \tag{4.269}$$

The gradient of a scalar quantity $T$ is a vector and its physical components are

$$\boldsymbol{\nabla} T = \text{grad } T = \frac{1}{a\sqrt{\cosh^2{(\xi)} - \sin^2{(\psi)}}}\left[\frac{\partial T}{\partial \xi}, \frac{\partial T}{\partial \psi}, \frac{\sqrt{\cosh^2{(\xi)} - \sin^2{(\psi)}}}{\cosh{(\xi)}\sin{(\psi)}}\frac{\partial T}{\partial \varphi}\right] \tag{4.270}$$

The divergence of a vector $\mathbf{V} = [\hat{V}_\xi, \hat{V}_\psi, \hat{V}_\varphi]$ is the scalar

$$\boldsymbol{\nabla} \cdot \mathbf{V} = \text{div}\mathbf{V} = \frac{1}{a\sqrt{\cosh^2{(\xi)} - \sin^2{(\psi)}}}\left[\frac{\partial \hat{V}_\xi}{\partial \xi} + \frac{2\cosh^2{(\xi)} - \sin^2{(\psi)}}{\cosh^2{(\xi)} - \sin^2{(\psi)}}\tanh{(\xi)}\,\hat{V}_\xi\right] + $$

$$+\frac{1}{a\sqrt{\cosh^2{(\xi)} - \sin^2{(\psi)}}}\left[\frac{\partial \hat{V}_\psi}{\partial \psi} + \frac{\cosh^2{(\xi)} - 2\sin^2{(\psi)}}{\cosh^2{(\xi)} - \sin^2{(\psi)}}\cot{(\psi)}\,\hat{V}_\psi\right] + $$

$$+\frac{1}{a\cosh{(\xi)}\sin{(\psi)}}\frac{\partial \hat{V}_\varphi}{\partial \varphi} \tag{4.271}$$

The curl of a vector $\mathbf{V} = [\hat{V}_\xi, \hat{V}_\psi, \hat{V}_\varphi]$ is the vector $\boldsymbol{\omega} = \text{curl}\mathbf{V}$

$$\begin{bmatrix} \hat{\omega}_\xi \\ \hat{\omega}_\psi \\ \hat{\omega}_\varphi \end{bmatrix} = \begin{bmatrix} \frac{1}{a[\cosh^2{(\xi)}-\sin^2{(\psi)}]^{1/2}}\left[\frac{\partial \hat{V}_\varphi}{\partial \psi} + \cot{(\psi)}\,\hat{V}_\varphi\right] - \frac{1}{a\cosh{(\xi)}\sin{(\psi)}}\frac{\partial \hat{V}_\psi}{\partial \varphi} \\ \frac{1}{a\cosh{(\xi)}\sin{(\psi)}}\frac{\partial \hat{V}_\xi}{\partial \varphi} - \frac{1}{a[\cosh^2{(\xi)}-\sin^2{(\psi)}]^{1/2}}\left[\frac{\partial \hat{V}_\varphi}{\partial \xi} + \tanh{(\xi)}\,\hat{V}_\varphi\right] \\ \frac{1}{a[\cosh^2{(\xi)}-\sin^2{(\psi)}]^{1/2}}\left[\frac{\partial \hat{V}_\psi}{\partial \xi} + \frac{\sinh{(\xi)}\cosh{(\xi)}}{\cosh^2{(\xi)}-\sin^2{(\psi)}}\hat{V}_\psi - \frac{\partial \hat{V}_\xi}{\partial \psi} + \frac{\sin{(\psi)}\cos{(\psi)}}{\cosh^2{(\xi)}-\sin^2{(\psi)}}\hat{V}_\xi\right] \end{bmatrix} \tag{4.272}$$

The Laplacian of the scalar $T$ is the scalar

$$\nabla^2 T = \frac{1}{a^2\left[\cosh^2{(\xi)} - \sin^2{(\psi)}\right]}\left[\frac{\partial^2 T}{\partial \xi^2} + \tanh{(\xi)}\frac{\partial T}{\partial \xi} + \frac{\partial^2 T}{\partial \psi^2} + \cot{(\psi)}\frac{\partial T}{\partial \psi}\right] + $$

$$+\frac{1}{a^2\cosh^2{(\xi)}\sin^2{(\psi)}}\frac{\partial^2 T}{\partial \varphi^2} \tag{4.273}$$

The Laplace and Helmholtz equations are *simple-separable* in this system, the corresponding Stäckel matrix is

$$\Phi = \begin{bmatrix} a^2 \cosh^2(\xi) & -1 & \frac{1}{\cosh^2(\xi)} \\ -a^2 \sin^2(\psi) & 1 & -\frac{1}{\sin^2(\psi)} \\ 0 & 0 & 1 \end{bmatrix} \tag{4.274}$$

the functions $f_k$ are

$$f_1 = f_\xi(\xi) = \cosh(\xi); \quad f_2 = f_\psi(\psi) = \sin(\psi); \quad f_3 = f_\varphi(\varphi) = a \tag{4.275}$$

the functions $q_k$ are

$$q_1 = q_\xi(\xi) = a^2 \cosh^2(\xi)\,\alpha_1 - \alpha_2 - \frac{\alpha_3}{\cosh^2(\xi)} \tag{4.276a}$$

$$q_2 = q_\psi(\psi) = -a^2 \sin^2(\psi)\,\alpha_1 + \alpha_2 - \frac{\alpha_3}{\sin^2(\psi)} \tag{4.276b}$$

$$q_3 = q_\varphi(\varphi) = \alpha_3 \tag{4.276c}$$

for the Laplace equation $\alpha_1 = 0$ and for the Helmholtz equation $\alpha_1 = K^2$.

The transformation matrix for physical components from oblate spheroidal to Cartesian coordinates, $\mathbf{M}^{u \to x}$ is

$$\mathbf{M}^{u \to x} = \begin{bmatrix} \frac{\sinh(\xi)\sin(\psi)}{\left[\cosh^2(\xi)-\sin^2(\psi)\right]^{1/2}}\cos(\varphi) & \frac{\cosh(\xi)\cos(\psi)}{\left[\cosh^2(\xi)-\sin^2(\psi)\right]^{1/2}}\cos(\varphi) & -\sin(\varphi) \\ \frac{\sinh(\xi)\sin(\psi)}{\left[\cosh^2(\xi)-\sin^2(\psi)\right]^{1/2}}\sin(\varphi) & \frac{\cosh(\xi)\cos(\psi)}{\left[\cosh^2(\xi)-\sin^2(\psi)\right]^{1/2}}\sin(\varphi) & \cos(\varphi) \\ \frac{\cosh(\xi)\cos(\psi)}{\left[\cosh^2(\xi)-\sin^2(\psi)\right]^{1/2}} & -\frac{\sinh(\xi)\sin(\psi)}{\left[\cosh^2(\xi)\;\sin^2(\psi)\right]^{1/2}} & 0 \end{bmatrix}$$

$$\tag{4.277}$$

## 4.4.7 Parabolic Coordinate System

Parabolic coordinates can be defined as

$$x = \mu\nu\cos(\varphi); \quad y = \mu\nu\sin(\varphi); \quad z = \frac{1}{2}\left(\mu^2 - \nu^2\right) \tag{4.278}$$

The coordinate space $(\mu, \nu, \varphi)$ is limited as: $0 \le \mu < +\infty$, $0 \le \nu < +\infty$, $0 \le \varphi < 2\pi$.

The surfaces $\mu = \mu_0$ and $\nu = \nu_0$ are paraboloids of revolution; the surface $\varphi = \varphi_0$ is an half-plane passing through the $z$-axis (see Fig. 4.7). The scale factors are

$$h_\mu = \sqrt{\mu^2 + \nu^2} = h_\nu; \quad h_\varphi = \mu\nu; \quad g^{1/2} = \mu\nu\left(\mu^2 + \nu^2\right) \tag{4.279}$$

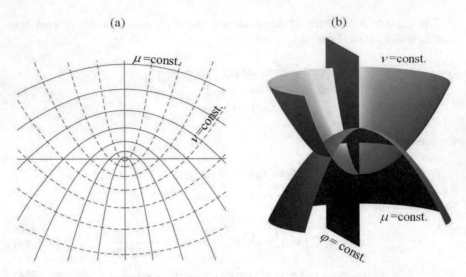

**Fig. 4.7** Parabolic coordinate system: **a** coordinates over a plane passing through the $z$-axis; **b** 3-D representation of the iso-surfaces

and the functions $W^{(k)}$ are

$$W^{(\mu)} = \mu\nu = W^{(\nu)}; \quad W^{(\varphi)} = \frac{\mu^2 + \nu^2}{\mu\nu} \tag{4.280}$$

The gradient of a scalar quantity $T$ is a vector and its physical components are

$$\nabla T = \text{grad } T = \left[ \frac{1}{\sqrt{\mu^2 + \nu^2}} \frac{\partial T}{\partial \mu}, \frac{1}{\sqrt{\mu^2 + \nu^2}} \frac{\partial T}{\partial \eta}, \frac{1}{\mu\nu} \frac{\partial T}{\partial \varphi} \right] \tag{4.281}$$

The divergence of a vector $\mathbf{V} = [\hat{V}_\mu, \hat{V}_\nu, \hat{V}_\varphi]$ is the scalar

$$\nabla \cdot \mathbf{V} = \text{div}\mathbf{V} = \frac{1}{\sqrt{\mu^2 + \nu^2}} \left[ \frac{\partial \hat{V}_\mu}{\partial \mu} + \frac{2\mu^2 + \nu^2}{\mu(\mu^2 + \nu^2)} \hat{V}_\mu + \frac{\partial \hat{V}_\nu}{\partial \nu} + \frac{\mu^2 + 2\nu^2}{\nu(\mu^2 + \nu^2)} \hat{V}_\nu \right] + \frac{1}{\mu\nu} \frac{\partial \hat{V}_\varphi}{\partial \varphi} \tag{4.282}$$

The curl of a vector $\mathbf{V} = [\hat{V}_\mu, \hat{V}_\nu, \hat{V}_\varphi]$ is the vector $\boldsymbol{\omega} = \text{curl}\mathbf{V}$

$$\begin{bmatrix} \hat{\omega}_\mu \\ \hat{\omega}_\nu \\ \hat{\omega}_\varphi \end{bmatrix} = \begin{bmatrix} \frac{1}{(\mu^2 + \nu^2)^{1/2}} \left( \frac{\partial \hat{V}_\varphi}{\partial \nu} + \frac{1}{\nu} \hat{V}_\varphi \right) - \frac{1}{\mu\nu} \frac{\partial \hat{V}_\nu}{\partial \varphi} \\ \frac{1}{\mu\nu} \frac{\partial \hat{V}_\mu}{\partial \varphi} - \frac{1}{(\mu^2 + \nu^2)^{1/2}} \left( \frac{\partial \hat{V}_\varphi}{\partial \mu} + \frac{1}{\mu} \hat{V}_\varphi \right) \\ \frac{1}{(\mu^2 + \nu^2)^{1/2}} \left( \frac{\partial \hat{V}_\nu}{\partial \mu} + \frac{\mu}{\mu^2 + \nu^2} \hat{V}_\nu - \frac{\partial \hat{V}_\mu}{\partial \nu} - \frac{\nu}{\mu^2 + \nu^2} \hat{V}_\mu \right) \end{bmatrix} \tag{4.283}$$

The Laplacian of the scalar $T$ is the scalar

$$\nabla^2 T = \frac{1}{\mu^2 + \nu^2} \left( \frac{\partial^2 T}{\partial \mu^2} + \frac{1}{\mu} \frac{\partial T}{\partial \mu} + \frac{\partial^2 T}{\partial \nu^2} + \frac{1}{\nu} \frac{\partial T}{\partial \nu} \right) + \frac{1}{\mu^2 \nu^2} \frac{\partial^2 T}{\partial \varphi^2} \qquad (4.284)$$

The Laplace and Helmholtz equations are *simple-separable* in this system, the corresponding Stäckel matrix is

$$\mathbf{\Phi} = \begin{bmatrix} \mu^2 & -1 & -\frac{1}{\mu^2} \\ \nu^2 & 1 & -\frac{1}{\nu^2} \\ 0 & 0 & 1 \end{bmatrix} \qquad (4.285)$$

the functions $f_k$ are

$$f_1 = f_\mu (\mu) = \mu; \quad f_2 = f_\nu (\nu) = \nu; \quad f_3 = f_\varphi (\varphi) = 1 \qquad (4.286)$$

the functions $q_k$ are

$$q_1 = q_\mu (\mu) = \mu^2 \alpha_1 - \alpha_2 - \frac{\alpha_3}{\mu^2}; \quad q_2 = q_\nu (\nu) = \nu^2 \alpha_1 + \alpha_2 - \frac{\alpha_3}{\nu^2}; \quad q_3 = q_\varphi (\varphi) = \alpha_3 \qquad (4.287)$$

for the Laplace equation $\alpha_1 = 0$ and for the Helmholtz equation $\alpha_1 = K^2$.

The transformation matrix for physical components from parabolic to Cartesian coordinates, $\mathbf{M}^{u \to x}$ is

$$\mathbf{M}^{u \to x} = \begin{bmatrix} \frac{\nu}{(\mu^2 + \nu^2)^{1/2}} \cos (\varphi) & \frac{\mu}{(\mu^2 + \nu^2)^{1/2}} \cos (\varphi) & -\sin (\varphi) \\ \frac{\nu}{(\mu^2 + \nu^2)^{1/2}} \sin (\varphi) & \frac{\mu}{(\mu^2 + \nu^2)^{1/2}} \sin (\varphi) & \cos (\varphi) \\ \frac{\mu}{(\mu^2 + \nu^2)^{1/2}} & -\frac{\nu}{(\mu^2 + \nu^2)^{1/2}} & 0 \end{bmatrix} \qquad (4.288)$$

## 4.4.8 Conical Coordinate System

Conical coordinates can be defined as

$$x^2 = \frac{R \mu \nu}{\beta \gamma}; \quad y^2 = \frac{R^2 \left( \mu^2 - \beta^2 \right) \left( \beta^2 - \nu^2 \right)}{\beta^2 \left( \gamma^2 - \beta^2 \right)}; \quad z^2 = \frac{R^2 \left( \gamma^2 - \mu^2 \right) \left( \gamma^2 - \nu^2 \right)}{\gamma^2 \left( \gamma^2 - \beta^2 \right)} \qquad (4.289)$$

The coordinate space $(R, \mu, \nu)$ is limited as: $0 \le R < +\infty$, $\beta < \mu < \gamma$, $0 < \nu < \beta$ and $\gamma > \beta > 0$.

The surface $R = R_0$ is a sphere; the surfaces $\mu = \mu_0$ and $\nu = \nu_0$ are elliptic cones (see Fig. 4.8). The scale factors are

**Fig. 4.8** Conical coordinate
system: 3-D representation
of the iso-surfaces

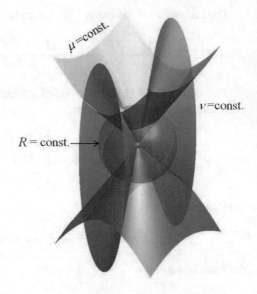

$$h_R = 1; \quad h_\mu = R\sqrt{\frac{\mu^2 - \nu^2}{\left(\mu^2 - \beta^2\right)\left(\gamma^2 - \mu^2\right)}}; \quad h_\nu = R\sqrt{\frac{\mu^2 - \nu^2}{\left(\beta^2 - \nu^2\right)\left(\gamma^2 - \nu^2\right)}}$$

(4.290)

$$g^{1/2} = \frac{R^2\left(\mu^2 - \nu^2\right)}{\sqrt{\left(\mu^2 - \beta^2\right)\left(\gamma^2 - \mu^2\right)\left(\beta^2 - \nu^2\right)\left(\gamma^2 - \nu^2\right)}}$$

(4.291)

and the functions $W^{(k)}$ are

$$W^{(R)} = \frac{R^2\left(\mu^2 - \nu^2\right)}{\sqrt{\left(\mu^2 - \beta^2\right)\left(\gamma^2 - \mu^2\right)\left(\beta^2 - \nu^2\right)\left(\gamma^2 - \nu^2\right)}}$$

(4.292)

$$W^{(\mu)} = \sqrt{\frac{\left(\mu^2 - \beta^2\right)\left(\gamma^2 - \mu^2\right)}{\left(\beta^2 - \nu^2\right)\left(\gamma^2 - \nu^2\right)}}; \quad W^{(\nu)} = \sqrt{\frac{\left(\beta^2 - \nu^2\right)\left(\gamma^2 - \nu^2\right)}{\left(\mu^2 - \beta^2\right)\left(\gamma^2 - \mu^2\right)}}$$

(4.293)

The gradient of a scalar quantity $T$ is a vector and its physical components are

$$\nabla T = \text{grad } T = \left[\frac{\partial T}{\partial R}, \frac{1}{R}\sqrt{\frac{\left(\mu^2 - \beta^2\right)\left(\gamma^2 - \mu^2\right)}{\mu^2 - \nu^2}}\frac{\partial T}{\partial \mu}, \frac{1}{R}\sqrt{\frac{\left(\beta^2 - \nu^2\right)\left(\gamma^2 - \nu^2\right)}{\mu^2 - \nu^2}}\frac{\partial T}{\partial \nu}\right]$$

(4.294)

The divergence of a vector $\mathbf{V} = [\hat{V}_R, \hat{V}_\mu, \hat{V}_\nu]$ is the scalar

$$\nabla \cdot \mathbf{V} = \mathrm{div}\mathbf{V} = \frac{\partial \hat{V}_R}{\partial R} + \frac{2}{R}\hat{V}_R + \frac{1}{R}\sqrt{\frac{(\mu^2 - \beta^2)(\gamma^2 - \mu^2)}{\mu^2 - \nu^2}}\left[\frac{\partial \hat{V}_\mu}{\partial \nu} + \frac{\mu}{\mu^2 - \nu^2}\hat{V}_\mu\right] +$$

$$+\frac{1}{R}\sqrt{\frac{(\beta^2 - \nu^2)(\gamma^2 - \nu^2)}{\mu^2 - \nu^2}}\left[\frac{\partial \hat{V}_\nu}{\partial \nu} - \frac{\nu}{\mu^2 - \nu^2}\hat{V}_\nu\right] \tag{4.295}$$

The curl of a vector $\mathbf{V} = [\hat{V}_R, \hat{V}_\mu, \hat{V}_\nu]$ is the vector $\boldsymbol{\omega} = \mathrm{curl}\mathbf{V}$

$$\begin{bmatrix} \hat{\omega}_R \\ \hat{\omega}_\mu \\ \hat{\omega}_\nu \end{bmatrix} = \begin{bmatrix} \frac{(\mu^2 - \beta^2)^{1/2}(\gamma^2 - \mu^2)^{1/2}}{R(\mu^2 - \nu^2)^{1/2}}\left[\frac{\partial \hat{V}_\nu}{\partial \mu} + \frac{\mu}{\mu^2 - \nu^2}\hat{V}_\nu\right] - \frac{(\beta^2 - \nu^2)^{1/2}(\gamma^2 - \nu^2)^{1/2}}{R(\mu^2 - \nu^2)^{1/2}}\left[\frac{\partial \hat{V}_\mu}{\partial \nu} - \frac{\nu}{\mu^2 - \nu^2}\hat{V}_\mu\right] \\ \frac{(\beta^2 - \nu^2)^{1/2}(\gamma^2 - \nu^2)^{1/2}}{R(\mu^2 - \nu^2)^{1/2}}\frac{\partial \hat{V}_R}{\partial \nu} - \frac{\partial \hat{V}_\nu}{\partial R} - \frac{1}{R}\hat{V}_\nu \\ \frac{\partial \hat{V}_\mu}{\partial R} + \frac{1}{R}\hat{V}_\mu - \frac{(\mu^2 - \beta^2)^{1/2}(\gamma^2 - \mu^2)^{1/2}}{R(\mu^2 - \nu^2)^{1/2}}\frac{\partial \hat{V}_R}{\partial \mu} \end{bmatrix} \tag{4.296}$$

The Laplacian of the scalar $T$ is the scalar

$$\nabla^2 T = \frac{\partial^2 T}{\partial R^2} + \frac{2}{R}\frac{\partial T}{\partial R} + \frac{(\mu^2 - \beta^2)(\gamma^2 - \mu^2)}{R^2(\mu^2 - \nu^2)}\frac{\partial^2 T}{\partial \mu^2} + \frac{\mu(\beta^2 + \gamma^2 - 2\mu^2)}{R^2(\mu^2 - \nu^2)}\frac{\partial T}{\partial \mu} +$$

$$+\frac{(\beta^2 - \nu^2)(\gamma^2 - \nu^2)}{R^2(\mu^2 - \nu^2)}\frac{\partial^2 T}{\partial \nu^2} - \frac{\nu(\beta^2 + \gamma^2 - 2\nu^2)}{R^2(\mu^2 - \nu^2)}\frac{\partial T}{\partial \nu} \tag{4.297}$$

The Laplace and Helmholtz equations are *simple-separable* in this system, the corresponding Stäckel matrix is

$$\Phi = \begin{bmatrix} 1 & -\frac{1}{R^2} & 0 \\ 0 & \frac{\mu^2}{(\mu^2 - \beta^2)(\gamma^2 - \mu^2)} & -\frac{1}{(\mu^2 - \beta^2)(\gamma^2 - \mu^2)} \\ 0 & -\frac{\nu^2}{(\beta^2 - \nu^2)(\gamma^2 - \nu^2)} & \frac{1}{(\beta^2 - \nu^2)(\gamma^2 - \nu^2)} \end{bmatrix} \tag{4.298}$$

the functions $f_k$ are

$$f_1 = f_R(R) = R^2; \quad f_2 = f_\mu(\mu) = \sqrt{(\mu^2 - \beta^2)(\gamma^2 - \mu^2)}; \quad f_3 = f_\nu(\nu) = \sqrt{(\beta^2 - \nu^2)(\gamma^2 - \nu^2)} \tag{4.299}$$

the functions $q_k$ are

$$q_1 = q_R(R) = \alpha_1 - \frac{\alpha_2}{R^2} \tag{4.300a}$$

$$q_2 = q_\mu(\mu) = \frac{\alpha_2\mu^2}{(\mu^2-\beta^2)(\gamma^2-\mu^2)} - \frac{\alpha_3}{(\mu^2-\beta^2)(\gamma^2-\mu^2)} \tag{4.300b}$$

$$q_3 = q_\nu(\nu) = -\frac{\alpha_2\nu^2}{(\beta^2-\nu^2)(\gamma^2-\nu^2)} + \frac{\alpha_3}{(\beta^2-\nu^2)(\gamma^2-\nu^2)} \tag{4.300c}$$

for the Laplace equation $\alpha_1 = 0$ and for the Helmholtz equation $\alpha_1 = K^2$.

The transformation matrix for physical components from conical to Cartesian coordinates, $\mathbf{M}^{u\to x}$ is

$$\mathbf{M}^{u\to x} = \begin{bmatrix} \dfrac{\mu\nu}{\beta\gamma} & \dfrac{\nu}{\beta\gamma}\dfrac{(\mu^2-\beta^2)^{1/2}(\gamma^2-\mu^2)^{1/2}}{(\mu^2-\nu^2)^{1/2}} & \dfrac{\mu}{\beta\gamma}\dfrac{(\beta^2-\nu^2)^{1/2}(\gamma^2-\nu^2)^{1/2}}{(\mu^2-\nu^2)^{1/2}} \\[2mm] \dfrac{1}{\beta}\dfrac{(\mu^2-\beta^2)^{1/2}(\beta^2-\nu^2)^{1/2}}{(\gamma^2-\beta^2)^{1/2}} & \dfrac{\mu}{\beta}\dfrac{(\gamma^2-\mu^2)^{1/2}(\beta^2-\nu^2)^{1/2}}{(\gamma^2-\beta^2)^{1/2}(\mu^2-\nu^2)^{1/2}} & -\dfrac{\nu}{\beta}\dfrac{(\mu^2-\beta^2)^{1/2}(\gamma^2-\nu^2)^{1/2}}{(\gamma^2-\beta^2)^{1/2}(\mu^2-\nu^2)^{1/2}} \\[2mm] \dfrac{1}{\gamma}\dfrac{(\gamma^2-\mu^2)^{1/2}(\gamma^2-\nu^2)^{1/2}}{(\gamma^2-\beta^2)^{1/2}} & -\dfrac{\mu}{\gamma}\dfrac{(\mu^2-\beta^2)^{1/2}(\gamma^2-\nu^2)^{1/2}}{(\gamma^2-\beta^2)^{1/2}(\mu^2-\nu^2)^{1/2}} & -\dfrac{\nu}{\gamma}\dfrac{(\gamma^2-\mu^2)^{1/2}(\beta^2-\nu^2)^{1/2}}{(\gamma^2-\beta^2)^{1/2}(\mu^2-\nu^2)^{1/2}} \end{bmatrix}$$
$$\tag{4.301}$$

### 4.4.9 Ellipsoidal Coordinate System

Ellipsoidal coordinates can be defined as

$$x^2 = \left(\frac{\xi\mu\nu}{\beta\gamma}\right)^2; \quad y^2 = \frac{(\xi^2-\beta^2)(\mu^2-\beta^2)(\beta^2-\nu^2)}{\beta^2(\gamma^2-\beta^2)}; \quad z^2 = \frac{(\xi^2-\gamma^2)(\gamma^2-\mu^2)(\gamma^2-\nu^2)}{\gamma^2(\gamma^2-\beta^2)} \tag{4.302}$$

The coordinate space $(\xi, \mu, \nu)$ is limited as: $\gamma^2 < \xi^2 < +\infty$, $\beta^2 < \mu^2 < \gamma^2$, $0 \le \nu^2 < \beta^2$.

The surface $\xi = \xi_0$ is an ellipsoid; the surfaces $\mu = \mu_0$ and $\nu = \nu_0$ are hyperboloids (see Fig. 4.9). The scale factors are

$$h_\xi = \sqrt{\frac{(\xi^2-\mu^2)(\xi^2-\nu^2)}{(\xi^2-\beta^2)(\xi^2-\gamma^2)}}; \quad h_\mu = \sqrt{\frac{(\mu^2-\nu^2)(\xi^2-\mu^2)}{(\mu^2-\beta^2)(\gamma^2-\mu^2)}}; \quad h_\nu = \sqrt{\frac{(\xi^2-\nu^2)(\mu^2-\nu^2)}{(\beta^2-\nu^2)(\gamma^2-\nu^2)}} \tag{4.303}$$

$$g^{1/2} = \frac{(\xi^2-\mu^2)(\xi^2-\nu^2)(\mu^2-\nu^2)}{\sqrt{(\xi^2-\beta^2)(\xi^2-\gamma^2)(\mu^2-\beta^2)(\gamma^2-\mu^2)(\beta^2-\nu^2)(\gamma^2-\nu^2)}} \tag{4.304}$$

and the functions $W^{(k)}$ are

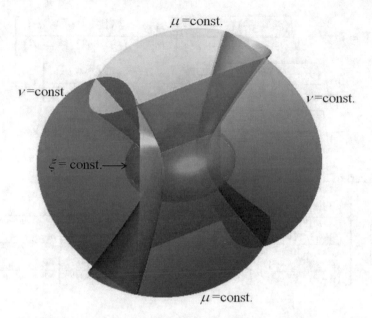

**Fig. 4.9** Ellipsoidal coordinate system: 3-D representation of the iso-surfaces

$$W^{(\xi)} = \left(\mu^2 - \nu^2\right)\sqrt{\frac{\left(\xi^2 - \beta^2\right)\left(\xi^2 - \gamma^2\right)}{\left(\mu^2 - \beta^2\right)\left(\gamma^2 - \mu^2\right)\left(\beta^2 - \nu^2\right)\left(\gamma^2 - \nu^2\right)}} \quad (4.305a)$$

$$W^{(\mu)} = \left(\xi^2 - \nu^2\right)\sqrt{\frac{\left(\mu^2 - \beta^2\right)\left(\gamma^2 - \mu^2\right)}{\left(\xi^2 - \beta^2\right)\left(\xi^2 - \gamma^2\right)\left(\beta^2 - \nu^2\right)\left(\gamma^2 - \nu^2\right)}} \quad (4.305b)$$

$$W^{(\nu)} = \left(\xi^2 - \mu^2\right)\sqrt{\frac{\left(\beta^2 - \nu^2\right)\left(\gamma^2 - \nu^2\right)}{\left(\xi^2 - \beta^2\right)\left(\xi^2 - \gamma^2\right)\left(\mu^2 - \beta^2\right)\left(\gamma^2 - \mu^2\right)}} \quad (4.305c)$$

The gradient of a scalar quantity $T$ is a vector and its physical components are

$$\nabla T = \operatorname{grad} T$$

$$= \left[\sqrt{\frac{\left(\xi^2 - \beta^2\right)\left(\xi^2 - \gamma^2\right)}{\left(\xi^2 - \mu^2\right)\left(\xi^2 - \nu^2\right)}}\frac{\partial T}{\partial \xi}, \sqrt{\frac{\left(\mu^2 - \beta^2\right)\left(\gamma^2 - \mu^2\right)}{\left(\mu^2 - \nu^2\right)\left(\xi^2 - \mu^2\right)}}\frac{\partial T}{\partial \mu}, \sqrt{\frac{\left(\beta^2 - \nu^2\right)\left(\gamma^2 - \nu^2\right)}{\left(\xi^2 - \nu^2\right)\left(\mu^2 - \nu^2\right)}}\frac{\partial T}{\partial \nu}\right]$$

$$(4.306)$$

The divergence of a vector $\mathbf{V} = [\hat{V}_\xi, \hat{V}_\mu, \hat{V}_\nu]$ is the scalar

$$\nabla \cdot \mathbf{V} = \mathrm{div}\,\mathbf{V} = \sqrt{\frac{\left(\xi^2 - \beta^2\right)\left(\xi^2 - \gamma^2\right)}{\left(\xi^2 - \mu^2\right)\left(\xi^2 - \nu^2\right)}}\left[\frac{\partial \hat{V}_\xi}{\partial \xi} + \frac{\xi\left(2\xi^2 - \nu^2 - \mu^2\right)}{\left(\xi^2 - \mu^2\right)\left(\xi^2 - \nu^2\right)}\right]\hat{V}_\xi +$$

$$+ \sqrt{\frac{\left(\mu^2 - \beta^2\right)\left(\gamma^2 - \mu^2\right)}{\left(\mu^2 - \nu^2\right)\left(\xi^2 - \mu^2\right)}}\left[\frac{\partial \hat{V}_\mu}{\partial \nu} + \frac{\mu\left(\xi^2 - 2\mu^2 + \nu^2\right)}{\left(\mu^2 - \nu^2\right)\left(\xi^2 - \mu^2\right)}\right]\hat{V}_\mu +$$

$$+ \sqrt{\frac{\left(\beta^2 - \nu^2\right)\left(\gamma^2 - \nu^2\right)}{\left(\xi^2 - \nu^2\right)\left(\mu^2 - \nu^2\right)}}\left[\frac{\partial \hat{V}_\nu}{\partial \nu} - \frac{\nu\left(\xi^2 + \mu^2 - 2\nu^2\right)}{\left(\xi^2 - \nu^2\right)\left(\mu^2 - \nu^2\right)}\right]\hat{V}_\nu \qquad (4.307)$$

The curl of a vector $\mathbf{V} = [\hat{V}_\xi, \hat{V}_\mu, \hat{V}_\nu]$ is the vector $\boldsymbol{\omega} = \mathrm{curl}\,\mathbf{V}$

$$\begin{bmatrix} \hat{\omega}_\xi \\ \hat{\omega}_\mu \\ \hat{\omega}_\nu \end{bmatrix} = \begin{bmatrix} \frac{(\mu^2-\beta^2)^{1/2}(\gamma^2-\mu^2)^{1/2}}{(\mu^2-\nu^2)^{1/2}(\xi^2-\mu^2)^{1/2}}\left[\frac{\partial \hat{v}_\nu}{\partial \mu} + \frac{\mu}{\mu^2-\nu^2}\hat{V}_\nu\right] - \frac{(\beta^2-\nu^2)^{1/2}(\gamma^2-\nu^2)^{1/2}}{(\xi^2-\nu^2)^{1/2}(\mu^2-\nu^2)^{1/2}}\left[\frac{\partial \hat{v}_\mu}{\partial \nu} - \frac{\nu}{\mu^2-\nu^2}\hat{V}_\mu\right] \\ \frac{(\beta^2-\nu^2)^{1/2}(\gamma^2-\nu^2)^{1/2}}{(\xi^2-\nu^2)^{1/2}(\mu^2-\nu^2)^{1/2}}\left[\frac{\partial \hat{v}_\xi}{\partial \nu} - \frac{\nu}{\xi^2-\nu^2}\hat{V}_\xi\right] - \frac{(\xi^2-\beta^2)^{1/2}(\xi^2-\gamma^2)^{1/2}}{(\xi^2-\mu^2)^{1/2}(\xi^2-\nu^2)^{1/2}}\left[\frac{\partial \hat{v}_\nu}{\partial \xi} + \frac{\xi}{\xi^2-\nu^2}\hat{V}_\nu\right] \\ \frac{(\xi^2-\beta^2)^{1/2}(\xi^2-\gamma^2)^{1/2}}{(\xi^2-\mu^2)^{1/2}(\xi^2-\nu^2)^{1/2}}\left[\frac{\partial \hat{v}_\mu}{\partial \xi} + \frac{\xi}{\xi^2-\mu^2}\hat{V}_\mu\right] - \frac{(\mu^2-\beta^2)^{1/2}(\gamma^2-\mu^2)^{1/2}}{(\mu^2-\nu^2)^{1/2}(\xi^2-\mu^2)^{1/2}}\left[\frac{\partial \hat{v}_\xi}{\partial \mu} - \frac{\mu}{\xi^2-\mu^2}\hat{V}_\xi\right] \end{bmatrix}$$
$$(4.308)$$

The Laplacian of the scalar $T$ is the scalar

$$\nabla^2 T = \frac{\left(\xi^2 - \beta^2\right)\left(\xi^2 - \gamma^2\right)}{\left(\xi^2 - \mu^2\right)\left(\xi^2 - \nu^2\right)}\frac{\partial^2 T}{\partial \xi^2} + \frac{\xi\left(2\xi^2 - \beta^2 - \gamma^2\right)}{\left(\xi^2 - \mu^2\right)\left(\xi^2 - \nu^2\right)}\frac{\partial T}{\partial \xi} +$$

$$+ \frac{\left(\mu^2 - \beta^2\right)\left(\gamma^2 - \mu^2\right)}{\left(\xi^2 - \mu^2\right)\left(\mu^2 - \nu^2\right)}\frac{\partial^2 T}{\partial \mu^2} - \frac{\mu\left(2\mu^2 - \beta^2 - \gamma^2\right)}{\left(\xi^2 - \mu^2\right)\left(\mu^2 - \nu^2\right)}\frac{\partial T}{\partial \mu} +$$

$$+ \frac{\left(\beta^2 - \nu^2\right)\left(\gamma^2 - \nu^2\right)}{\left(\xi^2 - \nu^2\right)\left(\mu^2 - \nu^2\right)}\frac{\partial^2 T}{\partial \nu^2} + \frac{\nu\left(2\nu^2 - \beta^2 - \gamma^2\right)}{\left(\xi^2 - \nu^2\right)\left(\mu^2 - \nu^2\right)}\frac{\partial T}{\partial \nu} \qquad (4.309)$$

The Laplace and Helmholtz equations are *simple-separable* in this system, the corresponding Stäckel matrix is

$$\boldsymbol{\Phi} = \begin{bmatrix} \frac{\xi^4}{(\xi^2-\beta^2)(\xi^2-\gamma^2)} & \frac{1}{(\xi^2-\beta^2)(\xi^2-\gamma^2)} & \frac{\xi^2}{(\xi^2-\beta^2)(\xi^2-\gamma^2)} \\ -\frac{\mu^4}{(\gamma^2-\mu^2)(\mu^2-\beta^2)} & -\frac{1}{(\gamma^2-\mu^2)(\mu^2-\beta^2)} & -\frac{\mu^2}{(\gamma^2-\mu^2)(\mu^2-\beta^2)} \\ \frac{\nu^4}{(\beta^2-\nu^2)(\gamma^2-\nu^2)} & \frac{1}{(\beta^2-\nu^2)(\gamma^2-\nu^2)} & \frac{\nu^2}{(\beta^2-\nu^2)(\gamma^2-\nu^2)} \end{bmatrix} \qquad (4.310)$$

the functions $f_k$ are

$$f_1 = f_\xi(\xi) = \sqrt{\left(\xi^2 - \beta^2\right)\left(\xi^2 - \gamma^2\right)} \qquad (4.311a)$$

$$f_2 = f_\mu(\mu) = \sqrt{\left(\gamma^2 - \mu^2\right)\left(\mu^2 - \beta^2\right)} \qquad (4.311b)$$

$$f_3 = f_\nu(\nu) = \sqrt{\left(\beta^2 - \nu^2\right)\left(\gamma^2 - \nu^2\right)} \qquad (4.311c)$$

the functions $q_k$ are

$$q_1 = q_\xi (\xi) = \frac{\xi^4 \alpha_1 + \alpha_2 + \xi^2 \alpha_3}{\left(\xi^2 - \beta^2\right)\left(\xi^2 - \gamma^2\right)} \tag{4.312a}$$

$$q_2 = q_\mu (\mu) = -\frac{\mu^4 \alpha_1 + \alpha_2 + \mu^2 \alpha_3}{\left(\gamma^2 - \mu^2\right)\left(\mu^2 - \beta^2\right)} \tag{4.312b}$$

$$q_3 = q_\nu (\nu) = \frac{\nu^4 \alpha_1 + \alpha_2 + \nu^2 \alpha_3}{\left(\beta^2 - \nu^2\right)\left(\gamma^2 - \nu^2\right)} \tag{4.312c}$$

for the Laplace equation $\alpha_1 = 0$ and for the Helmholtz equation $\alpha_1 = K^2$.

The transformation matrix for physical components from ellipsoidal to Cartesian coordinates, $\mathbf{M}^{u \to x}$ is

$$\mathbf{M}^{u \to x} = \begin{bmatrix} \frac{\mu\nu(\xi^2-\beta^2)^{1/2}(\xi^2-\gamma^2)^{1/2}}{\beta\gamma(\xi^2-\mu^2)^{1/2}(\xi^2-\nu^2)^{1/2}} & \frac{\xi\nu(\mu^2-\beta^2)^{1/2}(\gamma^2-\mu^2)^{1/2}}{\beta\gamma(\mu^2-\nu^2)^{1/2}(\xi^2-\mu^2)^{1/2}} & \frac{\xi\mu(\beta^2-\nu^2)^{1/2}(\gamma^2-\nu^2)^{1/2}}{\beta\gamma(\xi^2-\nu^2)^{1/2}(\mu^2-\nu^2)^{1/2}} \\ \frac{\xi[(\xi^2-\gamma^2)(\mu^2-\beta^2)(\beta^2-\nu^2)]^{1/2}}{\beta[(\xi^2-\mu^2)(\xi^2-\nu^2)(\gamma^2-\beta^2)]^{1/2}} & \frac{\mu[(\gamma^2-\mu^2)(\xi^2-\beta^2)(\beta^2-\nu^2)]^{1/2}}{\beta[(\mu^2-\nu^2)(\xi^2-\mu^2)(\gamma^2-\beta^2)]^{1/2}} & -\frac{\nu[(\gamma^2-\nu^2)(\xi^2-\beta^2)(\mu^2-\beta^2)]^{1/2}}{\beta[(\xi^2-\nu^2)(\mu^2-\nu^2)(\gamma^2-\beta^2)]^{1/2}} \\ \frac{\xi[(\xi^2-\beta^2)(\gamma^2-\mu^2)(\gamma^2-\nu^2)]^{1/2}}{\gamma[(\xi^2-\mu^2)(\xi^2-\nu^2)(\gamma^2-\beta^2)]^{1/2}} & -\frac{\mu[(\mu^2-\beta^2)(\xi^2-\gamma^2)(\gamma^2-\nu^2)]^{1/2}}{\gamma[(\mu^2-\nu^2)(\xi^2-\mu^2)(\gamma^2-\beta^2)]^{1/2}} & -\frac{\nu[(\beta^2-\nu^2)(\xi^2-\gamma^2)(\gamma^2-\mu^2)]^{1/2}}{\gamma[(\xi^2-\nu^2)(\mu^2-\nu^2)(\gamma^2-\beta^2)]^{1/2}} \end{bmatrix} \tag{4.313}$$

## 4.4.10  *Paraboloidal Coordinate System*

Paraboloidal coordinates can be defined as

$$x^2 = \frac{4}{\gamma - \beta}(\xi - \gamma)(\gamma - \mu)(\gamma - \nu); \quad y^2 = \frac{4}{\gamma - \beta}(\xi - \beta)(\beta - \mu)(\nu - \beta); \quad z = \xi + \mu + \nu - \beta - \gamma \tag{4.314}$$

The coordinate space $(\xi, \mu, \nu)$ is limited as: $\gamma \leq \xi < +\infty, 0 < \mu < \beta, \beta < \nu < \gamma$ and $\gamma > \beta > 0$.

The surfaces $\xi = \xi_0$ and $\mu = \mu_0$ are elliptic paraboloids; the surface $\nu = \nu_0$ is a hyperbolic paraboloid (see Fig. 4.10). The scale factors are

$$h_\xi = \sqrt{\frac{(\xi - \mu)(\xi - \nu)}{(\xi - \beta)(\xi - \gamma)}}; \quad h_\mu = \sqrt{\frac{(\xi - \mu)(\nu - \mu)}{(\beta - \mu)(\gamma - \mu)}}; \quad h_\nu = \sqrt{\frac{(\xi - \nu)(\nu - \mu)}{(\nu - \beta)(\gamma - \nu)}} \tag{4.315}$$

$$g^{1/2} = \frac{(\xi - \mu)(\xi - \nu)(\nu - \mu)}{\sqrt{(\xi - \beta)(\xi - \gamma)(\beta - \mu)(\gamma - \mu)(\nu - \beta)(\gamma - \nu)}} \tag{4.316}$$

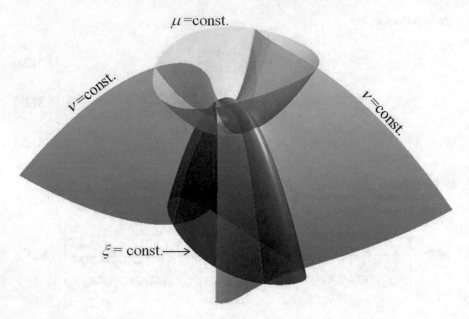

**Fig. 4.10**  Paraboloidal coordinate system: 3-D representation of the iso-surfaces

and the functions $W^{(k)}$ are

$$W^{(\xi)} = (\nu - \mu)\sqrt{\frac{(\xi - \beta)(\xi - \gamma)}{(\beta - \mu)(\gamma - \mu)(\nu - \beta)(\gamma - \nu)}} \tag{4.317a}$$

$$W^{(\mu)} = (\xi - \nu)\sqrt{\frac{(\beta - \mu)(\gamma - \mu)}{(\xi - \beta)(\xi - \gamma)(\nu - \beta)(\gamma - \nu)}} \tag{4.317b}$$

$$W^{(\nu)} = (\xi - \mu)\sqrt{\frac{(\nu - \beta)(\gamma - \nu)}{(\xi - \beta)(\xi - \gamma)(\beta - \mu)(\gamma - \mu)}} \tag{4.317c}$$

The gradient of a scalar quantity $T$ is a vector and its physical components are

$$\nabla T = \text{grad } T = \left[\sqrt{\frac{(\xi - \beta)(\xi - \gamma)}{(\xi - \mu)(\xi - \nu)}}\frac{\partial T}{\partial \xi}, \sqrt{\frac{(\beta - \mu)(\gamma - \mu)}{(\xi - \mu)(\nu - \mu)}}\frac{\partial T}{\partial \mu}, \sqrt{\frac{(\nu - \beta)(\gamma - \nu)}{(\xi - \nu)(\nu - \mu)}}\frac{\partial T}{\partial \nu}\right] \tag{4.318}$$

The divergence of a vector $\mathbf{V} = [\hat{V}_\xi, \hat{V}_\mu, \hat{V}_\nu]$ is the scalar

$$\nabla \cdot \mathbf{V} = \text{div}\mathbf{V} = \sqrt{\frac{(\xi - \beta)(\xi - \gamma)}{(\xi - \mu)(\xi - \nu)}}\left[\frac{\partial \hat{V}_\xi}{\partial \xi} + \frac{2\xi - \mu - \nu}{2(\xi - \mu)(\xi - \nu)}\hat{V}_\xi\right] +$$

$$+\sqrt{\frac{(\beta-\mu)\,(\gamma-\mu)}{(\xi-\mu)\,(\nu-\mu)}}\left[\frac{\partial\hat{V}_\mu}{\partial\nu}+\frac{2\mu-\xi-\nu}{2\,(\xi-\mu)\,(\nu-\mu)}\,\hat{V}_\mu\right]+$$

$$+\sqrt{\frac{(\gamma-\nu)\,(\nu-\beta)}{(\xi-\nu)\,(\nu-\mu)}}\left[\frac{\partial\hat{V}_\nu}{\partial\nu}-\frac{2\nu-\xi-\mu}{2\,(\xi-\nu)\,(\nu-\mu)}\,\hat{V}_\nu\right] \tag{4.319}$$

The curl of a vector $\mathbf{V}=[\hat{V}_\xi,\,\hat{V}_\mu,\,\hat{V}_\nu]$ is the vector $\boldsymbol{\omega}=\mathrm{curl}\mathbf{V}$

$$\begin{bmatrix}\hat{\omega}_\xi\\\hat{\omega}_\mu\\\hat{\omega}_\nu\end{bmatrix}=\begin{bmatrix}\frac{(\beta-\mu)^{1/2}(\gamma-\mu)^{1/2}}{(\xi-\mu)^{1/2}(\nu-\mu)^{1/2}}\left[\frac{\partial\hat{V}_\nu}{\partial\mu}-\frac{1}{2(\nu-\mu)}\hat{V}_\nu\right]-\frac{(\nu-\beta)^{1/2}(\gamma-\nu)^{1/2}}{(\xi-\nu)^{1/2}(\nu-\mu)^{1/2}}\left[\frac{\partial\hat{V}_\mu}{\partial\nu}+\frac{1}{2(\nu-\mu)}\hat{V}_\mu\right]\\\frac{(\nu-\beta)^{1/2}(\gamma-\nu)^{1/2}}{(\xi-\nu)^{1/2}(\nu-\mu)^{1/2}}\left[\frac{\partial\hat{V}_\xi}{\partial\nu}-\frac{1}{2(\xi-\nu)}\hat{V}_\xi\right]-\frac{(\xi-\beta)^{1/2}(\xi-\gamma)^{1/2}}{(\xi-\mu)^{1/2}(\xi-\nu)^{1/2}}\left[\frac{\partial\hat{V}_\nu}{\partial\xi}+\frac{1}{2(\xi-\nu)}\hat{V}_\nu\right]\\\frac{(\xi-\beta)^{1/2}(\xi-\gamma)^{1/2}}{(\xi-\mu)^{1/2}(\xi-\nu)^{1/2}}\left[\frac{\partial\hat{V}_\mu}{\partial\xi}+\frac{1}{2(\xi-\mu)}\hat{V}_\mu\right]-\frac{(\beta-\mu)^{1/2}(\gamma-\mu)^{1/2}}{(\xi-\mu)^{1/2}(\nu-\mu)^{1/2}}\left[\frac{\partial\hat{V}_\xi}{\partial\mu}-\frac{1}{2(\xi-\mu)}\hat{V}_\xi\right]\end{bmatrix} \tag{4.320}$$

The Laplacian of the scalar $T$ is the scalar

$$\nabla^2 T=\frac{(\xi-\beta)\,(\xi-\gamma)}{(\xi-\mu)\,(\xi-\nu)}\frac{\partial^2 T}{\partial\xi^2}+\frac{2\xi-\beta-\gamma}{2\,(\xi-\mu)\,(\xi-\nu)}\frac{\partial T}{\partial\xi}+\frac{(\beta-\mu)\,(\gamma-\mu)}{(\xi-\mu)\,(\nu-\mu)}\frac{\partial^2 T}{\partial\mu^2}+$$

$$+\frac{2\mu-\beta-\gamma}{2\,(\xi-\mu)\,(\nu-\mu)}\frac{\partial T}{\partial\mu}+\frac{(\nu-\beta)\,(\gamma-\nu)}{(\xi-\nu)\,(\nu-\mu)}\frac{\partial^2 T}{\partial\nu^2}-\frac{2\nu-\beta-\gamma}{2\,(\xi-\nu)\,(\nu-\mu)}\frac{\partial T}{\partial\nu} \tag{4.321}$$

The Laplace and Helmholtz equations are *simple-separable* in this system, the corresponding Stäckel matrix is

$$\boldsymbol{\Phi}=\begin{bmatrix}\frac{\xi^2}{(\xi-\gamma)(\xi-\beta)} & -\frac{1}{(\xi-\gamma)(\xi-\beta)} & \frac{\xi}{(\xi-\gamma)(\xi-\beta)}\\\frac{\mu^2}{(\gamma-\mu)(\beta-\mu)} & -\frac{1}{(\gamma-\mu)(\beta-\mu)} & \frac{\mu}{(\gamma-\mu)(\beta-\mu)}\\\frac{\nu^2}{(\nu-\beta)(\gamma-\nu)} & \frac{1}{(\nu-\beta)(\gamma-\nu)} & \frac{\nu}{(\nu-\beta)(\gamma-\nu)}\end{bmatrix} \tag{4.322}$$

the functions $f_k$ are

$$f_1=f_\xi\,(\xi)=\sqrt{(\xi-\gamma)\,(\xi-\beta)} \tag{4.323a}$$

$$f_2=f_\mu\,(\mu)=\sqrt{(\gamma-\mu)\,(\beta-\mu)} \tag{4.323b}$$

$$f_3=f_\nu\,(\nu)=\sqrt{(\nu-\beta)\,(\gamma-\nu)} \tag{4.323c}$$

the functions $q_k$ are

$$q_1 = q_\xi\,(\xi) = \frac{\xi^2\alpha_1 - \alpha_2 + \xi\alpha_3}{(\xi - \gamma)\,(\xi - \beta)} \qquad (4.324a)$$

$$q_2 = q_\mu\,(\mu) = \frac{\mu^2\alpha_1 - \alpha_2 + \mu\alpha_3}{(\gamma - \mu)\,(\beta - \mu)} \qquad (4.324b)$$

$$q_3 = q_\nu\,(\nu) = -\frac{\nu^2\alpha_1 - \alpha_2 + \nu\alpha_3}{(\nu - \beta)\,(\gamma - \nu)} \qquad (4.324c)$$

for the Laplace equation $\alpha_1 = 0$ and for the Helmholtz equation $\alpha_1 = K^2$.

The transformation matrix for physical components from paraboloidal to Cartesian coordinates, $\mathbf{M}^{u\to x}$ is

$$\mathbf{M}^{u\to x} = \begin{bmatrix} \frac{(\gamma-\mu)^{1/2}(\gamma-\nu)^{1/2}(\xi-\beta)^{1/2}}{(\gamma-\beta)^{1/2}(\xi-\mu)^{1/2}(\xi-\nu)^{1/2}} & -\frac{(\xi-\gamma)^{1/2}(\gamma-\nu)^{1/2}(\beta-\mu)^{1/2}}{(\gamma-\beta)^{1/2}(\xi-\mu)^{1/2}(\nu-\mu)^{1/2}} & -\frac{(\xi-\gamma)^{1/2}(\gamma-\mu)^{1/2}(\nu-\beta)^{1/2}}{(\gamma-\beta)^{1/2}(\xi-\nu)^{1/2}(\nu-\mu)^{1/2}} \\ \frac{(\beta-\mu)^{1/2}(\nu-\beta)^{1/2}(\xi-\gamma)^{1/2}}{(\gamma-\beta)^{1/2}(\xi-\mu)^{1/2}(\xi-\nu)^{1/2}} & -\frac{(\xi-\beta)^{1/2}(\nu-\beta)^{1/2}(\gamma-\mu)^{1/2}}{(\gamma-\beta)^{1/2}(\xi-\mu)^{1/2}(\nu-\mu)^{1/2}} & \frac{(\xi-\beta)^{1/2}(\beta-\mu)^{1/2}(\gamma-\nu)^{1/2}}{(\gamma-\beta)^{1/2}(\xi-\nu)^{1/2}(\nu-\mu)^{1/2}} \\ \frac{(\xi-\beta)^{1/2}(\xi-\gamma)^{1/2}}{(\xi-\mu)^{1/2}(\xi-\nu)^{1/2}} & \frac{(\beta-\mu)^{1/2}(\gamma-\mu)^{1/2}}{(\xi-\mu)^{1/2}(\nu-\mu)^{1/2}} & \frac{(\nu-\beta)^{1/2}(\gamma-\nu)^{1/2}}{(\xi-\nu)^{1/2}(\nu-\mu)^{1/2}} \end{bmatrix}$$

$$(4.325)$$

### 4.4.11  Tangent-Sphere Coordinate System

Tangent-sphere coordinates can be defined as

$$x = \frac{\mu}{\mu^2 + \nu^2}\cos(\varphi)\,;\quad y = \frac{\mu}{\mu^2 + \nu^2}\sin(\varphi)\,;\quad z = \frac{\nu}{\mu^2 + \nu^2} \qquad (4.326)$$

The coordinate space $(\mu, \nu, \varphi)$ is limited as: $0 < \mu < +\infty$, $-\infty < \nu < +\infty$, $0 \le \varphi < 2\pi$.

The surface $\mu = \mu_0$ is a toroid without hole in the centre; the surface $\nu = \nu_0$ is a sphere tangent to the $x - y$ plane at the origin; the surface $\varphi = \varphi_0$ is an half-plane passing through the $z$-axis (see Fig. 4.11). The scale factors are

$$h_\mu = \frac{1}{\mu^2 + \nu^2} = h_\nu\,;\quad h_\varphi = \frac{\mu}{\mu^2 + \nu^2}\,;\quad g^{1/2} = \frac{\mu}{\left(\mu^2 + \nu^2\right)^3} \qquad (4.327)$$

and the functions $W^{(k)}$ are

$$W^{(\mu)} = \frac{\mu}{\mu^2 + \nu^2} = W^{(\nu)}\,;\quad W^{(\varphi)} = \frac{1}{\mu\left(\mu^2 + \nu^2\right)} \qquad (4.328)$$

The gradient of a scalar quantity $T$ is a vector and its physical components are

$$\nabla T = \mathrm{grad}\,T = \left(\mu^2 + \nu^2\right)\left[\frac{\partial T}{\partial \mu}, \frac{\partial T}{\partial \nu}, \frac{1}{\mu}\frac{\partial T}{\partial \varphi}\right] \qquad (4.329)$$

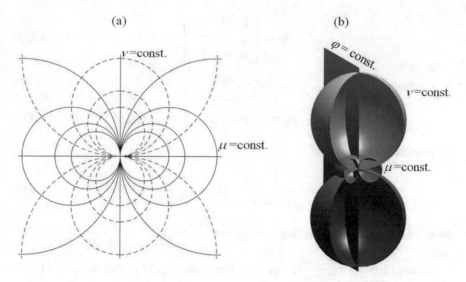

**Fig. 4.11** Tangent-sphere coordinate system: **a** coordinates over a plane passing through the $z$-axis; **b** 3-D representation of the iso-surfaces

The divergence of a vector $\mathbf{V} = [\hat{V}_\mu, \hat{V}_\nu, \hat{V}_\varphi]$ is the scalar

$$\nabla \cdot \mathbf{V} = \mathrm{div}\mathbf{V} = \left(\mu^2 + \nu^2\right)\frac{\partial \hat{V}_\mu}{\partial \mu} + \left(\frac{\nu^2 - 3\mu^2}{\mu}\right)\hat{V}_\mu + \left(\mu^2 + \nu^2\right)\frac{\partial \hat{V}_\nu}{\partial \nu} - 4\nu\hat{V}_\nu + \left(\frac{\mu^2 + \nu^2}{\mu}\right)\frac{\partial \hat{V}_\varphi}{\partial \varphi}$$

$$(4.330)$$

The curl of a vector $\mathbf{V} = [\hat{V}_\mu, \hat{V}_\nu, \hat{V}_\varphi]$ is the vector $\boldsymbol{\omega} = \mathrm{curl}\mathbf{V}$

$$\begin{bmatrix} \hat{\omega}_\mu \\ \hat{\omega}_\nu \\ \hat{\omega}_\varphi \end{bmatrix} = \begin{bmatrix} \left(\mu^2 + \nu^2\right)\frac{\partial \hat{V}_\varphi}{\partial \nu} - 2\nu\hat{V}_\varphi - \frac{\mu^2+\nu^2}{\mu}\frac{\partial \hat{V}_\nu}{\partial \varphi} \\ \frac{\mu^2+\nu^2}{\mu}\frac{\partial \hat{V}_\mu}{\partial \varphi} - \left(\mu^2 + \nu^2\right)\frac{\partial \hat{V}_\varphi}{\partial \mu} + \frac{\mu^2-\nu^2}{\mu}\hat{V}_\varphi \\ \left(\mu^2 + \nu^2\right)\frac{\partial \hat{V}_\nu}{\partial \mu} - 2\mu\hat{V}_\nu - \left(\mu^2 + \nu^2\right)\frac{\partial \hat{V}_\mu}{\partial \nu} + 2\nu\hat{V}_\mu \end{bmatrix}$$

$$(4.331)$$

The Laplacian of the scalar $T$ is the scalar

$$\nabla^2 T = \left(\mu^2 + \nu^2\right)\left[\left(\mu^2 + \nu^2\right)\frac{\partial^2 T}{\partial \mu^2} - \frac{\left(\mu^2 - \nu^2\right)}{\mu}\frac{\partial T}{\partial \mu} + \left(\mu^2 + \nu^2\right)\frac{\partial^2 T}{\partial \nu^2} - 2\nu\frac{\partial T}{\partial \nu}\right]$$
$$+ \frac{\left(\mu^2 + \nu^2\right)^2}{\mu^2}\frac{\partial^2 T}{\partial \varphi^2}$$

$$(4.332)$$

The Laplace equation is *R-separable* in this system, the corresponding Stäckel matrix is

$$\Phi = \begin{bmatrix} 1 & -1 & -\frac{1}{\mu^2} \\ 0 & 1 & 0 \\ 0 & 0 & 1 \end{bmatrix} \tag{4.333}$$

the functions $f_k$ are

$$f_1 = f_\mu(\mu) = \mu; \quad f_2 = f_\nu(\nu) = 1; \quad f_3 = f_\varphi(\varphi) = 1 \tag{4.334}$$

the functions $\tilde{R}$ and $\tilde{Q}$ are

$$\tilde{R} = \frac{1}{\sqrt{\mu^2 + \nu^2}}; \quad \tilde{Q} = \frac{1}{\left(\mu^2 + \nu^2\right)^2} \tag{4.335}$$

the constant $\alpha_1$ is nil and the functions $q_k$ are

$$q_1 = q_\mu(\mu) = -\alpha_2 - \frac{\alpha_3}{\mu^2}; \quad q_2 = q_\nu(\nu) = \alpha_2; \quad q_3 = q_\varphi(\varphi) = \alpha_3 \tag{4.336}$$

The Helmholtz equation is *not* separable in this system.

The transformation matrix for physical components from tangent-sphere to Cartesian coordinates, $M^{u \to x}$ is

$$M^{u \to x} = \begin{bmatrix} -\frac{\mu^2 - \nu^2}{\mu^2 + \nu^2} \cos(\varphi) & -\frac{2\mu\nu}{\mu^2 + \nu^2} \cos(\varphi) & -\sin(\varphi) \\ -\frac{\mu^2 - \nu^2}{\mu^2 + \nu^2} \sin(\varphi) & -\frac{2\mu\nu}{\mu^2 + \nu^2} \sin(\varphi) & \cos(\varphi) \\ -\frac{2\mu\nu}{\mu^2 + \nu^2} & \frac{\mu^2 - \nu^2}{\mu^2 + \nu^2} & 0 \end{bmatrix} \tag{4.337}$$

### 4.4.12   Cardioid Coordinate System

Cardioid coordinates can be defined as

$$x = \frac{\mu\nu}{\left(\mu^2 + \nu^2\right)^2} \cos(\varphi); \quad y = \frac{\mu\nu}{\left(\mu^2 + \nu^2\right)^2} \sin(\varphi); \quad z = \frac{\mu^2 - \nu^2}{2\left(\mu^2 + \nu^2\right)^2} \tag{4.338}$$

The coordinate space $(\mu, \nu, \varphi)$ is limited as: $0 \leq \mu < +\infty$, $0 \leq \nu < +\infty$, $0 \leq \varphi < 2\pi$.

The surfaces $\mu = \mu_0$ and $\nu = \nu_0$ are cardioids of revolution; the surface $\varphi = \varphi_0$ is an half-plane passing through the $z$-axis (see Fig. 4.12). The scale factors are

$$h_\mu = \frac{1}{\left(\mu^2 + \nu^2\right)^{3/2}} = h_\nu; \quad h_\varphi = \frac{\mu\nu}{\left(\mu^2 + \nu^2\right)^2}; \quad g^{1/2} = \frac{\mu\nu}{\left(\mu^2 + \nu^2\right)^5} \tag{4.339}$$

and the functions $W^{(k)}$ are

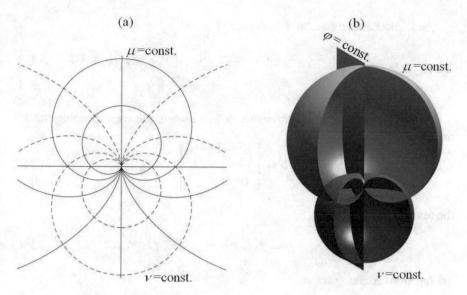

**Fig. 4.12** Cardioid coordinate system: **a** coordinates over a plane passing through the $z$-axis; **b** 3-D representation of the iso-surfaces

$$W^{(\mu)} = \frac{\mu\nu}{\left(\mu^2 + \nu^2\right)^2} = W^{(\nu)}; \quad W^{(\varphi)} = \frac{1}{\mu\nu\left(\mu^2 + \nu^2\right)} \tag{4.340}$$

The gradient of a scalar quantity $T$ is a vector and its physical components are

$$\nabla T = \text{grad } T = \left(\mu^2 + \nu^2\right)^{3/2} \left[\frac{\partial T}{\partial \mu}, \frac{\partial T}{\partial \nu}, \frac{\left(\mu^2 + \nu^2\right)^{1/2}}{\mu\nu} \frac{\partial T}{\partial \varphi}\right] \tag{4.341}$$

The divergence of a vector $\mathbf{V} = [\hat{V}_\mu, \hat{V}_\nu, \hat{V}_\varphi]$ is the scalar

$$\nabla \cdot \mathbf{V} = \text{div}\mathbf{V} = \left(\mu^2 + \nu^2\right)^{3/2} \left[\frac{\partial \hat{V}_\mu}{\partial \mu} - \frac{6\mu^2 - \nu^2}{\mu\left(\mu^2 + \nu^2\right)}\hat{V}_\mu + \frac{\partial \hat{V}_\nu}{\partial \nu} + \frac{\mu^2 - 6\nu^2}{\nu\left(\mu^2 + \nu^2\right)}\hat{V}_\nu\right]$$
$$+ \frac{\left(\mu^2 + \nu^2\right)^2}{\mu\nu} \frac{\partial \hat{V}_\varphi}{\partial \nu} \tag{4.342}$$

The curl of a vector $\mathbf{V} = [\hat{V}_\mu, \hat{V}_\nu, \hat{V}_\varphi]$ is the vector $\omega = \text{curl}\mathbf{V}$

$$\begin{bmatrix} \hat{\omega}_\mu \\ \hat{\omega}_\nu \\ \hat{\omega}_\varphi \end{bmatrix} = \begin{bmatrix} \left(\mu^2 + \nu^2\right)^{3/2}\frac{\partial \hat{V}_\varphi}{\partial \nu} + \frac{\left(\mu^2 - 3\nu^2\right)\left(\mu^2 + \nu^2\right)^{1/2}}{\nu}\hat{V}_\varphi - \frac{\left(\mu^2 + \nu^2\right)^2}{\mu\nu}\frac{\partial \hat{V}_\nu}{\partial \varphi} \\ \frac{\left(\mu^2 + \nu^2\right)^2}{\mu\nu}\frac{\partial \hat{V}_\mu}{\partial \varphi} - \left(\mu^2 + \nu^2\right)^{3/2}\frac{\partial \hat{V}_\varphi}{\partial \mu} + \frac{\left(3\mu^2 - \nu^2\right)\left(\mu^2 + \nu^2\right)^{1/2}}{\mu}\hat{V}_\varphi \\ \left(\mu^2 + \nu^2\right)^{3/2}\left[\frac{\partial \hat{V}_\nu}{\partial \mu} - \frac{3\mu}{\mu^2 + \nu^2}\hat{V}_\nu - \frac{\partial \hat{V}_\mu}{\partial \nu} + \frac{3\nu}{\mu^2 + \nu^2}\hat{V}_\mu\right] \end{bmatrix} \tag{4.343}$$

The Laplacian of the scalar $T$ is the scalar

$$\nabla^2 T = \left(\mu^2 + \nu^2\right)^2 \left[ \begin{array}{c} \left(\mu^2 + \nu^2\right) \frac{\partial^2 T}{\partial \mu^2} - \frac{3\mu^2 - \nu^2}{\mu} \frac{\partial T}{\partial \mu} + \left(\mu^2 + \nu^2\right) \frac{\partial^2 T}{\partial \nu^2} + \frac{\mu^2 - 3\nu^2}{\nu} \frac{\partial T}{\partial \nu} + \\ + \frac{\left(\mu^2 + \nu^2\right)^2}{\mu^2 \nu^2} \frac{\partial^2 T}{\partial \varphi^2} \end{array} \right]$$

(4.344)

The Laplace equation is *R-separable* in this system, the corresponding Stäckel matrix is

$$\Phi = \begin{bmatrix} \mu^2 & -1 & -\frac{1}{\mu^2} \\ \nu^2 & 1 & -\frac{1}{\nu^2} \\ 0 & 0 & 1 \end{bmatrix}$$

(4.345)

the functions $f_k$ are

$$f_1 = f_\mu(\mu) = \mu; \quad f_2 = f_\nu(\nu) = \nu; \quad f_3 = f_\varphi(\varphi) = 1$$

(4.346)

the functions $\tilde{R}$ and $\tilde{Q}$ are

$$\tilde{R} = \frac{1}{\mu^2 + \nu^2}; \quad \tilde{Q} = \frac{1}{\left(\mu^2 + \nu^2\right)^4}$$

(4.347)

the constant $\alpha_1$ is nil and the functions $q_k$ are

$$q_1 = q_\mu(\mu) = -\alpha_2 - \frac{\alpha_3}{\mu^2}; \quad q_2 = q_\nu(\nu) = \alpha_2 - \frac{\alpha_3}{\nu^2}; \quad q_3 = q_\varphi(\varphi) = \alpha_3 \quad (4.348)$$

The Helmholtz equation is *not* separable in this system.

The transformation matrix for physical components from cardioid to Cartesian coordinates, $\mathbf{M}^{u \to x}$ is

$$\mathbf{M}^{u \to x} = \begin{bmatrix} -\frac{\nu(3\mu^2 - \nu^2)}{(\mu^2 + \nu^2)^{3/2}} \cos(\varphi) & \frac{\mu(\mu^2 - 3\nu^2)}{(\mu^2 + \nu^2)^{3/2}} \cos(\varphi) & -\sin(\varphi) \\ -\frac{\nu(3\mu^2 - \nu^2)}{(\mu^2 + \nu^2)^{3/2}} \sin(\varphi) & \frac{\mu(\mu^2 - 3\nu^2)}{(\mu^2 + \nu^2)^{3/2}} \sin(\varphi) & \cos(\varphi) \\ -\frac{\mu(\mu^2 - 3\nu^2)}{(\mu^2 + \nu^2)^{3/2}} & -\frac{\nu(3\mu^2 - \nu^2)}{(\mu^2 + \nu^2)^{3/2}} & 0 \end{bmatrix}$$

(4.349)

### 4.4.13   Bispherical Coordinate System

Bispherical coordinates can be defined as

$$x = \frac{a \sin(\psi) \cos(\varphi)}{\Theta(\xi, \psi)}; \quad y = \frac{a \sin(\psi) \sin(\varphi)}{\Theta(\xi, \psi)}; \quad z = \frac{a \sinh(\xi)}{\Theta(\xi, \psi)}$$

(4.350)

(a)

(b)

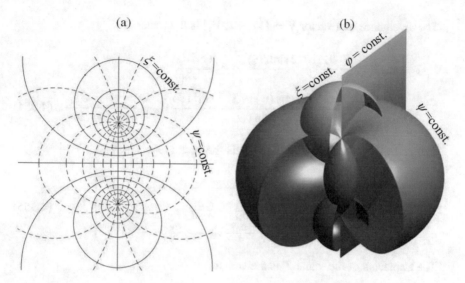

**Fig. 4.13** Bispherical coordinate system: **a** coordinates over a plane passing through the $z$-axis; **b** 3-D representation of the iso-surfaces

where $\Theta\,(\xi, \psi) = \cosh(\xi) - \cos(\psi)$.

The coordinate space $(\xi, \psi, \varphi)$ is limited as: $-\infty < \xi < +\infty$, $0 \le \psi < \pi$, $0 \le \varphi < 2\pi$.

The surface $\xi = \xi_0$ is a sphere; the surface $\psi = \psi_0$ is a self-intersecting toroid; the surface $\varphi = \varphi_0$ is an half-plane passing through the $z$-axis (see Fig. 4.13). The scale factors are

$$h_\xi = \frac{a}{\Theta} = h_\psi; \quad h_\varphi = \frac{a\sin(\psi)}{\Theta}; \quad g^{1/2} = \frac{a^3\sin(\psi)}{\Theta^3} \tag{4.351}$$

and the functions $W^{(k)}$ are

$$W^{(\xi)} = \frac{a\sin(\psi)}{\Theta} = W^{(\psi)}; \quad W^{(\varphi)} = \frac{a}{\Theta\sin(\psi)} \tag{4.352}$$

The gradient of a scalar quantity $T$ is a vector and its physical components are

$$\nabla T = \operatorname{grad} T = \frac{\Theta}{a}\left[\frac{\partial T}{\partial\xi}, \frac{\partial T}{\partial\psi}, \frac{1}{\sin(\psi)}\frac{\partial T}{\partial\varphi}\right] \tag{4.353}$$

The divergence of a vector $\mathbf{V} = [\hat{V}_\xi, \hat{V}_\psi, \hat{V}_\varphi]$ is the scalar

$$\nabla \cdot \mathbf{V} = \mathrm{div}\mathbf{V} = \frac{\Theta}{a}\frac{\partial \hat{V}_\xi}{\partial \xi} - \frac{2\sinh(\xi)}{a}\hat{V}_\xi + \frac{\Theta}{a}\frac{\partial \hat{V}_\psi}{\partial \psi}$$

$$+ \frac{\cosh(\xi)\cos(\psi) - 1 - \sin^2(\psi)}{a\sin(\psi)}\hat{V}_\psi + \frac{\Theta}{a\sin(\psi)}\frac{\partial \hat{V}_\varphi}{\partial \nu} \quad (4.354)$$

The curl of a vector $\mathbf{V} = [\hat{V}_\xi, \hat{V}_\psi, \hat{V}_\varphi]$ is the vector $\omega = \mathrm{curl}\mathbf{V}$

$$\begin{bmatrix} \hat{\omega}_\xi \\ \hat{\omega}_\psi \\ \hat{\omega}_\varphi \end{bmatrix} = \begin{bmatrix} \frac{\Theta}{a}\frac{\partial \hat{V}_\varphi}{\partial \psi} + \frac{\cosh(\xi)\cos(\psi)-1}{a\sin(\psi)}\hat{V}_\varphi - \frac{\Theta}{a\sin(\psi)}\frac{\partial \hat{V}_\psi}{\partial \varphi} \\ \frac{\Theta}{a\sin(\psi)}\frac{\partial \hat{V}_\xi}{\partial \varphi} - \frac{\Theta}{a}\frac{\partial \hat{V}_\varphi}{\partial \xi} + \frac{\sinh(\xi)}{a}\hat{V}_\varphi \\ \frac{\Theta}{a}\frac{\partial \hat{V}_\psi}{\partial \xi} - \frac{\sinh(\xi)}{a}\hat{V}_\psi - \frac{\Theta}{a}\frac{\partial \hat{V}_\xi}{\partial \psi} + \frac{\sin(\psi)}{a}\hat{V}_\xi \end{bmatrix} \quad (4.355)$$

The Laplacian of the scalar $T$ is the scalar

$$\nabla^2 T = \frac{\Theta^2}{a^2}\left[\frac{\partial^2 T}{\partial \xi^2} - \frac{\sinh(\xi)}{\Theta}\frac{\partial T}{\partial \xi}\right] + \frac{\Theta^2}{a^2}\left[\frac{\partial^2 T}{\partial \psi^2} + \frac{\cos(\psi)\cosh(\xi) - 1}{\sin(\psi)\Theta}\frac{\partial T}{\partial \psi}\right] + \frac{\Theta^2}{a^2\sin^2(\psi)}\frac{\partial^2 T}{\partial \varphi^2}$$
$$(4.356)$$

The Laplace equation is *R-separable* in this system, the corresponding Stäckel matrix is

$$\Phi = \begin{bmatrix} 1 & -1 & 0 \\ 0 & 1 & -\frac{1}{\sin^2(\psi)} \\ 0 & 0 & 1 \end{bmatrix} \quad (4.357)$$

the functions $f_k$ are

$$f_1 = f_\xi(\xi) = 1; \quad f_2 = f_\psi(\psi) = \sin(\psi); \quad f_3 = f_\varphi(\varphi) = a \quad (4.358)$$

the functions $\tilde{R}$ and $\tilde{Q}$ are

$$\tilde{R} = \frac{1}{\Theta^{1/2}}; \quad \tilde{Q} = \frac{a}{\Theta^2} \quad (4.359)$$

the constant $\alpha_1$ is equal to $-\frac{1}{4}$ and the functions $q_k$ are

$$q_1 = q_\xi(\xi) = -\frac{1}{4} - \alpha_2; \quad q_2 = q_\psi(\psi) = \alpha_2 - \frac{\alpha_3}{\sin^2(\psi)}; \quad q_3 = q_\varphi(\varphi) = \alpha_3$$
$$(4.360)$$

The Helmholtz equation is *not* separable in this system.

The transformation matrix for physical components from bispherical to Cartesian coordinates, $\mathbf{M}^{u \to x}$ is

$$\mathbf{M}^{u\to x} = \begin{bmatrix} -\frac{\sinh(\xi)\sin(\psi)}{\Theta}\cos(\varphi) & \frac{\cosh(\xi)\cos(\psi)-1}{\Theta}\cos(\varphi) & -\sin(\varphi) \\ -\frac{\sinh(\xi)\sin(\psi)}{\Theta}\sin(\varphi) & \frac{\cosh(\xi)\cos(\psi)-1}{\Theta}\sin(\varphi) & \cos(\varphi) \\ \frac{1-\cosh(\xi)\cos(\psi)}{\Theta} & -\frac{\sinh(\xi)\sin(\psi)}{\Theta} & 0 \end{bmatrix} \qquad (4.361)$$

### 4.4.14 Toroidal Coordinate System

Toroidal coordinates can be defined as

$$x = \frac{a\sinh(\xi)}{\Theta(\xi,\psi)}\cos(\varphi); \quad y = \frac{a\sinh(\xi)}{\Theta(\xi,\psi)}\sin(\varphi); \quad z = \frac{a}{\Theta(\xi,\psi)}\sin(\psi) \quad (4.362)$$

where $\Theta(\xi,\psi) = \cosh(\xi) - \cos(\psi)$.

The coordinate space $(\xi,\psi,\varphi)$ is limited as: $0 \le \xi < +\infty$, $-\pi < \psi \le \pi$, $0 \le \varphi < 2\pi$.

The surface $\xi = \xi_0$ is a toroid; the surface $\psi = \psi_0$ is a spherical bowl; the surface $\varphi = \varphi_0$ is an half-plane passing through the $z$-axis (see Fig. 4.14). The scale factors are

$$h_\xi = \frac{a}{\Theta} = h_\psi; \quad h_\varphi = \frac{a\sinh(\xi)}{\Theta}; \quad g^{1/2} = \frac{a^3\sinh(\xi)}{\Theta^3} \qquad (4.363)$$

and the functions $W^{(k)}$ are

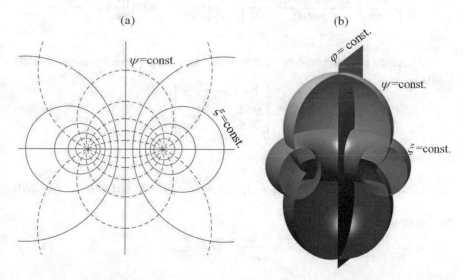

(a)     (b)

**Fig. 4.14** Toroidal coordinate system: **a** coordinates over a plane passing through the $z$-axis; **b** 3-D representation of the iso-surfaces

$$W^{(\xi)} = \frac{a \sinh(\xi)}{\Theta} = W^{(\psi)}; \quad W^{(\varphi)} = \frac{a}{\Theta \sinh(\xi)} \tag{4.364}$$

The gradient of a scalar quantity $T$ is a vector and its physical components are

$$\nabla T = \text{grad } T = \frac{\Theta}{a} \left[ \frac{\partial T}{\partial \xi}, \frac{\partial T}{\partial \psi}, \frac{1}{\sinh(\xi)} \frac{\partial T}{\partial \varphi} \right] \tag{4.365}$$

The divergence of a vector $\mathbf{V} = [\hat{V}_\xi, \hat{V}_\psi, \hat{V}_\varphi]$ is the scalar

$$\nabla \cdot \mathbf{V} = \text{div}\mathbf{V} = \frac{\Theta}{a} \frac{\partial \hat{V}_\xi}{\partial \xi} + \frac{\Theta \coth(\xi) - 2 \sinh(\xi)}{a} \hat{V}_\xi + \frac{\Theta}{a} \frac{\partial \hat{V}_\psi}{\partial \psi} - \frac{2 \sin(\psi)}{a} \hat{V}_\psi + \frac{\Theta}{a \sinh(\xi)} \frac{\partial \hat{V}_\varphi}{\partial \nu} \tag{4.366}$$

The curl of a vector $\mathbf{V} = [\hat{V}_\xi, \hat{V}_\psi, \hat{V}_\varphi]$ is the vector $\omega = \text{curl}\mathbf{V}$

$$\begin{bmatrix} \hat{\omega}_\xi \\ \hat{\omega}_\psi \\ \hat{\omega}_\varphi \end{bmatrix} = \begin{bmatrix} \frac{\Theta}{a} \frac{\partial \hat{V}_\varphi}{\partial \psi} - \frac{\sin(\psi)}{a} \hat{V}_\varphi - \frac{\Theta}{a \sinh(\xi)} \frac{\partial \hat{V}_\psi}{\partial \varphi} \\ \frac{\Theta}{a \sinh(\xi)} \frac{\partial \hat{V}_\xi}{\partial \varphi} - \frac{\Theta}{a} \frac{\partial \hat{V}_\varphi}{\partial \xi} + \frac{\cos(\psi) \cosh(\xi) - 1}{a \sinh(\xi)} \hat{V}_\varphi \\ \frac{\Theta}{a} \frac{\partial \hat{V}_\psi}{\partial \xi} - \frac{\sinh(\xi)}{a} \hat{V}_\psi - \frac{\Theta}{a} \frac{\partial \hat{V}_\xi}{\partial \psi} + \frac{\sin(\psi)}{a} \hat{V}_\xi \end{bmatrix} \tag{4.367}$$

The Laplacian of the scalar $T$ is the scalar

$$\nabla^2 T = \frac{\Theta}{a^2} \left[ \Theta \frac{\partial^2 T}{\partial \xi^2} + \frac{1 - \cosh(\xi) \cos(\psi)}{\sinh(\xi)} \frac{\partial T}{\partial \xi} \right] + \frac{\Theta}{a^2} \left[ \Theta \frac{\partial^2 T}{\partial \psi^2} - \sin(\psi) \frac{\partial T}{\partial \psi} \right] + \frac{\Theta^2}{a^2 \sinh^2(\xi)} \frac{\partial^2 T}{\partial \varphi^2} \tag{4.368}$$

The Laplace equation is *R-separable* in this system, the corresponding Stäckel matrix is

$$\boldsymbol{\Phi} = \begin{bmatrix} 1 & -1 & -\frac{1}{\sinh^2(\xi)} \\ 0 & 1 & 0 \\ 0 & 0 & 1 \end{bmatrix} \tag{4.369}$$

the functions $f_k$ are

$$f_1 = f_\xi(\xi) = \sinh(\xi); \quad f_2 = f_\psi(\psi) = 1; \quad f_3 = f_\varphi(\varphi) = a \tag{4.370}$$

the functions $\tilde{R}$ and $\tilde{Q}$ are

$$\tilde{R} = \frac{1}{\Theta^{1/2}}; \quad \tilde{Q} = \frac{a^2}{\Theta^2} \tag{4.371}$$

the constant $\alpha_1$ is equal to $\frac{1}{4}$ and the functions $q_k$ are

$$q_1 = q_\xi(\xi) = \frac{1}{4} - \alpha_2 - \frac{\alpha_3}{\sinh^2(\xi)}; \quad q_2 = q_\psi(\psi) = \alpha_2; \quad q_3 = q_\varphi(\varphi) = \alpha_3$$

$$(4.372)$$

The Helmholtz equation is *not* separable in this system.

The transformation matrix for physical components from toroidal to Cartesian coordinates, $\mathbf{M}^{u \to x}$ is

$$\mathbf{M}^{u \to x} = \begin{bmatrix} \frac{1-\cosh(\xi)\cos(\psi)}{\Theta}\cos(\varphi) & -\frac{\sinh(\xi)\sin(\psi)}{\Theta}\cos(\varphi) & -\sin(\varphi) \\ \frac{1-\cosh(\xi)\cos(\psi)}{\Theta}\sin(\varphi) & -\frac{\sinh(\xi)\sin(\psi)}{\Theta}\sin(\varphi) & \cos(\varphi) \\ -\frac{\sinh(\xi)\sin(\psi)}{\Theta} & \frac{\cosh(\xi)\cos(\psi)-1}{\Theta} & 0 \end{bmatrix}$$

$$(4.373)$$

### 4.4.15 Inverse Prolate Spheroidal Coordinate System

Inverse prolate spheroidal coordinates can be defined as

$$x = \frac{a\sqrt{\zeta^2-1}\sqrt{1-\eta^2}\cos(\varphi)}{\Theta}; \quad y = \frac{a\sqrt{\zeta^2-1}\sqrt{1-\eta^2}\sin(\varphi)}{\Theta}; \quad z = \frac{a\zeta\eta}{\Theta}$$

$$(4.374)$$

with $\Theta = \zeta^2 + \eta^2 - 1$.

The coordinate space $(\zeta, \eta, \varphi)$ is limited as: $1 \le \zeta < +\infty, -1 \le \eta \le 1, 0 \le \varphi < 2\pi$.

The surfaces $\zeta = \zeta_0$ is a rotational cyclide (inverse prolate spheroid); the surface $\eta = \eta_0$ is a rotational cyclide; the surface $\varphi = \varphi_0$ is a half plane passing through the $z$-axis (see Fig. 4.15). The scale factors are

$$h_\zeta = \frac{a}{\Theta}\sqrt{\frac{\zeta^2-\eta^2}{\zeta^2-1}}; \quad h_\eta = \frac{a}{\Theta}\sqrt{\frac{\zeta^2-\eta^2}{1-\eta^2}}; \quad h_\varphi = \frac{u\sqrt{(\zeta^2-1^2)(1-\eta^2)}}{\Theta}; \quad g^{1/2} = \frac{a^3(\zeta^2-\eta^2)}{\Theta^3}$$

$$(4.375)$$

and the functions $W^{(k)}$ are

$$W^{(\zeta)} = \frac{a(\zeta^2-1^2)}{\Theta}; \quad W^{(\eta)} = \frac{a(1-\eta^2)}{\Theta}; \quad W^{(\varphi)} = \frac{a(\zeta^2-\eta^2)}{\Theta(\zeta^2-1)(1-\eta^2)}$$

$$(4.376)$$

The gradient of a scalar quantity $T$ is a vector and its physical components are

$$\nabla T = \text{grad } T = \frac{\Theta}{a}\left[\sqrt{\frac{\zeta^2-1}{\zeta^2-\eta^2}}\frac{\partial T}{\partial\zeta}, \sqrt{\frac{1-\eta^2}{\zeta^2-\eta^2}}\frac{\partial T}{\partial\eta}, \frac{1}{\sqrt{(\zeta^2-1)(1-\eta^2)}}\frac{\partial T}{\partial\varphi}\right]$$

$$(4.377)$$

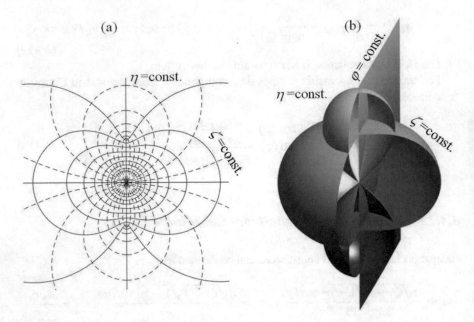

**Fig. 4.15** Inverse prolate spheroidal coordinate system: **a** coordinates over a plane passing through the $z$-axis; **b** 3-D representation of the iso-surfaces

The divergence of a vector $\mathbf{V} = [\hat{V}_\zeta, \hat{V}_\eta, \hat{V}_\varphi]$ is the scalar

$$
\nabla \cdot \mathbf{V} = \text{div}\mathbf{V} = \frac{\Theta}{a}\sqrt{\frac{\zeta^2-1}{\zeta^2-\eta^2}}\left[\frac{\partial \hat{V}_\zeta}{\partial \zeta} + \zeta\left(\frac{1}{\zeta^2-1} + \frac{1}{\zeta^2-\eta^2} - \frac{4}{\Theta}\right)\hat{V}_\zeta\right] +
$$

$$
+\frac{\Theta}{a}\sqrt{\frac{1-\eta^2}{\zeta^2-\eta^2}}\left[\frac{\partial \hat{V}_\eta}{\partial \eta} - \eta\left(\frac{1}{1-\eta^2} + \frac{1}{\zeta^2-\eta^2} + \frac{4}{\Theta}\right)\hat{V}_\eta\right] +
$$

$$
+\frac{\Theta}{a}\frac{1}{\sqrt{(1-\eta^2)(\zeta^2-1)}}\frac{\partial \hat{V}_\varphi}{\partial \varphi} \tag{4.378}
$$

The curl of a vector $\mathbf{V} = [\hat{V}_\zeta, \hat{V}_\eta, \hat{V}_\varphi]$ is the vector $\boldsymbol{\omega} = \text{curl}\mathbf{V}$

$$
\begin{bmatrix}\hat{\omega}_\zeta \\ \hat{\omega}_\eta \\ \hat{\omega}_\varphi\end{bmatrix} =
\begin{bmatrix}
\frac{\Theta(1-\eta^2)^{1/2}}{a(\zeta^2-\eta^2)^{1/2}}\frac{\partial \hat{V}_\varphi}{\partial \eta} - \frac{\eta(1-\eta^2+\zeta^2)}{a(1-\eta^2)^{1/2}(\zeta^2-\eta^2)^{1/2}}\hat{V}_\varphi - \frac{\Theta}{a(\zeta^2-1)^{1/2}(1-\eta^2)^{1/2}}\frac{\partial \hat{V}_\eta}{\partial \varphi} \\
\frac{\Theta}{a(\zeta^2-1)^{1/2}(1-\eta^2)^{1/2}}\frac{\partial \hat{V}_\zeta}{\partial \varphi} - \frac{\Theta(\zeta^2-1)^{1/2}}{a(\zeta^2-\eta^2)^{1/2}}\frac{\partial \hat{V}_\varphi}{\partial \zeta} - \frac{\zeta(1+\eta^2-\zeta^2)}{a(\zeta^2-1)^{1/2}(\zeta^2-\eta^2)^{1/2}}\hat{V}_\varphi \\
\frac{\Theta(\zeta^2-1)^{1/2}}{a(\zeta^2-\eta^2)^{1/2}}\frac{\partial \hat{V}_\eta}{\partial \zeta} - \frac{\zeta(1-3\eta^2+\zeta^2)(\zeta^2-1)^{1/2}}{a(\zeta^2-\eta^2)^{3/2}}\hat{V}_\eta - \frac{\Theta(1-\eta^2)^{1/2}}{a(\zeta^2-\eta^2)^{1/2}}\frac{\partial \hat{V}_\zeta}{\partial \eta} - \frac{\eta(1+\eta^2-3\zeta^2)(1-\eta^2)^{1/2}}{a(\zeta^2-\eta^2)^{3/2}}\hat{V}_\zeta
\end{bmatrix}
\tag{4.379}
$$

The Laplacian of the scalar $T$ is the scalar

$$\nabla^2 T = \frac{\Theta^2}{a^2 \left(\zeta^2 - \eta^2\right)} \left[ \left(\zeta^2 - 1\right) \frac{\partial^2 T}{\partial \zeta^2} + \frac{2\zeta \eta^2}{\Theta} \frac{\partial T}{\partial \zeta} + \left(1 - \eta^2\right) \frac{\partial^2 T}{\partial \eta^2} - \frac{2\zeta^2 \eta}{\Theta} \frac{\partial T}{\partial \eta} \right.$$

$$\left. + \frac{\left(\zeta^2 - \eta^2\right)}{\left(\zeta^2 - 1\right)\left(1 - \eta^2\right)} \frac{\partial^2 T}{\partial \varphi^2} \right] \tag{4.380}$$

The Laplace equation is *R-separable* in this system, the corresponding Stäckel matrix is

$$\mathbf{\Phi} = \begin{bmatrix} a^2 & -\dfrac{1}{a\left(\zeta^2 - 1^2\right)} & -\dfrac{1}{\left(\zeta^2 - 1\right)^2} \\[2mm] a^2 & \dfrac{1}{a\left(1 - \eta^2\right)} & -\dfrac{1}{\left(1 - \eta^2\right)^2} \\[2mm] 0 & 0 & 1 \end{bmatrix} \tag{4.381}$$

the functions $f_k$ are

$$f_1 = f_\zeta \left(\zeta\right) = a \left(\zeta^2 - 1\right); \quad f_2 = f_\eta \left(\eta\right) = a \left(1 - \eta^2\right); \quad f_3 = f_\varphi \left(\varphi\right) = 1 \tag{4.382}$$

the functions $\tilde{R}$ and $\tilde{Q}$ are

$$\tilde{R} = \frac{1}{\Theta^{1/2}}; \quad \tilde{Q} = \frac{1}{\Theta^2} \tag{4.383}$$

the constant $\alpha_1$ is nil and the functions $q_k$ are

$$q_1 = q_\zeta \left(\zeta\right) = -\frac{\alpha_2}{a \left(\zeta^2 - 1\right)} - \frac{\alpha_3}{\left(\zeta^2 - 1\right)^2}; \quad q_2 = q_\eta \left(\eta\right) = \frac{\alpha_2}{a \left(1 - \eta^2\right)} - \frac{\alpha_3}{\left(1 - \eta^2\right)^2}; \quad q_3 = q_\varphi \left(\varphi\right) = \alpha_3 \tag{4.384}$$

The Helmholtz equation is *not* separable in this system.

The transformation matrix for physical components from inverse prolate spheroid to Cartesian coordinates, $\mathbf{M}^{u \to x}$ is

$$\mathbf{M}^{u \to x} = \begin{bmatrix} \dfrac{\zeta\left(1 - \eta^2\right)^{1/2}\left(\eta^2 - \zeta^2 + 1\right)}{\Theta\left(\zeta^2 - \eta\right)^{1/2}} \cos\left(\varphi\right) & \dfrac{\eta\left(\zeta^2 - 1\right)^{1/2}\left(\eta^2 - \zeta^2 - 1\right)}{\Theta\left(\zeta^2 - \eta\right)^{1/2}} \cos\left(\varphi\right) & -\sin\left(\varphi\right) \\[4mm] \dfrac{\zeta\left(1 - \eta^2\right)^{1/2}\left(\eta^2 - \zeta^2 + 1\right)}{\Theta\left(\zeta^2 - \eta\right)^{1/2}} \sin\left(\varphi\right) & \dfrac{\eta\left(\zeta^2 - 1\right)^{1/2}\left(\eta^2 - \zeta^2 - 1\right)}{\Theta\left(\zeta^2 - \eta\right)^{1/2}} \sin\left(\varphi\right) & \cos\left(\varphi\right) \\[4mm] \dfrac{\eta\left(\zeta^2 - 1\right)^{1/2}\left(\eta^2 - \zeta^2 - 1\right)}{\Theta\left(\zeta^2 - \eta\right)^{1/2}} & -\dfrac{\zeta\left(1 - \eta^2\right)^{1/2}\left(\eta^2 - \zeta^2 + 1\right)}{\Theta\left(\zeta^2 - \eta\right)^{1/2}} & 0 \end{bmatrix} \tag{4.385}$$

The inverse prolate spheroidal coordinate system can also be defined using the coordinates $(\xi, \psi, \varphi)$ as

$$x = \frac{a \sinh\left(\xi\right) \sin\left(\psi\right) \cos\left(\varphi\right)}{\Theta}; \quad y = \frac{a \sinh\left(\xi\right) \sin\left(\psi\right) \sin\left(\varphi\right)}{\Theta}; \quad z = \frac{a \cosh\left(\xi\right) \cos\left(\psi\right)}{\Theta} \tag{4.386}$$

with $\Theta = \cosh^2\left(\xi\right) - \sin^2\left(\psi\right)$.

The coordinate space $(\xi, \psi, \varphi)$ is limited as: $0 \leq \xi < +\infty, 0 \leq \psi \leq \pi, 0 \leq \varphi < 2\pi$ and the relation with the system (4.374) is given by: $\zeta = \cosh(\xi)$ and $\eta = \cos(\psi)$. The scale factors are

$$h_\xi = a\frac{\Pi^{1/2}}{\Theta} = h_\psi; \quad h_\varphi = a\frac{\sinh(\xi)\sin(\psi)}{\Theta} \tag{4.387}$$

$$g^{1/2} = a^3\frac{\Pi \sinh(\xi)\sin(\psi)}{\Theta^3} \tag{4.388}$$

where $\Pi = \sinh^2(\xi) + \sin^2(\psi) = \cosh^2(\xi) - \cos^2(\psi)$ and the functions $W^{(k)}$ are

$$W^{(\xi)} = a\frac{\sinh(\xi)\sin(\psi)}{\Theta} = W^{(\psi)}; \quad W^{(\varphi)} = a\frac{\Pi}{\Theta \sinh(\xi)\sin(\psi)} \tag{4.389}$$

The gradient of a scalar quantity $T$ is a vector

$$\nabla T = \text{grad } T = \frac{\Theta}{a\Pi^{1/2}}\left[\frac{\partial T}{\partial \xi}, \frac{\partial T}{\partial \psi}, \frac{\Pi^{1/2}}{\sinh(\xi)\sin(\psi)}\frac{\partial T}{\partial \varphi}\right] \tag{4.390}$$

The divergence of a vector $\mathbf{V} = [\hat{V}_\xi, \hat{V}_\psi, \hat{V}_\varphi]$ is the scalar

$$\nabla \cdot \mathbf{V} = \text{div}\mathbf{V} = \frac{\Theta}{a\Pi^{1/2}}\left[\frac{\partial \hat{V}_\xi}{\partial \xi} - \sinh(\xi)\cosh(\xi)\left(\frac{4}{\Theta} - \frac{1}{\sinh^2(\xi)} - \frac{1}{\Pi}\right)\hat{V}_\xi\right] +$$

$$+\frac{\Theta}{a\Pi^{1/2}}\left[\frac{\partial \hat{V}_\psi}{\partial \psi} + \sin(\psi)\cos(\psi)\left(\frac{4}{\Theta} + \frac{1}{\sin^2(\psi)} + \frac{1}{\Pi}\right)\hat{V}_\psi\right] +$$

$$+\frac{\Theta}{a\sinh(\xi)\sin(\psi)}\frac{\partial \hat{V}_\varphi}{\partial \varphi} \tag{4.391}$$

The curl of a vector $\mathbf{V} = [\hat{V}_\xi, \hat{V}_\psi, \hat{V}_\varphi]$ is the vector $\omega = \text{curl}\mathbf{V}$

$$\begin{bmatrix} \hat{\omega}_\xi \\ \hat{\omega}_\psi \\ \hat{\omega}_\varphi \end{bmatrix} = \begin{bmatrix} \frac{\Theta}{a\Pi^{1/2}}\left(\frac{\partial \hat{V}_\varphi}{\partial \psi} + \frac{\sin^2(\psi)+\cosh^2(\xi)}{\Theta\tan(\psi)}\hat{V}_\varphi\right) - \frac{\Theta}{a\sinh(\xi)\sin(\psi)}\frac{\partial \hat{V}_\psi}{\partial \varphi} \\ \frac{\Theta}{a\sinh(\xi)\sin(\psi)}\frac{\partial \hat{V}_\xi}{\partial \varphi} - \frac{\Theta}{a\Pi^{1/2}}\left(\frac{\partial \hat{V}_\varphi}{\partial R} + \frac{\cos^2(\psi)-\sinh^2(\xi)}{\Theta\tanh(\xi)}\hat{V}_\varphi\right) \\ \frac{\Theta}{a\Pi^{1/2}}\left(\begin{array}{l}\frac{\partial \hat{V}_\psi}{\partial \xi} + \frac{(3\cos^2(\psi)-1-\cosh^2(\xi))\sinh(\xi)\cosh\xi}{\Theta\Pi}\hat{V}_\psi+ \\ -\frac{\partial \hat{V}_\xi}{\partial \psi} + \frac{(\cos^2(\psi)-3\cosh^2(\xi)+1)\sin(\psi)\cos(\psi)}{\Theta\Pi}\hat{V}_\xi\end{array}\right) \end{bmatrix} \tag{4.392}$$

The Laplacian of the scalar $T$ is the scalar

$$\nabla^2 T = \frac{\Theta^2}{a^2 \Pi} \left[ \frac{\partial^2 T}{\partial \xi^2} + \frac{\cos^2(\psi) - \sinh^2(\xi)}{\Theta \tanh(\xi)} \frac{\partial T}{\partial \xi} \right] +$$

$$+ \frac{\Theta^2}{a^2 \Pi^{1/2}} \left[ \frac{\partial^2 T}{\partial \psi^2} + \frac{\sin^2(\psi) + \cosh^2(\xi)}{\Theta \tan(\psi)} \frac{\partial T}{\partial \psi} \right] +$$

$$+ \frac{\Theta^2}{a^2 \sinh^2(\xi) \sin^2(\psi)} \frac{\partial^2 T}{\partial \varphi^2} \tag{4.393}$$

The Laplace equation is *R-separable* in this system, the corresponding Stäckel matrix is

$$\Phi = \begin{bmatrix} a^2 \sinh^2(\xi) & -1 & -\frac{1}{\sinh^2(\xi)} \\ a^2 \sin^2(\psi) & 1 & -\frac{1}{\sin^2(\psi)} \\ 0 & 0 & 1 \end{bmatrix} \tag{4.394}$$

the functions $f_k$ are

$$f_1 = f_\xi(\xi) = \sinh(\xi); \quad f_2 = f_\psi(\psi) = \sin(\psi); \quad f_3 = f_\varphi(\varphi) = a \tag{4.395}$$

the functions $\tilde{R}$ and $\tilde{Q}$ are

$$\tilde{R} = \frac{1}{\Theta^{1/2}}; \quad \tilde{Q} = \frac{1}{\Theta^2} \tag{4.396}$$

the constant $\alpha_1$ is nil and the functions $q_k$ are

$$q_1 = q_\xi(\xi) = -\alpha_2 - \frac{\alpha_3}{\sinh^2(\xi)}; \quad q_2 = q_\psi(\psi) = \alpha_2 - \frac{\alpha_3}{\sin^2(\theta)}; \quad q_3 = q_\varphi(\varphi) = \alpha_3 \tag{4.397}$$

The Helmholtz equation is *not* separable in this system.

The transformation matrix for physical components from inverse prolate spheroidal to Cartesian coordinates, $\mathbf{M}^{u \to x}$ is

$$\mathbf{M}^{u \to x} = \begin{bmatrix} \frac{\cosh(\xi)\sin(\psi)\left[\cos^2(\psi)-\sinh^2(\xi)\right]}{\Theta\Pi^{1/2}} \cos(\varphi) & \frac{\sinh(\xi)\cos(\psi)\left[\cosh^2(\xi)+\sin^2(\psi)\right]}{\Theta\Pi^{1/2}} \cos(\varphi) & -\sin(\varphi) \\ \frac{\cosh(\xi)\sin(\psi)\left[\cos^2(\psi)-\sinh^2(\xi)\right]}{\Theta\Pi^{1/2}} \sin(\varphi) & \frac{\sinh(\xi)\cos(\psi)\left[\cosh^2(\xi)+\sin^2(\psi)\right]}{\Theta\Pi^{1/2}} \sin(\varphi) & \cos(\varphi) \\ -\frac{\sinh(\xi)\cos(\psi)\left[\cosh^2(\xi)+\sin^2(\psi)\right]}{\Theta\Pi^{1/2}} & \frac{\cosh(\xi)\sin(\psi)\left[\cos^2(\psi)-\sinh^2(\xi)\right]}{\Theta\Pi^{1/2}} & 0 \end{bmatrix} \tag{4.398}$$

### 4.4.16 *Inverse Oblate Spheroidal Coordinate System*

Inverse oblate spheroidal coordinates can be defined as

$$x = \frac{a\sqrt{\zeta^2+1}\sqrt{1-\eta^2}\cos(\varphi)}{\Theta}; \quad y = \frac{a\sqrt{\zeta^2+1}\sqrt{1-\eta^2}\sin(\varphi)}{\Theta}; \quad z = \frac{a\zeta\eta}{\Theta}$$
(4.399)

with $\Theta = \zeta^2 - \eta^2 + 1$.

The coordinate space $(\zeta, \eta, \varphi)$ is limited as: $0 \le \zeta < +\infty, -1 \le \eta \le 1, 0 \le \varphi < 2\pi$.

The surfaces $\zeta = \zeta_0$ is a rotational cyclide (inverse oblate spheroid); the surface $\eta = \eta_0$ is a rotational cyclide; the surface $\varphi = \varphi_0$ is a half plane passing through the $z$-axis (see Fig. 4.16). The scale factors are

$$h_\zeta = \frac{a}{\Theta}\sqrt{\frac{\zeta^2+\eta^2}{\zeta^2+1}}; \quad h_\eta = \frac{a}{\Theta}\sqrt{\frac{\zeta^2+\eta^2}{1-\eta^2}}; \quad h_\varphi = \frac{a\sqrt{(\zeta^2+1)(1-\eta^2)}}{\Theta}; \quad g^{1/2} = \frac{a^3(\zeta^2+\eta^2)}{\Theta^3}$$
(4.400)

and the functions $W^{(k)}$ are

$$W^{(\zeta)} = \frac{a(\zeta^2+1)}{\Theta}; \quad W^{(\eta)} = \frac{a(1-\eta^2)}{\Theta}; \quad W^{(\varphi)} = \frac{a(\zeta^2+\eta^2)}{\Theta(\zeta^2+1)(1-\eta^2)}$$
(4.401)

The gradient of a scalar quantity $T$ is a vector and its physical components are

(a)                                                        (b)

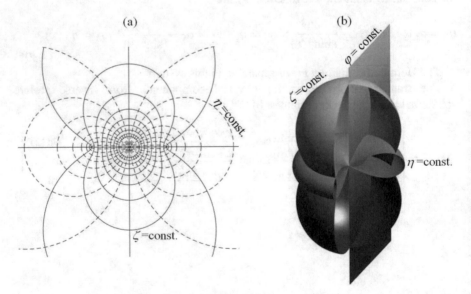

**Fig. 4.16** Inverse oblate spheroidal coordinate system: **a** coordinates over a plane passing through the $z$-axis; **b** 3-D representation of the iso-surfaces

$$\nabla T = \operatorname{grad} T = \frac{\Theta}{a}\left[\sqrt{\frac{\zeta^2+1}{\zeta^2+\eta^2}}\frac{\partial T}{\partial\zeta},\ \sqrt{\frac{1-\eta^2}{\zeta^2+\eta^2}}\frac{\partial T}{\partial\eta},\ \frac{1}{\sqrt{(\zeta^2+1)(1-\eta^2)}}\frac{\partial T}{\partial\varphi}\right]$$

(4.402)

The divergence of a vector $\mathbf{V} = [\hat{V}_\zeta,\ \hat{V}_\eta,\ \hat{V}_\varphi]$ is the scalar

$$\nabla\cdot\mathbf{V} = \operatorname{div}\mathbf{V} = \frac{\Theta}{a}\sqrt{\frac{\zeta^2+1}{\zeta^2+\eta^2}}\left[\frac{\partial\hat{V}_\zeta}{\partial\zeta}+\zeta\left(\frac{1}{\zeta^2+\eta^2}+\frac{1}{\zeta^2+1}-\frac{4}{\Theta}\right)\hat{V}_\zeta\right]+$$

$$+\frac{\Theta}{a}\sqrt{\frac{1-\eta^2}{\zeta^2+\eta^2}}\left[\frac{\partial\hat{V}_\eta}{\partial\eta}+\eta\left(\frac{1}{\zeta^2+\eta^2}-\frac{1}{1-\eta^2}+\frac{4}{\Theta}\right)\hat{V}_\eta\right]+$$

$$+\frac{\Theta}{a}\frac{1}{\sqrt{(1-\eta^2)(\zeta^2+1)}}\frac{\partial\hat{V}_\varphi}{\partial\varphi}$$

(4.403)

The curl of a vector $\mathbf{V} = [\hat{V}_\zeta,\ \hat{V}_\eta,\ \hat{V}_\varphi]$ is the vector $\boldsymbol{\omega} = \operatorname{curl}\mathbf{V}$

$$\begin{bmatrix}\hat{\omega}_\zeta\\\hat{\omega}_\eta\\\hat{\omega}_\varphi\end{bmatrix}=\begin{bmatrix}\dfrac{\Theta(1-\eta^2)^{1/2}}{a(\zeta^2+\eta^2)^{1/2}}\dfrac{\partial\hat{V}_\varphi}{\partial\eta}+\dfrac{\eta(1-\eta^2-\zeta^2)}{a(1-\eta^2)^{1/2}(\zeta^2+\eta^2)^{1/2}}\hat{V}_\varphi-\dfrac{\Theta}{a(\zeta^2+1)^{1/2}(1-\eta^2)^{1/2}}\dfrac{\partial\hat{V}_\eta}{\partial\varphi}\\[12pt]\dfrac{\Theta}{a(\zeta^2+1)^{1/2}(1-\eta^2)^{1/2}}\dfrac{\partial\hat{V}_\zeta}{\partial\varphi}-\dfrac{\Theta(\zeta^2+1)^{1/2}}{a(\zeta^2+\eta^2)^{1/2}}\dfrac{\partial\hat{V}_\varphi}{\partial\zeta}+\dfrac{\zeta(1+\eta^2+\zeta^2)}{a(\zeta^2+1)^{1/2}(\zeta^2+\eta^2)^{1/2}}\hat{V}_\varphi\\[12pt]\dfrac{\Theta(\zeta^2+1)^{1/2}}{a(\zeta^2+\eta^2)^{1/2}}\dfrac{\partial\hat{V}_\eta}{\partial\zeta}+\dfrac{\zeta(1-3\eta^2-\zeta^2)(\zeta^2+1)^{1/2}}{a(\zeta^2+\eta^2)^{3/2}}\hat{V}_\eta-\dfrac{\Theta(1-\eta^2)^{1/2}}{a(\zeta^2+\eta^2)^{1/2}}\dfrac{\partial\hat{V}_\zeta}{\partial\eta}-\dfrac{\eta(1+\eta^2+3\zeta^2)(1-\eta^2)^{1/2}}{a(\zeta^2+\eta^2)^{3/2}}\hat{V}_\zeta\end{bmatrix}$$

(4.404)

The Laplacian of the scalar $T$ is the scalar

$$\nabla^2 T = \frac{\Theta^2}{a^2(\zeta^2+\eta^2)}\left[(\zeta^2+1)\frac{\partial^2 T}{\partial\zeta^2}-\frac{2\zeta\eta^2}{\Theta}\frac{\partial T}{\partial\zeta}+(1-\eta^2)\frac{\partial^2 T}{\partial\eta^2}-\frac{2\zeta^2\eta}{\Theta}\frac{\partial T}{\partial\eta}\right]+$$

$$+\frac{\Theta^2}{a^2(\zeta^2+\eta^2)(1-\eta^2)}\frac{\partial^2 T}{\partial\varphi^2}$$

(4.405)

The Laplace equation is *R-separable* in this system, the corresponding Stäckel matrix is

$$\Phi = \begin{bmatrix}a^2 & -\dfrac{1}{a(\zeta^2+1^2)} & \dfrac{1}{(\zeta^2+1)^2}\\[8pt]-a^2 & \dfrac{1}{a(1-\eta^2)} & -\dfrac{1}{(1-\eta^2)^2}\\[8pt]0 & 0 & 1\end{bmatrix}$$

(4.406)

the functions $f_k$ are

$$f_1 = f_\zeta(\zeta) = a\left(\zeta^2 + 1\right); \quad f_2 = f_\eta(\eta) = a\left(1 - \eta^2\right); \quad f_3 = f_\varphi(\varphi) = 1$$

$$(4.407)$$

the functions $\tilde{R}$ and $\tilde{Q}$ are

$$\tilde{R} = \frac{1}{\Theta^{1/2}}; \quad \tilde{Q} = \frac{1}{\Theta^2} \tag{4.408}$$

the constant $\alpha_1$ is nil and the functions $q_k$ are

$$q_1 = q_\zeta(\zeta) = -\frac{\alpha_2}{a\left(\zeta^2 + 1\right)} + \frac{\alpha_3}{\left(\zeta^2 + 1\right)^2}; \quad q_2 = q_\eta(\eta) = \frac{\alpha_2}{a\left(1 - \eta^2\right)} - \frac{\alpha_3}{\left(1 - \eta^2\right)^2}; \quad q_3 = q_\varphi(\varphi) = \alpha_3$$

$$(4.409)$$

The Helmholtz equation is *not* separable in this system.

The transformation matrix for physical components from inverse oblate spheroid to Cartesian coordinates, $\mathbf{M}^{u \to x}$ is

$$\mathbf{M}^{u \to x} = \begin{bmatrix} -\frac{\zeta\left(1 - \eta^2\right)^{1/2}\left(1 + \eta^2 + \zeta^2\right)}{\Theta\left(\zeta^2 + \eta\right)^{1/2}}\cos(\varphi) & \frac{\eta\left(\zeta^2 + 1\right)^{1/2}\left(1 - \eta^2 - \zeta^2\right)}{\Theta\left(\zeta^2 + \eta\right)^{1/2}}\cos(\varphi) & -\sin(\varphi) \\ -\frac{\zeta\left(1 - \eta^2\right)^{1/2}\left(1 + \eta^2 + \zeta^2\right)}{\Theta\left(\zeta^2 + \eta\right)^{1/2}}\sin(\varphi) & \frac{\eta\left(\zeta^2 + 1\right)^{1/2}\left(1 - \eta^2 - \zeta^2\right)}{\Theta\left(\zeta^2 + \eta\right)^{1/2}}\sin(\varphi) & \cos(\varphi) \\ \frac{\eta\left(\zeta^2 + 1\right)^{1/2}\left(1 - \eta^2 - \zeta^2\right)}{\Theta\left(\zeta^2 + \eta\right)^{1/2}} & \frac{\zeta\left(1 - \eta^2\right)^{1/2}\left(1 + \eta^2 + \zeta^2\right)}{\Theta\left(\zeta^2 + \eta\right)^{1/2}} & 0 \end{bmatrix}$$

$$(4.410)$$

The inverse prolate spheroidal coordinate system can also be defined using the coordinates $(\xi, \psi, \varphi)$ as

$$x = \frac{a\cosh(\xi)\sin(\psi)\cos(\varphi)}{\Theta}; \quad y = \frac{a\cosh(\xi)\sin(\psi)\sin(\varphi)}{\Theta}; \quad z = \frac{a\sinh(\xi)\cos(\psi)}{\Theta}$$

$$(4.411)$$

with $\Theta = \cosh^2(\xi) - \cos^2(\psi)$.

The coordinate space $(\xi, \psi, \varphi)$ is limited as: $0 \leq \xi < +\infty, 0 \leq \psi \leq \pi, 0 \leq \varphi < 2\pi$ and the relation with the system (4.399) is given by: $\zeta = \sinh(\xi)$ and $\eta = \cos(\psi)$.

The scale factors are

$$h_\xi = a\frac{\Pi^{1/2}}{\Theta} = h_\psi; \quad h_\varphi = a\frac{\cosh(\xi)\sin(\psi)}{\Theta} \tag{4.412}$$

$$g^{1/2} = a^3\frac{\Pi\cosh(\xi)\sin(\psi)}{\Theta^3} \tag{4.413}$$

where $\Pi = \cosh^2(\xi) - \sin^2(\psi) = \sinh^2(\xi) + \cos^2(\psi)$ and the functions $W^{(k)}$ are

$$W^{(\xi)} = a\frac{\cosh(\xi)\sin(\psi)}{\Theta} = W^{(\psi)}; \quad W^{(\varphi)} = a\frac{\Pi}{\Theta\cosh(\xi)\sin(\psi)} \tag{4.414}$$

The gradient of a scalar quantity $T$ is a vector

$$\nabla T = \operatorname{grad} T = \frac{\Theta}{a\Pi^{1/2}} \left[ \frac{\partial T}{\partial \xi}, \frac{\partial T}{\partial \psi}, \frac{\Pi^{1/2}}{\cosh(\xi)\sin(\psi)} \frac{\partial T}{\partial \varphi} \right] \tag{4.415}$$

The divergence of a vector $\mathbf{V} = [\hat{V}_\xi, \hat{V}_\psi, \hat{V}_\varphi]$ is the scalar

$$\begin{aligned}
\nabla \cdot \mathbf{V} = \operatorname{div}\mathbf{V} = {}& \frac{\Theta}{a\Pi^{1/2}} \frac{\partial \hat{V}_\xi}{\partial \xi} + \\
&+ \frac{-10\cos(2\psi)\sinh(2\xi) - [3 + \cos(4\psi) + 6\cosh(2\xi) + 2\cosh(4\xi)]\tanh(\xi)}{2\sqrt{2}a\,[\cos(2\psi) + \cosh(2\xi)]^{3/2}} \hat{V}_\xi + \\
&+ \frac{\Theta}{a\Pi^{1/2}} \frac{\partial \hat{V}_\psi}{\partial \psi} + \frac{\cot(\psi)\left[3 - 6\cos(2\psi) + 2\cos(4\psi) + \cosh(4\xi) - 20\cosh(2\xi)\sin^2(\psi)\right]}{2\sqrt{2}a\,[\cos(2\psi) + \cosh(2\xi)]^{3/2}} \hat{V}_\psi + \\
&+ \frac{\Theta}{a\cosh(\xi)\sin(\psi)} \frac{\partial \hat{V}_\varphi}{\partial \varphi}
\end{aligned} \tag{4.416}$$

The curl of a vector $\mathbf{V} = [\hat{V}_\xi, \hat{V}_\psi, \hat{V}_\varphi]$ is the vector $\boldsymbol{\omega} = \operatorname{curl}\mathbf{V}$

$$\begin{bmatrix} \hat{\omega}_\xi \\ \hat{\omega}_\psi \\ \hat{\omega}_\varphi \end{bmatrix} = \begin{bmatrix} \frac{\Theta}{a\Pi^{1/2}}\left( \frac{\partial \hat{V}_\varphi}{\partial \psi} + \frac{\sinh^2(\xi) - \sin^2(\psi)}{\Theta\tan(\psi)} \hat{V}_\varphi \right) - \frac{\Theta}{a\cosh(\xi)\sin(\psi)} \frac{\partial \hat{V}_\psi}{\partial \varphi} \\ \frac{\Theta}{a\cosh(\xi)\sin(\psi)} \frac{\partial \hat{V}_\xi}{\partial \varphi} - \frac{\Theta}{a\Pi^{1/2}}\left( \frac{\partial \hat{V}_\varphi}{\partial \xi} + \frac{\cosh^2(\xi) + \cos^2(\psi)}{\Theta\coth(\xi)} \hat{V}_\varphi \right) \\ \frac{\Theta}{a\Pi^{1/2}}\left( \frac{\partial \hat{V}_\psi}{\partial \xi} + \frac{[3\cos^2(\psi) + \cosh^2(\xi) - 2]\sinh(\xi)\cosh(\xi)}{\Pi^{1/2}\Theta} \hat{V}_\psi + -\frac{\partial \hat{V}_\xi}{\partial \psi} + \frac{[3\cosh^2(\xi) + \cos^2(\psi) - 2]\sin(\psi)\cos(\psi)}{\Pi^{1/2}\Theta} \hat{V}_\xi \right) \end{bmatrix} \tag{4.417}$$

The Laplacian of the scalar $T$ is the scalar

$$\begin{aligned}
\nabla^2 T = {}& \frac{\Theta^2}{a^2\Pi} \left[ \frac{\partial^2 T}{\partial \xi^2} - \frac{\cosh^2(\xi) + \cos^2(\psi)}{\Theta\coth(\xi)} \frac{\partial T}{\partial \xi} \right] + \\
&+ \frac{\Theta^2}{a^2\Pi^{1/2}} \left[ \frac{\partial^2 T}{\partial \psi^2} + \frac{\sinh^2(\xi) - \sin^2(\psi)}{\Theta\tan(\psi)} \frac{\partial T}{\partial \psi} \right] + \\
&+ \frac{\Theta^2}{a^2\cosh^2(\xi)\sin^2(\psi)} \frac{\partial^2 T}{\partial \varphi^2}
\end{aligned} \tag{4.418}$$

The Laplace equation is *R-separable* in this system, the corresponding Stäckel matrix is

$$\Phi = \begin{bmatrix} a^2\cosh^2(\xi) & -1 & \frac{1}{\cosh^2(\xi)} \\ -a^2\sin^2(\psi) & 1 & -\frac{1}{\sin^2(\psi)} \\ 0 & 0 & 1 \end{bmatrix} \tag{4.419}$$

the functions $f_k$ are

$$f_1 = f_\xi(\xi) = \cosh(\xi); \quad f_2 = f_\psi(\psi) = \sin(\psi); \quad f_3 = f_\varphi(\varphi) = a \tag{4.420}$$

the functions $\tilde{R}$ and $\tilde{Q}$ are

$$\tilde{R} = \frac{1}{\Theta^{1/2}}; \quad \tilde{Q} = \frac{1}{\Theta^2} \tag{4.421}$$

the constant $\alpha_1$ is nil and the functions $q_k$ are

$$q_1 = q_\xi\,(\xi) = -\alpha_2 + \frac{\alpha_3}{\cosh^2(\xi)}; \quad q_2 = q_\psi\,(\psi) = \alpha_2 - \frac{\alpha_3}{\sin^2(\theta)}; \quad q_3 = q_\varphi\,(\varphi) = \alpha_3 \tag{4.422}$$

The Helmholtz equation is *not* separable in this system.

The transformation matrix for physical components from inverse oblate spheroidal to Cartesian coordinates, $\mathbf{M}^{u \to x}$ is

$$\mathbf{M}^{u \to x} = \begin{bmatrix} -\frac{\sinh(\xi)\sin(\psi)\left[\cosh^2(\xi)+\cos^2(\psi)\right]}{\Theta\Pi^{1/2}}\cos(\varphi) & \frac{\cosh(\xi)\cos(\psi)\left[\sinh^2(\xi)-\sin^2(\psi)\right]}{\Theta\Pi^{1/2}}\cos(\varphi) & -\sin(\varphi) \\ -\frac{\sinh(\xi)\sin(\psi)\left[\cosh^2(\xi)+\cos^2(\psi)\right]}{\Theta\Pi^{1/2}}\sin(\varphi) & \frac{\cosh(\xi)\cos(\psi)\left[\sinh^2(\xi)-\sin^2(\psi)\right]}{\Theta\Pi^{1/2}}\sin(\varphi) & \cos(\varphi) \\ -\frac{\cosh(\xi)\cos(\psi)\left[\sinh^2(\xi)-\sin^2(\psi)\right]}{\Theta\Pi^{1/2}} & -\frac{\sinh(\xi)\sin(\psi)\left[\cosh^2(\xi)+\cos^2(\psi)\right]}{\Theta\Pi^{1/2}} & 0 \end{bmatrix} \tag{4.423}$$

# References

1. Helms, L.L.: Potential, Theory, Universitext. Springer, London (2009)
2. Moon, P., Spencer, D.E.: Field Theory Handbook, 2nd edn. Springer, Berlin (1971)

# Chapter 5
# Sturm–Liouville Problems

In Chap. 3 we have seen how the separability of PDEs leads to ordinary differential equations problems, usually of second order. The problem is complemented with B.C.s and the reduction of the initial PDE to second order ODEs often yield a so-called Sturm–Liouville (SL) problem (named after the French mathematicians Jacques Charles François Sturm, 1803–1855, and Joseph Liouville, 1809–1882).

Since the solution of such a problem is obtained in terms of orthogonal functions, the first part of this chapter will give to the reader some elements of function analysis needed to deal with them. Concepts like orthogonality among functions, basis of a functional space and generalised Fourier series will be analysed. The fundamentals of the Sturm–Liouville theory is given at the end of the chapter together with some details about the properties of solution of a SL problem.

## 5.1 Orthogonal Functions

The concept of orthogonality comes from geometry as a generalisation of the concept of perpendicularity, which is the property of two lines that intersect at a right angle. Two vectors are said to be orthogonal when they form a right angle and in vector algebra formalism this property can be simply expressed by saying that the scalar product (the dot product) of the two vectors is nil. The orthogonality among functions is a property that can be expressed as a generalisation of this last definition. The interested reader can refer to [1] for a deeper treatment.

### 5.1.1 Linear Spaces, Hilbert Space and the Inner Product

A linear space (also called *vector* space) is a collection of objects complemented with two operations: sum and scalar multiplication. The sum is an operation that takes two

© Springer Nature Switzerland AG 2021
G. E. Cossali and S. Tonini, *Drop Heating and Evaporation: Analytical Solutions in Curvilinear Coordinate Systems*, Mathematical Engineering,
https://doi.org/10.1007/978-3-030-49274-8_5

elements of the space (which will be called *vectors*) and associate to them a third element of the space, while the scalar multiplication takes any element of a field (also called *scalar*) and any vector and yields another vector. These two operations must satisfy a number of requirements (in what follows $u$, $v$ and $w$ are *vectors* and $a, b, \ldots$ are *scalars*), precisely:

(a) the sum must be commutative ($u + v = v + u$) and associative ($u + (v + w) = (u + v) + w$);
(b) a *zero* element ($0$) must exist, such that: $v + 0 = v$ for all the vectors;
(c) for each vector $u$ an *inverse* element ($-u$) exists such that their sum yields the zero vector;
(d) the scalar multiplication must be distributive: $a(u + v) = au + av$
(e) the scalar identity ($1$) exists such that: $1 u = u$.

The definition of linear space is quite general and it may apply to a vast variety of objects. For example, the vectors used in physics to represents forces, velocities, etc. form a linear space, and the terminology used in linear spaces theory is borrowed from vector algebra. Another example, which is the most important for us, is that of a set of functions defined over a given domain, and taking a real or complex number as the element of the field in scalar multiplication. The *sum* operation is the usual sum between two functions

$$f(x) + g(x) = h(x) \tag{5.1}$$

and the *scalar multiplication* is the usual multiplication between a function and a constant

$$a f(x) = g(x) \tag{5.2}$$

To make the above defined space more interesting for the applications that we have in mind, a further operation must be added, which transforms a linear space in a so-called Hilbert space (named after the German mathematician David Hilbert, 1862–1943). This operation is the *inner product*, which associates any pair of vectors to an either real or complex scalar, and it is formally written as

$$\langle u, v \rangle = a \tag{5.3}$$

The inner product must satisfy the following requirements:

(a) it must be linear: $\langle u + v, w \rangle = \langle u, w \rangle + \langle v, w \rangle$ and $\langle au, v \rangle = a \langle u, v \rangle$
(b) it must be symmetric : $\langle u, v \rangle = \overline{\langle v, u \rangle}$
(c) it must be positively defined: $\langle u, u \rangle \geq 0$ and if $\langle u, u \rangle = 0$ then $u = 0$, and vice versa.

To notice that the linearity and the symmetry requirements yield

$$\langle u, bv \rangle = \overline{\langle bv, u \rangle} = \overline{b \langle v, u \rangle} = \overline{b} \langle u, v \rangle \tag{5.4}$$

while, if the inner product is always a real number, the symmetry property (b) yields: $\langle u, v \rangle = \langle v, u \rangle$.

This definition of inner product is quite general and it can be applied to many cases. For example, the usual scalar product defined in vector algebra satisfies these requirements; in fact, defining

$$\langle u, v \rangle = \sum_j u_j v_j \qquad (5.5)$$

where $u_j$ and $v_j$ are the components of the vectors $u$ and $v$, it is easy to prove that all the above reported requirements are satisfied. The inner product satisfies an important relation, the *Schwarz inequality* (named after the German mathematician Karl Hermann Amandus Schwarz, 1843–1921)

$$|\langle u, v \rangle|^2 \leq \langle u, u \rangle \langle v, v \rangle \qquad (5.6)$$

and it can be proven as follows. First notice that, if $\langle u, v \rangle = 0$, the inequality clearly holds, and it also holds when $u$ is a multiple of $v$ (or vice versa) or when $u$ or $v$ are the nil vector. Consider now the vector $u + av$, where $a$ is a non nil complex number to be defined, the positive definiteness of the inner product assures that $\langle u + av, u + av \rangle \geq 0$, and expanding it yields, thanks to the linearity and symmetry properties

$$\langle u + av, u + av \rangle = \langle u, u \rangle + \langle u, av \rangle + a \langle v, u \rangle + a \langle v, av \rangle$$
$$= \langle u, u \rangle + \overline{a} \langle u, v \rangle + a \overline{\langle u, v \rangle} + a\overline{a} \langle v, v \rangle \qquad (5.7)$$

Having excluded the previously mentioned particular cases, taking $a = - \langle u, v \rangle / \langle v, v \rangle$ yields

$$\langle u + av, u + av \rangle = \langle u, u \rangle - \frac{\overline{\langle u, v \rangle} \langle u, v \rangle}{\langle v, v \rangle} \geq 0 \qquad (5.8)$$

that proves the thesis.

## 5.1.2 Orthogonality of Functions

Consider now the Hilbert space built by taking as vectors the functions defined over a given domain $[a, b]$ (where $b > a$, which can be finite or infinite), and as the scalar field that of the complex numbers. The inner product can be defined as

$$\langle f, g \rangle = \int_a^b f(x) \, \overline{g}(x) \, w(x) \, dx \qquad (5.9)$$

where $\bar{g}(x)$ is the complex conjugate of $g(x)$ and $w(x) > 0$ is a real function. This product is linear

$$\int_a^b [f(x) + h(x)] \, \bar{g}(x) \, w(x) \, dx = \int_a^b f(x) \, \bar{g}(x) \, w(x) \, dx + \int_a^b h(x) \, \bar{g}(x) \, w(x) \, dx \quad (5.10)$$

$$\int_a^b a\, f(x) \, \bar{g}(x) \, w(x) \, dx = a \int_a^b f(x) \, \bar{g}(x) \, w(x) \, dx \quad (5.11)$$

it is symmetric

$$\int_a^b f(x) \, \bar{g}(x) \, w(x) \, dx = \int_a^b \bar{g}(x) \, f(x) \, w(x) \, dx = \overline{\int_a^b g(x) \, \bar{f}(x) \, w(x) \, dx} \quad (5.12)$$

and it is positively defined

$$\int_a^b f(x) \, \bar{f}(x) \, w(x) \, dx = \int_a^b |f(x)|^2 \, w(x) \, dx \geq 0 \quad (5.13)$$

and if $f(x) \equiv 0$ then $\int_a^b f(x) \, \bar{f}(x) \, w(x) \, dx = 0$, while $\int_a^b f(x) \, \bar{f}(x) \, w(x) \, dx = 0$ only if $f(x) \equiv 0$.

As stated in the introduction, orthogonality is a property that can be defined in a simple way whenever a scalar product is defined. It is then straightforward to define the orthogonality of two functions defined over a given domain.

**Definition** Two functions $f(x)$ and $g(x)$ are said to be orthogonal when their inner product is nil

$$\langle f, g \rangle = \int_a^b f(x) \, \bar{g}(x) \, w(x) \, dx = 0 \quad (5.14)$$

As an example, the two functions $f = \cos(x)$ and $g = \cos(2x)$ are orthogonal over the domain $[a, b] = [0, 2\pi]$, with weighting function $w(x) = 1$, in fact

$$\langle f, g \rangle = \int_0^{2\pi} \cos(x) \cos(2x) \, dx = \left[ \frac{1}{2} \sin(x) + \frac{1}{3} \sin(3x) \right]_0^{2\pi} = 0 \quad (5.15)$$

As a second example, the two *probabilists'* (named after the French mathematician Charles Hermite, 1822–1901)

$$He_1(x) = x; \quad He_3(x) = x^3 - 3x \quad (5.16)$$

are orthogonal with weighting function $w(x) = e^{-x^2/2}$ over the domain $[a, b] = [-\infty, \infty]$ in fact

$$\langle He_1, He_3 \rangle = \int_{-\infty}^{\infty} He_1(x) \, He_3(x) \, e^{-x^2} \, dx = \int_{-\infty}^{\infty} (x^4 - 3x^2) \, e^{-x^2/2} dx = -\left[ e^{-x^2/2} x^3 \right]_{-\infty}^{+\infty} = 0 \quad (5.17)$$

Also $He_4(x) = x^4 - 6x^2 + 3$ and $He_1(x) = x$ are orthogonal

$$\langle He_1, He_4 \rangle = \int_{-\infty}^{\infty} He_1(x) \, He_4(x) \, e^{-x^2} dx = \int_{-\infty}^{\infty} \left( x^5 - 6x^3 + 3x \right) e^{-x^2/2} dx = \quad (5.18)$$
$$= \left[ e^{-x^2/2} \left( 1 + 2x^2 - x^4 \right) \right]_{-\infty}^{+\infty} = 0$$

and generally for all the couples $He_n(x)$, $He_m(x)$ with $n \neq m$

$$\langle He_n, He_m \rangle = 0 \qquad (5.19)$$

It should be stressed at this point that the concept of inner product of two functions is not at all restricted to functions of one variable, although this last case will be the main subject of the following discussion.

The existence of a scalar product allows to generalise some important concepts usually encountered in vector algebra. Thanks to the fact that $\langle f, f \rangle$ is non-negative, it is possible to introduce a *norm* for each vector, defined as the square root of $\langle f, f \rangle$

$$\| f \| = \langle f, f \rangle^{1/2} \qquad (5.20)$$

In the present case the norm of $f$ is the generalisation of the concept of *length* of a vector. For the sake of completeness it should be said that the concept of norm can be unrelated to that of an inner product. A norm must satisfy the following fundamental properties:

(a) it must be always non-negative: $\| f \| \geq 0$ and it is nil if and only if the vector is the nil vector;
(b) it is linear for scalar multiplication, in the sense that: $\| af \| = |a| \; \| f \|$;
(c) it must satisfy the *triangle inequality*: $\| f + g \| \leq \| f \| + \| g \|$.

It is easy to show that these properties are satisfied by the definition (5.20). The first one is a direct consequence of the positive definiteness of the inner product (Eq. 5.13). The second one is a direct consequence of (5.11). The third one can be demonstrated as follows: consider the norm of $f + g$, its square is

$$\| f + g \|^2 = \langle f + g, f + g \rangle = \langle f, f \rangle + \langle f, g \rangle + \overline{\langle f, g \rangle} + \langle g, g \rangle = \| f \|^2 + 2 \operatorname{Re} \{ \langle f, g \rangle \} + \| g \|^2 \quad (5.21)$$

where the symmetry of the scalar product has been used to yield $\langle f, g \rangle + \overline{\langle f, g \rangle} = 2 \operatorname{Re} \{ \langle f, g \rangle \}$. Since $\operatorname{Re} \{ \langle f, g \rangle \} \leq \sqrt{\operatorname{Re}^2 \{ \langle f, g \rangle \} + \operatorname{Im}^2 \{ \langle f, g \rangle \}} = |\langle f, g \rangle|$, the Schwarz inequality (5.6) yields

$$\| f + g \|^2 \leq \| f \|^2 + 2 |\langle f, g \rangle| + \| g \|^2 \leq \| f \|^2 + 2 \| f \| \| g \| + \| g \|^2 = (\| f \| + \| g \|)^2 \quad (5.22)$$

which proves the thesis. This inequality is sometime referred as Minkowski inequality (named after the German mathematician Hermann Minkowski, 1864–1909). To notice that Eq. (5.21) applied to two orthogonal vectors ($\langle f, g \rangle = 0$) yields: $\| f + g \|^2 = \| f \|^2 + \| g \|^2$ which is the generalisation of the Pythagorean theorem.

Another important concept that can be generalised is that of *projection*. When dealing with physical vectors, the projection $\mathbf{w}$ of a vector $\mathbf{u}$ on another vector $\mathbf{v}$ is

**Fig. 5.1** Projection of a
vector **u** on a vector **v**

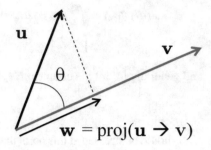

$$\mathbf{w} = \mathrm{proj}(\mathbf{u} \rightarrow \mathbf{v})$$

defined as the dot product between the vector and the versor $\phi = \frac{\mathbf{v}}{|\mathbf{v}|}$ (i.e. the unit
vector parallel to the vector **v**), multiplied by the versor itself:

$$\mathbf{w} = (\mathbf{u} \cdot \phi) \ \phi = \frac{(\mathbf{u} \cdot \mathbf{v})}{\|\mathbf{v}\|^2} \mathbf{v} \qquad (5.23)$$

This concept can be easily generalised as follows: the projection of the function $f$
on the function $g$ is

$$\mathrm{proj}\,(f \rightarrow g) = \frac{\langle f, g \rangle}{\|g\|^2} g = \langle f, \phi \rangle \, \phi \qquad (5.24)$$

where $\phi = \frac{g}{\|g\|}$ is the *versor*; to notice that $\mathrm{proj}\,(f \rightarrow g)$ is still a vector. The pro-
jection has some interesting properties, among them the following is of a certain
importance: the vector $f - \mathrm{proj}\,(f \rightarrow g)$ is orthogonal to $g$. Figure 5.1 shows this
for vectors, and for the general case the proof is simple

$$\langle f - \mathrm{proj}\,(f \rightarrow g)\,, g \rangle = \langle f, g \rangle - \frac{\langle f, g \rangle}{\|g\|^2} \langle g, g \rangle = 0 \qquad (5.25)$$

## 5.2  Orthogonal Sequences

Consider a sequence of functions $f_n\,(x)$, defined over a given domain, characterised
by the following property

$$\langle f_n, f_k \rangle = 0 \qquad (5.26)$$

whenever $k \neq n$; this sequence is called an *orthogonal sequence*. Since for many
applications, and in particular those related to the use of generalised Fourier series
(see Sect. 5.3), orthogonal sequences are of paramount importance, we will empha-
sise here some of the most important properties.

   A finite sequence of functions $f_1, \ldots,\ f_n$ is said to be linearly independent if the
only linear combination that is identically nil is that with all coefficients equal to zero,

i.e. $a_1 f_1 + a_2 f_2 + \cdots + a_n f_n = 0$ only if all $a_j = 0$. This concept can be extended also to infinite sequences of functions: an infinite sequence of functions is linearly independent if and only if any finite sub-sequence is linear independent. Given a finite, linearly independent sequence of functions $f_1, \ldots, f_n$, a n-dimensional linear space can be defined as the set of all linear combinations of elements of the sequence $f_1, \ldots, f_n$. The sequence $f_1, \ldots, f_n$ is called a *basis* of this space, since clearly each element of the space can be obtained by a linear combination of the elements $f_k$. The idea can then be reversed, saying that a set of functions form a n-dimensional linear space if there exists a linear independent set of functions $f_1, \ldots, f_n$ such that each element of the space can be obtained as a linear combination of $f_k$, which is called *basis* of the space. It should be stressed at this point that the basis of a space is not at all unique. Moreover, the fundamental theorem of linear algebra (see [1, 2] for a more complete discussion) states that if the sequence $f_1, \ldots, f_n$ is taken in a definite order, then each element of the space can be obtained by a linear combination in a unique way, i.e. there exists only one set of numbers $a_j$ that is associated to each element of the space. The set $(a_1, \ldots, a_n)$ is sometimes defined as the *coordinate sequence* of the element, with respect to the basis $f_1, \ldots, f_n$.

If the space has a basis that is formed by an infinite, linearly independent sequence of functions, the space is said to be infinite-dimensional. An example of a similar space is that of all functions of one real variable continuous over an interval $[a, b]$, usually indicated by $C[a, b]$. However, in what follows, we will need a *wider* space, precisely that of all functions that are square-integrable over a domain $[a, b]$, i.e. all functions defined over $[a, b]$ such that the integral $\int_a^b |f|^2 \, dx$ is finite, this space is called $L^2([a, b])$.

Let now assume that a finite sequence of functions $\phi_1, \ldots, \phi_n$, all belonging to the same space $S$ (say for example $L^2([a, b])$), is given and that they are mutually orthogonal (i.e. $\langle \phi_m, \phi_k \rangle = 0$ whenever $m \neq k$) and with unitary norm (i.e. $\langle \phi_n, \phi_n \rangle^{1/2} = \|\phi_n\| = 1$). This sequence is said to be *orthonormal* and it is linearly independent; in fact, suppose that a linear combination $\sum_{j=1}^{n} a_j \phi_n = f$ exists that is identically nil, the scalar products $\langle f, \phi_k \rangle = \sum_j a_j \langle \phi_j, \phi_k \rangle = a_k$ are then necessarily all nil. The sequence $\phi_1, \ldots, \phi_n$ is a basis of a n-dimensional linear space $H_n$, which is called a *sub-space* of $S$, formed by all the linear combinations of $\phi_1, \ldots, \phi_n$. Taking now an element $f \in S$, the following definition of projection over a sub-space can be given.

**Definition** The projection of $f$ on the subspace $H_n$ is defined as

$$\text{proj}\,(f \to H_n) = \sum_{k=1}^{n} \langle f, \phi_k \rangle \, \phi_k \tag{5.27}$$

It is easy to see that proj $(f \to H_n)$ belongs to $H_n$ since it is a linear combination of the basis.

**Fig. 5.2** Projection of a
vector in 3D space on a plane

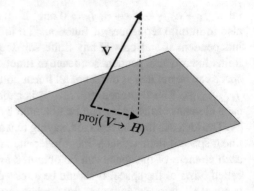

Moreover the following statements hold:

(a) If $f \in H_n$ then the projection of $f$ on $H_n$ is $f$ itself.
(b) the vector $g = f - \text{proj}\,(f \to H_n)$ is orthogonal to any element of $H_n$.

In fact, $g$ is orthogonal to each vector $\phi_j$

$$\langle g, \phi_j \rangle = \langle f - \text{proj}\,(f \to H_n)\,, \phi_j \rangle = \left\langle f - \sum_{k=1}^{n} \langle f, \phi_k \rangle\, \phi_k, \phi_j \right\rangle \tag{5.28}$$

$$= \langle f, \phi_j \rangle - \sum_{k=1}^{n} \langle f, \phi_k \rangle\, \langle \phi_k, \phi_j \rangle = \langle f, \phi_j \rangle - \langle f, \phi_j \rangle \langle \phi_j, \phi_j \rangle = 0$$

and since a generic element of $H_n$ is a linear combination of $\phi_j$ $(h = \sum_{k=1}^{n} h_k \phi_k)$
then

$$\langle g, h \rangle = \sum_{k=1}^{n} h_k \langle g, \phi_k \rangle = 0 \tag{5.29}$$

It is interesting to notice that the above defined projection on a sub-space can be
seen as a generalisation of the concept of projection of a vector in 3-D space on a
plane (see Fig. 5.2).

It is useful to introduce the concept of orthogonalisation of a finite sequence of
functions. Consider the sequence of linearly independent functions $f_1, \ldots, f_n$, we
know that this forms a basis of the sub-space of the linear combinations of $f_1, \ldots, f_n$.
Due to its special properties, it is sometimes useful to search for an orthonormal basis,
i.e. a basis made by a sequence of mutually orthogonal functions of unitary norm. To
this end, procedures exist to transform a linearly independent set of functions into a
set of orthonormal functions. The Gram–Schmidt procedure (named after the Danish
mathematician Jørgen Pedersen Gram, 1850–1916, and the German mathematician

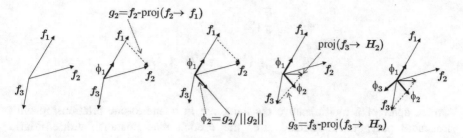

**Fig. 5.3** Orthogonalisation of the sequence of function $f_1$, $f_2$ and $f_3$

Erhard Schmidt 1876–1959) is here briefly described. Starting with the ordered sequence $f_1, \ldots, f_n$ an ordered sequence of orthonormal functions is obtained by the following process

$$\phi_1 = \frac{1}{\|f_1\|} f_1$$

$$\phi_2 = \frac{1}{\|g_2\|} g_2 \text{ where } g_2 = f_2 - \langle f_2, \phi_1 \rangle \phi_1$$

$$\phi_3 = \frac{1}{\|g_2\|} g_2 \text{ where } g_3 = f_3 - \langle f_3, \phi_1 \rangle \phi_1 - \langle f_3, \phi_2 \rangle \phi_2 \qquad (5.30)$$

$$\ldots$$

$$\phi_n = \frac{1}{\|g_n\|} g_n \text{ where } g_n = f_n - \sum_{k=1}^{n-1} \langle f_n, \phi_k \rangle \phi_k$$

At each step, a new vector $g_k$ is found by subtracting to $f_k$ its projection on the sub-space $H_{k-1}$ (spanned by the first $k-1$ orthonormal vectors $\phi_j$); this new vector is orthogonal to $H_{k-1}$ and after normalisation it becomes the $k-th$ element of the orthonormal sequence $\phi_k$ (see Fig. 5.3 for a sketch of the procedure).

As an example, take as linearly independent sequence the following: $1, x, x^2, \ldots,$ $x^n$, which is a basis of the sub-space of polynomials of maximum degree equal to $n$. The Gram–Schmidt procedure can be used to generate all the known orthogonal polynomials (see [3] for further details), by choosing a proper domain $[a, b]$ and a weighting function $w(x)$.

If the domain is $[-1, 1]$ and the weighting function is $w(x) = 1$, the Gram–Schmidt procedure yields the following sequence of orthogonal polynomials

$$\phi_0 = \frac{1}{\sqrt{2}}$$

$$\phi_1 = \sqrt{\frac{3}{2}} x$$

$$\phi_2 = \sqrt{\frac{5}{8}} \left(3x^2 - 1\right) \qquad (5.31)$$

$$\phi_3 = \sqrt{\frac{7}{8}} \left(5x^3 - 3x\right)$$

$$\phi_4 = \frac{3}{8\sqrt{2}} \left(35x^4 - 30x^2 + 3\right)$$

$$\cdots$$

which, apart of a multiplicative constant, are in a one-to-one relationship with Legendre polynomials (named after the French mathematician Adrien-Marie Legendre, 1752–1833)

$$\phi_n = \sqrt{\frac{2n+1}{2}} P_n(x) \tag{5.32}$$

The appearance of the constant $\sqrt{\frac{2n+1}{2}}$ is due to the fact that Legendre polynomials are not normalised since $\langle P_n, P_n \rangle = \|P_n\|^2 = \frac{2}{2n+1}$ [3]. An interesting property of $P_n(x)$ is the following: $P_n$ is orthogonal to any polynomial of order $m < n$. This can be shown as follows. Let $H_m$ be the space of all polynomials of degree $\leq m$, a basis for this space is $P_0, \ldots, P_m$ in fact each $P_j$ belong to $H_m$ since it is a polynomial and the sequence is mutually orthogonal. Then any element of $H_m$ can be written as $h = \sum_{k=0}^{m} a_k P_k$, and each $P_n$ with $n > m$ is orthogonal to $H_m$, in fact

$$\langle P_n, h \rangle = \sum_{k=0}^{m} a_k \langle P_n, P_k \rangle = 0 \tag{5.33}$$

Since any polynomial of order $m < n$ belongs to $H_m$ the thesis is proven.

Choosing the same domain and $w(x) = \frac{1}{\sqrt{1-x^2}}$ as the weighting function, the following polynomials are obtained

$$\phi_0 = \frac{1}{\sqrt{\pi}}$$

$$\phi_1 = \sqrt{\frac{2}{\pi}} x$$

$$\phi_2 = \sqrt{\frac{2}{\pi}} \left(2x^2 - 1\right) \tag{5.34}$$

$$\phi_3 = \sqrt{\frac{2}{\pi}} \left(4x^3 - 3x\right)$$

$$\phi_4 = \sqrt{\frac{2}{\pi}} \left(8x^4 - 8x^2 + 1\right)$$

$$\cdots$$

which are one-to-one related to the Chebyshev polynomials of the 1st kind (named after the Russian mathematicia Pafnuty Lvovich Chebyshev, 1821–1894) [3]

**Table 5.1** Orthogonal polynomials

| Name | Domain | $w(x)$ | Symbol |
|---|---|---|---|
| Legendre | $[-1, 1]$ | $1$ | $P_n(x)$ |
| Chebishev 1st kind | $[-1, 1]$ | $\frac{1}{\sqrt{1-x^2}}$ | $T_n(x)$ |
| Chebishev 2nd kind | $[-1, 1]$ | $\sqrt{1-x^2}$ | $U_n(x)$ |
| Laguerre | $[0, \infty)$ | $e^{-x}$ | $L_n(x)$ |
| *probabilists'* Hermite | $(-\infty, \infty)$ | $e^{-\frac{1}{2}x^2}$ | $He_n(x)$ |
| *physicists'* Hermite | $(-\infty, \infty)$ | $e^{-x^2}$ | $H_n(x)$ |

$$\phi_0 = \sqrt{\frac{1}{\pi}} T_0(x); \quad \phi_n = \sqrt{\frac{2}{\pi}} T_n(x) \tag{5.35}$$

Choosing the domain $[0, \infty)$ and $w(x) = e^{-x}$ the application of the Gram–Schmidt procedure yields

$$\begin{aligned}
\phi_0 &= 1 \\
\phi_1 &= x - 1 \\
\phi_2 &= x^2 - 4x + 2 \\
\phi_3 &= x^3 - 9x^2 + 18x - 6 \\
\phi_4 &= x^4 - 16x^3 + 72x^2 - 96x + 24
\end{aligned} \tag{5.36}$$

which are the Laguerre polynomials (named after the French mathematician Edmond Nicolas Laguerre, 1834–1886): $\phi_n = L_n(x)$ [3]. Table 5.1. reports the domain and the weighting function that can be used to obtain some well known orthogonal polynomials. A more exhaustive table can be found in [4].

It should be pointed out that often the commonly used sequences of orthogonal polynomials are not normalised but *standardised*. Standardisation, differently to normalisation, is based on the multiplication of each polynomial by a constant (that may depend on $n$) chosen by common practice or accepted conventions. For example, Legendre polynomials are standardised by setting $P_n(1) = 1$, the same happen with Chebishev polynomial of first kind ($T_n(1) = 1$) while for Chebishev polynomial of second kind $U_n(1) = n + 1$.

To conclude the section, it is easy to see that since the sequence $f_1, \ldots, f_n$ is linearly independent and it is a basis for the space of all linear combinations of $f_n$, also the orthonormal sequence $\phi_1, \ldots, \phi_n$ is a basis for the same space. In fact, the sequence $\phi_k$ is linearly independent and each vector $\phi_k$ can be written as

$$\phi_j = \frac{f_j - \sum_{k=1}^{j-1} \langle f_j, \phi_k \rangle \phi_k}{\left\| f_j - \sum_{k=1}^{n-1} \langle f_j, \phi_k \rangle \phi_k \right\|} \tag{5.37}$$

then each function $f_j$ is a linear combination of $\phi_k$

$$f_j = b_j \phi_j + \sum_{k=1}^{j-1} \langle f_j, \phi_k \rangle \phi_k \tag{5.38}$$

where $b_j = \left\| f_j - \sum_{k=1}^{n-1} \langle f_j, \phi_k \rangle \phi_k \right\|$. Each element of the space can be written as a linear combination of $f_j$ and then it can be written as a linear combination of $\phi_j$, then $\phi_1, \ldots, \phi_n$ is a basis for that space.

## 5.3  Generalised Fourier Series

The existence of the norm induced by the inner product allows the generalisation of the concept of *distance* between two functions. Given the functions $f$ and $g$ the norm of their difference $\| f - g \|$ is the equivalent of the *distance* between the two *points* in a space. For example, given three different points in an Euclidean space, say $A$, $B$ and $C$, the distance between two of them $\overline{AB}$ is never larger than the sum of the distances $\overline{AC}$ and $\overline{BC}$.

Given three functions $f$, $g$ and $h$ the *distances* among them are: $\| f - g \|$, $\| f - h \|$ and $\| g - h \|$ and the following inequality is always true

$$\| f - g \| \le \| f - h \| + \| g - h \| \tag{5.39}$$

in fact, since $f - g = (f - h) + (h - g)$ the inequality (5.39) is a simple consequence of the triangular inequality (see Sect. 5.1.2).

Another matching with a result of elementary geometry is the parallelogram equality (see Fig. 5.4). From elementary geometry we know that the sum of the squares of the lengths of the four sides of a parallelogram equals the sum of the squares of the lengths of the two diagonals: $A^2 + B^2 = 2a^2 + 2b^2$.

The equivalent identity (also called *parallelogram equality*) is the following: given any pair of functions $f$ and $g$

$$\| f + g \|^2 + \| f - g \|^2 = 2 \| f \|^2 + 2 \| g \|^2 \tag{5.40}$$

In fact

$$\| f + g \|^2 = \langle f + g, f + g \rangle = \| f \|^2 + \langle f, g \rangle + \langle g, f \rangle + \| g \|^2 \tag{5.41}$$
$$\| f - g \|^2 = \langle f - g, f - g \rangle = \| f \|^2 - \langle f, g \rangle - \langle g, f \rangle + \| g \|^2$$

and the sum of both sides of these two equations yields the wanted equality.

The concept of distance $\| \|$ can be used to give a geometrical interpretation to the following result. Consider again the orthonormal sequence $\phi_1, \ldots, \phi_n$ which is a basis for the n-dimensional sub-space $H_n$ (see Sect. 5.2), then take a function $f$ and search for an element $h$ of $H_n$ that minimises the distance $\| f - h \|$. The answer is the following: the element $h$ is the projection of $f$ on the sub-space $H_n$, i.e.

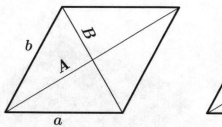
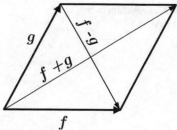

**Fig. 5.4** Parallelogram equality

$$h = \sum_{k=1}^{n} \langle f, \phi_k \rangle \, \phi_k \tag{5.42}$$

and it can be proven as follows. A general element of $H_n$ can be written as $h = \sum_{k=1}^{n} a_k \phi_k$, since $\phi_1, \ldots, \phi_n$ is a basis for $H_n$; now

$$\|f - h\|^2 = \langle f - h, f - h \rangle = \|f\|^2 - \langle f, h \rangle - \langle h, f \rangle + \|h\|^2 \tag{5.43}$$

and, from the properties of the inner product (see Sect. 5.1)

$$\|h\|^2 = \sum_{k=1}^{n} |a_k|^2; \quad \langle h, f \rangle = \sum_{k=1}^{n} a_k \overline{\langle f, \phi_k \rangle}; \quad \langle f, h \rangle = \sum_{k=1}^{n} \overline{a}_k \langle f, \phi_k \rangle \tag{5.44}$$

Substitution into Eq. (5.43) yields

$$\|f - h\|^2 = \|f\|^2 - \sum_{k=1}^{n} \overline{a}_k \langle f, \phi_k \rangle - \sum_{k=1}^{n} a_k \overline{\langle f, \phi_k \rangle} + \sum_{k=1}^{n} |a_k|^2 = \tag{5.45}$$

$$= \|f\|^2 + \sum_{k=1}^{n} |a_k - \langle f, \phi_k \rangle|^2 - \sum |\langle f, \phi_k \rangle|^2$$

where the last equality is obtained by observing that:

$$\sum_{k=1}^{n} |a_k - \langle f, \phi_k \rangle|^2 = \sum_{k=1}^{n} [a_k - \langle f, \phi_k \rangle] \left[ \overline{a}_k - \overline{\langle f, \phi_k \rangle} \right] =$$

$$= \sum_{k=1}^{n} |a_k|^2 - \sum_{k=1}^{n} a_k \overline{\langle f, \phi_k \rangle} - \sum_{k=1}^{n} \overline{a}_k \langle f, \phi_k \rangle + \sum_{k=1}^{n} |\langle f, \phi_k \rangle|^2$$

$$\tag{5.46}$$

**Fig. 5.5** Minimisation of the distance $AA\prime$

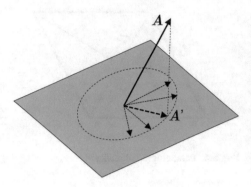

Since the coefficients $a_k$ in (5.45) have to be chosen to minimise $\| f - h \|$, the only choice is $a_k = \langle f, \phi_k \rangle$ that proves the statement, and as a consequence

$$\left\| f - \sum_{k=1}^{n} \langle f, \phi_k \rangle\, \phi_k \right\|^2 = \| f \|^2 - \sum |\langle f, \phi_k \rangle|^2 \qquad (5.47)$$

Since $h$ is the projection of $f$ on the sub-space $H_n$, the vector $f - h$ is orthogonal to $H_m$ (see Eq. 5.29). The analogy with geometry is clear: taking a plane and a point $A$ out of the plane, the projection $A'$ of the point onto the plane is the one that minimises the distance between $A$ and the plane, and the segment $AA'$ is orthogonal to the plane (see Fig. 5.5).

This fundamental result prompts also the following consideration: if we take the distance $\| f - h \|$ as a measure of the *closeness* between $f$ and $h$, we can say that $h = \sum_{k=1}^{n} \langle f, \phi_k \rangle\, \phi_k$ is the element of $H_n$ *closer* to $f$, in other words $h$ is the best approximation of $f$ built with elements of $H_n$.

The use of linear combination of orthonormal sequences to approximate a function is then fully justified by the previous result, which anyway raises a new problem. Consider the orthonormal sequence $\phi_1, \ldots, \phi_n, \phi_{n+1}$ built by adding a new element $\phi_{n+1}$, orthogonal to all the other $n$ elements. A new sub-space $H_{n+1}$ is spanned by all the linear combinations of this sequence, and it is straightforward to see that this new sub-space contains $H_n$, in the sense that each element of $H_n$ belongs to $H_{n+1}$, but not vice versa. Given again the same function $f$ as before, the approximation that can be built choosing an element of $H_{n+1}$ is at least not worse than the previous one. In fact $h = \sum_{k=1}^{n} \langle f, \phi_k \rangle\, \phi_k$ is an element of $H_{n+1}$, but it may be different from the projection $h'$ of $f$ over $H_{n+1}$, which we already know that minimises the distance $\| f - h' \|$. The interesting fact is that the new approximation: $h' = \sum_{k=1}^{n+1} \langle f, \phi_k \rangle\, \phi_k$ can be built as

$$h' = \sum_{k=1}^{n+1} \langle f, \phi_k \rangle\, \phi_k = h + \langle f, \phi_{n+1} \rangle\, \phi_{n+1} \qquad (5.48)$$

i.e. it is necessary to find just the $n + 1 - th$ coefficient $\langle f, \phi_{n+1} \rangle$ to improve the approximation, since all the others are exactly the same as before. This result is sometimes expressed by saying that the coefficients of the approximation satisfy the *condition of finality*. It can be shown (see [1] for more details) that this condition is satisfied if and only if the sequence $\phi_j$ is orthogonal (not necessarily orthonormal), while it is not true if a general linearly independent sequence is used.

As an example, take $f$ as the fourth degree polynomial $f = 3 + 2x + 5x^2 + 7x^3 + x^4$ in the domain $[-1, 1]$ and search an approximation on the space $H_2$ spanned by the basis $1, x, x^2$, refer to Fig. 5.6. $H_2$ is the space of all the polynomials of degree $m \le 2$. The sequence $1, x, x^2$ is a basis in $H_2$, but is not orthonormal. The sequence $\phi_0, \phi_1, \phi_2$, where

$$\phi_n = \sqrt{\frac{2n+1}{2}} P_n(x) \tag{5.49}$$

and $P_n(x)$ is a Legendre polynomial, is an orthonormal basis on $H_2$ (see section Eqs. 5.31). The *best* approximation of $f$ is

$$h^{(2)} = \sum_{k=0}^{2} \langle f, \phi_k \rangle \phi_k = \frac{73\sqrt{2}}{15} \phi_0 + \frac{31}{5}\sqrt{\frac{2}{3}}\phi_1 + \frac{82}{21}\sqrt{\frac{2}{5}}\phi_2 \tag{5.50}$$

that can be written as a polynomial (using Eqs. 5.31 and 5.49), i.e. using the basis $1, x, x^2$, as

$$h^{(2)} = \frac{102}{35} + \frac{31}{5}x + \frac{41}{7}x^2 \tag{5.51}$$

Let now take the space $H_3$ of all the polynomials of degree $m \le 3$ and search for a better approximation of $f$. We know that this is given by

$$h^{(3)} = \sum_{k=0}^{3} \langle f, \phi_k \rangle \phi_k = \frac{73\sqrt{2}}{15} \phi_0 + \frac{31}{5}\sqrt{\frac{2}{3}}\phi_1 + \frac{82}{21}\sqrt{\frac{2}{5}}\phi_2 + \frac{2}{5}\sqrt{14}\phi_3 \tag{5.52}$$

where the first three coefficients are exactly the same as before, since, using an orthonormal basis, they satisfy the *condition of finality*. It can be noticed that also using as a basis $\phi_j = P_j(x)$, which is orthogonal but not orthonormal, the same results are obtained. But when the same function is written using the basis $1, x, x^2, x^3$, i.e. as a polynomial

$$h^{(3)} = \frac{102}{35} + 2x + \frac{41}{7}x^2 + 7x^3 \tag{5.53}$$

it can be seen that the coefficient of the linear term has changed, and this is because the basis $1, x, x^2, x^3$ is not an orthogonal basis and the condition of finality is not satisfied.

This property is the reason why the use of orthonormal sequences to approximate a function should be preferred, since it is clearly more efficient, also from a computational viewpoint. It is also clear that increasing the dimension of $H_n$ does not worsen the approximation of a function $f$ by the sum $\sum_{k=1}^{n} \langle f, \phi_k \rangle \phi_k$, thus

**Fig. 5.6** Approximation of a
$4^{th}$ degree polynomial

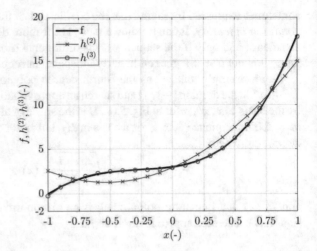

suggesting that wider bases would improve the result. The use of infinite sequences
is then the obvious next step, but this must be considered with some caution.

Given an infinite sequence of orthonormal functions belonging to a given space
(as for example the normalised Legendre polynomials, Eq. 5.49), it is in general
not true that any linear combination of all these elements yield an element of the
same space, since it may not even yield a function. When dealing with infinite series
like $\sum_{k=1}^{\infty} a_k \phi_k = f$ it is necessary to consider their convergence. The definition
of convergence that is needed here is based on the following consideration. If we
take a sequence of functions $s_n$ such that the distance $\|s - s_n\|$ from a function $s$
approaches zero when $n$ approaches infinity, we say that the sequence $s_n$ converges
*in the mean* to $s$ (which should not be confused with pointwise convergence). Then,
considering the sequence of partial sums $\sum_{k=1}^{n} a_k \phi_k = s_n$ the expression

$$f = \sum_{k=1}^{\infty} a_k \phi_k \tag{5.54}$$

is meaningful if the sequence $s_n$ converges *in the mean* to $f$, i.e. if

$$\lim_{n \to \infty} \left\| f - \sum_{k=1}^{n} a_k \phi_k \right\| = 0 \tag{5.55}$$

An example of such a case is given by the Fourier series. When dealing with
series of functions, *completeness* of the functional space is an important issue. The
convergence of a sequence of functions $f_n$ to a function $f$ does not mean that the
function $f$ belongs to the same space of $f_n$. A classical example is the following:
the sequence of functions

**Fig. 5.7** Convergence in the mean of series of functions

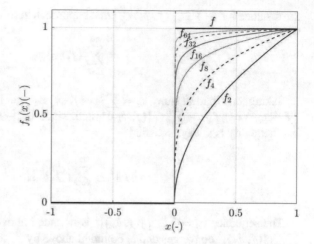

$$f_n = \begin{cases} 0 & \text{if } x \leq 0 \\ x^{1/n} & \text{if } x > 0 \end{cases} \tag{5.56}$$

all belonging to $C\left([-1, 1]\right)$ converges in the mean to the function (see Fig. 5.7)

$$f_n = \begin{cases} 0 \text{ if } x \leq 0 \\ 1 \text{ if } x > 0 \end{cases} \tag{5.57}$$

which does not belong to $C\left([-1, 1]\right)$. The space $C\left([-1, 1]\right)$ is not complete. However, the space $L^2\left([a, b]\right)$ contains all the functions of $C\left([a, b]\right)$ (i.e. continuous) and also those that are limits of converging sequences of continuous functions, thus $L^2\left([a, b]\right)$ in this sense is complete.

In the following we will assume that we can always find an orthonormal sequence $\phi_1, \phi_2, \ldots$ such that for every function $f$ there exists a series $\sum_{k=1}^{\infty} a_k \phi_k$ that converges in the mean to $f$, and we will call $\phi_1, \phi_2, \ldots$ an *orthonormal basis*. A little digression is needed at this point. The term *basis* for an orthonormal sequence is correctly used for a finite dimension space (like the $H_m$ sub-spaces encountered above), while for an infinite dimension space the term *basis* is used in functional analysis for something more general (see for example [5]). There is no agreement on how to call the above defined sequence $\phi_1, \phi_2, \ldots$, for example [1] uses the term *approximating basis*, while [5] uses *total orthonormal sequence* and others (like [6]) use *complete orthonormal sequence* (and sometime *closed* is used with the meaning of complete). Also the existence of *basis* for general infinite dimensions spaces is a delicate matter and we will not go further in details on these important questions; the reader that is interested should refer to classical books on functional analysis like [5, 7]. In this book we will use the term *orthonormal basis* as above defined.

When such sequence exists, it can be shown that $a_k = \langle f, \phi_k \rangle$. In fact, the convergence in mean assures that $\left\| f - \sum_{k=1}^{n} a_k \phi_k \right\|$ approaches to zero when $n$ goes to infinity, and, since $\left\| f - \sum_{k=1}^{n} a_k \phi_k \right\|$ is minimised by choosing $a_k = \langle f, \phi_k \rangle$, the distance $\left\| f - \sum_{k=1}^{n} \langle f, \phi_k \rangle \phi_k \right\|$ must approach zero when $n \to \infty$, then we can write

$$f = \sum_{k=1}^{\infty} \langle f, \phi_k \rangle \phi_k \qquad (5.58)$$

Taking any partial sum $s_n = \sum_{k=1}^{n} \langle f, \phi_k \rangle \phi_k$ we have seen that (Eq. 5.47): $\| f - s_n \|^2 = \| f \|^2 - \sum_{k=1}^{n} |\langle f, \phi_k \rangle|^2$ and, since $\| f - s_n \|^2 \geq 0$, then the following inequality is always satisfied

$$\| f \|^2 \geq \sum_{k=1}^{n} |\langle f, \phi_k \rangle|^2 \qquad (5.59)$$

The sequence $c_n = \sum_{k=1}^{n} |\langle f, \phi_k \rangle|^2$ is bounded above by $\| f \|^2$, then the series $\sum_{k=1}^{\infty} |\langle f, \phi_k \rangle|^2$ converges and is bounded above by $f$, thus

$$\| f \|^2 \geq \sum_{k=1}^{\infty} |\langle f, \phi_k \rangle|^2 \qquad (5.60)$$

which is knows as *Bessel's inequality* (named after the German mathematician and astronomer Friedrich Wilhelm Bessel, 1784–1846). To notice that the convergence of $\sum_{k=1}^{\infty} |\langle f, \phi_k \rangle|^2$ also implies that

$$\lim_{n \to \infty} \langle f, \phi_k \rangle = 0 \qquad (5.61)$$

For the present case we assume that the series given by Eq. (5.58) converges in the mean to $f$, the LHS of Eq. (5.47) approaches zero when $n \to \infty$, and so does the RHS, then Bessel's inequality reduces to the *Parseval's equality* (named after the French mathematician Marc-Antoine Parseval des Chênes, 1755–1836)

$$\| f \|^2 = \sum_{k=1}^{\infty} |\langle f, \phi_k \rangle|^2 \qquad (5.62)$$

To summarise, under the hypotheses above expressed, given an orthonormal basis $\phi_1, \phi_2, \ldots$ and a function $f \in L^2([a, b])$ it is always possible to write $f$ as a finite or infinite (series) linear combination of $\phi_j$, where the coefficients are $\langle f, \phi_k \rangle$. Moreover, in the case of $\langle f, \phi_k \rangle = 0$ for every $k$, the function $f$ is the nil element. It must be added that the nil element consists of a function that is equal zero *almost everywhere*, i.e. a function that is nil everywhere except on a countable set of points.

We conclude this section with some examples of series of orthogonal functions.

The most famous is the Fourier series. In this case, given any finite interval $[a, b]$ every function $f \in L^2([a, b])$ can be written as a series

$$f = \sum_{n=-\infty}^{\infty} a_n \phi_n \tag{5.63}$$

where the orthonormal basis is

$$\phi_n = \frac{1}{\sqrt{P}} e^{i \frac{2\pi n}{P} x}; \quad P = b - a \tag{5.64}$$

which is more often written as

$$f = \phi_0 + \sum_{n=1}^{\infty} a_n \phi_n^{(s)} + \sum_{n=1}^{\infty} b_n \phi_n^{(s)} \tag{5.65}$$

where the orthonormal basis is

$$\phi_0 = 0; \quad \phi_n^{(c)} = \sqrt{\frac{2}{P}} \cos\left(\frac{2\pi n}{P} x\right); \quad \phi_n^{(s)} = \sqrt{\frac{2}{P}} \sin\left(\frac{2\pi n}{P} x\right) \tag{5.66}$$

However, the obvious consequences of the discussion reported in this section is that there can be an infinity of such series, depending on the domain $[a, b]$ and the orthonormal sequence that is chosen.

Choosing the domain $[-1, 1]$ (but any other finite interval would work, since it can be reduced to this one by a simple linear transformation of $x$) and the orthonormal basis

$$\phi_j = \sqrt{\frac{2n + 1}{2}} P_n(x) \tag{5.67}$$

where $P_n(x)$ are the Legendre polynomials, allows to write any function $f$ as

$$f = \sum_{n=0}^{\infty} a_n \phi_n(x) \tag{5.68}$$

where $a_n = \langle f, \phi_n \rangle$, that is called Fourier-Legendre series. It can be noticed that the use of orthogonal (but not orthonormal) basis will work as well, since we can always multiply the elements of the sequence $\phi_n$ by any scalar without changing their reciprocal orthogonality, for example we can write

$$f = \sum_{n=0}^{\infty} a_n \phi_n(x) = \sum_{n=0}^{\infty} \sqrt{\frac{2n + 1}{2}} a_n \frac{\phi_n(x)}{\sqrt{\frac{2n+1}{2}}} = \sum_{n=0}^{\infty} c_n P_n(x) \tag{5.69}$$

where $c_n = \sqrt{\frac{2n+1}{2}} a_n$. In fact, the only change is the way the coefficients are calculated

$$c_n = \frac{\langle f, P_n \rangle}{\langle P_n, P_n \rangle} = \sqrt{\frac{2n+1}{2}} \frac{\langle f, \phi_n \rangle}{\langle \phi_n, \phi_n \rangle} = \sqrt{\frac{2n+1}{2}} a_n \qquad (5.70)$$

There is no need to have an orthonormal basis to expand a function in a series like the one given by Eq. (5.68), but it is sufficient that the basis is orthogonal (i.e. all the functions $\phi_j$ must be mutually orthogonal).

It must always be reminded that the inner product $\langle \ , \ \rangle$ is defined by choosing the domain and the weighting function $w(x)$. Consider the domain $[0, 1]$ and the Bessel functions of the first kind and order $p \geq 0$, $J_p(x)$. Defining $\alpha_1, \alpha_2, \dots$ as the positive roots of $J_p(x)$ (i.e. $J_p(\alpha_k) = 0$ for any $k$) the following sequence of orthogonal functions can be defined

$$\phi_k = J_p(\alpha_k x) \qquad (5.71)$$

where the orthogonality is defined by taking $w(x) = x$. The orthogonality of the sequence (5.71) is proven by the properties of $J_p(\alpha_k x)$, see [4]). It is then possible to expand a function on $[0, 1]$ as

$$f = \sum_{k=1}^{\infty} a_k J_p(\alpha_k x) \qquad (5.72)$$

where the coefficients $a_k$ are found as

$$a_k = \frac{\langle f, J_p(\alpha_k x) \rangle}{\| J_p(\alpha_k x) \|^2} = \frac{\int_0^1 f \, J_p(\alpha_k x) \, x \, dx}{\int_0^1 |J_p(\alpha_k x)|^2 \, x \, dx} = \frac{2 \int_0^1 f \, J_p(\alpha_k x) \, x \, dx}{|J_{p+1}(\alpha_k)|^2} \qquad (5.73)$$

and the last equality holds due to the property of $J_p(x)$: $\| J_p(\alpha_k x) \|^2 = \frac{1}{2} |J_{p+1}(\alpha_k)|^2$.

All the series $\sum_{k=1}^{\infty} a_k \phi_k$ share with the Fourier series the most important properties and that is why they are all called *generalised Fourier series*.

## 5.4    Elements of Sturm–Liouville Theory

The general Sturm–Liouville theory in $L^2$, that will be briefly sketched below, shows that for a linear second order differential operator that satisfy some conditions (self-adjointness) the solutions can be written as an orthogonal sequence of function that spans the space $L^2$, and this is the reason for the two previous sections. A rather complete treatment of SL problems can be found in [8, 9].

A linear second order ODE has the general form

$$a_2(x) \frac{d^2 y}{dx^2} + a_1(x) \frac{dy}{dx} + a_0(x) y = f(x) \qquad (5.74)$$

where $a_j(x)$ and $f(x)$ are complex functions on the domain $[a, b]$ (which may also be infinite); when $f(x) \equiv 0$ the equation is said to be *homogeneous*. A compact way to write this equation is by introducing the linear differential operator

$$\mathcal{L} = a_2(x)\frac{d^2}{dx^2} + a_1(x)\frac{d}{dx} + a_0(x) \tag{5.75}$$

then Eq. (5.74) can be written as: $\mathcal{L}y = f$. Linearity of $\mathcal{L}$ means that $\mathcal{L}(a_1 y_1 + a_2 y_2) = a_1\mathcal{L}y_1 + a_2\mathcal{L}y_2$, and this is also a restatement of the *superposition principle* that holds for linear homogeneous differential equations: a linear combination of two solutions is still a solution of the homogeneous linear ODE. Equation (5.74) with $f(x) \equiv 0$ (homogeneous equation) has two independent solutions $y_1(x)$ and $y_2(x)$, and their linear combination is the general solution. If $y_0(x)$ is any particular solution of the non-homogeneous ($f(x) \neq 0$) equation the general solution of (5.74) is $a_1 y_1(x) + a_2 y_2(x) + y_0(x)$. Equation (5.74) can be written in a more convenient form, if $a_2(x)$ does not vanish at any point of the domain, i.e.

$$\frac{d^2 y}{dx^2} + q(x)\frac{dy}{dx} + r(x)y = h(x) \tag{5.76}$$

where $q(x) = a_1(x)/a_2(x)$, $r(x) = a_0(x)/a_2(x)$, $h(x) = f(x)/a_2(x)$. If there is a point $s_0 \in [a, b]$ where $a_2(s_0) = 0$, the equation is said to be *singular* ($s_0$ is the singular point) otherwise is said to be *regular*. The existence and uniqueness theorem for linear ODE states that (see [10, 11]) given the *initial* conditions

$$y(x_0) = c_0; \quad \left(\frac{dy}{dx}\right)_{x=x_0} = c_1 \tag{5.77}$$

where $x_0 \in [a, b]$, if $q$, $r$, and $h$, are continuous in $[a, b]$, then there is a unique solution of Eq. (5.76) that satisfies Eqs. (5.77). More often one has to face a different problem: find a solution that satisfies conditions on both boundaries (a boundary value problem). The quite general form of linear boundary conditions (B.C.) is the following

$$\alpha_0 y(a) + \alpha_1 \left(\frac{dy}{dx}\right)_{x=a} + \alpha_2 y(b) + \alpha_3 \left(\frac{dy}{dx}\right)_{x=b} = c_0 \tag{5.78}$$

$$\beta_0 y(b) + \beta_1 \left(\frac{dy}{dx}\right)_{x=b} + \beta_2 y(a) + \beta_3 \left(\frac{dy}{dx}\right)_{x=a} = c_1$$

where not all $\alpha_j$ or $\beta_j$ are nil, while $c_0$ and $c_1$ can be nil (in such case the B.C. are said to be *homogeneous*). However the most common form of B.C. are those where $\alpha_2 = \alpha_3 = \beta_2 = \beta_3 = 0$, i.e.

$$\alpha_0 y(a) + \alpha_1 \left(\frac{dy}{dx}\right)_{x=a} = c_0; \quad \beta_0 y(b) + \beta_1 \left(\frac{dy}{dx}\right)_{x=b} = c_1 \tag{5.79}$$

that are said to be *separated*, while sometimes the following form is found

$$y\,(a) = y\,(b)\,;\qquad \left(\frac{dy}{dx}\right)_{x=a} = \left(\frac{dy}{dx}\right)_{x=b} \tag{5.80}$$

then B.C. are not separated and they are said to be *periodic*. This last form may be connected to the so-called *degeneracy* of eigensolutions (see below). The Wronskian determinant (named after the Polish scientist Józef Maria Hoene-Wroński, 1776–1853)

$$W\,(x) = \begin{vmatrix} y_1\,(x) & y_2\,(x) \\ \frac{dy_1(x)}{dx} & \frac{dy_2(x)}{dx} \end{vmatrix} \tag{5.81}$$

where $y_1$ and $y_2$ are two solutions of the homogeneous ODE (5.76), is an important function since the two solutions are independent if and only if $W\,(x) \neq 0$ on the domain [10, 11].

In some applications, the points where the solutions of

$$\frac{d^2y}{dx^2} + q\,(x)\,\frac{dy}{dx} + r\,(x)\,y = 0 \tag{5.82}$$

are nil, which are called *zeroes* of the solutions, assume a certain importance. The characteristics of those points can be deduced by the properties of the functions $q\,(x)$ and $r\,(x)$, and one of the most striking is known as *Sturm separation theorem*: two linearly independent solutions of Eq. (5.82) have distinct zeroes and the two sequences of zeroes alternate (see [8] for a simple proof).

As an example, when $q = 0$ and $r = 1$, Eq. (5.82) becomes

$$\frac{d^2y}{dx^2} + y = 0 \tag{5.83}$$

and the two linearly independent solutions are

$$y_1 = \sin\,(x)\,;\quad y_2 = \cos\,(x) \tag{5.84}$$

which have zeroes on $x_n = n\pi$ and on $x_m = \frac{\pi}{2} + m\pi$, where $m$ and $n$ are integer, and the two sequences are alternated.

It can also be shown that the numerosity of the zeroes depends on the functions $g$ and $r$. Precisely, defining the function $s\,(x) = r\,(x) - \frac{1}{4}q^2\,(x) - \frac{1}{2}q'\,(x)$, the following cases can be found (see [8, 9]):

(a) if $s\,(x) > k^2 > 0$ the solution is *oscillatory*, i.e. it has an infinite number of zeroes distributed between the zeroes of the solution of the equation $y'' + k^2y = 0$.

(b) if $s\,(x) \leq 0$ on the domain, the solution has at most one zero in $[a, b]$.

Equation (5.83) is an example of case (a) while as an example of case (b) consider the equation

$$\frac{d^2y}{dx^2} + \frac{dy}{dx} - \frac{3}{4}y = 0 \tag{5.85}$$

where $r = -\frac{3}{4}, q = 1$ then: $s(x) = -1 \le 0$ and the solution is

$$y = c_1 e^{\frac{1}{2}x} + c_2 e^{-\frac{3}{2}x} \tag{5.86}$$

which may have at most one zero, depending to the values of $c_j$.

### 5.4.1  Eigenvalue Problems

We have seen that separation of PDE may lead to ODE of the form

$$\mathcal{L}y = \lambda y \tag{5.87}$$

As an example, separation of Helmholtz equation $\nabla^2 \Psi + K \Psi = 0$ in spherical coordinates yielded equation (3.37), which can be written as

$$x^2 y'' + 2xy' + K^2 x^2 y - \alpha_2 y = 0 \tag{5.88}$$

i.e. like Eq. (5.87) with $a_2(x) = x^2$; $a_1(x) = 2x$; $a_0(x) = K^2 x^2$ and $\lambda = \alpha_2$, where for sake of simplicity the first and second derivatives of a function $y$ are written using the compact notation: $y'$ and $y''$, respectively, and the same notation will be used below.

Equation (5.87) is an *eigenvalue problem* for the operator $\mathcal{L}$, which means that, apart of the trivial solution $y = 0$ (satisfied for any value of $\lambda$), the equation may be satisfied for certain values of the constant $\lambda$, which are then called *eigenvalues* of $\mathcal{L}$, while any function non identically nil over the given domain that satisfies the equation for that values of $\lambda$ is called *eigenfunction* of $\mathcal{L}$. It will be seen that this problem shares some properties with the well known problem of linear algebra of finding vectors $u$ that satisfy the algebraic equation

$$\mathbf{A}u = \lambda u \tag{5.89}$$

where $\mathbf{A}$ is a square matrix; in that environment, the vectors $u$ are called *eigenvectors* and the corresponding values of $\lambda$ are the *eigenvalues*. It is a result of linear algebra that if $\mathbf{A}$ is a self-adjoint matrix (i.e. $\mathbf{A} = \overline{\mathbf{A}}^T$, where $\overline{\mathbf{A}}$ is the conjugate of $\mathbf{A}$ and $\mathbf{A}^T$ is the transpose of $\mathbf{A}$) then the eigenvalues of $\mathbf{A}$ are real numbers and the eigenvector corresponding to different eigenvalues are orthogonal and form a basis of the vector space.

A strikingly similar result applies to the problem (5.87). It should be remarked that below, the space of the solution is not $L^2([a, b])$ but its intersection with the space of the continuous double differentiable functions, $C^2$ ($[a, b]$), since each term of Eq. (5.74) must be continuous.

The adjoint of the operator $\mathcal{L}$ (which we will call $\hat{\mathcal{L}}$) is defined by the equation

$$\langle \mathcal{L}f, g \rangle = \left\langle f, \hat{\mathcal{L}}g \right\rangle \tag{5.90}$$

where $\langle \, , \, \rangle$ is the inner product on $L^2$ ($[a, b]$). The *formal adjoint* of the operator $\mathcal{L}$ is defined as

$$\mathcal{L}^*g = (\overline{a_2}g)'' - (\overline{a_1}g)' + \overline{a_0}g \tag{5.91}$$

and it should be pointed out that the adjective *formal* is needed since $\mathcal{L}^*$ may be different form $\hat{\mathcal{L}}$. In fact, it can be shown using integration by part, that

$$\langle \mathcal{L}f, g \rangle = \langle f, \mathcal{L}^*g \rangle + \left[ (a_1 - a_2') f\overline{g} - a_2 \left( f'\overline{g} - f\overline{g}' \right) \right]_a^b \tag{5.92}$$

The operator $\mathcal{L}$ is said to be *formally self-adjoint* when $\mathcal{L} = \mathcal{L}^*$, and this may happen only when

$$a_2 g'' + a_1 g' + a_0 = (\overline{a_2}g)'' - (\overline{a_1}g)' + \overline{a_0}g \tag{5.93}$$

i.e. when

$$\overline{a_2} = a_2 \tag{5.94}$$
$$2\overline{a_2}' - \overline{a_1} = a_1$$
$$\overline{a_2}'' + \overline{a_0} - \overline{a_1}' = a_0$$

The first equation says that $a_2$ is real, and if $a_2' = a_1$ the second one is satisfied and $a_1$ is real, thus the third one implies that $a_0$ is real too ($a_2'' = a_1'$ is a consequence of $a_2' = a_1$). A *formally self adjoint* operator $\mathcal{L}$ has then the form

$$\mathcal{L} = \frac{d}{dx} \left( a_2 \frac{d}{dx} \right) + a_0 \tag{5.95}$$

where $a_2$ and $a_1$ are real, and $\mathcal{L}$ is also *self-adjoint* when

$$\left[ a_2 \left( f'\overline{g} - f\overline{g}' \right) \right]_a^b = 0 \tag{5.96}$$

for any pair of function $f$ and $g$. Self-adjointness of $\mathcal{L}$ is a quite important property since for this case the solutions achieve quite interesting properties. However, It may appear that the form (5.95) may be too specific with respect to the more general Eq. (5.75) to be really useful. Nevertheless, it is easy to see that an operator of the form of Eq. (5.75) (with $a_j$ real) can be put into the form of Eq. (5.95) by a simple transformation. In fact, supposing that $a_2$ does not vanish on $[a, b]$, define

$$w(x) = \frac{c}{a_2(x)} e^{\int_a^x \frac{a_1(x)}{a_2(x)} dx} \tag{5.97}$$

where $c$ is a constant, then the operator $\mathcal{L}_w = w\mathcal{L} = wa_2 \frac{d^2}{dx^2} + wa_1 \frac{d}{dx} + wa_0$ is formally self-adjoint when

$$(wa_2)' = wa_1 \tag{5.98}$$

which is identically satisfied by Eq. (5.97). Then when we have an equation like Eq. (5.87) with $\mathcal{L}$ given by Eq. (5.75), i.e.

$$a_2(x) \frac{d^2 y}{dx^2} + a_1(x) \frac{dy}{dx} + a_0(x) y = \lambda y \tag{5.99}$$

we can reduce $\mathcal{L}$ to the form of Eq. (5.95) when $a_2' = a_1$ while in the other cases we can multiply both sides by $w(x)$ given by Eq. (5.97) to obtain

$$\frac{d}{dx}\left(wa_2(x) \frac{dy}{dx}\right) + wa_0(x) y = \lambda wy \tag{5.100}$$

where the differential operator has the form of Eq. (5.95) and we can write

$$\mathcal{L}_w y = \lambda w(x) \, y \tag{5.101}$$

This last form is then the most general one, since each 2nd order ODE with real coefficients can be reduced to it. Suppose now that we have an ODE of the form (5.101) where $\mathcal{L}_w$ has been reduced to the form (5.95) ($\mathcal{L}_w$ is formally self-adjoint), the following fundamental result holds: when Eq. (5.96) holds, the eigenvalues of (5.101) are all real and any pair of eigenfunctions associated with different values of eigenvalues are orthogonal, where the inner product is defined as

$$\langle f, g \rangle_w = \int_a^b f(x)\, \overline{g}(x)\, w(x)\, dx \tag{5.102}$$

The proof is as follows:
(a) suppose $\lambda_1$ is an eigenvalue, then a function $f_1$ exists such that

$$\mathcal{L}_w f_1 = \lambda_1 w(x) \, f_1 \tag{5.103}$$

and, since $w(x)$ is a real function and defining $\langle f, g \rangle_1 = \int_a^b f(x)\, \overline{g}(x)\, dx$

$$\lambda_1 \|f_1\|_w^2 = \langle \lambda_1 f_1, f_1 \rangle_w = \langle \lambda_1 w f_1, f_1 \rangle_1 = \langle \mathcal{L}_w f_1, f_1 \rangle_1 =$$
$$= \langle f_1, \mathcal{L}_w f_1 \rangle_1 = \langle f_1, \lambda_1 w \, f_1 \rangle_1 = \overline{\lambda}_1 \langle f_1, w \, f_1 \rangle_1 = \overline{\lambda}_1 \|f_1\|_w^2 \tag{5.104}$$

which shows that $\lambda_1$ is real;

(b) take another eigenvalue $\lambda_2$ (which now we know it is real) and the associate eigenfunction $f_2$

$$(\lambda_1 - \lambda_2)\langle f_1, f_2 \rangle_w = \langle \lambda_1 f_1, f_2 \rangle_w - \langle f_1, \lambda_2 f_2 \rangle_w = \tag{5.105}$$
$$= \langle \mathcal{L}_w f_1, f_2 \rangle_1 - \langle f_1, \mathcal{L}_w f_2 \rangle_1 = \langle \mathcal{L}_w f_1, f_2 \rangle_1 - \langle \mathcal{L}_w f_1, f_2 \rangle_1 = 0 \tag{5.106}$$

and since $\lambda_1 \neq \lambda_2$ then $\langle f_1, f_2 \rangle_w = 0$, i.e. $f_1$ and $f_2$ are orthogonal. To notice that when $w(x) \neq 1$, the space of the solution is $L_w^2([a,b])$, where the subscript $w$ is used to indicate the weighting function when the square integrability is checked.

Let now turn to the main point of this section: the Sturm–Liouville problem.

### 5.4.2  The Sturm–Liouville System

Consider the following quite general form of a second order ODE

$$\frac{d}{dx}\left(a_2(x)\frac{dy}{dx}\right) + a_0 y - \lambda w(x) y = 0 \tag{5.107}$$

where the coefficients are real and, without loss of generality, $a_2(x)$ can be taken positive. Equation (5.107) shall be solved with the following boundary conditions

$$\alpha_1 y(a) + \alpha_2 y'(a) = 0 \tag{5.108a}$$
$$\beta_1 y(b) + \beta_2 y'(b) = 0 \tag{5.108b}$$

with $\alpha_j$, $\beta_j$ are real constants. This is called a Sturm–Liouville (SL) problem.

Due to the given boundary conditions (5.108), the following equation always holds for any couple of solutions $f$ and $g$

$$\left[a_2\left(f'\bar{g} - f\bar{g}'\right)\right]_a^b = 0 \tag{5.109}$$

In fact, taken two solutions $y_1$ and $y_2$, from Eqs. (5.108) it stems that

$$y_1(a) y_2'(a) - y_2(a) y_1'(a) = 0 \tag{5.110}$$
$$y_1(b) y_2'(b) - y_2(b) y_1'(b) = 0 \tag{5.111}$$

and Eq. (5.109) is satisfied, thus the operator $\mathcal{L} = \frac{d}{dx}\left(a_2(x)\frac{dy}{dx}\right) + a_0 y$ is self-adjoint. From the previous results we know that if eigenvalues exist they are real and the eigenfunctions are orthogonal with respect to the inner product $\langle\,,\,\rangle_w$.

There are two different types of SL problems: when the domain is bounded (i.e. both $a$ and $b$ are finite) and the function $a_2$ is non-nil on $[a, b]$ the problem is said *regular*, otherwise is said to be *singular*.

The general theory on Sturm–Liouville problems assures (see for example [8, 9]) that for a *regular* SL problem, the solutions of Eq. (5.107) exist for a countable number of $\lambda$ values and they form a *basis* for $L_w^2([a, b])$. Precisely, the SL problem has an infinite sequence of eigenvalues $\lambda_0, \lambda_1, \ldots$; to each eigenvalue $\lambda_j$ it corresponds a single eigenfunction $\phi_j$, and $\phi_0, \phi_1, \ldots$ are mutually orthogonal. Then, given a function $f \in L_w^2([a, b])$, it can be expanded in a unique way as $f = \sum_{k=0}^{\infty} c_k \phi_k$.

**Example: The Harmonic Oscillator Equation**
The harmonic oscillator equation has the for

$$y'' - \lambda y = 0 \tag{5.112}$$

and the general solution if $\lambda \neq 0$ is

$$y = c_1 e^{\sqrt{\lambda} x} + c_2 e^{-\sqrt{\lambda} x} \tag{5.113}$$

Consider now the following homogeneous boundary value problem on $[a, b] = [0, 1]$

$$y(0) = 0; \quad y(1) = 0 \tag{5.114}$$

The operator $L = \frac{d^2}{dx^2}$ is formally self-adjoint since it is a particular case of the general form given by Eq. (5.107)

$$L = \frac{d}{dx}\left(a_2(x)\frac{dy}{dx}\right) + a_0 y \tag{5.115}$$

where $a_2(x) = 1$ and $a_0(x) = 0$. The B.C. are a particular case of Eq. (5.108a) where $\alpha_2 = \beta_2 = 0$, then Eq. (5.109) is satisfied and since $a_2(x)$ is not vanishing on the domain, this is a regular SL problem.

The B.C. (5.114) are satisfied when

$$c_1 = -c_2 \tag{5.116}$$

$$c_1\left(e^{\sqrt{\lambda}} - e^{-\sqrt{\lambda}}\right) = 0 \tag{5.117}$$

The case $c_1 = 0$ yields the trivial solution ($y \equiv 0$), while for $c_1 = -c_2 \neq 0$ the B.C. are satisfied for

$$e^{2\sqrt{\lambda}} = 1 \tag{5.118}$$

that, when $\lambda \neq 0$, yields an infinity of values for $\lambda$

$$\lambda_n = -n^2 \pi^2 \tag{5.119}$$

for $n = 0, \pm 1, \pm 2$ and the solution, apart for a multiplying constant, is

$$y = c_1 \left( e^{in\pi x} - e^{-in\pi x} \right) = \sin(n\pi x) \tag{5.120}$$

The values $\lambda_n = -n^2\pi^2$ are the eigenvalues, and $f_n = \sin(n\pi x)$ are the corresponding eigenfunctions. It is worth to notice that the values $n$ and $-n$ would yield the same eigenvalue, then only $n = 0, 1, 2 \ldots$ can be used to distinguish eigenvalues and eigenfunctions. The functions $f_n$ form an orthogonal sequence

$$\langle f_n, f_m \rangle = 0 \text{ if } n \neq m \tag{5.121}$$

in fact

$$\langle f_n, f_m \rangle = \int_0^1 \sin(n\pi x) \sin(n\pi x) = \begin{cases} 0 \text{ if } m \neq n \\ \frac{1}{2} \text{ if } m = n \end{cases} \tag{5.122}$$

The general results about a regular SL problem assures that the sequence $f_0, f_1, \ldots$ (or the normalised one $\phi_j = \sqrt{2}f_j$) is a basis in $L^2([0,1])$ and then any function $g \in L^2([0,1])$ can be written as

$$g = \sum_{k=0}^{\infty} c_k f_k = \sum_{k=0}^{\infty} c_k \sin(k\pi x) \tag{5.123}$$

where the convergence of the series is intended to be *in the mean* and

$$c_k = \frac{\langle g, f_k \rangle}{\|f_k\|^2} = 2 \int_0^1 g(x) \sin(k\pi x)\, dx \tag{5.124}$$

This is a results that can be obtained from the Fourier series theory, in fact Eq. (5.123) is the expansion of $g(x)$ in a Fourier series of sine functions.

A word should be said about the case $\lambda = 0$. In such case the general solution is

$$y = c_1 x + c_2 \tag{5.125}$$

but the boundary conditions would yield $c_1 = c_2 = 0$ i.e. the trivial solution.

### 5.4.3   The Periodic Sturm–Liouville Problem

In some problems the boundary conditions (5.108a) may be replaced by the following

$$y(a) = y(b); \quad y'(a) = y'(b) \tag{5.126}$$

which are not *separated*. Equation (5.109) is still verified if $a_2(a) = a_2(b)$ and the corresponding SL problem is said to be *periodic*. In such case the operator $\mathcal{L}$ is again self-adjoint and all the previous results are valid with the only exception of the uniqueness of the solution: at each eigenvalue may correspond different eigensolutions. This is a quite important difference and the existence of different eigensolutions for the same eigenvalue is sometimes called *degeneracy* of the eigenfunctions in physics. We can see this case with an example.

**Example: The Fourier Series**
Consider again the equation

$$y'' - \lambda y = 0 \tag{5.127}$$

on the domain $[-1, 1]$ and now set the periodic conditions

$$y(-1) = y(+1); \quad y'(-1) = y'(+1) \tag{5.128}$$

Since $a_2(-1) = a_2(+1) = 1$ this set a periodic SL problem (see also the previous example). The general solution of equation (5.127) is

$$y = \begin{cases} c_1 e^{\sqrt{\lambda}x} + c_2 e^{-\sqrt{\lambda}x} & \text{if } \lambda > 0 \\ c_1 x + c_2 & \text{if } \lambda = 0 \\ c_1 \sin\left(\sqrt{|\lambda|}x\right) + c_2 \cos\left(\sqrt{|\lambda|}x\right) & \text{if } \lambda < 0 \end{cases} \tag{5.129}$$

and setting the B.C. (5.128) yields

$$c_1 = c_2 = 0 \text{ if } \lambda > 0 \text{ (i.e. the trivial solution)} \tag{5.130}$$
$$c_1 = 0 \text{ if } \lambda = 0 \text{ (i.e. the solution } y = c_2)$$

while for $\lambda < 0$

$$2c_1 \sin\left(\sqrt{|\lambda|}\right) - 0 \tag{5.131}$$

$$2\sqrt{\lambda}c_2 \sin\left(\sqrt{|\lambda|}\right) = 0 \tag{5.132}$$

i.e. either the trivial solution ($c_1 = c_2 = 0$) or

$$\lambda_n = -\pi^2 n^2 \tag{5.133}$$

where $n = 0, \pm1, \pm2 \ldots$, and again, since $n$ and $-n$ yield the same eigenvalue, we can consider only $n = 0, 1, 2, \ldots$. Since the B.C do not yield a value for $c_1, c_2$ they can be chosen at will, and then, for each eigenvalue, two linearly independent functions are the solution: $\sin(n\pi x)$ and $\cos(n\pi x)$, i.e. to one eigenvalue ($\lambda_n = -\pi^2 n^2$) correspond two eigenfunctions. However, the results about a SL problem still holds, i.e. the eigenfunctions are all mutually orthogonal, and this holds also for those corresponding to the same eigenvalue

$$\int_{-1}^{1} \sin(n\pi x) \cos(n\pi x) = 0 \tag{5.134}$$

Again, given a function $g \in L^2 ([-1, 1])$ it can be expanded as a series of eigenfunctions

$$g = \sum_{k=0}^{\infty} a_k \sin (k\pi x) + b_k \cos (k\pi x) \tag{5.135}$$

and

$$a_k = \frac{\langle g, \sin (k\pi x) \rangle}{\| \sin (k\pi x) \|^2}; \quad b_k = \frac{\langle g, \cos (k\pi x) \rangle}{\| \cos (k\pi x) \|^2} \tag{5.136}$$

Equation (5.135) is the well known Fourier series expansion of a periodic function.
To conclude this section we have to deal with another exception from what above
discussed, the case of a *singular* SL problem.

### 5.4.4   The Singular Sturm–Liouville Problem

When the domain is not bounded or when $a_2 (x)$ vanishes somewhere on the domain,
the SL problem is a *singular* one. When the singularity is caused by the vanishing
of $a_2 (x)$, we will suppose that this happens on the boundaries of the domain. The
results obtained above for regular SL problems extend to these cases (see [8, 9] for
more details). As a special case it is worth mentioning that when $a_2 (a) = 0$ and
$a_2 (b) = 0$ the condition (5.109) is always satisfied and $\mathcal{L}$ is then self-adjoint, as in
the following case.

**Example: The Legendre Polynomials**
Consider the Legendre equation

$$\left(1 - x^2\right) y'' - 2xy' + \lambda y = 0 \tag{5.137}$$

over the domain $[-1, +1]$, the differential operator $\mathcal{L} = \left(1 - x^2\right) y'' - 2xy'$ can be
put into the form

$$\mathcal{L} = \frac{d}{dx} \left[ \left(1 - x^2\right) \frac{d}{dx} \right] \tag{5.138}$$

which shows that $\mathcal{L}$ is formally self-adjoint. Since $a_2 (x) = 1 - x^2$ is nil at both ends
of the domain, the problem is *singular*. In this case Eq. (5.96) is always verified and
$\mathcal{L}$ is self-adjoint, whatever boundary condition we set, and the results we have seen
for self-adjoint operators hold. Without loss of generality we can put $\lambda = \nu (\nu + 1)$
and the solution of (5.137) are the so-called Legendre functions $P_\nu (x)$ and $Q_\nu (x)$.
The only case of solutions bounded over all the domain $[-1, +1]$ (and in particular
on $x = -1$) is obtained when $\nu$ is an integer. In such case $P_n (x)$   are the so-called
Legendre polynomials

$$P_n (x) = \frac{1}{2^n n!} \frac{d^n \left(x^2 - 1\right)^n}{dx^n} \tag{5.139}$$

To notice that the second linear independent solution $Q_n(x)$, the Legendre functions of second kind

$$Q_n(x) = \frac{1}{2} P_n \ln \frac{1+x}{1-x} - \sum_{m=1}^{n} \frac{1}{m} P_{m-1}(x) P_{n-m}(x) \qquad (5.140)$$

is unbounded on $x = \pm 1$. See Appendix for a sketch of some of both kind of functions.

Since $\mathcal{L}$ is self-adjoint, all the results above obtained hold, in particular the functions $P_n(x)$ are all mutually orthogonal.

We also know from the general results that the sequence $P_0, P_1, \ldots$ is a basis for $L^2([-1, +1])$, then given any function $g \in L^2([-1, 1])$ the following series expansion holds

$$g = \sum_{n=0}^{\infty} c_n P_n(x) \qquad (5.141)$$

and

$$c_n = \frac{\langle g, P_n \rangle}{\| P_n \|^2} \qquad (5.142)$$

The case of a singular Sturm-Liouvile problem due to an unbounded domain is also common in the applications, consider as an example the case of Laguerre or Hermite polynomials. The Laguerre polynomials are defined over the interval $[0, \infty)$ and are the solutions of the Laguerre equation

$$x \frac{d^2 y}{dx} + (1 - x) \frac{dy}{dx} + \lambda y = 0 \qquad (5.143)$$

corresponding to the eigenvalues $\lambda = n$ where $n$ is an integer. Their definition through the Rodrigues' formula (named after the French mathematician Benjamin Olinde Rodrigues, 1795–1851) is

$$L_n(x) = \frac{e^x}{n!} \frac{d^n e^{-x} x^n}{dx^n} \qquad (5.144)$$

Multiplying all term of Eq. (5.143) by $e^{-x}$, it can be recasted into the form of Eq. (5.100), and in this case the inner product is defined by using the weighting function $w(x) = e^{-x}$, and the functions $L_0, L_1, \ldots$ are mutually orthogonal

$$\int_0^{\infty} L_n(x) L_k(x) e^{-x} dx = 0, \text{ if } n \neq k \qquad (5.145)$$

The Hermite polynomials, are defined over the unbounded interval $(-\infty, \infty)$. There are actually two kinds of Hermite polynomials, solutions of the two eigenvalue equations

$$\frac{d^2y}{dx^2} - x\frac{dy}{dx} + \lambda y = 0 \tag{5.146a}$$

$$\frac{d^2y}{dx^2} - 2x\frac{dy}{dx} + 2\lambda y = 0 \tag{5.146b}$$

The solutions of the first one corresponding to the eigenvalues $\lambda = n$, where $n$ is a non-negative integer, are called $He_n(x)$ and sometime named *probabilists' Hermite polynomials*, to distinguish them from the eigensolutions of the second equation (again corresponding to $\lambda$ equal to a non-negative integer) $H_n(x)$ named *physicists' Hermite polynomial*, and related to the previous ones by the equation

$$H_n(x) = 2^{n/2} He_n\left(\sqrt{2}x\right) \tag{5.147}$$

Apart of this details, the polynomials $H_n$ can be written as

$$H_n(x) = (-1)^n e^{x^2} \frac{d^n e^{-x^2}}{dx^n} \tag{5.148}$$

Multiplying all term of Eq. (5.146b) by $e^{-x^2}$, it can be recasted into the form of Eq. (5.100), and then the inner product is defined using $w(x) = e^{-x^2}$ and each element of the sequence $H_0, H_1 \ldots$ is orthogonal to each other

$$\int_0^\infty H_n(x) H_k(x) e^{-x^2} dx = 0, \text{ if } n \neq k \tag{5.149}$$

Clearly polynomials are not the only functions that have the above mentioned properties, see for example the above mentioned Legendre functions of second kind, $Q_n(x)$, which are not polynomials but share with $P_n(x)$ all the orthogonality properties as well as many others. A full and exhaustive treatment of many kind of special functions that are generated from solutions of Sturm–Liouville problems can be found in [3, 4, 12].

# References

1. Davis, H.F.: Fourier Series and Orthogonal Functions. Dover, New York (1989)
2. Strang, G.: The fundamental theorem of linear algebra. Am. Math. Month. **100**(9), 848–855 (1993)
3. Olver, F.W., Lozier, D.W., Boisvert, R.F., Clark, C.W. (eds.): NIST Handbook of Mathematical Functions. Cambridge University Press, Cambridge (2010)
4. Abramowitz, M., Stegun, I.A.: Handbook of Mathematical Functions with Formulas, Graphs, and Mathematical Tables, 10th edn. Dover Publications Inc., United States (1972)
5. Kreyszig, E.: Introductory Functional Analysis with Applications. Wiley, New York (1978)
6. Arfken, G.B., Weber, H.J.: Mathematical Methods for Physicists. Elsevier Academic Press, Cambridge (2005)

7. Rudin, W.: Functional Analysis. Mc Graw Hill, New York (1991)
8. Al-Gwaiz, M.A.: Sturm-Liouville Theory and its Applications. Springer, Berlin (2008)
9. Zettl, A.: Sturm-Liouville Theory. American Mathematical Society, Providence (2010)
10. Ince, L.: Ordinary Differential Equations. Dover, Illinois (1956)
11. Tenenbaum, M., Pollard, H.: Ordinary Differential Equations. Dover, Illinois (1985)
12. Lebedev, N.N.: Special Functions and Their Applications. Dover, Illinois (1972)

# Part II
# Mass, Momentum and Energy Conservation Equations in Curvilinear Coordinates

The second part of this book is devoted to the equations that govern the transport of mass, momentum and energy in a fluid. A particular attention is paid to the case of multi-component fluids and to the presence of interfaces between different phases, since the evaporation of liquid drops is the main target of the analytical approaches that will be reported in Part III, and multi-component mixtures and interfaces are always present in such applications. Conservation of mass, chemical species, momentum and energy will be derived from first principles, as well as the balances across an interface between phases. Constitutive equations for multi-component fluids will be also treated in some details. As for the previous part, there is no claim of completeness and the interested reader is often referred to fundamental books on these subjects.

In Chaps. 6 and 7 the equations will be written in Cartesian coordinates only and, since in such a system there is no difference between covariant, contravariant and physical components of vectors and tensors, only subscripts will be used to indicate tensor components; the repeated index convention is applied when necessary, also between indices at the same level (subscripts). In Chap. 8 the conservation and constitutive equations will be written in the most general tensorial form and examples in different coordinate systems will be reported. Then the usual convention about use of indices for tensor components will be recovered.

The last chapter reports and discusses the most common simplifying assumptions, which are used to transform the governing equations into a form that may allow analytical solutions.

# Chapter 6
# Conservation Equations

Conservation laws can be seen as statements about the fact that some measurable quantities do not change with time when the system under investigation is isolated. In fluid mechanics there are at least two ways to mathematically state the conservation of quantities like mass, chemical species, momentum and energy. The first one states the conservation law over any finite portion of fluid in a so-called *integral form*; the second one states the law for any point inside the fluid volume and yields the *differential form* of the conservation law. The two forms are just two different ways to state the same law and they can be alternatively used, although for analytical approaches to the solution of thermo-fluid problems the second one is certainly the most convenient. The tools for switching from one form to the other rely on two fundamental theorems: the divergence theorem and the transport theorem, and both will be revisited in the first sections of this chapter. Conservation equations can be derived from first principles, and in this respect the integral forms are the easiest way to do it; in this chapter the conservation of mass, chemical species, momentum and energy will be formulated in both forms with one important simplification: since chemical reactions will not be considered in this book, the source terms in the species and energy conservation equations will be dropped.

## 6.1 The Divergence Theorem

Given any sufficiently smooth vector field $F_k$, the divergence theorem states a relation between the integral of the divergence of this field over a finite volume and the integral of the normal component of the vector over the surface that encloses the volume, precisely

$$\int_V \nabla_k F_k dV = \int_S F_k \check{N}_k dS \qquad (6.1)$$

© Springer Nature Switzerland AG 2021
G. E. Cossali and S. Tonini, *Drop Heating and Evaporation: Analytical Solutions in Curvilinear Coordinate Systems*, Mathematical Engineering,
https://doi.org/10.1007/978-3-030-49274-8_6

where $V$ is the volume of space surrounded by the closed surface $S$, $\nabla_k F_k$ is the divergence of the vector $\mathbf{F}$ and $\check{N}_k$ are the components of the normal to the surface. The reader should notice that, as remarked in the preface to this part of the book, since the equations are written in Cartesian coordinates, there is no distinction among covariant, contravariant and physical components and the indices are always lower, while the summation convention on repeated indices is used, unless otherwise stated. The demonstration of this theorem in general form is due to the Russian mathematician Mikhail Vasilyevich Ostrogradsky (1801–1862). However, the use of a particular form of this equation is found in a work of Lagrange (Joseph-Louis Lagrange, an Italian mathematician, born as Giuseppe Lodovico Lagrangia, 1736–1813) in 1762 [1] and the theorem was first stated by Gauss in 1813. A demonstration of this theorem can be found in [2]. In the next chapters this theorem will be used many times, particularly in conjunction with the transport theorem (see next section), to transform surface integral into volume integrals and vice-versa.

Although stated for a vector field, this theorem can be extended to tensor fields. In an Euclidean 3D space, using Cartesian coordinates, the extension of this theorem to second order tensors $H_{jk}$ is

$$\int_V \nabla_k H_{jk} dV = \int_S H_{jk} \check{N}_k dS \qquad (6.2)$$

The proof of this version is simple: take any constant vector field $B_j$, then

$$B_j \int_S H_{jk} \check{N}_k dS = \int_S (B_j H_{jk}) \check{N}_k dS \qquad (6.3)$$

since $(B_j H_{jk})$ is a vector, the application of Eq. (6.1) yields

$$B_j \int_S H_{jk} \check{N}_k dS = \int_V \nabla_k (B_j H_{jk}) dV = \int_V B_j \nabla_k H_{jk} dV = B_j \int_V \nabla_k H_{jk} dV \qquad (6.4)$$

and the arbitrariness of the vector $B_j$ yields Eq. (6.2).

Some implications of this theorem are the following. Consider the curl of a field $F_k$, i.e. $\omega_k = \varepsilon_{kjl} \nabla_j F_l$, then the following rule holds

$$\int_V \omega_k dV = \int_S \varepsilon_{kjl} F_l \check{N}_j dV \qquad (6.5)$$

In fact, taking again a constant vector $B_k$,

$$B_k \int_V \omega_k dV = B_k \int_V \varepsilon_{kjl} \nabla_j F_l dV = \int_V \nabla_j (B_k \varepsilon_{kjl} F_l) dV = \int_S (B_k \varepsilon_{kjl} F_l) \check{N}_j dS = B_k \int_S \varepsilon_{kjl} F_l \check{N}_j dS \qquad (6.6)$$

where Eq. (6.1) was used to obtain the third equality since $(B_k \varepsilon_{kjl} F_l)$ is a vector; the arbitrariness of the vector $B_k$ proves the result.

Now, consider the derivative of a vector field $F_k$, i.e. $u_{kj} = \frac{\partial F_k}{\partial x_j} = \nabla_j F_k$, taking again an arbitrary constant vector $B_j$,

$$B_j \int_V \nabla_j F_k dV = \int_V \nabla_j (B_j F_k) dV = \int_S (B_j F_k) \check{N}_j dS = B_j \int_S F_k \check{N}_j dS \quad (6.7)$$

and then

$$\int_V \nabla_j F_k dV = \int_S F_k \check{N}_j dS \quad (6.8)$$

## 6.2 The Transport Theorem

The following is sometime called *Reynolds transport theorem* (named after the British engineer Osborne Reynolds, 1842–1912), a statement of this theorem in the form given by Reynolds can be found in [3]. This theorem can also be seen as a generalisation to higher dimensions of the Leibniz's rule (named after the German mathematician and philosopher Gottfried Wilhelm Leibniz, 1646–1716) for differentiation under the integral sign.

Consider a portion of space defined by a closed surface $S$, which may deform with time; then also the volume $V$ enclosed by the surface may depend on time. Given any function of coordinates and time, $B(\mathbf{x}, t)$, the integral of $B$ over the volume $V$ is a function of time only

$$A(t) = \int_{V(t)} B(\mathbf{x}, t) dV \quad (6.9)$$

The time derivative of $A(t)$ can then be calculated as

$$\frac{dA(t)}{dt} = \frac{\int_{V(t+dt)} B(\mathbf{x}, t+dt) dV - \int_{V(t)} B(\mathbf{x}, t) dV}{dt} = \quad (6.10)$$

$$= \frac{\int_{V(t+dt)} B(\mathbf{x}, t+dt) dV - \int_{V(t+dt)} B(\mathbf{x}, t) dV}{dt} + \frac{\int_{V(t+dt)} B(\mathbf{x}, t) dV - \int_{V(t)} B(\mathbf{x}, t) dV}{dt}$$

where the quantity $\int_{V(t+dt)} B(\mathbf{x}, t) dV$ is the integral of $B$ as evaluated at time $t$, over the portion of space occupied at time $t + dt$. The first term on the most RHS of Eq. (6.10) can be transformed as

$$\frac{\int_{V(t+dt)} B(\mathbf{x}, t+dt) dV - \int_{V(t+dt)} B(\mathbf{x}, t) dV}{dt} = \int_{V(t+dt)} \frac{B(\mathbf{x}, t+dt) - B(\mathbf{x}, t)}{dt} dV = \int_{V(t+dt)} \frac{\partial B(\mathbf{x}, t)}{\partial t} dV \quad (6.11)$$

To transform the second term on the RHS of Eq. (6.10) let suppose that the volume $V(t)$ is contained into the volume $V(t + dt)$ (see Fig. 6.1) then

$$\int_{V(t+dt)} B(\mathbf{x}, t) dV - \int_{V(t)} B(\mathbf{x}, t) dV = \int_{\delta V} B(\mathbf{x}, t) dV \quad (6.12)$$

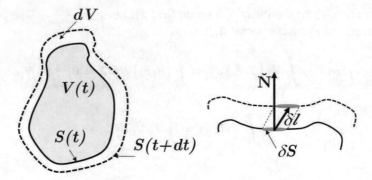

**Fig. 6.1** Portion of space enclosed by the time deforming surface $S$ (left) and volume $dV$ swept infinitesimal area $dS$ during the time interval $dt$ (right)

where $\delta V$ represents the infinitesimal portion of space between the two surfaces. Performing the integration over $\delta V$, we can also consider that the element of volume $dV$ can be calculated as the product of the element of area $dS$ multiplied by the distance between the surface measured along the normal direction, i.e. $dV = dS \, dl_k \check{N}_k$ (see again Fig. 6.1), then the second term on the RHS of Eq. (6.10) becomes

$$\frac{1}{dt} \int_{\delta V} B\left(\mathbf{x}, t\right) dV = \int_{S} B\left(\mathbf{x}, t\right) \frac{dl_k}{dt} \check{N}_k dS = \int_{S} B\left(\mathbf{x}, t\right) V_{s,k} \check{N}_k dS \qquad (6.13)$$

where $V_{s,k} = \frac{dl_k}{dt}$ are the components of the surface velocity vector $\mathbf{V}_s$. Substituting Eqs. (6.11) and (6.13) into Eq. (6.10), (in the limit $dt \to 0$), yields

$$\frac{d}{dt} \int_{V(t)} B\left(\mathbf{x}, t\right) dV = \int_{V(t)} \frac{\partial B\left(\mathbf{x}, t\right)}{\partial t} dV + \int_{S(t)} B\left(\mathbf{x}, t\right) V_{s,k} \check{N}_k dS \qquad (6.14)$$

which is the statement of the transport theorem. Similar arguments can be used also when the surfaces $S(t)$ and $S(t + dt)$ intersect. For a more formal demonstration the interested reader can refer to [4]. It should be noticed that the statement holds for any quantity $B$ (scalar, vector, tensor) and the velocity appearing in the surface integral is that of the surface, which can be different from that of a fluid that may occupy the space, i.e. the surface does not need to be a material surface.

The importance of Eq. (6.14) will become evident in the following sections, particularly in the next one, where it will be used to derive the local form of the conservation equations from the integral ones.

## 6.3   The Conservation Equations

From a physical point of view, the most significant form of a conservation equation is the integral one, since it translates in the simplest way the conservation principles. However, from a mathematical point of view, the differential form is certainly the most useful, particularly when analytical solutions are sought. The two forms are clearly strictly connected and the tools that allow the conversion of one to the other are the two theorems discussed in the previous sections.

### *6.3.1   The Mass Conservation Equation*

The conservation of mass can be expressed in a simple form using the integral approach. Taking any closed volume $V$, the mass contained into this volume is

$$m = \int_V \rho dV \tag{6.15}$$

where $\rho$ is the mass density. Let now assume that the contour surface of $V$ is a material surface, i.e. a surface that moves with the fluid, meaning that any point on the surface has the velocity of the fluid at that point, i.e. $V_{s,j} = v_j$ where $\mathbf{v}$ is the fluid velocity. This is enough to state that the fluid cannot pass through the surface and the mass $m$ does not change with time, i.e.

$$\frac{d \int_V \rho dV}{dt} = 0 \tag{6.16}$$

This integro-differential equation states the mass conservation law in integral form.

Applying now the transport theorem (Eq. 6.14) with $B \equiv \rho$ results in

$$0 = \frac{d \int_V \rho dV}{dt} = \int_V \frac{\partial \rho}{\partial t} dV + \int_S \rho v_j \check{N}_j dS = \int_V \frac{\partial \rho}{\partial t} dV + \int_V \nabla_j (\rho v_j) dV = \int_V \left[ \frac{\partial \rho}{\partial t} + \nabla_j (\rho v_j) \right] dV \tag{6.17}$$

where the second equality is obtained applying the divergence theorem to the surface integral. Since this equation must hold for any volume enclosed in a material surface, the integrand must be nil, thus yielding the differential form of the mass conservation law, also called *continuity equation*

$$\frac{\partial \rho}{\partial t} + \nabla_j (\rho v_j) = 0 \tag{6.18}$$

When the density of the fluid can be assumed constant, like it is often done when dealing with liquids, the continuity equation assumes the simpler form

$$\nabla_j v_j = 0 \tag{6.19}$$

and the velocity field is said to be *divergenceless* or *solenoidal*.

## 6.3.2 The Species Conservation Equations

Let now consider the case of a multi-component fluid, i.e. a fluid that is a mixture of $n + 1$ species. To characterise the flow and the composition of a multi-component fluid we need to introduce some new quantities and to specify the relationships among them. Let us start with the composition of a multi-component fluid, which can be defined in two distinct ways. The *mass concentration* of component $p$ , $\rho^{(p)}$, is defined as the mass of species $p$ per unit volume of the mixture. The $n + 1$ quantities $\rho^{(p)}$ $(p = 0, \ldots, n)$ satisfy the trivial relation

$$\rho = \sum_{p=0}^{n} \rho^{(p)} \tag{6.20}$$

where $\rho$ is the mass density. The composition of the mixture can be defined also in term of *molar concentrations* $c^{(p)}$, i.e. the number of moles of species $p$ per unit volume. Again they satisfy a relation analogous to Eq. (6.20)

$$c = \sum_{p=0}^{n} c^{(p)} \tag{6.21}$$

where $c$ is the *molar density* of the mixture. The *mass fraction* of each component is then defined as the ratio

$$\chi^{(p)} = \frac{\rho^{(p)}}{\rho} \tag{6.22}$$

and analogously, the *molar fractions* are

$$y^{(p)} = \frac{c^{(p)}}{c} \tag{6.23}$$

There are simple relations among the two sets (mass and molar) of variables, precisely

$$c^{(p)} = \frac{\rho^{(p)}}{Mm^{(p)}}; \quad y^{(p)} = \frac{\rho}{cMm^{(p)}} \chi^{(p)} \tag{6.24}$$

where $Mm^{(p)}$ is the *molar mass* of component $p$, and also

$$c = \sum_{p=0}^{n} \frac{\rho^{(p)}}{Mm^{(p)}} = \rho \sum_{p=0}^{n} \frac{\chi^{(p)}}{Mm^{(p)}} \tag{6.25}$$

$$\rho = c \sum_{p=0}^{n} Mm^{(p)} y^{(p)} \tag{6.26}$$

and the average molar mass of the mixture can be defined as

$$\overline{Mm} = \frac{\rho}{c} = \sum_{p=0}^{n} Mm^{(p)} y^{(p)} \tag{6.27}$$

In a mixture of different components it is necessary to define the fluxes of each component, and again the fluxes can be defined on a molar or mass basis. The mass of component $p$ that crosses the unit of surface in a unit of time is the *mass flux* of component $p$, and it will be written $\mathbf{n}^{(p)}$ since it is a vector of components $n_j^{(p)}$. The number of moles of component $p$ that crosses the unit of surface in a unit of time is the *molar flux* of component $p$, and it will be written $\mathbf{N}^{(p)}$, a vector of components $N_j^{(p)}$. These two vectors are related to each other by

$$n_j^{(p)} = N_j^{(p)} Mm^{(p)} \tag{6.28}$$

and they allow a simple definition of the *species velocity vector* $\mathbf{v}^{(p)}$

$$v_j^{(p)} = \frac{n_j^{(p)}}{\rho^{(p)}} = \frac{N_j^{(p)}}{c^{(p)}} \tag{6.29}$$

This quantity admits a simple physical interpretation (see [5]) as the average velocity of the molecules of component $p$ at a given point. The average mass, $\bar{\mathbf{v}}$, and molar, $\bar{\mathbf{v}}^*$, velocity vectors of the mixture are defined as

$$\bar{v}_j = \frac{\sum_{p=0}^{n} \rho^{(p)} v_j^{(p)}}{\sum_{p=0}^{n} \rho^{(p)}} \tag{6.30a}$$

$$\bar{v}_j^* = \frac{\sum_{p=0}^{n} c^{(p)} v_j^{(p)}}{\sum_{p=0}^{n} c^{(p)}} \tag{6.30b}$$

and since $n_j^T = \sum_{p=0}^{n} n_j^{(p)}$ and $N_j^T = \sum_{p=0}^{n} N_j^{(p)}$ are the total mass and molar fluxes, respectively, the following result holds

$$n_j^T = \sum_{p=0}^{n} \rho^{(p)} v_j^{(p)} = \rho \bar{v}_j \tag{6.31a}$$

$$N_j^T = \sum_{p=0}^{n} c^{(p)} v_j^{(p)} = c \bar{v}_j^* \tag{6.31b}$$

Let now assume that the reference system moves with a velocity equal to the mass average velocity $\bar{\mathbf{v}}$ at a given point, then the species velocity vectors with respect to this new reference system are $\mathbf{v}^{(p)} - \bar{\mathbf{v}}$ and the corresponding species mass fluxes in this system are

$$j_j^{(p)} = \rho^{(p)} \left( v_j^{(p)} - \bar{v}_j \right) \tag{6.32}$$

These fluxes are in general non nil although the fluid is at rest in that particular system at the chosen point. These fluxes represent the transport of species due to molecular motion and they are called *molecular* (or *diffusive*) *mass* fluxes. The same holds if we consider the molar fluxes in a reference system moving with the molar average velocity $\bar{\mathbf{v}}^*$, and the *molecular* (diffusive) *molar* fluxes are

$$J_j^{(p)} = c^{(p)} \left( v_j^{(p)} - \bar{v}_j^* \right) \tag{6.33}$$

It is worth to notice, and straightforward to see, that the following relations among the diffusive fluxes always hold

$$\sum_{p=0}^{n} j_j^{(p)} = 0 \tag{6.34a}$$

$$\sum_{p=0}^{n} J_j^{(p)} = 0 \tag{6.34b}$$

and that the mass and molar fluxes can be always split like

$$n_j^{(p)} = \rho^{(p)} v_j^{(p)} = \rho^{(p)} \bar{v}_j + j_j^{(p)} = \rho \bar{v}_j \chi^{(p)} + j_j^{(p)} = n_j^T \chi^{(p)} + j_j^{(p)} \tag{6.35a}$$

$$N_j^{(p)} = c^{(p)} v_j^{(p)} = c^{(p)} \bar{v}_j^* + J_j^{(p)} = c \bar{v}_j^* y^{(p)} + J_j^{(p)} = N_j^T y^{(p)} + J_j^{(p)} \tag{6.35b}$$

where

$$\rho \bar{v}_j \chi^{(p)} = n_j^T \chi^{(p)} \tag{6.36a}$$

$$c \bar{v}_j^* y^{(p)} = N_j^T y^{(p)} \tag{6.36b}$$

are the *convective* parts of the respective fluxes (being $\mathbf{j}^{(p)}$ and $\mathbf{J}^{(p)}$ the *diffusive* parts) and they are sometime called *convective* (mass or molar) fluxes. As for other quantities that we will encounter later, the diffusive part of the species fluxes ($\mathbf{j}^{(p)}$ and

$\mathbf{J}^{(p)}$) need to be evaluated as a function of the concentration field, i.e. a *constitutive law* is necessary, and this will be discussed in the next chapter.

It is interesting to notice that the relationships between $\mathbf{j}^{(p)}$ and $\mathbf{J}^{(p)}$ are usually quite cumbersome, with the exception of the case of binary mixtures. In such case, set $p = 1$ and from Eqs. (6.35a) and (6.35b)

$$\mathbf{N}^T y^{(1)} + \mathbf{J}^{(1)} = \mathbf{N}^{(1)} = \frac{1}{Mm^{(1)}} \mathbf{n}^{(1)} = \frac{1}{Mm^{(1)}} \mathbf{n}^T \chi^{(1)} + \frac{1}{Mm^{(1)}} \mathbf{j}^{(1)} \qquad (6.37)$$

and then

$$\mathbf{J}^{(1)} = \frac{1}{Mm^{(1)}} \mathbf{n}^T \chi^{(1)} - \mathbf{N}^T y^{(1)} + \frac{1}{Mm^{(1)}} \mathbf{j}^{(1)} \qquad (6.38)$$

Here, and often later on, when dealing with species fluxes, we have used the *vector notation* in Cartesian coordinates, to increase the readability of formulas that contain also superscripts to indicate the different species.

From the definitions of $\mathbf{N}^T$ and $\mathbf{n}^T$

$$\mathbf{N}^T = \mathbf{N}^{(1)} + \mathbf{N}^{(0)} = \frac{1}{Mm^{(1)}} \mathbf{n}^{(1)} + \frac{1}{Mm^{(0)}} \mathbf{n}^{(0)} = \qquad (6.39)$$

$$= \frac{1}{Mm^{(1)}} \left( \mathbf{n}^{(T)} \chi^{(1)} + \mathbf{j}^{(1)} \right) + \frac{1}{Mm^{(0)}} \left( \mathbf{n}^{(T)} \chi^{(0)} + \mathbf{j}^{(0)} \right)$$

$$= \mathbf{n}^{(T)} \left[ \frac{1}{Mm^{(1)}} \chi^{(1)} + \frac{1}{Mm^{(0)}} \chi^{(0)} \right] + \mathbf{j}^{(1)} \frac{1}{Mm^{(1)}} + \mathbf{j}^{(0)} \frac{1}{Mm^{(0)}}$$

$$= \mathbf{n}^{(T)} \frac{c}{\rho} + \mathbf{j}^{(1)} \left( \frac{Mm^{(0)} - Mm^{(1)}}{Mm^{(1)} Mm^{(0)}} \right)$$

where the last equality is obtained using Eqs. (6.25) and (6.34a). Substituting Eq. (6.39) into Eq. (6.38), using Eq. (6.24) to eliminate $y^{(1)}$ in favour of $\chi^{(1)}$, yields

$$\mathbf{J}^{(1)} = \mathbf{j}^{(1)} \left[ \frac{1}{Mm^{(1)}} - \left( \frac{Mm^{(0)} - Mm^{(1)}}{Mm^{(1)} Mm^{(0)}} \right) \frac{\rho}{cMm^{(1)}} \chi^{(1)} \right] \qquad (6.40)$$

Finally, observing that from Eq. (6.25)

$$\chi^{(1)} \left( \frac{Mm^{(0)} - Mm^{(1)}}{Mm^{(1)} Mm^{(0)}} \right) = \frac{c}{\rho} - \frac{1}{Mm^{(0)}} \qquad (6.41)$$

and substituting into Eq. (6.40) yields a simple relationship between molar and mass diffusive fluxes

$$\mathbf{J}^{(1)} = \mathbf{j}^{(1)} \frac{\rho}{cMm^{(0)} Mm^{(1)}} \qquad (6.42)$$

The same holds for component $p = 0$

$$\mathbf{J}^{(0)} = \mathbf{j}^{(0)} \frac{\rho}{cMm^{(0)} Mm^{(1)}} \qquad (6.43)$$

It is worth to stress again that these simple relations hold only for binary mixtures.

It is now time to write the species conservation equations, and this can be done following the same method seen in the previous section. Taking again a volume of fluid surrounded by a material surface, we can evaluate the mass of component $p$ contained into the volume $V$ at any time as

$$m^{(p)} = \int_V \rho^{(p)} dV \tag{6.44}$$

We have seen above that the diffusive flux of component $p$ may be different from zero also when the average fluid velocity is nil. Since the material surface moves with velocity $\bar{\mathbf{v}}$, then the flux of component $p$ that crosses the moving material surface at any time is just the diffusive part of the flux: $\mathbf{j}^{(p)}$, and the total mass of component $p$ that exits the whole surface per unit of time can be calculated as $\int_S j_j^{(p)} \check{N}_j dS$, where $\check{\mathbf{N}}$ is the normal to the surface directed outside.

Generally, the mass of component $p$ inside the volume $V$ may change also due to chemical reactions, and a source term should be considered. However, such term does not add any complexity to the problem and since this book will not deal with reacting flows, production rates due to chemical or nuclear reactions will be assumed to be nil. The mass balance over the volume $V$ can then be written as

$$\frac{dm^{(p)}}{dt} = -\int_S j_j^{(p)} \check{N}_j dS \tag{6.45}$$

where the minus sign is needed since the surface normal is assumed to point outside. The application of the transport theorem to the LHS of Eq. (6.45), after substituting Eq. (6.44), and the use of the divergence theorem yields

$$\int_V \frac{\partial \rho^{(p)}}{\partial t} dV + \int_V \nabla_j \left( \rho^{(p)} \bar{v}_j \right) dV = -\int_V \nabla_j j_j^{(p)} dV \tag{6.46}$$

and thanks to the arbitrariness of the volume $V$ one obtains

$$\frac{\partial \rho^{(p)}}{\partial t} + \nabla_j \left( \rho^{(p)} \bar{v}_j \right) = -\nabla_j j_j^{(p)} \tag{6.47}$$

This equation can be written (see Eq. 6.35a) as

$$\frac{\partial \rho^{(p)}}{\partial t} + \nabla_j n_j^{(p)} = 0 \tag{6.48}$$

A similar discussion based on the mole conservation yields the analogous equation

$$\frac{\partial c^{(p)}}{\partial t} + \nabla_j N_j^{(p)} = 0 \tag{6.49}$$

The $n + 1$ Eqs. (6.48) (and analogously for Eqs. 6.49) are not all independent, in fact summing both sides over the index $p$ yields

$$\text{LHS} \sum_{p=0}^{n} \frac{\partial \rho^{(p)}}{\partial t} = \frac{\partial \sum_{p=0}^{n} \rho^{(p)}}{\partial t} = \frac{\partial \rho}{\partial t} \tag{6.50}$$

$$\text{RHS} \sum_{p=0}^{n} -\nabla_j n_j^{(p)} = -\nabla_j \sum_{p=0}^{n} n_j^{(p)} = -\nabla_j n_j^T = -\nabla_j \rho \bar{v}_j \tag{6.51}$$

and the continuity equation is recovered

$$\frac{\partial \rho}{\partial t} + \nabla_j \left( \rho \bar{v}_j \right) = 0 \tag{6.52}$$

It should be noticed that the velocity $v_j$ appearing in Eq. (6.18) is the average mass velocity $\bar{v}_j$; in the next sections $v_j$ will always be assumed identical to $\bar{v}_j$ unless otherwise stated.

To notice that, using the molar forms (6.49), would yield an equivalent *molar form* of the continuity equation

$$\frac{\partial c}{\partial t} + \nabla_j \left( c \bar{v}_j^* \right) = 0 \tag{6.53}$$

where $\bar{v}^*$ is the *molar* averaged velocity vector, which is not the one that would appear in the momentum (and energy) equation, and that is why the mass form of the continuity equation is preferred.

### 6.3.3 The Momentum Conservation Equation

Consider again a volume $V$ surrounded by a material surface $S$, the momentum of this fluid lump, **p**, can be easily calculated as

$$p_j = \int_V \rho v_j dV \tag{6.54}$$

where hereinafter, as above pointed out, the velocity **v** has to be understood as the average mass velocity $\bar{v}$.

Newton's second law of dynamics (named after the celebrated English scientist Isaac Newton, 1642–1726) states that the time derivative of the momentum of a given mass is equal to the resultant force applied to the mass

$$\frac{dp_j}{dt} = F_j \tag{6.55}$$

The forces applied to the lump of fluids are of two types: i) external forces acting on each fluid particle belonging to the fluid lump, the so-called *volume forces*, $\mathbf{F}^V$, and ii) the forces acting on the surface, which are due to the interaction with the surrounding fluid, also called *surface forces*, $\mathbf{F}^S$. Assuming that a single field acts on the fluid (for example, the gravity field), the volume force can be evaluated as

$$F_j^V = \int_V \rho \, f_j dV \tag{6.56}$$

where $\mathbf{f}$ is the external force per unit of mass. Writing Eq. (6.56), we assume that each species is acted on by the same external force (like for gravity). There exist cases (which will not be treated in this book) where each species may be acted on by different forces, like for example when some species are ions and the force is generated by an electrical field; in such cases the term $\rho \mathbf{f}$ in Eq. (6.56) should be substituted by $\sum_{p=0}^n \rho^{(p)} \mathbf{f}^{(p)}$. The surface forces can instead be evaluated by introducing the stress tensor $\mathcal{T}_{jk}$ as

$$F_j^S = \int_S \mathcal{T}_{jk} \check{N}_k dS \tag{6.57}$$

In fluid mechanics, the stress expresses the internal forces exerted on each other by neighbouring fluid particles. Newton's second law (6.55) then becomes

$$\frac{d \int_V \rho v_j dV}{dt} = \int_S \mathcal{T}_{jk} \check{N}_k dS + \int_V \rho \, f_j dV \tag{6.58}$$

Using the transport theorem to transform the LHS yields

$$\frac{d \int_V \rho v_j dV}{dt} = \int_V \frac{\partial \rho v_j}{\partial t} dV + \int_S \rho v_j v_k \check{N}_k dS \tag{6.59}$$

and applying the divergence theorem to transform the surface integrals into volume integrals allow to re-write Eq. (6.58) as

$$\int_V \frac{\partial \rho v_j}{\partial t} dV + \int_V \nabla_k \left( \rho v_j v_k \right) dV = \int_V \nabla_k \mathcal{T}_{jk} dV + \int_V \rho f_j dV \tag{6.60}$$

that, recalling that the equation must hold for any volume $V$, yields the differential form of the momentum conservation equation

$$\frac{\partial \rho v_j}{\partial t} + \nabla_k \left( \rho v_j v_k \right) = \nabla_k \mathcal{T}_{jk} + \rho f_j \tag{6.61}$$

The stress tensor can be split into two parts

$$T_{jk} = -P\delta_{jk} + \tau_{jk} \tag{6.62}$$

where $P$ is the pressure and the tensor $\tau_{jk}$ is called the *deviatoric* stress tensor. The form of $\tau_{jk}$ depends on the velocity field and on the properties of the fluid and this will be discussed in Chap. 7.

### 6.3.4 The Energy Conservation Equation

The first law of thermodynamics states that the variation of the energy of a close system equals the heat and work exchanged with the environment. Consider the system made by the matter contained into the material surface $S$ as above. The internal plus the kinetic energy of this system can be evaluated as

$$E = \int_V \rho \left( \hat{u} + \frac{1}{2} |\mathbf{v}|^2 \right) dV \tag{6.63}$$

where $\hat{u}$ is the internal energy per unit mass and $|\mathbf{v}|^2 = v_j v_j$. The variation of the energy per unit time must then equal the energy exchanged per unit time as work and heat. As above, we will not consider any *heat generation*, which is a way to generally define all the transformation of different forms of energy (electromagnetic, chemical, nuclear) into internal energy, consistently to the previous assumption of absence of reactions. The *exchanged work* is relatively easy to define, since it can be taken as the work done by all forces acting on the volume, which as previously mentioned, can be subdivided into volume (external) forces and surface forces. The work done on the system by the external field $\mathbf{f}$ (cfr Sect. 6.3.3) per unit of time can be easily calculated as

$$\dot{W}_V = \int_V \rho \, f_j v_j dV \tag{6.64}$$

while the work done on the system by the surface forces can be written as

$$\dot{W}_S = \int_S T_{jk} v_j \check{N}_k dS \tag{6.65}$$

In absence of *heat generation*, the heat exchange between the system and the environment can be assumed to take place at the surface of the system and, defining the heat flux vector $\mathbf{q}$, the heat exchanged per unit of time (again positive when entering the volume) is

$$\dot{Q} = -\int_S q_j \check{N}_j dS \tag{6.66}$$

The first law of thermodynamics can be generally written as

$$\frac{dE}{dt} = \dot{Q} + \dot{W}_S + \dot{W}_V \tag{6.67}$$

and, after substituting Eqs. (6.63), (6.64), (6.65) and (6.66), it becomes

$$\frac{d \int_V \rho \left( \hat{u} + \frac{1}{2} |\mathbf{v}|^2 \right) dV}{dt} = - \int_S q_j \check{N}_j dS + \int_S T_{jk} v_j \check{N}_k dS + \int_V \rho f_j v_j dV \tag{6.68}$$

The LHS of Eq. (6.68) can be transformed using the transport theorem

$$\frac{d \int_V \rho \left( \hat{u} + \frac{1}{2} |\mathbf{v}|^2 \right) dV}{dt} = \int_V \frac{\partial \rho \left( \hat{u} + \frac{1}{2} |\mathbf{v}|^2 \right)}{\partial t} dV + \int_S \rho \left( \hat{u} + \frac{1}{2} |\mathbf{v}|^2 \right) v_j \check{N}_j dS \tag{6.69}$$

and, using the divergence theorem to transform all the surface integrals into volume integrals, yields the integral-differential equation

$$\int_V \frac{\partial \rho \left( \hat{u} + \frac{1}{2} |\mathbf{v}|^2 \right)}{\partial t} dV + \int_V \nabla_j \left[ \rho \left( \hat{u} + \frac{1}{2} |\mathbf{v}|^2 \right) v_j \right] dV = \int_V \left[ -\nabla_j q_j + \nabla_j \left( T_{jk} v_k \right) + \rho f_j v_j \right] dV \tag{6.70}$$

and, due to the arbitrariness of the volume $V$, the differential form of the energy conservation equation is

$$\frac{\partial \rho \left( \hat{u} + \frac{1}{2} |\mathbf{v}|^2 \right)}{\partial t} + \nabla_j \left[ \rho \left( \hat{u} + \frac{1}{2} |\mathbf{v}|^2 \right) v_j \right] = -\nabla_j q_j + \nabla_j \left( T_{jk} v_k \right) + \rho f_j v_j \tag{6.71}$$

The momentum equation can be used to eliminate the kinetic energy terms from Eq. (6.71); taking the dot product of all terms of Eq. (6.61) by $\mathbf{v}$ yields

$$v_j \frac{\partial \rho v_j}{\partial t} + v_j \nabla_k \left( \rho v_j v_k \right) = v_j \nabla_k T_{jk} + \rho \, f_j v_j \tag{6.72}$$

then, transforming the LHS terms as

$$v_j \frac{\partial \rho v_j}{\partial t} = \rho v_j \frac{\partial v_j}{\partial t} + v_j v_j \frac{\partial \rho}{\partial t} = \rho \frac{\partial \frac{1}{2} |\mathbf{v}|^2}{\partial t} + |\mathbf{v}|^2 \frac{\partial \rho}{\partial t} \tag{6.73}$$

$$v_j \nabla_k \left( \rho v_j v_k \right) = \rho v_k v_j \nabla_k \left( v_j \right) + v_j v_j \nabla_k \left( \rho v_k \right) = \rho v_k \nabla_k \left( \frac{1}{2} |\mathbf{v}|^2 \right) + |\mathbf{v}|^2 \nabla_k \left( \rho v_k \right) \tag{6.74}$$

and making use of the continuity Eqs. (6.18) and (6.72) becomes

$$\rho \frac{\partial \frac{1}{2} |\mathbf{v}|^2}{\partial t} + \rho v_k \nabla_k \left( \frac{1}{2} |\mathbf{v}|^2 \right) = v_j \nabla_k T_{jk} + \rho \, f_j v_j \tag{6.75}$$

sometimes called the *equation of change of kinetic energy* [5].

Substracting Eq. (6.75) from Eq. (6.71), we obtain the following more compact form

$$\frac{\partial \rho \hat{u}}{\partial t} + \nabla_j \left( \rho v_j \hat{u} \right) = -\nabla_j q_j + T_{jk} \nabla_k v_j \tag{6.76}$$

which contains only the contribution of the internal energy. It is worth to point out that Eqs. (6.76) and (6.71) are perfectly equivalent, since one can be obtained from the other making use of Eq. (6.61), as it is done above.

The energy conservation equation can be written in different forms, some of them more convenient than others for certain problems. For example, introducing the enthalpy per unit mass as $\hat{h} = \hat{u} + \frac{P}{\rho}$, Eq. (6.76) transforms to

$$\frac{\partial \rho \hat{h}}{\partial t} + \nabla_j \left( \rho v_j \hat{h} \right) = -\nabla_j q_j + \tau_{jk} \nabla_k v_j + v_j \nabla_j P + \frac{\partial P}{\partial t} \tag{6.77}$$

where Eq. (6.62) was used. See [5] and [6] for a complete collection of different forms of this equation.

A final remark on the heat flux $\mathbf{q}$ is necessary. As for the diffusive fluxes $\mathbf{j}^{(p)}$, the heat flux $\mathbf{q}$ needs to be defined as a function of the fluid properties, and this will be done in Chap. 7, where constitutive equations are treated. For multi-component systems the form of $\mathbf{q}$ may be relatively complex, due to the many mechanisms of molecular transport of energy involved. In Chap. 10 we will see how simplifying assumptions may be used to transform the conservation equations into forms that may allow an analytical approach to the solution.

## 6.4 The Balances at Interfaces

Drop heating and evaporation is a typical multiphase problem and the main distinction with single-phase problems is the existence of *interfaces* between different phases. The phase interface may be considered as a *singular* surface with respect to specific extensive properties (like specific volume, energy, etc.), i.e. the value of such properties approaching the surface from one side may be different from that reached by approaching the surface from the other side. The presence of interfaces requires to re-derive the balances of mass, chemical species, momentum and energy, encompassing both phases and the interface and taking into account the above mentioned discontinuous (jump) variations across the sharp interface. The results of this analysis are the so-called *jump conditions*, which are unique to multiphase flows.

In Sect. 6.2 we studied a fundamental mathematical tool, the transport theorem, that we used to derive the conservation equations for a single-phase flow, making the implicit assumption that the flow and fluid properties were subject to a smooth variation inside any given control volume. In this section, we will start developing the analogous tool necessary to deal with multiphase systems, and then we will apply it to derive the proper forms of *balances at interfaces* for mass, chemical species, momentum and energy.

### 6.4.1  The Transport Theorem in the Presence of Interfaces

Let us consider a finite region of space characterised by the presence of a sharp interface, i.e. a surface (which will be assumed sufficiently smooth) across which the quantities defining the fluid characteristics and the velocity may experience a discontinuous variation. We will refer to Fig. 6.2 for the nomenclature.

The region of space is divided into two subregions where all properties and velocity assume values that will be indicated by an index ($A$ or $B$). At each point of the interface two normal versors are considered, $\check{\mathbf{N}}^{(A)} = -\check{\mathbf{N}}^{(B)}$, the versor $\check{\mathbf{N}}^{(A)}$ pointing toward region $A$ and similarly for $\check{\mathbf{N}}^{(B)}$ (see Fig. 6.2). The surface is assumed to move with velocity $\mathbf{V}_I$, i.e. given a point particle on the surface its velocity is $\mathbf{V}_I$; this velocity may be different from that of the fluids at the same locations, i.e. the interface may not be a material surface. Consider now any quantity $\Psi$ continuous and differentiable inside each subregion $A$ or $B$, but possibly discontinuous across the surface. Its integral over the volume $V$ can be split into two parts

$$\int_V \Psi dV = \int_{V^{(A)}} \Psi dV + \int_{V^{(B)}} \Psi dV \qquad (6.78)$$

The time derivative is

$$\frac{d}{dt} \int_V \Psi dV = \frac{d}{dt} \int_{V^{(A)}} \Psi dV + \frac{d}{dt} \int_{V^{(B)}} \Psi dV \qquad (6.79)$$

and applying to each term on the RHS of Eq. (6.79) the usual transport theorem (i.e. Eq. 6.14), considering that the closed surface that defines the region $A$ is made by a part of the contour of the whole region $S^{(A)}$ and by the interface $S^{(I)}$, and similarly for the region $B$ yields

**Fig. 6.2** Sketch of volume of fluid crossed by an interface

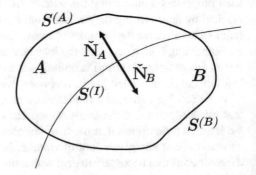

$$\frac{d}{dt} \int_{V^{(A)}} \Psi dV = \int_{V^{(A)}} \frac{\partial \Psi}{\partial t} dV + \int_{S^{(A)}} \Psi \, \mathbf{V}_s \cdot \check{\mathbf{N}} dS - \int_{S^{(I)}} \Psi^{(A)} \mathbf{V}_I \cdot \check{\mathbf{N}}^{(A)} dS$$

(6.80)

$$\frac{d}{dt} \int_{V^{(B)}} \Psi dV = \int_{V^{(B)}} \frac{\partial \Psi}{\partial t} dV + \int_{S^{(B)}} \Psi \, \mathbf{V}_s \cdot \check{\mathbf{N}} dS - \int_{S^{(I)}} \Psi^{(B)} \mathbf{V}_I \cdot \check{\mathbf{N}}^{(B)} dS$$

(6.81)

where $\mathbf{V}_s$ and $\check{\mathbf{N}}$ are the velocity and normal (pointing outside) on the external surface $S$, $\Psi^{(A)}$ and $\Psi^{(B)}$ are the limiting values of $\Psi$ when approaching the interface from side $A$ and $B$, respectively, and the minus sign in the last term is due to the choice of the orientation of $\check{\mathbf{N}}^{(A)}$ and $\check{\mathbf{N}}^{(B)}$. Substitution into Eq. (6.79) yields

$$\frac{d}{dt} \int_V \Psi dV = \int_{V^{(A)}} \frac{\partial \Psi}{\partial t} dV + \int_{S^{(A)}} \Psi \, \mathbf{V}_s \cdot \check{\mathbf{N}} dS - \int_{S^{(I)}} \Psi^{(A)} \mathbf{V}_I \cdot \check{\mathbf{N}}^{(A)} dS \quad (6.82)$$

$$+ \int_{V^{(B)}} \frac{\partial \Psi}{\partial t} dV + \int_{S^{(B)}} \Psi \, \mathbf{V}_s \cdot \check{\mathbf{N}} dS - \int_{S^{(I)}} \Psi^{(B)} \mathbf{V}_I \cdot \check{\mathbf{N}}^{(B)} dS =$$

$$= \int_V \frac{\partial \Psi}{\partial t} dV + \int_S \Psi \, \mathbf{V}_s \cdot \check{\mathbf{N}} dS - \int_{S^{(I)}} \{\Psi V_{In}\} \, dS$$

where the curly brackets are used to represent

$$\{\Psi V_{In}\} = \Psi^{(A)} V_{In}^{(A)} + \Psi^{(B)} V_{In}^{(B)}$$

(6.83)

and $V_{In}^{(A)} = \mathbf{V}_I \cdot \check{\mathbf{N}}^{(A)}$; $V_{In}^{(B)} = \mathbf{V}_I \cdot \check{\mathbf{N}}^{(B)}$. Equation (6.82) is the transport theorem for regions containing an interface. It is interesting to notice that Eq. (6.83) contains only the normal component of the interface velocity, $V_{In}^{(A)} = \mathbf{V}_I \cdot \check{\mathbf{N}}^{(A)}$; $V_{In}^{(B)} = \mathbf{V}_I \cdot \check{\mathbf{N}}^{(B)}$ and in Chap. 2 we have seen that all the information about the motion of the interface is contained in this quantity.

### 6.4.2 The Jump Mass Balance at Interface

We can now start to re-derive the conservation equations for regions of space containing an interface, starting with the mass balance. Choosing the external surface to be *material*

$$\frac{d}{dt} \int_V \rho dV = 0$$

(6.84)

applying Eq. (6.82) yields

$$\int_V \frac{\partial \rho}{\partial t} dV + \int_S \rho \, \mathbf{v} \cdot \check{\mathbf{N}} dS - \int_{S^{(I)}} \{\rho V_{In}\} \, dS = 0$$

(6.85)

where the fluid velocity $\mathbf{v}$ is used in the second integral since the surface is *material* and $\mathbf{v} \equiv \mathbf{V}_s$. Let now consider each region singularly, where the continuity equation holds, then necessarily

$$\int_{V^{(A)}} \left( \frac{\partial \rho}{\partial t} + \nabla \cdot (\rho \mathbf{v}) \right) dV = 0 \tag{6.86}$$

$$\int_{V^{(B)}} \left( \frac{\partial \rho}{\partial t} + \nabla \cdot (\rho \mathbf{v}) \right) dV = 0 \tag{6.87}$$

applying the divergence theorem, the two equations can be written as

$$\int_{V^{(A)}} \frac{\partial \rho}{\partial t} dV + \int_{S^{(A)}} \rho \mathbf{v} \cdot \check{\mathbf{N}} dS - \int_{S^{(I)}} \rho \mathbf{v}^{(A)} \cdot \check{\mathbf{N}}^{(A)} dS = 0 \tag{6.88}$$

$$\int_{V^{(B)}} \frac{\partial \rho}{\partial t} dV + \int_{S^{(B)}} \rho \mathbf{v} \cdot \check{\mathbf{N}} dS - \int_{S^{(I)}} \rho \mathbf{v}^{(B)} \cdot \check{\mathbf{N}}^{(B)} dS = 0 \tag{6.89}$$

where $v_j^{(A)}$ and $v_j^{(B)}$ are the *fluid* velocity at the interface (which is in general different from the interface velocity at that point, since the interface is not *material*). Summing Eqs. (6.88) and (6.89) yields

$$\int_V \frac{\partial \rho}{\partial t} dV + \int_S \rho \mathbf{v} \cdot \check{\mathbf{N}} dS - \int_{S^{(I)}} \{ \rho v_n \} dS = 0 \tag{6.90}$$

where $v_n = \mathbf{v} \cdot \check{\mathbf{N}}$. Subtracting Eq. (6.90) from (6.85) yields

$$\int_{S^{(I)}} \{ \rho \left( v_n - V_{In} \right) \} dS = 0 \tag{6.91}$$

and since it must be true for any portion of fluid

$$\{ \rho \left( v_n - V_{In} \right) \} = 0 \tag{6.92}$$

which is the *jump mass balance* at the interface.

### 6.4.3   The Jump Species Balance at the Interface

Let now consider a single species $p$ in a multi-component mixture, the species conservation equation, taking a volume defined by a closed material surface, is

$$\frac{d}{dt} \int_V \rho^{(p)} dV = - \int_S \mathbf{j}^{(p)} \cdot \check{\mathbf{N}} dS \tag{6.93}$$

applying Eq. (6.82) to the LHS yields

$$\int_V \frac{\partial \rho^{(p)}}{\partial t} dV + \int_S \rho^{(p)} \mathbf{v} \cdot \check{\mathbf{N}} dS - \int_{S^{(I)}} \left\{ \rho^{(p)} V_{In} \right\} dS = - \int_S \mathbf{j}^{(p)} \cdot \check{\mathbf{N}} dS \quad (6.94)$$

Let now consider each region singularly, where the conservation equation

$$\frac{\partial \rho^{(p)}}{\partial t} + \nabla \cdot \mathbf{n}^{(p)} = 0 \quad (6.95)$$

holds, then necessarily

$$\int_{V^{(A)}} \left( \frac{\partial \rho^{(p)}}{\partial t} + \nabla \cdot \mathbf{n}^{(p)} \right) dV = 0 \quad (6.96)$$

$$\int_{V^{(B)}} \left( \frac{\partial \rho^{(p)}}{\partial t} + \nabla \cdot \mathbf{n}^{(p)} \right) dV = 0 \quad (6.97)$$

applying the divergence theorem the two equations can be written as

$$\int_{V^{(A)}} \frac{\partial \rho^{(p)}}{\partial t} dV + \int_{S^{(A)}} \mathbf{n}^{(p)} \cdot \check{\mathbf{N}} dS - \int_{S^{(I)}} \mathbf{n}_{(A)}^{(p)} \cdot \check{\mathbf{N}}^{(A)} dS = 0 \quad (6.98a)$$

$$\int_{V^{(B)}} \frac{\partial \rho^{(p)}}{\partial t} dV + \int_{S^{(B)}} \mathbf{n}^{(p)} \cdot \check{\mathbf{N}} dS - \int_{S^{(I)}} \mathbf{n}_{(B)}^{(p)} \cdot \check{\mathbf{N}}^{(B)} dS = 0 \quad (6.98b)$$

where $\mathbf{n}_{(A)}^{(p)}$ and $\mathbf{n}_{(B)}^{(p)}$ are the mass fluxes at the interface. Summing Eqs. (6.98a) and (6.98b) yields

$$\int_V \frac{\partial \rho^{(p)}}{\partial t} dV + \int_S \mathbf{n}^{(p)} \cdot \check{\mathbf{N}} dS - \int_{S^{(I)}} \left\{ n_n^{(p)} \right\} dS = 0 \quad (6.99)$$

Subtracting this equation from Eq. (6.94) and recalling Eq. (6.35a) yields

$$\int_{S^{(I)}} \left\{ \rho^{(p)} \left( v_n^{(p)} - V_{In} \right) \right\} dS = 0 \quad (6.100)$$

and since it must be true for any portion of fluid, then

$$\left\{ \rho^{(p)} \left( v_n^{(p)} - V_{In} \right) \right\} = 0 \quad (6.101)$$

which is the *jump species balance* at the interface.

### 6.4.4   The Jump Momentum Balance at the Interface

Let us take again a volume containing an interface and write the integral form of the momentum balance (Eq. 6.58)

$$\frac{d}{dt} \int_V \rho v_j dV = \int_S T_{jk} \check{N}_k dS + \int_V \rho f_j dV \tag{6.102}$$

since we have to deal with the stress tensor now, it is more convenient to switch to the indicial representation in a Cartesian coordinate system. Applying Eq. (6.82) yields the following equation

$$\int_V \frac{\partial \rho v_j}{\partial t} dV + \int_S \rho v_j v_j \check{N}_k dS - \int_{S^{(I)}} \{\rho v_j V_{In}\} \, dS - \int_S T_{jk} \check{N}_k dS - \int_V \rho f_j dV = 0 \tag{6.103}$$

The differential momentum conservation Eq. (6.61)

$$\frac{\partial \rho v_j}{\partial t} + \nabla_k (\rho v_i v_k) - \nabla_k T_{jk} - \rho f_j = 0 \tag{6.104}$$

must hold in both regions $A$ and $B$; taking the integrals over the two regions and transforming the volume integrals of the second and third term into a surface integral yields

$$\int_{V(A)} \frac{\partial \rho v_j}{\partial t} dV + \int_{S(A)} (\rho v_j v_k - T_{jk}) \check{N}_k dS - \int_{S(I)} (\rho v_j v_k - T_{jk}) \check{N}_k^{(A)} dS - \int_{V(A)} \rho f_j dV = 0 \tag{6.105}$$

$$\int_{V(B)} \frac{\partial \rho v_j}{\partial t} dV + \int_{S(B)} (\rho v_j v_k - T_{jk}) \check{N}_k dS - \int_{S(I)} (\rho v_j v_k - T_{jk}) \check{N}_k^{(B)} dS - \int_{V(B)} \rho f_j dV = 0 \tag{6.106}$$

and summation yields

$$\int_V \frac{\partial \rho v_j}{\partial t} dV + \int_S (\rho v_j v_k - T_{jk}) \check{N}_k dS - \int_V \rho f_j dV - \int_{S^{(I)}} \{\rho v_j v_n\} dS + \int_{S^{(I)}} \left\{ T_{jk} \check{N}_k \right\} dS = 0 \tag{6.107}$$

Subtracting this equation from Eq. (6.103) and considering the arbitrariness of the chosen volume yields the *jump momentum balance*

$$\left\{ \rho v_j (v_n - V_{In}) - T_{jk} \check{N}_k \right\} = 0 \tag{6.108}$$

### 6.4.5  The Jump Energy Balance at the Interface

Finally, the same procedure can be applied to the energy balance, starting with the integral form (Chap. 6)

$$\frac{d \int_V \rho \left( \hat{u} + \frac{1}{2} |v|^2 \right) dV}{dt} = - \int_S q_j \check{N}_j dS + \int_S T_{jk} v_j \check{N}_k dS + \int_V \rho f_j v_j dV \quad (6.109)$$

and applying the transport theorem (6.82) the LHS becomes

$$\int_V \frac{\partial \rho \left( \hat{u} + \frac{1}{2} |v|^2 \right) dV}{\partial t} + \int_S \rho \left( \hat{u} + \frac{1}{2} |v|^2 \right) v_k \check{N}_k dS - \int_{S^{(I)}} \left\{ \rho \left( \hat{u} + \frac{1}{2} |v|^2 \right) V_{In} \right\} dS \quad (6.110)$$

Consider now the differential energy Eq. (6.71), then taking the integral over both volumes $A$ and $B$, and using the divergence theorem to transform the integrals of the divergence terms into surface integrals yields

$$\int_{V^{(A)}} \frac{\partial \rho \left( \hat{u} + \frac{1}{2} |v|^2 \right)}{\partial t} dV + \int_{S^{(A)}} \rho \left( \hat{u} + \frac{1}{2} |v|^2 \right) v_j \check{N}_j dS - \int_{S^{(I)}} \rho \left( \hat{u} + \frac{1}{2} |v|^2 \right) v_j \check{N}_j^{(A)} dS = \quad (6.111)$$

$$= - \int_{S^{(A)}} (q_j - T_{jk} v_j) \check{N}_j dV + \int_{S^{(I)}} (q_k - T_{jk} v_j) \check{N}_k^{(A)} dS + \int_{V^{(A)}} \rho f_j v_j dV$$

$$\int_{V^{(B)}} \frac{\partial \rho \left( \hat{u} + \frac{1}{2} |v|^2 \right)}{\partial t} dV + \int_{S^{(B)}} \rho \left( \hat{u} + \frac{1}{2} |v|^2 \right) v_j \check{N}_j dS - \int_{S^{(I)}} \rho \left( \hat{u} + \frac{1}{2} |v|^2 \right) v_j \check{N}_j^{(B)} dS = \quad (6.112)$$

$$= - \int_{S^{(B)}} (q_j - T_{jk} v_j) \check{N}_j dV + \int_{S^{(I)}} (q_k - T_{jk} v_j) \check{N}_k^{(B)} dS + \int_{V^{(B)}} \rho f_j v_j dV$$

and summing yields

$$\int_V \frac{\partial \rho \left( \hat{u} + \frac{1}{2} |v|^2 \right)}{\partial t} dV + \int_S \rho \left( \hat{u} + \frac{1}{2} |v|^2 \right) v_j \check{N}_j dS - \int_{S^{(I)}} \left\{ \rho \left( \hat{u} + \frac{1}{2} |v|^2 \right) v_j \check{N}_j \right\} dS \quad (6.113)$$

$$= - \int_S (q_k - T_{jk} v_j) \check{N}_k dS + \int_V \rho f_j v_j dV + \int_{S^{(I)}} \left\{ (q_k - T_{jk} v_j) \check{N}_k \right\} dS$$

Subtraction from Eq. (6.109) after substituting the LHS with Eq. (6.110) yields

$$\int_{S^{(I)}} \left\{ \rho \left( \hat{u} + \frac{1}{2} |v|^2 \right) (v_n - V_{In}) + (q_k - T_{jk} v_j) \check{N}_k \right\} dS = 0 \quad (6.114)$$

and the *jump energy balance* is

$$\left\{ \rho \left( \hat{u} + \frac{1}{2} |v|^2 \right) (v_n - V_{In}) + (q_k - T_{jk} v_j) \check{N}_k \right\} = 0 \quad (6.115)$$

Other forms for this jump energy balance may also be obtained, for example considering that

$$\rho\hat{u} = \rho\widehat{h} - P \tag{6.116}$$

then

$$0 = \left\{ \rho\left(\hat{u} + \frac{1}{2}|v|^2\right)(v_n - V_{In}) + (q_k - T_{jk}v_j)\,\check{N}_k \right\} =$$

$$= \left\{ \rho\left(\widehat{h} + \frac{1}{2}|v|^2\right)(v_n - V_{In}) - Pv_k\check{N}_k + PV_{In} + (q_k - T_{jk}v_j)\,\check{N}_k \right\} =$$

$$= \left\{ \rho\left(\widehat{h} + \frac{1}{2}|v|^2\right)(v_n - V_{In}) + PV_{In} + (q_k - Pv_k - [-P\delta_{jk} + \tau_{jk}]\,v_j)\,\check{N}_k \right\} =$$

$$= \left\{ \rho\left(\widehat{h} + \frac{1}{2}|v|^2\right)(v_n - V_{In}) + PV_{In} + (q_k - \tau_{jk}v_j)\,\check{N}_k \right\} \tag{6.117}$$

## References

1. Serret, J.A. (ed.): Oeuvres de Lagrange (in French), vol. 1, pp. 151–316. Gauthier-Villars, Paris (1867)
2. Kellogg, O.D.: Foundations of Potential Theory. Springer, Berlin (1967)
3. Reynolds, O.: Collected Papers on Mechanical and Physical Subjects, vol. 3. Cambridge University Press, Cambridge (1903)
4. Leal, L.G.: Advanced Transport Phenomena: Fluid Mechanics and Convective Transport Processes. Cambridge University Press, Cambridge (2007)
5. Bird, R.B., Stewart, W.E., Lightfoot, E.N.: Transport Phenomena, 2nd edn. Wiley, Hoboken (2002)
6. Slattery, C.: Momentum, Energy and Mass Transfer In continua, 2nd edn. Edition R. Krieger Publishing, New York (1981)

# Chapter 7
# Introduction to Constitutive Equations

When the conservation equations for mass, chemical species, momentum and energy were derived in the previous chapter, it became soon evident that the number of unknown functions was far larger than that of the equations. To allow the closure of the problem some quantities need to be related to others and to the properties of matter, and these are the diffusive mass fluxes, $\mathbf{j}^{(p)}$, the deviatoric stress tensor, $\tau_{jk}$, the internal energy per unity of mass, $\hat{u}$ (or the specific enthalpy, $\hat{h}$) and the heat flux, $\mathbf{q}$. The laws that describe these quantities are known as *constitutive equations*, and in thermo-fluids they are inherently empirical, although they must satisfy some requirement based upon first principles, like the condition of *material objectivity* (material properties must be independent of observer), the symmetry properties of a material body and the law of thermodynamics (particularly, the entropy inequality).

The derivation of the most complete forms of the constitutive relations is out of the scope of this book, and the interested reader is invited to refer to specialised books on transport phenomena like [1, 2]. In the next sections we will discuss the most used forms of the constitutive equations with the aim of clarifying the conditions of applicability and the assumptions often introduced to simplify these equations.

## 7.1 Generalised Maxwell–Stefan Equations

The species conservation equations (6.48) and (6.49) are written in terms of species (mass or molar) fluxes, and the species mass fluxes are defined as

$$\mathbf{n}^{(p)} = \mathbf{n}^T \chi^{(p)} + \mathbf{j}^{(p)} \qquad (p = 0, 1, \ldots, n) \tag{7.1}$$

The diffusive mass fluxes $\mathbf{j}^{(p)}$ are unknown and need to be related to the properties of the fluid. An approach based on the kinetic theory of dilute gases was proposed

© Springer Nature Switzerland AG 2021
G. E. Cossali and S. Tonini, *Drop Heating and Evaporation: Analytical Solutions in Curvilinear Coordinate Systems*, Mathematical Engineering,
https://doi.org/10.1007/978-3-030-49274-8_7

by [3] (see also [4]), to express the diffusional fluxes as a function of a set of vectors $\mathbf{d}^{(p)}$, called *diffusional driving forces* [1], (see also [2], page 476)

$$\mathbf{j}^{(p)} = -D_T^{(p)} \nabla \ln T + \rho^{(p)} \sum_{q=0}^{n} \tilde{D}_{qp} \mathbf{d}^{(q)} \tag{7.2}$$

In these equations, $D_T^{(p)}$ are the multi-component *thermal* diffusion coefficients, and $\tilde{D}_{qp}$ are the *multi-component* diffusion coefficients ([2], page 476); to notice that $\tilde{D}_{qp} = \tilde{D}_{pq}$, but $\tilde{D}_{qp}$ are not the *binary diffusion coefficients* and they depend on the concentration of all species. The multi-component thermal diffusion coefficient $D_T^{(p)}$ defines the diffusion of molecules due to thermal gradients. There exists experimental evidence showing that for binary mixtures of dilute gases the species with larger molecular weight migrates to regions of lower temperatures, or, when the molecular weights are similar, the molecules of larger size move to colder regions [1].

Since the sum of all diffusional fluxes must be nil (Eq. 6.34a) the following relations among the coefficients must hold

$$\sum_{p=0}^{n} D_T^{(p)} = 0 \tag{7.3}$$

$$\sum_{p=0}^{n} \rho^{(p)} \sum_{q=0}^{n} \tilde{D}_{qp} d_j^{(q)} = \sum_{q=0}^{n} \left( \sum_{p=0}^{n} \rho^{(p)} \tilde{D}_{qp} \right) d_j^{(q)} = 0 \tag{7.4}$$

and Eq. (7.4) implies

$$\sum_{p=0}^{n} \chi^{(p)} \tilde{D}_{qp} = 0 \tag{7.5}$$

which shows that, as above pointed out, the multicomponent diffusion coefficients must depend on composition.

Equations (7.2) can be re-written by inverting $\mathbf{d}^{(p)}$ with $\mathbf{j}^{(p)}$ (see [5] for a derivation), obtaining the equations

$$\mathbf{d}^{(p)} = \sum_{\substack{q=0 \\ q \neq p}}^{n} \frac{y^{(p)} y^{(q)}}{D_{pq}} \left( \frac{D_T^{(q)}}{\rho^{(q)}} - \frac{D_T^{(p)}}{\rho^{(p)}} \right) \nabla \ln T + \sum_{\substack{q=0 \\ q \neq p}}^{n} \frac{y^{(p)} y^{(q)}}{D_{pq}} \left( \frac{\mathbf{j}^{(q)}}{\rho^{(q)}} - \frac{\mathbf{j}^{(p)}}{\rho^{(p)}} \right) \tag{7.6}$$

These equations are the *generalised Maxwell–Stefan equations*, also called generalised Stefan–Maxwell equations (named after the Scottish physicist James Clerk Maxwell, 1831–1879 and the Slovene physicist Josef Stefan, 1835–1893). It must be noticed that $D_{pq}$ in Eq. (7.6) are different from $\tilde{D}_{pq}$ in Eq. (7.2) and they are generally empirical coefficients referred as *Maxwell–Stefan diffusion coefficients*. However, for dilute gases they are the *binary diffusion coefficients*, and this is why the approach based on Eq. (7.6) should be preferred with respect to that based on Eq. (7.2), when dealing with dilute gas mixtures, since in that case $D_{pq}$ are relatively easier to measure or estimate and they are practically independent of mixture composition. Clearly $\tilde{D}_{pq}$ and $D_{pq}$ are related to each other, but the relations are quite cumbersome; in [2], examples for binary and ternary mixtures are reported.

The last term in Eq. (7.6) can be written, using Eqs. (7.1), under the following alternative forms

$$
\sum_{\substack{q=0 \\ q \neq p}}^{n} \frac{y^{(p)} y^{(q)}}{D_{pq}} \left( \frac{\mathbf{j}^{(q)}}{\rho^{(q)}} - \frac{\mathbf{j}^{(p)}}{\rho^{(p)}} \right) = \sum_{\substack{q=0 \\ q \neq p}}^{n} \frac{y^{(p)} y^{(q)}}{D_{pq}} \left( \frac{\mathbf{n}^{(q)}}{\rho^{(q)}} - \frac{\mathbf{n}^{(p)}}{\rho^{(p)}} \right) = \tag{7.7}
$$

$$
= \sum_{\substack{q=0 \\ q \neq p}}^{n} \frac{1}{c D_{pq}} \left( y^{(p)} \mathbf{N}^{(q)} - y^{(q)} \mathbf{N}^{(p)} \right) =
$$

$$
= \sum_{\substack{q=0 \\ q \neq p}}^{n} \frac{1}{c D_{pq}} \left( y^{(p)} \mathbf{J}^{(q)} - y^{(q)} \mathbf{J}^{(p)} \right)
$$

## 7.2  The Diffusional Forces

The general form of the diffusional forces $\mathbf{d}^{(p)}$ can be obtained by a thermodynamic analysis, which will not be reported here; the interested reader is referred to [2] for a detailed derivation. That analysis shows that $\mathbf{d}^{(q)}$ can be written as (see [2] Sect. 8.3.6-4 page 469, where the terms that account for changes in the kinetic energy of diffusion have been neglected, following [5])

$$
\mathbf{d}^{(p)} = \frac{\rho^{(p)}}{cRT} \left( s^{(p)} \nabla T + \nabla \hat{\mu}^{(p)} - \frac{1}{\rho} \nabla P - \mathbf{f}^{(p)} + \sum_{q=0}^{n} \chi^{(p)} \mathbf{f}^{(p)} \right) \tag{7.8}
$$

where $s^{(p)}$ is the partial entropy, on mass basis, of the component $p$, and $\hat{\mu}^{(p)}$, the chemical potential on mass basis, is defined as

$$
\hat{\mu}^{(p)} = -T \left( \frac{\partial \bar{S}}{\partial M^{(p)}} \right)_{\bar{U}, V, M^{(k \neq p)}} \tag{7.9}
$$

where $\bar{S} = \bar{S} \left( \bar{U}, V, \mathbf{M} \right)$ is the fundamental relation of the mixture ($\bar{U}$ is the internal energy and $\mathbf{M} \equiv \left[ M^{(0)}, \ldots, M^{(n)} \right]$ are the masses of each component). These diffusional forces satisfy the relation $\sum_{p=0}^{n} \mathbf{d}^{(p)} = 0$. In fact, first notice that

$$\sum_{p=0}^{n} \frac{\rho^{(p)}}{cRT} \left( \mathbf{f}^{(p)} - \sum_{q=0}^{n} \chi^{(q)} \mathbf{f}^{(q)} \right) = \sum_{p=0}^{n} \frac{\rho^{(p)}}{cRT} \mathbf{f}^{(p)} - \frac{\rho}{cRT} \sum_{p=0}^{n} \chi^{(p)} \mathbf{f}^{(p)} = 0 \quad (7.10)$$

Take now the Gibbs–Duhem relation (on a mass basis)

$$\sum_{p=0}^{n} d\hat{\mu}^{(p)} \chi^{(p)} + \hat{s}\, dT - \hat{v}\, dP = 0 \quad (7.11)$$

where $\hat{s}$ and $\hat{v}$ are the entropy and volume of the mixture for unit of mass; multiplying all terms by $\rho$ yields

$$\sum_{p=0}^{n} \nabla \hat{\mu}^{(p)} \rho^{(p)} + \rho \hat{s}\, \nabla_j T - \nabla_j P = 0 \quad (7.12)$$

thus

$$\sum_{p=0}^{n} \frac{\rho^{(p)}}{cRT} \left\{ s^{(p)} \nabla_j T + \nabla_j \hat{\mu}^{(p)} - \frac{1}{\rho} \nabla_j P \right\} = \frac{1}{cRT} \left\{ \begin{array}{l} \sum_{p=0}^{n} \rho^{(p)} s^{(p)} \nabla_j T + \\ + \sum_{p=0}^{n} \rho^{(p)} \nabla_j \hat{\mu}^{(p)} - \sum_{p=0}^{n} \rho^{(p)} \frac{1}{\rho} \nabla_j P \end{array} \right\} =$$

$$= \frac{1}{cRT} \left\{ \rho \hat{s} \nabla_j T + \sum_{p=0}^{n} \rho^{(p)} \nabla_j \hat{\mu}^{(p)} - \nabla_j P \right\} = 0$$

$$(7.13)$$

where $\sum_{p=0}^{n} \rho^{(p)} s^{(p)} = \rho \hat{s}$ has been used and the last equality comes from Eq. (7.12).

Equation (7.8) can be re-written in a more useful form observing the following. To calculate the gradient of the chemical potential $\hat{\mu}^{(p)}$ that appears in Eq. (7.8) we need to calculate the partial derivative with respect to $T$, $P$, and all the $\chi^{(q)}$. However, when calculating the partial derivatives of $\hat{\mu}^{(p)}$, it must be remembered that only $n$ mass fractions $\chi^{(q)}$ are independent, since $\chi^{(p)} = 1 - \sum_{\substack{k=0 \\ k \neq p}}^{n} \chi^{(k)}$, then the partial derivative $\left( \frac{\partial \hat{\mu}^{(p)}}{\partial \chi^{(k)}} \right)_{T,P\chi^{(m \neq k)}}$ must be substituted by $\left( \frac{\partial \hat{\mu}^{(p)}}{\partial \chi^{(k)}} \right)_{T,P\chi^{(m \neq k,p)}}$.

The gradient of $\hat{\mu}^{(p)}$ can then be expanded as

$$\nabla_j \hat{\mu}^{(p)} = \sum_{q=0}^{n} \left( \frac{\partial \hat{\mu}^{(p)}}{\partial \chi^{(q)}} \right)_{T,P,\chi^{(k \neq q,p)}} \nabla_j \chi^{(q)} + \left( \frac{\partial \hat{\mu}^{(p)}}{\partial T} \right)_{P,\chi^{(m)}} \nabla_j T + \left( \frac{\partial \hat{\mu}^{(p)}}{\partial P} \right)_{T,\chi^{(m)}} \nabla_j P =$$

$$= \sum_{q=0}^{n} \left( \frac{\partial \hat{\mu}^{(p)}}{\partial \chi^{(q)}} \right)_{T,P,\chi^{(k \neq q,p)}} \nabla_j \chi^{(q)} - s^{(p)} \nabla_j T + v^{(p)} \nabla_j P \quad (7.14)$$

since (see also [6], page 273)

$$\left(\frac{\partial \hat{\mu}^{(p)}}{\partial T}\right)_{P,\chi^{(k)}} = -s^{(p)} \tag{7.15}$$

$$\left(\frac{\partial \hat{\mu}^{(p)}}{\partial P}\right)_{T,\chi^{(k)}} = v^{(p)} \tag{7.16}$$

When performing the partial derivatives with respect to $\chi^{(q)}$, while keeping the other $\chi^{(k)}$ constant with the exception of $\chi^{(q)}$ and $\chi^{(p)}$, like $\left(\frac{\partial \hat{\mu}^{(p)}}{\partial \chi^{(q)}}\right)_{T,P,\chi^{(k\neq q,p)}}$, the condition $\sum_{k=0}^{n} \chi^{(k)} = 1$ has also another implication. Since

$$\chi^{(q)} + \chi^{(p)} = 1 - \sum_{\substack{k=0 \\ k\neq p,q}}^{n} \chi^{(k)} \tag{7.17}$$

and the sum on the RHS is kept constant when performing the derivation, then $\chi^{(q)} = -\chi^{(p)} + const$ and

$$\left(\frac{\partial \hat{\mu}^{(p)}}{\partial \chi^{(q)}}\right)_{\chi^{(k\neq p,q)}} = -\left(\frac{\partial \hat{\mu}^{(p)}}{\partial \chi^{(p)}}\right)_{\chi^{(k\neq p,q)}} \tag{7.18}$$

Using Eqs. (7.18) and (7.14), Eq. (7.8) yields the following form for the diffusional forces

$$\mathbf{d}^{(p)} = \frac{\rho^{(p)}}{cRT} \left\{ -\sum_{\substack{r=0 \\ r\neq p}}^{n} \left(\frac{\partial \hat{\mu}^{(p)}}{\partial \chi^{(p)}}\right)_{T,P,\chi^{(k\neq r,p)}} \nabla \chi^{(r)} + \left(v^{(p)} - \frac{1}{\rho}\right) \nabla P - \mathbf{f}^{(p)} + \sum_{r=0}^{n} \chi^{(r)} \mathbf{f}^{(r)} \right\} \tag{7.19}$$

This equation can also be written on molar basis, i.e. using molar fractions and the chemical potential on molar basis $\breve{\mu}^{(p)}$ (see also [2] Sect. 8.4.5-4) as

$$\mathbf{d}^{(p)} = \frac{y^{(p)}}{RT} \left\{ \begin{array}{l} -\sum_{\substack{r=0 \\ r\neq p}}^{n} \left(\frac{\partial \breve{\mu}^{(p)}}{\partial y^{(p)}}\right)_{T,P,y^{(k\neq p,r)}} \nabla y^{(r)} + Mm^{(p)} \left(v^{(p)} - \frac{1}{\rho}\right) \nabla P + \\ -Mm^{(p)} \left(\mathbf{f}^{(p)} - \sum_{r=0}^{n} \chi^{(r)} \mathbf{f}^{(r)}\right) \end{array} \right\} \tag{7.20}$$

## 7.3 Components of the Diffusive Fluxes

When Eq. (7.19) or Eq. (7.20) are substituted into Eq. (7.2), the diffusive fluxes can be split into four components

$$\mathbf{j}^{(k)} = \mathbf{j}_f^{(k)} + \mathbf{j}_P^{(k)} + \mathbf{j}_T^{(k)} + \mathbf{j}_o^{(k)} \tag{7.21}$$

where

$$
\begin{aligned}
\mathbf{j}_f^{(p)} &= -\rho^{(p)} \sum_{q=0}^n \tilde{D}_{qp} \frac{y^{(q)}}{RT} M m^{(q)} \left( \mathbf{f}^{(q)} - \sum_{r=0}^n \chi^{(r)} \mathbf{f}^{(r)} \right) & \text{force} \\
\mathbf{j}_P^{(p)} &= \rho^{(p)} \sum_{q=0}^n \tilde{D}_{qp} \frac{y^{(q)}}{RT} M m^{(q)} \left( v^{(q)} - \frac{1}{\rho} \right) \nabla P & \text{pressure} \\
\mathbf{j}_T^{(p)} &= -D_T^{(p)} \nabla \ln T & \text{thermal (Soret)} \\
\mathbf{j}_o^{(p)} &= \rho^{(p)} \sum_{q=0}^n \tilde{D}_{qp} \frac{y^{(q)}}{RT} \sum_{\substack{r=0 \\ r \neq q}}^n \left( \frac{\partial \breve{\mu}^{(q)}}{\partial y^{(r)}} \right)_{T,P,y^{(k \neq q,r)}} \nabla y^{(r)} & \text{ordinary}
\end{aligned}
\tag{7.22}
$$

The *force* diffusion component $\mathbf{j}_f^{(p)}$ is caused by external forces and only exists when the external force acting over the species depends on the species; this may happen for example for electrolytes in an electric field, while for neutral species it is nil. The *pressure* component $\mathbf{j}_P^{(p)}$ becomes important only when strong pressure gradients are present; for most of the ordinary conditions it is negligible. The *Soret* component $\mathbf{j}_T^{(p)}$ is the diffusion caused by temperature gradients inside the mixture. It is not frequent, but for combustion conditions, where temperature gradients may be high, it may have an effect. Finally, $\mathbf{j}_o^{(p)}$ is the *ordinary* diffusive flux, caused by the existence of concentration gradients inside the mixture, which is the most important in evaporative processes.

## 7.4   Dilute Gas Mixtures

A quite common assumption when dealing with gases is that of ideal mixture, which implies that (refer to [6])

$$
\breve{\mu}^{(p)} = \breve{\mu}^{0(p)} + RT \ln y^{(p)} \tag{7.23}
$$

where $\breve{\mu}^{0(p)}$ is the corresponding value of the chemical potential for the pure component ($y^{(p)} = 1$) and

$$
\left( \frac{\partial \breve{\mu}^{(p)}}{\partial y^{(q)}} \right)_{T,P,y^{(r \neq q,p)}} = \frac{RT}{y^{(p)}} \delta_{qp} \tag{7.24}
$$

where $\delta_{qp}$ is the Kronecker delta. Substituting Eq. (7.24) in Eq. (7.20) and neglecting the *pressure* and *force* components in Eq. (7.20) yields the simple expression

$$
\mathbf{d}^{(p)} = \nabla y^{(p)} \tag{7.25}
$$

Equation (7.2) then assumes the simple form

$$
\mathbf{j}^{(p)} = -D_T^{(p)} \nabla \ln T + \rho^{(p)} \sum_{q=0}^n \tilde{D}_{qp} \nabla y^{(q)} \tag{7.26}
$$

that, for negligible temperature gradients, becomes

$$\mathbf{j}^{(p)} = \rho^{(p)} \sum_{q=0}^{n} \tilde{D}_{qp} \nabla y^{(q)} \tag{7.27}$$

and, under the same assumptions, Eqs. (7.6) yields

$$\mathbf{d}^{(p)} = \nabla y^{(p)} = \sum_{\substack{q=0 \\ q \neq p}}^{n} \frac{y^{(p)} y^{(q)}}{D_{pq}} \left( \frac{\mathbf{j}^{(q)}}{\rho^{(q)}} - \frac{\mathbf{j}^{(p)}}{\rho^{(p)}} \right) \tag{7.28}$$

which are known as Maxwell–Stefan (or Stefan–Maxwell) equations. These equations appeared first in a work of Maxwell in 1866 [7], which was further developed by Stefan in 1871 [8]; they were experimentally tested by [9–11], yielding a good agreement between theory and experiments. Refer to [12] for a simplified derivation of Eqs. (7.28) from molecular dynamics.

## 7.5 Binary Mixtures and Fick's Law

Let now consider the simplest case of mixtures: a binary mixture. Assuming negligible Soret effect, the Maxwell–Stefan equations (7.6) simplify, and since there is only ($n = 1$) independent equation and $\mathbf{j}^{(0)} = -\mathbf{j}^{(1)}$ one obtains

$$\mathbf{d}^{(1)} = -\frac{y^{(1)} y^{(0)}}{D_{10}} \left( \frac{\rho}{\rho^{(0)} \rho^{(1)}} \right) \mathbf{j}^{(1)} \tag{7.29}$$

or

$$\mathbf{j}^{(1)} = -\frac{c^2}{\rho} D_{10} M m^{(0)} M m^{(1)} \mathbf{d}^{(1)} \tag{7.30}$$

where Eqs. (6.24) have been used.

Assuming only ordinary diffusion, i.e. no external force or pressure gradient effects, Eq. (7.20) yields (to notice that $\nabla y^{(1)} = -\nabla y^{(0)}$)

$$\mathbf{d}^{(1)} = \frac{y^{(1)}}{RT} \left( \frac{\partial \breve{\mu}^{(1)}}{\partial y^{(1)}} \right)_{T,P} \nabla y^{(1)} \tag{7.31}$$

and Eq. (7.30) becomes

$$\mathbf{j}^{(1)} = -\frac{c^2}{\rho} M m^{(0)} M m^{(1)} \frac{y^{(1)}}{RT} \left( \frac{\partial \breve{\mu}^{(1)}}{\partial y^{(1)}} \right)_{T,P} D_{10} \nabla y^{(1)} \tag{7.32}$$

Now, from Eq. (6.24)

$$\nabla y^{(1)} = \frac{\rho}{cMm^{(1)}}\nabla\chi^{(1)} + \frac{\chi^{(1)}}{Mm^{(1)}}\nabla\frac{\rho}{c} = \frac{\rho}{cMm^{(1)}}\nabla\chi^{(1)} + \frac{\chi^{(1)}}{Mm^{(1)}}\left(Mm^{(1)} - Mm^{(0)}\right)\nabla y^{(1)}$$
(7.33)

where the last equality has been obtained from Eq. (6.26), then

$$\nabla y^{(1)} = \frac{\rho}{cMm^{(1)}\left(1 + \frac{\chi^{(1)}}{Mm^{(1)}}\left(Mm^{(0)} - Mm^{(1)}\right)\right)}\nabla\chi^{(1)}$$
(7.34)

and since from Eq. (6.25) it stems that

$$Mm^{(0)}\frac{c}{\rho} = \left[1 + \frac{\chi^{(1)}\left(Mm^{(0)} - Mm^{(1)}\right)}{Mm^{(1)}}\right]$$
(7.35)

then

$$\nabla y^{(1)} = \frac{\rho^2}{c^2 Mm^{(1)}Mm^{(0)}}\nabla\chi^{(1)}$$
(7.36)

Substituting Eq. (7.36) into Eq. (7.32) yields

$$\mathbf{j}^{(1)} = -\frac{y^{(1)}}{RT}\left(\frac{\partial\check{\mu}^{(1)}}{\partial y^{(1)}}\right)_{T,P}\rho D_{10}\nabla\chi^{(1)}$$
(7.37)

and since for a mixture of ideal gases equation (7.24) holds, i.e. $\frac{y^{(1)}}{RT}\left(\frac{\partial\check{\mu}^{(1)}}{\partial y^{(1)}}\right)_{T,P} = 1$, then

$$\mathbf{j}^{(1)} = -\rho D_{10}\nabla\chi^{(1)}$$
(7.38)

which is commonly referred as *Fick's law* or, more precisely, *first Fick's law* of binary diffusion (named after the German physician and physiologist Adolf Eugen Fick, 1829–1901).

From the relations between molar and mass fluxes (Eq. 6.42), another expression for the Fick's law can be obtained

$$\mathbf{J}^{(1)} = \mathbf{j}^{(1)}\frac{\rho}{cMm^{(0)}Mm^{(1)}} = -c\frac{y^{(1)}}{RT}\left(\frac{\partial\check{\mu}^{(1)}}{\partial y^{(1)}}\right)_{T,P}D_{10}\nabla y^{(1)}$$
(7.39)

that for a mixture of ideal gases yields

$$\mathbf{J}^{(1)} = -c D_{10}\nabla y^{(1)}$$
(7.40)

It is worth to notice that for binary mixtures equation (7.38) and (7.40) are perfectly equivalent, and the use of one or the other form is just a matter of choice (see also the discussion in [13]).

## 7.6 Fick's Law in Multi-component Mixtures

In this section we assume that *pressure*, *force* and *thermal* (*Soret*) effects can be neglected. The simplicity of Eqs. (7.38) and (7.40), which hold for binary mixtures, i.e. when one component is diffusing into another, has hinted the possibility of describing diffusional fluxes in a multi-component mixture in a similar simple manner through the use of an approximate equation like

$$\mathbf{J}^{(p)} = -cD_p^{(m)}\nabla y^{(p)} \tag{7.41}$$

where $D_p^{(m)}$ must be considered as the *diffusion coefficient of component p into the mixture*. This approach is considered acceptable when dealing with the diffusion in a mixture where the component $p$ is present as trace [2], although it is often used also for conditions where this assumption may not hold.

To notice that Eq. (7.41) is often substituted by the analogous

$$\mathbf{j}^{(p)} = -\rho D_p^{(m)}\nabla \chi^{(p)} \tag{7.42}$$

With the exception of the binary mixture case, Eqs. (7.41) and (7.42) are not equivalent. In fact, from Eq. (6.24)

$$\nabla y^{(p)} = \frac{\rho}{cMm^{(p)}}\nabla \chi^{(p)} + \frac{\chi^{(p)}}{Mm^{(p)}}\nabla\frac{\rho}{c} \tag{7.43}$$

then using (7.41)

$$\mathbf{j}^{(p)} = Mm^{(p)}\mathbf{J}^{(p)} = -Mm^{(p)}cD_p^{(m)}\nabla y^{(p)} = -\rho D_p^{(m)}\nabla \chi^{(p)} - cD_p^{(m)}\chi^{(p)}\nabla\frac{\rho}{c} \tag{7.44}$$

and Eq. (7.42) can be obtained only assuming that the gradients of $\frac{\rho}{c} = \sum_{p=0}^{n} Mm^{(p)}y^{(p)} = \left(\sum_{p=0}^{n} \frac{\chi^{(p)}}{Mm^{(p)}}\right)^{-1}$ (see Eqs. 6.26, 6.25) can be neglected.

The diffusion coefficient $D_p^{(m)}$ is usually a function of the concentration of all components and there exist empirical relations used to evaluate it from the binary diffusion coefficients.

Blanc [14], studying the mobility of ions in binary gaseous mixtures, found that the reciprocal of ion mobility was linearly dependent on the partial pressures of each gas; since it was also found [15] that the mobility is proportional to the diffusion coefficient, the following rule, called *Blanc's law*, can be stated as a general outcome of such investigation

$$\frac{1}{D_j^{(m)}} = \sum_{\substack{k=0 \\ k \neq j}}^{n} \frac{y^{(k)}}{D_{jk}} \tag{7.45}$$

Assuming that the component $p = n$ diffuses in a stagnant mixture, Wilke [16] derived the following law to evaluate the effective diffusion coefficient

$$D_p^{(m)} = \frac{1 - y^{(p)}}{\displaystyle\sum_{\substack{j=0 \\ j \neq p}}^{n} \frac{y^{(j)}}{D_{pj}}} \tag{7.46}$$

Other rules were proposed to evaluate $D_p^{(m)}$, for example in [17] (see also [12]) the following rule was proposed

$$\frac{1}{D_n^{(m)}} = \frac{y^{(n)}}{D_{n0}} + \sum_{p=0}^{n-1} \frac{y^{(p)}}{D_{np}} \tag{7.47}$$

The interested reader can refer to [12] for a more extended analysis.

All these formulas can be obtained by comparing Eq. (7.41) with the Stefan–Maxwell equations (7.28) and applying some simplifying assumptions. Consider the Stefan–Maxwell equations for dilute gas mixtures, which can be written for the component $p = n$ as

$$\nabla y^{(n)} = y^{(n)} \sum_{p=0}^{n-1} \frac{y^{(p)}}{cD_{np}} \left( \frac{\mathbf{J}^{(p)}}{y^{(p)}} \right) - \left( \sum_{p=0}^{n-1} \frac{y^{(p)}}{cD_{np}} \right) \mathbf{J}^{(n)} \tag{7.48}$$

eliminating $\nabla y^{(n)}$ between this equation and Eq. (7.41) yields the following equation

$$y^{(n)} \sum_{p=0}^{n-1} \frac{1}{D_{np}} \mathbf{J}^{(p)} - \left( \sum_{p=0}^{n-1} \frac{y^{(p)}}{D_{np}} \right) \mathbf{J}^{(n)} = -\frac{1}{D_n^{(m)}} \mathbf{J}^{(n)} \tag{7.49}$$

Now, assuming that the component $p = n$ is present in traces, Eq. (7.49) can be simplified by the condition $y^{(n)} \to 0$, yielding

$$\left( \sum_{p=0}^{n-1} \frac{y^{(p)}}{D_{np}} \right) \mathbf{J}^{(n)} = \frac{1}{D_n^{(m)}} \mathbf{J}^{(n)} \tag{7.50}$$

or

$$\sum_{p=0}^{n-1} \frac{y^{(p)}}{D_{np}} = \frac{1}{D_n^{(m)}} \tag{7.51}$$

i.e. the Blanc's law. To notice that since (see Eqs. 6.24 and 6.27)

$$y^{(p)} = \frac{\rho}{cMm^{(p)}}\chi^{(p)}; \quad \overline{Mm} = \frac{\rho}{c} = \sum_{p=0}^{n} Mm^{(p)} y^{(p)} \tag{7.52}$$

then

$$\frac{1}{D_n^{(m)}} = \frac{\rho}{c}\sum_{p=0}^{n-1} \frac{1}{D_{np}}\frac{\chi^{(p)}}{Mm^{(p)}} = \sum_{p=0}^{n-1} \frac{1}{D_{np}}\frac{\overline{Mm}}{Mm^{(p)}}\chi^{(p)} \tag{7.53}$$

and when $\frac{\overline{Mm}}{Mm^{(p)}} \sim 1$ the Blanc's law can also be written in the form

$$\frac{1}{D_n^m} = \sum_{p=0}^{n-1} \frac{1}{D_{np}}\chi^{(p)} \tag{7.54}$$

which then holds only when the species $p = n$ diffuses into a mixture with similar average molar weight.

Let instead assume that the component $p = n$ is diffusing in a *stagnant* mixture, i.e. that $\mathbf{N}^{(q)} = 0$ for $q \neq n$; as a consequence: $\mathbf{N}^{(n)} = \mathbf{N}^T$; this assumption implies that

$$0 = \mathbf{N}^{(q)} = \mathbf{N}^T y^{(q)} + \mathbf{J}^{(q)} \Rightarrow \mathbf{J}^{(q)} = -\mathbf{N}^T y^{(q)} \tag{7.55}$$

$$\mathbf{N}^T = \mathbf{N}^{(n)} = \mathbf{N}^T y^{(n)} + \mathbf{J}^{(n)} \Rightarrow \mathbf{J}^{(n)} = \mathbf{N}^T \left(1 - y^{(n)}\right) \tag{7.56}$$

Substituting into Eq. (7.49) yields

$$-y^{(n)}\sum_{p=0}^{n-1}\frac{1}{D_{np}}\mathbf{N}^T y^{(p)} - \left(\sum_{p=0}^{n-1}\frac{y^{(p)}}{D_{np}}\right)\mathbf{N}^T\left(1 - y^{(n)}\right) = -\frac{1}{D_n^{(m)}}\mathbf{N}^T\left(1 - y^{(n)}\right) \tag{7.57}$$

and simplifying

$$D_n^{(m)} = \frac{1 - y^{(n)}}{\sum_{q=0}^{n-1}\frac{y^{(q)}}{D_{nq}}} \tag{7.58}$$

i.e. Wilke's formula. To notice that in this case the molar fraction of the $p = n$ component does not need to be small; however Blanc's formula is recovered when $y^{(n)} \to 0$.

Finally, since

$$\mathbf{J}^{(0)} = -\sum_{p=1}^{n-1} \mathbf{J}^{(p)} - \mathbf{J}^{(n)} \tag{7.59}$$

equation (7.49) can be written, eliminating $\mathbf{J}^{(0)}$, as

$$y^{(n)} \sum_{p=1}^{n-1} \left( \frac{1}{D_{np}} - \frac{1}{D_{n0}} \right) \mathbf{J}^{(p)} - \left[ \frac{y^{(n)}}{D_{n0}} + \left( \sum_{p=0}^{n-1} \frac{y^{(p)}}{D_{np}} \right) \right] \mathbf{J}^{(n)} = -\frac{1}{D_n^{(m)}} \mathbf{J}^{(n)} \tag{7.60}$$

and the rule given by Eq. (7.47) holds if the first term of the LHS can be assumed to be negligible.

All the above mentioned rules may be used to evaluate $D_n^{(m)}$ and the interested reader can again refer to [12] for a quantitative comparison among them.

It is worth to notice that when components $p = 1, \ldots, n$ are present in traces into the mixture, while $p = 0$ is the main component, the use of Eq. (7.41) should be restricted to the evaluation of the fluxes $\mathbf{J}^{(1)}, \ldots, \mathbf{J}^{(n)}$, since the use of a similar relation for $\mathbf{J}^{(0)}$ would yield a contradiction when trying to satisfy the relation

$$\sum_{p=0}^{n} \mathbf{J}^{(p)} = 0 \tag{7.61}$$

Despite of this problem, Eq. (7.41) has been commonly used to predict vapour diffusion in models for multi-component drop evaporation [18–20]. To avoid inconsistencies when modelling multi-component drop evaporation, the use of the Maxwell–Stefan equations is advised (see also [21] for details).

## 7.7 Constitutive Equations for the Heat Flux

The most common constitutive equation for the heat flux in an isotropic medium is the so-called Fourier's law

$$q_j = -k\nabla_j T \tag{7.62}$$

where the thermal conductivity $k$ is a property of the matter, that may depend on pressure, temperature and composition. Equation (7.62) shows that the vectors $\mathbf{q}$ and $\nabla T$ are parallel and they point in the opposite direction, which is a necessary condition for respecting the second law of thermodynamics. If the material is anisotropic, this equation is substituted by

$$q_j = -\sum_{p=1}^{3} k_{jp} \nabla_p T \tag{7.63}$$

where $k_{jp}$ is the thermal conductivity tensor, symmetric with respect to the indexes (see [22], page 236), and in this case the vector $\mathbf{q}$ and the vector $\nabla T$ are not necessarily parallel. Many solids with particular structure can be anisotropic, and also some liquids have this property. For example, polymeric liquids may exhibit non-isotropic heat transport behaviour when not at rest (see [23, 24]).

Equation (7.62), combined with the energy equation for rigid bodies, leads to a diffusion equation that has the rather unphysical property of yielding instantaneous propagation of thermal pulses. The first to propose a correction to Fourier's law, based on the kinetic theory of gases, was Cattaneo [25, 26] and later Vernotte [27] proposed independently a similar correction, that can be written, in analogy of Maxwell model of linear viscoelasticity [28] as

$$q_j + \frac{\partial q_j}{\partial t} t_0 = -k \nabla_j T \tag{7.64}$$

where $t_0$ is called *relaxation time*; this equation is sometimes called Cattaneo–Vernotte equation and Fourier's law is obtained by setting $t_0 = 0$. The combination of this equation with the energy equation leads, under the assumption of constant properties, to a partial differential equation that is hyperbolic instead of parabolic, which assures a finite propagation velocity of thermal pulses. Equation (7.64) was later derived in different ways by many authors (see for example [29–32], the latter reporting an accurate bibliographic chronology). Further modifications of equation (7.64) were also proposed (see for example [32] for a *Jeffrey-type* conduction equation and a deep analysis of its implications). These modifications of the Fourier's law, while solving the above mentioned paradox of instantaneous propagation, are not without drawbacks, and criticisms were raised by different authors [33–35]. Moreover, the relaxation time $t_0$ is expected to be very small and no fully reliable measurements appear to be available in the literature.

In a multi-component mixture, the energy *diffusional flux*, i.e. the heat flux, must depend also on diffusional fluxes and the most general constitutive equation is the following [1, 2, 36]

$$\mathbf{q} = -k\nabla T + \sum_{p=0}^{n} \mathbf{j}^{(p)} h^{(p)} + cRT \sum_{p=0}^{n} \sum_{\substack{q=0 \\ q \neq p}}^{n} \frac{D_T^{(p)} y^{(p)} y^{(q)}}{\rho^{(p)} D_{pq}} \left( \frac{\mathbf{j}^{(p)}}{\rho^{(p)}} - \frac{\mathbf{j}^{(q)}}{\rho^{(q)}} \right) \tag{7.65}$$

where $h^{(p)}$ is the partial enthalpy per unit of mass of component $p$ that is defined as [6]

$$h^{(p)} = \left( \frac{\partial \bar{H}}{\partial M^{(p)}} \right)_{T, P, M^{(q \neq p)}} \tag{7.66}$$

where $M^{(p)}$ is the mass of the species $p$ and $\bar{H}$ is the enthalpy of the mixture.

It can be observed that $\mathbf{q}$ is the sum of three components, a *conductive* flux, proportional to the temperature gradient, a *diffusive* flux, that accounts for the molecular transport of enthalpy, and a contribution that depends on diffusional forces and it accounts for the *Dufour* effect (named after the Swiss physicist Louis Dufour, 1832–1892), which is usually small and often neglected; the expression for this last term reported in Eq. (7.65) was given by [36] and it can be found in [1].

## 7.8 Constitutive Equations for the Deviatoric Stress Tensor

The form of the deviatoric stress tensor (also called *viscous stress tensor*) depends on the fluid properties and in this respect fluids are often divided into two categories: *Newtonian* fluids (named after the celebrated physicist and mathematician Isaac Newton, 1642–1727), like gases and liquids of relatively low molecular weight, which follows a generalised form of Newton's law of viscosity, first proposed by Newton in his Principia Mathematica [37], and *non-Newtonian* fluids, with complex structure like polymeric solutions, suspensions, emulsions, etc., which follow more complex laws.

Newton's law of viscosity postulates that in a steady state 2-D shear flow the viscous stress is proportional to the gradient of the velocity. A generalisation of this empirical observation is the following. First observe that the generalisation of the 2-D shear flow velocity gradient is the so-called *deformation tensor* $\frac{\partial v_k}{\partial x_j}$, and its symmetric, $\epsilon_{kj}$, and skew-symmetric, $\Omega_{kj}$, parts

$$\epsilon_{kj} = \frac{1}{2}\left(\frac{\partial v_k}{\partial x_j} + \frac{\partial v_j}{\partial x_k}\right) \tag{7.67a}$$

$$\Omega_{kj} = \frac{1}{2}\left(\frac{\partial v_k}{\partial x_j} - \frac{\partial v_j}{\partial x_k}\right) \tag{7.67b}$$

are called *rate of strain* tensor and *spin* tensor, respectively. The spin tensor is fully defined by only three non-nil values and it is directly related to the flow vorticity $\omega_k = \varepsilon_{klm}\frac{\partial v_m}{\partial x_l}$ by the relation

$$\Omega_{kj} = -\frac{1}{2}\varepsilon_{kjm}\omega_m \tag{7.68}$$

The generalisation of the proportionality between stress and velocity gradient, postulated by the Newton's law, is a proportionality between the viscous stress tensor and the deformation tensor, and the most general form is

$$\tau_{jk} = a_{jkpq}\frac{\partial v_p}{\partial x_q} \tag{7.69}$$

that involves 81 components of the tensor $a_{jkpq}$. However, some requirements on the properties of $a_{jkpq}$ allow to simplify the problem. A fluid like a gas or a simple liquid is isotropic, i.e. its properties do not depend on direction. Isotropy of $a_{jkpq}$ is a consequence of this requirement and the most general isotropic fourth order tensor must have the following form [38]

$$a_{jkpq} = a\delta_{jk}\delta_{pq} + b\delta_{jp}\delta_{kq} + c\delta_{jq}\delta_{kp} \tag{7.70}$$

The symmetry of $\tau_{jk}$ imposes a further restriction: $a_{jkpq} = a_{kjpq}$, which implies

$$a\delta_{jk}\delta_{pq} + b\delta_{jp}\delta_{kq} + c\delta_{jq}\delta_{kp} = a\delta_{kj}\delta_{pq} + b\delta_{kp}\delta_{jq} + c\delta_{kq}\delta_{jp} = a\delta_{jk}\delta_{pq} + c\delta_{jp}\delta_{kq} + b\delta_{jq}\delta_{kp} \tag{7.71}$$

where the most RHS is obtained from the second one using the symmetry of $\delta_{pq}$ and rearranging the terms; the obvious consequence is $b = c$ and

$$a_{jkpq} = a\delta_{jk}\delta_{pq} + b\left(\delta_{jp}\delta_{kq} + \delta_{jq}\delta_{kp}\right) \tag{7.72}$$

Substituting Eq. (7.72) into Eq. (7.69) yields

$$\tau_{jk} = a\delta_{jk}\delta_{pq}\frac{\partial v_p}{\partial x_q} + b\left(\delta_{jp}\delta_{kq} + \delta_{jq}\delta_{kp}\right)\frac{\partial v_p}{\partial x_q} = a\delta_{jk}\frac{\partial v_p}{\partial x_p} + b\left(\frac{\partial v_j}{\partial x_k} + \frac{\partial v_k}{\partial x_j}\right) =$$
$$= \lambda\delta_{jk}\epsilon_{pp} + 2\mu\epsilon_{jk} \tag{7.73}$$

where $\mu$ is called *shear viscosity* and $\lambda$ is called *dilatational viscosity* (sometimes also *volume viscosity* or *bulk viscosity*). Differently form the shear viscosity, the dilatational viscosity of many fluids is often poorly known, although in many cases its value is larger that the shear viscosity [38]. However, in many applications, the viscous stress tensor becomes important under conditions where the velocity divergence $\nabla \cdot \mathbf{v}$ is small and the relative term in the stress tensor formula becomes neglectful. This is also true in liquids, where incompressibility assumption ($\nabla \cdot \mathbf{v} = 0$) is usually justified. It should also be remembered that *transport properties* like $\mu$ and $\lambda$ are expected to depend also on the flow characteristics (like velocity gradients); only the shear viscosity for Newtonian fluids is independent of flow characteristics.

The final form of the viscous stress tensor $\tau_{jk}$ for *Newtonian* fluids is then

$$\tau_{jk} = \mu\left(\frac{\partial v_j}{\partial x_k} + \frac{\partial v_k}{\partial x_j}\right) + \lambda\delta_{jk}\frac{\partial v_p}{\partial x_p} \tag{7.74}$$

and the stress tensor (see Eq. 6.62) becomes

$$T_{jk} = -P\delta_{jk} + \mu\left(\frac{\partial v_j}{\partial x_k} + \frac{\partial v_k}{\partial x_j}\right) + \lambda\delta_{jk}\frac{\partial v_p}{\partial x_p} \tag{7.75}$$

A little discussion about the meaning of the pressure term in Eq. (7.75) is required. Sometimes the pressure is defined from a mechanical point of view, based on the trace of the stress tensor as

$$\hat{P} = -\frac{1}{3}T_{kk} \tag{7.76}$$

Taking this definition for the pressure appearing in Eq. (7.75) yields

$$\hat{P} = -\frac{1}{3}T_{jk} = \frac{1}{3}P\delta_{jk} - \frac{2}{3}\mu\frac{\partial v_p}{\partial x_p} - \lambda\frac{\partial v_p}{\partial x_p} \tag{7.77}$$

and the statement $P = \hat{P}$ yields the relation

$$\lambda = -\frac{2}{3}\mu \tag{7.78}$$

This relation holds for ideal mono-atomic gases, and in such cases $\hat{P} = -\frac{1}{3}T_{kk}$ is equal to the thermodynamic pressure $P$, but in general $\hat{P}$ differs from the thermodynamic pressure, since the stress tensor depends on the flow condition of the fluid while $P$ depends on the thermodynamic state.

As said at the beginning of this section, not all fluids follow the constitutive equation (7.74). Fluids that deviate from the behaviour described by Eq. (7.74) are called *non-Newtonian* fluids. Those fluids can exhibit different unusual behaviour, for example (i) Bingham fluids (named after the American chemist Eugene Cook Bingham, 1878–1945) behave like rigid solids at low stress, but they flow when the stress becomes larger, like mud or toothpaste; (ii) some liquids (shear-thickening fluid) show an increase in viscosity with the rate of strain, this happens for example when a liquid contains suspended solid particles; (iii) polymer solutions usually exhibit a decrease of viscosity when the rate of strain increases, and are called shear-thinning fluids. There are many other deviations from the Newtonian behaviour, but it is out of the scope of this book to go into details on non-Newtonian fluids, and the interested reader can refer to [39–43] for more information.

# References

1. Bird, R.B., Stewart, W.E., Lightfoot, E.N.: Transport Phenomena, 2nd edn. Wiley, Hoboken (2002)
2. Slattery, C.: Momentum, Energy and Mass Transfer in Continua, 2nd edn. R. Krieger Publishing, New York (1981)
3. Curtiss, C.F.: Symmetric gaseous diffusion coefficients. J. Chem. Phys. **49**, 2917–2919 (1968)
4. Condiff, D.W.: On symmetric multicomponent diffusion coefficients. J. Chem. Phys. **51**, 4209–4212 (1969)
5. Hirschfelder, J.O., Curtiss, C.F., Bird, R.B.: Molecular Theory of Gases and Liquids. Wiley, New York (1964)

6. Gyftopoulos, E.P., Beretta, G.P.: Thermodynamics: Foundations and Applications. Macmillan, New York (1991)
7. Maxwell, J.C.: On the dynamical theory of gases. Philos. Trans. R. Soc. **157**, 49–88 (1866)
8. Stefan, J.: Uber das Gleichgewicht und die Bewegung, insbesondere die Diffusion von Gasmengen (in German). Sitzungsber. Akad. Wiss. Wien **63**, 63–124 (1871)
9. Hesse, D., Hugo, P.: Untersuchung der Mehrkomponenten-Diffusion im Gemisch Wasserstoff, Athylen und Athan mit einem Stationarem Messverfahren (in German). Chem. Ing. Tech. **44**, 1312–1318 (1972)
10. Carty, R., Schrodt, J.T.: Concentration profiles in ternary gaseous diffusion. Ind. Eng. Chem. Fundam. **14**, 276–278 (1975)
11. Bres, M., Hatzfeld, C.: Three gas diffusion-experimental and theoretical studies. Pflügers Arch. **371**, 227–233 (1977)
12. Taylor, R., Krishna, R.: Multicomponent Mass Transfer. Wiley, Hoboken (1993)
13. Tonini, S., Cossali, G.E.: On molar- and mass-based approaches to single component drop evaporation modelling. Int. Commun. Heat Mass Transf. **77**, 87–93 (2016)
14. Blanc, A.: Recherches sur les mobilités des ions dans les gaz (in French). J. Phys. **7**, 825 (1908)
15. Townsend, J.S.E.: The diffusion of ions in gases. Philos. Trans. R. Soc. Lond. Ser. A **193**, 129–158 (1900)
16. Wilke, C.R.: Diffusional properties of multicomponent gases. Chem. Eng. Prog. **46**, 95–104 (1950)
17. Burghardt, A., Krupiczka, R.: Wnikanie Masy W Ukladach Wielaskladnikowych-Teortyczna Analiza Zagadnienia i Okreslenie Wspolczynnikow Wnikania Masy (in Polish). Inz. Chem. **5**, 487–510 (1975)
18. Brenn, G., Deviprasath, L.J., Durst, F., Fink, C.: Evaporation of acoustically levitated multicomponent liquid droplets. Int. J. Heat Mass Transf. **50**, 5073–5086 (2007)
19. Zhang, L., Kong, S.-C.: Multicomponent vaporization modeling of bio-oil and its mixtures with other fuels. Fuel **95**, 471–480 (2012)
20. Tonini, S., Cossali, G.E.: A novel formulation of multi-component drop evaporation models for spray applications. Int. J. Therm. Sci. **89**, 245–253 (2015)
21. Tonini, S., Cossali, G.E.: A multi-component drop evaporation model based on analytical solution of Stefan-Maxwell equations. Int. J. Heat Mass Transf. **92**, 184–189 (2016)
22. DeGrot, S.R., Mazur, P.: Non-equilibrium Thermodynamics. Dover, New York (1984)
23. van den Brule, B.H.A.A.: A network theory for the thermal conductivity of an amorphous polymeric material. Rheol. Acta **28**, 257–266 (1989)
24. Curtiss, C.F., Bird, R.B.: Statistical mechanics of transport phenomena: polymeric liquid mixtures. Adv. Polym. Sci. **125**, 1–101 (1996)
25. Cattaneo, C.: Sulla conduzione del calore (in Italian). Atti Semin. Mat. Fis. Univ. Modena **3**(3) (1948)
26. Cattaneo, C.: Sur une forme de l'equation de la chaleur eliminant le paradoxe d'une propagation instanteneé (in French). Comptes Rendus Acad. Sci. **247**, 431–433 (1958)
27. Vernotte, P.: Les paradoxes de la theorie continue de l'equation de la chaleur (in French). Comptes Rendus Acad. Sci. **246**, 3154–3155 (1958)
28. Christensen, R.M.: Theory of Viscoelasticity, 2nd edn. Dover Publications, Inc., New York (1982)
29. Tavernier, J.: Sur l'equation de conduction de la chaleur (in French). Comptes Rendus Acad. Sci. **254**, 69–71 (1962)
30. Muller, I.: Toward relativistic thermodynamics. Arch. Ration. Mech. Anal. **34**, 259–282 (1969)
31. Meixner, J.: On the linear theory of heat conduction. Arch. Ration. Mech. Anal. **39**, 108–130 (1970)
32. Joseph, D.D., Preziosi, L.: Heat waves. Rev. Mod. Phys. **61**(1), 41–73 (1989)
33. Bai, C., Lavine, A.S.: On hyperbolic heat conduction and second law of thermodynamics. J. Heat Transf. **117**, 256–263 (1995)
34. Körner, C., Bergmann, H.W.: The physical defect of the hyperbolic heat conduction equation. Appl. Phy. A **67**, 397–401 (1998)

35. Guillemet, P., Bardon, J.-P.: Conduction de la chaleur aux temps courts: les limites spatio-temporelles des modèles parabolique et hyperbolique (in French). Int. J. Therm. Sci. **39**, 968–982 (2000)
36. Curtiss, C.F., Bird, R.B.: Diffusion multicomponent. Ind. Eng. Chem. Res. **38**, 2515–2522 (1999)
37. Newton, I.: The Principia - Mathematical Principles of Natural Philosophy, translation by Cohen, I.B., Whitman, A. University of California Press, Berkeley (1999)
38. Landau, L.D., Lifshitz, E.M.: Fluid Mechanics, 2nd edn. Pergamon, Oxford (1987)
39. Chabbra, R.P.: Bubbles, Drops, and Particles in Non-Newtonian Fluids. Taylor & Francis, Boca Raton (2007)
40. Chabbra, R.P.: Non-Newtonian fluids: an introduction. In: Krishnan, J., Deshpande, A., Kumar, P. (eds.) Rheology of Complex Fluids. Springer, New York (2010)
41. Witten, T.A., Pincus, P.A.: Structured Fluids, Polymers, Colloids, Surfactants. Oxford University Press, Oxford (2010)
42. Irgens, F.: Rheology and Non-Newtonian Fluids. Springer International Publishing, Berlin (2014)
43. Wilkinson, W.L.: Non-Newtonian Fluids: Fluid Mechanics, Mixing and Heat Transfer. Pergamon, Oxford (1960)

# Chapter 8
# Conservation and Constitutive Equations in Curvilinear Coordinates

The formulation of the conservation and constitutive differential equations derived in the previous chapters was obtained under the implicit assumption that the coordinate system was a Cartesian one. In practical problems it is sometime useful to switch to more *natural* coordinate systems, where the actual form of the differential equations may be simplified, thanks to some symmetry properties of the problem. For example, when dealing with the heating and evaporation of a spherical drop, the *natural* coordinate system is the spherical one, since in such a system the governing differential equations may assume a much simpler form. The scope of this chapter is to reformulate the conservation and constitutive differential equations in an invariant form, that can be used in any coordinate system. We will restrict again the formulation to the usual 3D Euclidean space, starting from the equations written in a Cartesian coordinate system. The summation convention will be used, unless specifically said. Some examples will help the reader to appreciate on one hand the compactness and simplicity of the invariant formulation and, on the other hand, the straightforwardness of writing these equations in explicit form for any curvilinear coordinate system.

## 8.1 The Conservation Equations

The generalisation to curvilinear coordinate systems of the conservation equations can be obtained by observing that the differential operators appearing in these equations can be written in a straightforward way following the definition of covariant derivative given in Chap. 1. Since in curvilinear systems covariant, contravariant and physical components are different, care should be taken in positioning the indices as superscripts or subscripts.

© Springer Nature Switzerland AG 2021
G. E. Cossali and S. Tonini, *Drop Heating and Evaporation: Analytical Solutions in Curvilinear Coordinate Systems*, Mathematical Engineering,
https://doi.org/10.1007/978-3-030-49274-8_8

### 8.1.1   The Mass and Species Conservation Equations

Let start with the mass conservation equation (6.18); it can be noticed that it contains the divergence operator of the contravariant vector $\rho v^j$ and since in general curvilinear coordinate $u^j$ the divergence is given by Eq. (4.153), then the covariant form of the mass conservation equation is

$$\frac{\partial \rho}{\partial t} + \frac{1}{\sqrt{g}} \frac{\partial \left(\sqrt{g}\rho v^s\right)}{\partial u^s} = 0 \qquad (8.1)$$

or in a more compact form

$$\frac{\partial \rho}{\partial t} + \left(\rho v^s\right)_{/s} = 0 \qquad (8.2)$$

Analogously, the species conservation equations (6.48) can be now written as

$$\frac{\partial \rho^{(p)}}{\partial t} + \frac{1}{\sqrt{g}} \frac{\partial \left(\sqrt{g}n^{(p)s}\right)}{\partial u^s} = 0 \qquad (8.3)$$

where $n^{(p)s}$ is the $s$th contravariant component of the mass flux vector $\mathbf{n}^{(p)}$. The same holds for the molar form (6.49)

$$\frac{\partial c^{(p)}}{\partial t} + \frac{1}{\sqrt{g}} \frac{\partial \left(\sqrt{g}N^{(p)s}\right)}{\partial u^s} = 0 \qquad (8.4)$$

#### 8.1.1.1   Example: The Species Conservation Equations in Spherical, Spheroidal and Inverse Spheroidal Coordinates

Consider the invariant form of the species conservation equation (8.3), the following are the explicit forms of this equation in spherical, oblate spheroidal, prolate spheroidal, inverse oblate spheroidal and inverse prolate spheroidal coordinate systems (refer to Chap. 4 for details). These coordinate systems are used when modelling evaporation from deformed droplets, in particular for the case of oscillating droplets (see Chap. 12). Figure 8.1 shows the general shape of such kind of droplets.

The *physical* components of the fluxes, $\hat{n}_k^{(p)} = n^{(p)k}h_k = n_k^{(p)}\frac{1}{h_k}$ (where $h_k$ are the scale factors), will be used and the summation convention is suspended in these examples, then the general form is

$$\frac{\partial \rho^{(p)}}{\partial t} = -\sum_{k=1}^{3} \frac{1}{\sqrt{g}} \frac{\partial \left(\frac{\sqrt{g}}{h_k}\hat{n}_k^{(p)}\right)}{\partial u^k} \qquad (8.5)$$

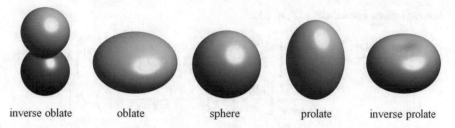

inverse oblate      oblate      sphere      prolate      inverse prolate

**Fig. 8.1** Examples of inverse oblate, oblate, spherical, prolate and inverse prolate shapes

and the explicit forms in the above mentioned coordinate systems can be obtained by substituting the explicit forms of the scale factors and of $\sqrt{g} = h_1 h_2 h_3$.

*Spherical coordinates* $(R, \eta, \varphi)$:

$$-\frac{\partial \rho^{(p)}}{\partial t} = \frac{\partial \hat{n}_R^{(p)}}{\partial R} + \frac{2}{R}\hat{n}_R^{(p)} + \frac{\sqrt{1-\eta^2}}{R}\frac{\partial \hat{n}_\eta^{(p)}}{\partial \eta} - \frac{\eta}{R\sqrt{1-\eta^2}}\hat{n}_\eta^{(p)} + \frac{1}{R\sqrt{1-\eta^2}}\frac{\partial \hat{n}_\varphi^{(p)}}{\partial \varphi} \quad (8.6)$$

*Spherical coordinates* $(R, \theta, \varphi)$:

$$-\frac{\partial \rho^{(p)}}{\partial t} = \frac{\partial \hat{n}_R^{(p)}}{\partial R} + \frac{2}{R}\hat{n}_R^{(p)} + \frac{1}{R}\frac{\partial \hat{n}_\theta^{(p)}}{\partial \theta} + \frac{\cot(\theta)}{R}\hat{n}_\theta^{(p)} + \frac{1}{R\sin(\theta)}\frac{\partial \hat{n}_\varphi^{(p)}}{\partial \varphi} \quad (8.7)$$

*Oblate spheroidal coordinates* $(\zeta, \eta, \varphi)$:

$$-\frac{\partial \rho^{(p)}}{\partial t} = \frac{\sqrt{\zeta^2+1}}{a\sqrt{\zeta^2+\eta^2}}\frac{\partial \hat{n}_\zeta^{(p)}}{\partial \zeta} + \frac{\zeta\left(2\zeta^2+1+\eta^2\right)}{a\left(\zeta^2+\eta^2\right)^{3/2}\sqrt{\zeta^2+1}}\hat{n}_\zeta^{(p)} + \frac{\sqrt{1-\eta^2}}{a\sqrt{\zeta^2+\eta^2}}\frac{\partial \hat{n}_\eta^{(p)}}{\partial \eta} +$$

$$+ \frac{\eta\left(1-\zeta^2-2\eta^2\right)}{a\left(\zeta^2+\eta^2\right)^{3/2}\sqrt{1-\eta^2}}\hat{n}_\eta^{(p)} + \frac{1}{a\sqrt{\zeta^2+1}\sqrt{1-\eta^2}}\frac{\partial \hat{n}_\varphi^{(p)}}{\partial \varphi}$$

$$\quad (8.8)$$

*Prolate spheroidal coordinates* $(\zeta, \eta, \varphi)$:

$$-\frac{\partial \rho^{(p)}}{\partial t} = \frac{\sqrt{\zeta^2-1}}{a\sqrt{\zeta^2-\eta^2}}\frac{\partial \hat{n}_\zeta^{(p)}}{\partial \zeta} + \frac{\zeta\left(2\zeta^2-\eta^2-1\right)}{a\left(\zeta^2-\eta^2\right)^{3/2}\sqrt{\zeta^2-1}}\hat{n}_\zeta^{(p)} + \frac{\sqrt{1-\eta^2}}{a\sqrt{\zeta^2-\eta^2}}\frac{\partial \hat{n}_\eta^{(p)}}{\partial \eta} +$$

$$- \frac{\eta\left(1+\zeta^2-2\eta^2\right)}{a\left(\zeta^2-\eta^2\right)^{3/2}\sqrt{1-\eta^2}}\hat{n}_\eta^{(p)} + \frac{1}{a\sqrt{\zeta^2-1}\sqrt{1-\eta^2}}\frac{\partial \hat{n}_\varphi^{(p)}}{\partial \varphi}$$

$$\quad (8.9)$$

*Inverse oblate coordinates* $(\zeta, \eta, \varphi)$:

$$-\frac{\partial \rho^{(p)}}{\partial t} = \frac{\Theta\sqrt{\zeta^2+1}}{a\sqrt{\zeta^2+\eta^2}}\left[\frac{\partial \hat{n}_\zeta^{(p)}}{\partial \zeta} + \zeta\left(\frac{1}{\zeta^2+\eta^2} + \frac{1}{\zeta^2+1} - \frac{4}{\Theta}\right)\hat{n}_\zeta^{(p)}\right] + \tag{8.10}$$

$$+\frac{\Theta\sqrt{1-\eta^2}}{a\sqrt{\zeta^2+\eta^2}}\left[\frac{\partial \hat{n}_\eta^{(p)}}{\partial \eta} + \eta\left(\frac{1}{\zeta^2+\eta^2} - \frac{1}{1-\eta^2} + \frac{4}{\Theta}\right)\hat{n}_\eta^{(p)}\right] +$$

$$+\frac{\Theta}{a\sqrt{\zeta^2+1}\sqrt{1-\eta^2}}\frac{\partial \hat{n}_\varphi^{(p)}}{\partial \varphi}$$

$$\Theta = \zeta^2 + 1 - \eta^2 \tag{8.11}$$

*Inverse prolate coordinates* $(\zeta, \eta, \varphi)$:

$$-\frac{\partial \rho^{(p)}}{\partial t} = \frac{\Theta\sqrt{\zeta^2-1}}{a\sqrt{\zeta^2-\eta^2}}\left[\frac{\partial \hat{n}_\zeta^{(p)}}{\partial \zeta} + \zeta\left(\frac{1}{\zeta^2-1} + \frac{1}{\zeta^2-\eta^2} - \frac{4}{\Theta}\right)\hat{n}_\zeta^{(p)}\right] + \tag{8.12}$$

$$+\frac{\sqrt{1-\eta^2}\Theta}{a\sqrt{\zeta^2-\eta^2}}\left[\frac{\partial \hat{n}_\eta^{(p)}}{\partial \eta} - \eta\left(\frac{1}{1-\eta^2} + \frac{1}{\zeta^2-\eta^2} + \frac{4}{\Theta}\right)\hat{n}_\eta^{(p)}\right] +$$

$$+\frac{\Theta}{a\sqrt{\zeta^2-1}\sqrt{1-\eta^2}}\frac{\partial \hat{n}_\varphi^{(p)}}{\partial \varphi}$$

$$\Theta = \zeta^2 - 1 + \eta^2 \tag{8.13}$$

These examples can help the reader to appreciate the compactness of the indicial notation and the convenience of using the invariant form of the equations when developing a problem in curvilinear coordinates. However, when numerical evaluations are needed, the implementation of the equations in explicit forms, like those of these examples, cannot be avoided.

## 8.1.2  The Momentum Conservation Equation

The momentum conservation equation (6.61) contains the divergence of the stress tensor $T_{jk}$ and the dyadic product $\rho v_j v_k$ that is also a tensor

$$\frac{\partial \rho v_j}{\partial t} + \nabla_k\left(\rho v_j v_k\right) = \nabla_k T_{jk} + \rho f_j \tag{8.14}$$

The generalisation of this equation is obtained observing that these terms can be written using the covariant derivatives as

$$\nabla_k T_{jk} \rightarrow T_{/k}^{jk} \tag{8.15}$$

$$\text{Cartesian} \rightarrow \text{curvilinear}$$

and then

$$T^{jk}_{/k} = T^{jk}_{,k} + \left\{ \begin{matrix} j \\ m\ k \end{matrix} \right\} T^{mk} + \left\{ \begin{matrix} k \\ m\ k \end{matrix} \right\} T^{jm} = T^{jk}_{,k} + \left\{ \begin{matrix} j \\ m\ k \end{matrix} \right\} T^{mk} + \frac{1}{\sqrt{g}} \frac{\partial \sqrt{g}}{\partial u^m} T^{jm}$$

(8.16)

where Eq. (1.194) has been used to substitute $\left\{ \begin{matrix} k \\ m\ k \end{matrix} \right\}$. The same can be done for the term $\nabla_k \left( \rho v_j v_k \right)$

$$\nabla_k \left( \rho v_j v_k \right) \rightarrow \left( \rho v^j v^k \right)_{/k}$$

(8.17)

Cartesian → curvilinear

and the covariant form of the momentum conservation equation becomes:

$$\frac{\partial \rho v^j}{\partial t} + \left( \rho v^j v^k \right)_{/k} = T^{jk}_{/k} + \rho f^j$$

(8.18)

or, in extended form

$$\frac{\partial \rho v^j}{\partial t} + \left( \rho v^j v^k \right)_{,k} + \rho v^m \left[ \left\{ \begin{matrix} j \\ m\ k \end{matrix} \right\} v^k + \frac{1}{\sqrt{g}} \frac{\partial \sqrt{g}}{\partial u^m} v^j \right] =$$

$$= T^{jk}_{,k} + \left\{ \begin{matrix} j \\ m\ k \end{matrix} \right\} T^{mk} + \frac{1}{\sqrt{g}} \frac{\partial \sqrt{g}}{\partial u^m} T^{jm} + \rho f^j$$

(8.19)

It is interesting to notice that, since

$$\frac{\partial \rho v^j}{\partial t} + \left( \rho v^j v^k \right)_{,k} = \rho \frac{\partial v^j}{\partial t} + v^j \frac{\partial \rho}{\partial t} + \rho v^k v^j_{,k} + v^j \left( \rho v^k \right)_{,k}$$

(8.20)

and from (8.1)

$$\left( \rho v^k \right)_{,k} = -\frac{1}{\sqrt{g}} \rho v^k \frac{\partial \left( \sqrt{g} \right)}{\partial u^k} - \frac{\partial \rho}{\partial t}$$

(8.21)

then

$$\frac{\partial \rho v^j}{\partial t} + \left( \rho v^j v^k \right)_{,k} = \rho \frac{\partial v^j}{\partial t} + \rho v^k v^j_{,k} - \frac{1}{\sqrt{g}} \rho v^k v^j \frac{\partial \left( \sqrt{g} \right)}{\partial u^k}$$

(8.22)

and the LHS of Eq. (8.19) transforms to

$$\rho \frac{\partial v^j}{\partial t} + \rho v^k \left[ v^j_{,k} + v^m \left\{ \begin{matrix} j \\ m\ k \end{matrix} \right\} \right] = \rho \frac{\partial v^j}{\partial t} + \rho v^k v^j_{/k}$$

(8.23)

thus yielding a second form of the momentum conservation equation

$$\rho \frac{\partial v^j}{\partial t} + \rho v^k \left[ v^j_{,k} + v^m \left\{ \begin{matrix} j \\ m \ k \end{matrix} \right\} \right] = T^{jk}_{,k} + \left\{ \begin{matrix} j \\ m \ k \end{matrix} \right\} T^{mk} + \frac{1}{\sqrt{g}} \frac{\partial \sqrt{g}}{\partial u^m} T^{jm} + \rho g^j$$

(8.24)

or in a more compact form

$$\rho \frac{\partial v^j}{\partial t} + \rho v^k v^j_{/k} = T^{jk}_{/k} + \rho g^j$$

(8.25)

## 8.1.3  The Energy Conservation Equation

Following the method used above, the divergence terms in Eq. (6.76) become

$$\nabla_j \left( \rho v_j \hat{u} \right) = \frac{1}{\sqrt{g}} \frac{\partial \left( \sqrt{g} \rho v^s \hat{u} \right)}{\partial u^s}$$

(8.26)

$$\nabla_j q_j = \frac{1}{\sqrt{g}} \frac{\partial \left( \sqrt{g} q^s \right)}{\partial u^s}$$

(8.27)

and using covariant derivatives

$$\frac{\partial \rho \hat{u}}{\partial t} + \frac{1}{\sqrt{g}} \left( \sqrt{g} \rho v^s \hat{u} \right)_{,s} = -\frac{1}{\sqrt{g}} \left( \sqrt{g} q^s \right)_{,s} + g_{mj} T^{mk} v^j_{/k}$$

(8.28)

where the rising and lowering index procedure has been used in the stress tensor term. In explicit form

$$\frac{\partial \rho \hat{u}}{\partial t} + \frac{1}{\sqrt{g}} \left( \sqrt{g} \rho v^s \hat{u} \right)_{,s} = -\frac{1}{\sqrt{g}} \left( \sqrt{g} q^s \right)_{,s} + g_{mj} T^{mk} v^j_{,k} + g_{mj} T^{mk} v^s \left\{ \begin{matrix} j \\ sk \end{matrix} \right\}$$

(8.29)

For any other form of the energy equation, the same procedure can be followed; for example for the enthalpy form (6.77)

$$\frac{\partial \rho \hat{h}}{\partial t} + \frac{1}{\sqrt{g}} \left( \sqrt{g} \rho v^s \hat{h} \right)_{,s} = -\frac{1}{\sqrt{g}} \left( \sqrt{g} q^s \right)_{,s} + g_{mj} \tau^{mk} v^j_{/k} + v^s P_{,s} + \frac{\partial P}{\partial t}$$

(8.30)

To notice that Eqs. (8.29) and (8.30) are written in such a way to evidence the actual operations to perform, however when one has to manipulate on these equations it may be useful to write them in the most compact form

$$\frac{\partial \rho \hat{u}}{\partial t} + \left(\rho v^s \hat{u}\right)_{/s} = -q^s_{/s} + g_{mk} T^{mk} v^j_{/k} \qquad (8.31)$$

$$\frac{\partial \rho \hat{h}}{\partial t} + \left(\rho v^s \hat{h}\right)_{/s} = -q^s_{/s} + g_{mk} \tau^{mk} v^j_{/k} + v^s P_{,s} + \frac{\partial P}{\partial t} \qquad (8.32)$$

## 8.2 The Constitutive Equations

Constitutive equations for mass diffusion are quite straightforward to generalise, since they mainly contain gradients and relations among vectors, then no real modifications are needed. For example, the Maxwell–Stefan equations (7.28) are

$$d_s^{(p)} = y_{,s}^{(p)} = \sum_{\substack{q=0 \\ q \neq p}}^{n} \frac{y^{(p)} y^{(q)}}{D_{pq}} \left( \frac{j_s^{(q)}}{\rho^{(q)}} - \frac{j_s^{(p)}}{\rho^{(p)}} \right) \qquad (8.33)$$

and the Fick's law (7.40) is

$$J_s^{(p)} = -c D_{10} y_{,s}^{(p)} \qquad (8.34)$$

The same holds for the heat flux constitutive equation (7.65)

$$q_s = -k T_{,s} + \sum_{p=1}^{n} j_s^{(p)} h^{(p)} + cRT \sum_{p=0}^{n} \sum_{\substack{q=0 \\ q \neq p}}^{n} \frac{D_T^{(p)} y^{(p)} y^{(q)}}{\rho^{(p)} D_{pq}} \left( \frac{j_s^{(p)}}{\rho^{(p)}} - \frac{j_s^{(q)}}{\rho^{(q)}} \right) \qquad (8.35)$$

A little more attention must be paid to the stress tensor constitutive equations (for Newtonian fluids) since it contains tensor derivatives. In particular, the rate of strain tensor is the symmetric part of the tensor obtained from derivatives of the velocity field, in covariant form

$$v_{p/m} = \frac{1}{2} \left( v_{p/m} + v_{m/p} \right) + \frac{1}{2} \left( v_{p/m} - v_{m/p} \right) = \epsilon_{pk} + \Omega_{pk} \qquad (8.36)$$

Using the rising and lowering indexes procedure

$$\epsilon_{pm} = \frac{1}{2} \left( v_{p/m} + v_{m/p} \right) = \frac{1}{2} \left( (v^q g_{qp})_{/m} + (v^q g_{qm})_{/p} \right) = \frac{1}{2} \left( v^q_{/m} g_{qp} + v^q_{/p} g_{qm} \right) \qquad (8.37)$$

then rising both indexes yields

$$\epsilon^{jk} = g^{jp} g^{km} \epsilon_{pm} = g^{jp} g^{km} \frac{1}{2} \left( v_{p/m} + v_{m/p} \right) = \frac{1}{2} \left( v^q_{/m} g_{qp} g^{jp} g^{km} + v^q_{/p} g_{mq} g^{jp} g^{km} \right) =$$
$$= \frac{1}{2} \left( v^q_{/m} \delta^j_q g^{km} + v^q_{/p} \delta^k_q g^{jp} \right) = \frac{1}{2} \left( v^j_{/m} g^{km} + v^k_{/p} g^{jp} \right) \qquad (8.38)$$

Equation (7.74) can be written as

$$T^{jk} = -Pg^{jk} + \mu \left( v^j_{/s} g^{sk} + v^k_{/s} g^{sj} \right) + \lambda g^{jk} v^p_{/p} \tag{8.39}$$

or in extended form

$$T^{jk} = -Pg^{jk} + \mu \left( v^j_{,s} g^{sk} + v^k_{,s} g^{sj} + v^s \frac{1}{2} \left( g^{rj} g^{mk} + g^{rk} g^{mj} \right) g_{mr,s} \right) + \lambda \frac{g^{jk}}{\sqrt{g}} \left( \sqrt{g} \rho v^p \right)_{,p} \tag{8.40}$$

where the identity

$$\left[ \left\{ \begin{matrix} j \\ m\ s \end{matrix} \right\} g^{mk} + \left\{ \begin{matrix} k \\ m\ s \end{matrix} \right\} g^{mj} \right] = \frac{1}{2} \left( g^{rj} g^{mk} + g^{rk} g^{mj} \right) g_{mr,s} \tag{8.41}$$

was used.

## 8.2.1   Example: The Steady State Energy Equation

Consider the steady state form of the energy conservation equation (8.32) under the simplifying assumptions of uniform and constant pressure and negligible viscous dissipation

$$\left( \rho v^s \hat{h} \right)_{/s} = -q^s_{/s} \tag{8.42}$$

Assuming Fourier law as constitutive equation for the heat flux ($q_k = -kT_{,k}$), proportionality of specific enthalpy with temperature $\left( \hat{h} = \hat{c}_p T \right)$ and constant thermophysical properties, yields

$$\frac{1}{\sqrt{g}} \frac{\partial}{\partial u^s} \left( \sqrt{g} T v^s \right) = \alpha \frac{1}{\sqrt{g}} \frac{\partial}{\partial u^s} \left( \sqrt{g} g^{sk} T_{,k} \right) \tag{8.43}$$

where $\alpha = \frac{k}{\rho \hat{c}_p}$ is the thermal diffusivity and the divergences have been transformed using Eq. (1.195). This equation can be written in a simpler form considering that, with the above mentioned assumption, the continuity equation is

$$\frac{1}{\sqrt{g}} \frac{\partial}{\partial u^s} \left( \sqrt{g} v^s \right) = 0 \tag{8.44}$$

and then Eq. (8.43) becomes

$$\sqrt{g} v^s \frac{\partial T}{\partial u^s} = \alpha \frac{\partial}{\partial u^s} \left( \sqrt{g} g^{sk} T_{,k} \right) \tag{8.45}$$

In the next part of this example the summation convention will be suspended and recalling Eq. (4.165) the energy equation becomes, in *orthogonal* curvilinear coordinates

$$\sqrt{g} \sum_{s=1}^{3} v^s \frac{\partial T}{\partial u^s} = \alpha \sum_{s=1}^{3} \frac{\partial}{\partial u^s} \left( W^{(s)} \frac{\partial T}{\partial u^s} \right) \tag{8.46}$$

where $W^{(s)} = \frac{\sqrt{g}}{h_s^2}$. As an example, let us now write the explicit form of this equation in spherical and bispherical (see Chap. 4) coordinate systems using the physical components of the velocity vector: $\hat{v}_s = v^s h_s = v_s \frac{1}{h_s}$.

*Spherical coordinates* $(R, \eta, \varphi)$:

$$\hat{v}_R \frac{\partial T}{\partial R} + \frac{\sqrt{1-\eta^2}}{R} \hat{v}_\eta \frac{\partial T}{\partial \eta} + \frac{1}{R\sqrt{1-\eta^2}} \hat{v}_\varphi \frac{\partial T}{\partial \varphi} = \alpha \left[ \begin{array}{c} \frac{\partial^2 T}{\partial R^2} + \frac{2}{R} \frac{\partial T}{\partial R} + \frac{(1-\eta^2)}{R^2} \frac{\partial^2 T}{\partial \eta^2} + \\ - \frac{2\eta}{R^2} \frac{\partial T}{\partial \eta} + \frac{1}{R^2(1-\eta^2)} \frac{\partial^2 T}{\partial \varphi^2} \end{array} \right] \tag{8.47}$$

*Spherical coordinates* $(R, \theta, \varphi)$:

$$\hat{v}_R \frac{\partial T}{\partial R} + \frac{1}{R} \hat{v}_\theta \frac{\partial T}{\partial \theta} + \frac{1}{R \sin(\theta)} \hat{v}_\varphi \frac{\partial T}{\partial \varphi} = \alpha \left[ \begin{array}{c} \frac{\partial^2 T}{\partial R^2} + \frac{2}{R} \frac{\partial T}{\partial R} + \frac{1}{R^2} \frac{\partial^2 T}{\partial \theta^2} + \\ + \frac{\cot(\theta)}{R^2} \frac{\partial T}{\partial \theta} + \frac{1}{R^2 \sin^2(\theta)} \frac{\partial^2 T}{\partial \varphi^2} \end{array} \right] \tag{8.48}$$

Bispherical *coordinates* $(\xi, \psi, \varphi)$:

$$\hat{v}_\xi \frac{\partial T}{\partial \xi} + \hat{v}_\psi \frac{\partial T}{\partial \psi} + \frac{1}{\sin(\psi)} \hat{v}_\varphi \frac{\partial T}{\partial \varphi} = \frac{\alpha}{a} \left[ \begin{array}{c} \Theta \frac{\partial^2 T}{\partial \xi^2} - \sinh \xi \frac{\partial T}{\partial \xi} + \Theta \frac{\partial^2 T}{\partial \psi^2} + \\ + \frac{\cosh(\xi)\cos(\psi)-1}{\sin(\psi)} \frac{\partial T}{\partial \psi} + \frac{\Theta}{\sin^2(\psi)} \frac{\partial^2 T}{\partial \varphi^2} \end{array} \right] \tag{8.49}$$

where $\Theta = \cosh(\xi) - \cos(\psi)$.

# Chapter 9
# Modelling Heat and Mass Transfer from an Evaporating Drop

The heat and mass transfer mechanisms taking place when a drop evaporates in a gaseous atmosphere can be accurately described making use of the equations developed in the previous chapters of this part of the book. However, when analytical tools are to be used to find solutions to such equations, some simplifying assumptions must be adopted to transform the governing equation to a more manageable form. In this chapter we will analyse the most common assumptions adopted when modelling the heating and evaporation of a liquid drop, and we will apply them to the conservation and constitutive equations to obtain differential equations that may admit analytical solutions.

The relevant differential equations will be generally written in Cartesian coordinates, since the simplifications are easy to be dealt with in this system, and the methods reported in Chap. 8 allow to re-write the simplified equations in invariant form, and then in any coordinate system.

The simplifications described in this chapter will be used in Part III of this book to investigate the heating and evaporation of liquid drops under different simplifying assumptions and with different geometrical characteristics, using the more *natural* curvilinear coordinate systems.

## 9.1 Mass Transfer

Starting with the species conservation equations (6.52, 6.53), the most common assumption is that of quasi-steadiness: all parameters and functions are considered independent of time. Under such assumptions the following equivalent species conservation equations are obtained

$$\nabla_k n_k^{(p)} = 0 \tag{9.1a}$$

$$\nabla_k N_k^{(p)} = 0 \tag{9.1b}$$

© Springer Nature Switzerland AG 2021
G. E. Cossali and S. Tonini, *Drop Heating and Evaporation: Analytical Solutions in Curvilinear Coordinate Systems*, Mathematical Engineering,
https://doi.org/10.1007/978-3-030-49274-8_9

where

$$\mathbf{n}^{(p)} = \mathbf{n}^T \chi^{(p)} + \mathbf{j}^{(p)} \tag{9.2a}$$
$$\mathbf{N}^{(p)} = \mathbf{N}^T y^{(p)} + \mathbf{J}^{(p)} \tag{9.2b}$$

Since the mass conservation equation (that can be obtained by summing Eqs. (9.1a) and (9.1b) over $p$) are

$$\nabla_k n_k^T = 0; \quad \nabla_k N_k^T = 0 \tag{9.3}$$

then Eqs. (9.1a) and (9.1b) can be written as

$$n_k^T \nabla_k \chi^{(p)} + \nabla_k j_k^{(p)} = 0 \tag{9.4a}$$
$$N_k^T \nabla_k y^{(p)} + \nabla_k J_k^{(p)} = 0 \tag{9.4b}$$

The diffusive fluxes $\mathbf{j}^{(p)}$ and $\mathbf{J}^{(p)}$ are modelled through the constitutive equations (see Chap. 7), that also need simplifications. Let us start with the most general case of multi-component mixtures, and consider the generalised Maxwell-Stefan equations (7.6) that link the diffusive fluxes to the diffusional forces. The thermal (Soret) component is often neglected, since high temperature gradients are needed to see a detectable effect, although this assumption may become questionable for some applications, like drop evaporation in combustion environments. When evaluating the diffusional forces (see Eq. 7.8), the pressure gradient effects are commonly neglected, as well as the force effects. Under these assumptions the Maxwell-Stefan equations yield the following set of ODEs

$$\nabla_k y^{(p)} = \sum_{\substack{q=0 \\ q \neq p}}^{n} \frac{1}{cD_{pq}} \left( y^{(p)} J_k^{(q)} - y^{(q)} J_k^{(p)} \right) \tag{9.5}$$

These equations can be written also using mass variables ($\chi^{(p)}$, $\rho$) and mass fluxes ($j_k^{(p)}$), in such case the RHS transforms easily while the LHS assumes a more complex form, yielding

$$\nabla_k \left[ \frac{\rho}{c} \chi^{(p)} \right] = \sum_{\substack{q=0 \\ q \neq p}}^{n} \frac{\rho}{Mm^{(q)} D_{pq} c^2} \left( \chi^{(p)} j_k^{(q)} - \chi^{(q)} j_k^{(p)} \right) \tag{9.6}$$

which only simplifies under the assumption that $\frac{\rho}{c}$ is constant, and since (see Eq. 6.26)

$$\frac{\rho}{c} = \sum_{p=0}^{n} Mm^{(p)} y^{(p)} = \left[ \sum_{p=0}^{n} \frac{\chi^{(p)}}{Mm^{(p)}} \right]^{-1} \tag{9.7}$$

this can be acceptable only when the mixture composition depends weakly on the position.

For binary mixtures, it has been shown (Eqs. 7.38, 7.40) that the Maxwell-Stefan equations yield the well known Fick's law

$$\mathbf{J}^{(1)} = -cD_{10}\nabla y^{(1)} \tag{9.8}$$
$$\mathbf{j}^{(1)} = -\rho D_{10}\nabla_k \chi^{(1)} \tag{9.9}$$

which is the most used form. To notice that Eq. (9.9), which is equivalent to Eq. (9.8), since they are both a direct consequence of Eq. (9.5), are often used to model also multi-component drop evaporation (see for example [1–3]), but in such case they must be considered an approximation of (9.5), valid for dilute mixtures, and the two forms (molar (9.8) and mass (9.9)) are not equivalent. Equation (9.9) is often used together with the assumption of constant mass density and diffusivity, since in such case Eq. (9.4a) yields

$$n_k^T \nabla_k \chi^{(1)} - \rho D_{10}\nabla^2 \chi^{(1)} = 0 \tag{9.10}$$

a form that, under further assumptions (see Chap. 10), allows simple solutions for the vapour concentration distribution, $\chi^{(1)}$, and consequently a simple evaluation of the evaporation rate. Also Eq. (9.8) can be used with the assumption of constant molar density and diffusivity, and Eq. (9.4b) then yields a form similar to (9.10)

$$N_k^T \nabla_k y^{(1)} - cD_{10}\nabla^2 y^{(1)} = 0 \tag{9.11}$$

However the two Eqs. (9.10) and (9.11) are not anymore equivalent, since they are obtained under different assumptions (constant *mass* density for Eq. 9.10 and constant *molar* density for Eq. 9.11), and this may yield substantial differences when modelling mass transfer in a gas mixture. A detailed analysis of this point can be found in [4].

Drop evaporation is a two-phase problem and jump balance conditions must be taken into account. For multi-component problems the mass jump balances for each component can be written as (see Eq. 6.101)

$$\rho_G^{(p)}\left(v_{Gn}^{(p)} - V_{In}\right) = \rho_L^{(p)}\left(v_{Ln}^{(p)} - V_{In}\right) \tag{9.12}$$

where the subscripts $G$ and $L$ are used to indicate the gaseous and the liquid phases, respectively. When a drop evaporates in air, this is usually assumed to behave as a single gas, and its diffusion into the liquid phase is assumed to be negligible. In the following the superscript $p = 0$ will refer to the component that is not present in the liquid phase. Under this assumption, the following approximations hold

$$\rho_L^{(0)} = 0; \quad v_{Ln}^{(0)} = 0 \tag{9.13}$$

Then, Eq. (9.12) for $p = 0$ yields

$$v_{Gn}^{(0)} = V_{In} \tag{9.14}$$

i.e. the velocity of component $p = 0$ in the gas mixture at drop surface is equal to the interface velocity. The interface velocity is generally non nil, since the drop can deform with time both due to evaporation (shrinking) and possibly time-dependent deformation (like shape oscillation). Depending on further assumptions about the drop shape variation, Eq. (9.12) can yield different results. To notice that the liquid density $\rho_L$ is that of the pure component $p = 1$, i.e. $\rho_L^{(1)} = \rho_L$, and the velocity of the liquid phase is $v_{Ln} = v_{Ln}^{(1)}$; these results are used in the next paragraphs.

Often, when developing analytical models for single component drop evaporation, the drop is assumed to be spherical with constant radius $R_d$; this assumption is clearly unphysical but it allows to treat the problem as quasi-steady, with consequent great simplifications. In this case

$$V_{In} = \dot{R}_d = 0 \tag{9.15}$$

and Eq. (9.12) yield

$$v_{Gn}^{(0)} = 0 \tag{9.16}$$

$$\rho_G^{(1)} v_{Gn}^{(1)} = \rho_L v_{Ln} \tag{9.17}$$

showing that the velocity of the component $p = 0$ in the gaseous phase is nil, while the velocity of the liquid phase (which is made by component $p = 1$ only) is not nil; this last results shows that the quasi-steady assumption implies the existence of a mass source inside the drop.

Let consider again a single component spherical droplet and relieve the quasi-steady assumption, then $V_{In} = \dot{R}_d \neq 0$; on the other hand, the velocity of the liquid phase must be nil (if no recirculation inside the drop is assumed): $v_{Ln} = 0$. In this case Eq. (9.12) yield

$$v_{Gn}^{(0)} = \dot{R}_d \tag{9.18}$$

$$\rho_G^{(1)} v_{Gn}^{(1)} = \left( \rho_G^{(1)} - \rho_L \right) \dot{R}_d \tag{9.19}$$

and the flux of component $p = 1$ on the drop surface is given by

$$n_s^{(1)} = \rho_G^{(1)} v_{Gn}^{(1)} = \left( \rho_G^{(1)} - \rho_L \right) \dot{R}_d \tag{9.20}$$

When dealing with mass transfer from an evaporating drop, the term *evaporation flux* is sometimes used to indicate the flux of components $p$ (with $p \neq 0$) that leaves the drop, which is correct when the drop surface is fixed. However, when dealing with a moving interface care should be taken to give an univocal definition of the *evaporation* flux. Consider again the case of an evaporating spherical drop, the evaporation rate can be calculated from the time variation of the drop mass

$$m_{ev} = -\frac{dM}{dt} = -4\pi \rho_L R_d^2 \dot{R}_d \tag{9.21}$$

where constancy of the liquid density $\rho_L$ is assumed. The evaporation flux in this case is simply obtained as

$$n_{ev} = \frac{m_{ev}}{4\pi R_d^2} = -\rho_L \dot{R}_d \tag{9.22}$$

while the flux of component $p = 1$ is given by Eq. (9.20), then $n_{ev} \neq n_s^{(1)}$.

Consider now the most general case, where the drop surface is deforming with time, for example as in an evaporating and oscillating drop. The local evaporation flux must be linearly dependent on the four velocities $v_{Gn}^{(1)}$, $v_{Gn}^{(0)}$, $v_{Ln}$ and $V_{In}$, but for a single component drop $v_{Gn}^{(0)} = V_{In}$ and $V_{In}$ can be written as a linear combination of $v_{Gn}^{(1)}$, $v_{Ln}$ using Eq. (9.12) for $p = 1$

$$V_{In} = \frac{\rho_G^{(1)}}{\left(\rho_G^{(1)} - \rho_L\right)} v_{Gn}^{(1)} - \frac{\rho_L}{\left(\rho_G^{(1)} - \rho_L\right)} v_{Ln} \tag{9.23}$$

Moreover, since $\rho_G v_{Gn} = \rho_G^{(1)} v_{Gn}^{(1)} + \rho_G^{(0)} v_{Gn}^{(0)}$, then

$$v_{Gn}^{(1)} = \frac{\rho_G}{\rho_G^{(1)}} \frac{\left(\rho_G^{(1)} - \rho_L\right)}{(\rho_G - \rho_L)} v_{Gn} + \frac{\rho_G^{(0)}}{\rho_G^{(1)}} \frac{\rho_L}{(\rho_G - \rho_L)} v_{Ln} \tag{9.24}$$

where $\rho_G = \rho_G^{(0)} + \rho_G^{(1)}$, and we can generally write

$$n_{ev} = a v_{Ln} + b v_{Gn} \tag{9.25}$$

For a deforming *non evaporating* drop, the normal components of the velocity of the gaseous mixture and that of the liquid phase at the interface must be equal to the interface normal velocity, i.e.

$$v_{Gn} = v_{Ln} = V_{In} \tag{9.26}$$

and since in this case the flux must be nil, the following condition must be satisfied

$$0 = a v_{Ln} + b v_{Gn} \tag{9.27}$$

i.e. $a = -b$.

For a spherical *evaporating* drop, Eq. (9.20) must be verified when $v_{Ln} = 0$ and the second condition

$$- \rho_L V_{I\,n} = b v_{G\,n} \tag{9.28}$$

must also be satisfied, and substituting Eq. (9.23) for $V_{I\,n}$, with the condition $v_{Ln} = 0$, i.e. $V_{In} = \frac{\rho_G}{(\rho_G - \rho_L)} v_{Gn}$, yields

$$b = - \frac{\rho_L \rho_G}{(\rho_G - \rho_L)} \tag{9.29}$$

The evaporation flux can then be consistently written as

$$n_{ev} = \frac{\rho_L \rho_G}{(\rho_G - \rho_L)} (v_{Ln} - v_{Gn}) \tag{9.30}$$

Alternative forms can be found using the relationships between the four velocities $v_{Gn}^{(1)}, v_{Gn}^{(0)}, v_{Ln}, V_{In}$ (Eqs. 9.23 and 9.24), yielding, for example

$$n_{ev} = \rho_L v_{Ln} - \rho_L V_{In} \tag{9.31}$$

and then

$$v_{Ln} = \frac{n_{ev}}{\rho_L} + V_{In} \tag{9.32}$$

Equation (9.30) comprises both the extreme cases of a shrinking spherical drop (Eq. 9.22) and that of a non-evaporating deforming drop and it can be taken as a consistent definition of the *evaporating flux*.

## 9.2 Energy Transfer

The energy conservation equation in the forms reported in Chap. 6 (Eqs. 6.76 and 6.77) are quite complex to be solved analytically, moreover some of the terms (like the heat flux) can be modelled in a relatively complex way (see Sect. 7.7), then some simplifying assumptions are usually necessary. We will refer here to the enthalpy form (6.77) of the energy conservation equation

$$\frac{\partial \rho \hat{h}}{\partial t} + \nabla_j \left( \rho v_j \hat{h} \right) = -\nabla q_j + \tau_{jk} \nabla_k v_j + v_j \nabla_j P + \frac{\partial P}{\partial t} \tag{9.33}$$

It should be remarked that when modelling drop evaporation the dissipative term $(\tau_{jk} \nabla_k v_j)$ can usually be neglected on the basis of relatively low flow velocities involved. A dimensional analysis (see for example [5]) shows that this term is usually negligible when the Eckert number (named after the Austro-Hungarian scientist Ernst Rudolf Georg Eckert, 1904–2004) $Ec = \frac{|v|^2}{\hat{c}_p \Delta T}$ (where $v$ is a reference velocity and $\hat{c}_p$ is the constant pressure heat capacity on mass basis) is much lower than one. Under conditions typical of drop evaporation modelling also the pressure terms appearing on the RHS are usually assumed negligible, and the energy equation then becomes

$$\frac{\partial \rho \hat{h}}{\partial t} + \nabla_j \left( \rho v_j \hat{h} \right) + \nabla_j q_j = 0 \tag{9.34}$$

This equation can be conveniently transformed using Eq. (7.65). The term modelling the Dufour effect is usually neglected in drop evaporation, and the heat flux is then simply

$$q_k = -k \nabla_k T + \sum_{p=0}^{n} j_k^{(p)} h^{(p)} \tag{9.35}$$

When modelling drop evaporation, the gas mixture is often assumed to be a *Gibbs-Dalton mixture*, (named after the American scientist Josiah Willard Gibbs, 1839–1903, and the English scientist John Dalton, 1766–1844) [6], also called *ideal* mixture, and in this case the relationship between the mixture enthalpy per unit of mass and the partial enthalpies of the component is given by the simple relation

$$\hat{h} = \sum_{p=0}^{n} \frac{\rho^{(p)}}{\rho} h^{(p)} = \sum_{p=0}^{n} \chi^{(p)} h^{(p)} \tag{9.36}$$

The two divergence terms in Eq. (9.34) can be conveniently written as $\nabla_j \left( \rho v_j \hat{h} + q_j \right)$; the first term inside the divergence operator can be transformed, assuming an *ideal* mixture behaviour, to

$$\rho v_k \hat{h} = n_k^T \hat{h} = n_k^T \sum_{p=0}^{n} \chi^{(p)} h^{(p)} = \sum_{p=0}^{n} n_k^T \chi^{(p)} h^{(p)} = \sum_{p=0}^{n} \left( n_k^{(p)} - j_k^{(p)} \right) h^{(p)} \tag{9.37}$$

where Eq. (6.35a) has been used, then

$$\rho v_k \hat{h} + q_k = \sum_{p=0}^{n} \left( n_k^{(p)} - j_k^{(p)} \right) h^{(p)} - k \nabla_k T + \sum_{p=0}^{n} j_k^{(p)} h^{(p)} = \sum_{p=0}^{n} n_k^{(p)} h^{(p)} - k \nabla_k T \tag{9.38}$$

Taking the divergence of this term and assuming constant thermal conductivity of the mixture, yields

$$\nabla_k \left( \rho v_k \hat{h} + q_k \right) = \sum_{p=0}^{n} \nabla_k \left( n_k^{(p)} h^{(p)} \right) - k \nabla^2 T \tag{9.39}$$

and the energy equation can be written as

$$\frac{\partial \rho \hat{h}}{\partial t} + \sum_{p=0}^{n} \nabla_k \left( n_k^{(p)} h^{(p)} \right) = k \nabla^2 T \tag{9.40}$$

Another quite common assumption is that of ideal gas behaviour of each component in a gas mixture, since pressure is often low enough and temperature high enough to make this a safe assumption, then the partial enthalpies can be written as

$$h^{(p)} = \hat{c}_p^{(p)} T + h_0^{(p)} \tag{9.41}$$

where $\hat{c}_p^{(p)}$ is the constant pressure heat capacity on mass basis of component $p$ and $h_0^{(p)}$ is the partial specific enthalpy on mass basis of component $p$ in a reference state. The LHS of Eq. (9.40) becomes

$$\frac{\partial \rho \hat{h}}{\partial t} + \sum_{p=0}^{n} \nabla_k \left( n_k^{(p)} h^{(p)} \right) = \sum_{p=0}^{n} \left[ \frac{\partial \rho^{(p)} h^{(p)}}{\partial t} + \nabla_k \left( n_k^{(p)} h^{(p)} \right) \right] = \tag{9.42}$$

$$= \sum_{p=0}^{n} \left\{ \frac{\partial \rho^{(p)} \hat{c}_p^{(p)} T}{\partial t} + \nabla_k \left( n_k^{(p)} \hat{c}_p^{(p)} T \right) + h_0^{(p)} \left[ \frac{\partial \rho^{(p)}}{\partial t} + \nabla_k n_k^{(p)} \right] \right\} = \tag{9.43}$$

$$= \sum_{p=0}^{n} \left[ \frac{\partial \rho^{(p)} \hat{c}_p^{(p)} T}{\partial t} + \nabla_k \left( n_k^{(p)} \hat{c}_p^{(p)} T \right) \right] = \frac{\partial \rho \hat{c}_p T}{\partial t} + \sum_{p=0}^{n} \nabla_k \left( n_k^{(p)} \hat{c}_p^{(p)} T \right) \tag{9.44}$$

where $\hat{c}_p = \sum_{p=0}^{n} \frac{\rho^{(p)} \hat{c}_p^{(p)}}{\rho} = \sum_{p=0}^{n} \chi^{(p)} \hat{c}_p^{(p)}$ and the third equality is obtained using the species conservation equations (6.48). Then Eq. (9.40) becomes

$$\frac{\partial \rho \hat{c}_p T}{\partial t} + \sum_{p=0}^{n} \nabla_k \left( n_k^{(p)} \hat{c}_p^{(p)} T \right) = k \nabla^2 T \tag{9.45}$$

### 9.2.1    The Heat Transfer inside a Single-Component Drop

The energy equation (9.33) can also be used to study the transient heat transfer inside an evaporating drop (and the case will be analysed in detail in Chap. 11). Considering the case of a single component drop, the interdiffusional terms and the terms describing the Dufour effect in Eq. (7.65) are nil. Also the pressure terms are neglectful and when the drop internal recirculation is neglected, as for example when considering a drop floating in a steady environment, the velocity vector is nil everywhere inside the drop and the convective terms in Eq. (9.45) disappear as well as the dissipative terms. Finally, for a pure liquid the enthalpy per unit of mass can be accurately approximated by the relation $\hat{h} = \hat{c}_p T + \hat{h}_0$, with $\hat{c}_p$ and $\hat{h}_0$ constant.

The energy equation is this case simplifies to

$$\frac{\partial \rho \hat{c}_p T}{\partial t} = \nabla_j \left( k \nabla_j T \right) \tag{9.46}$$

Often, further simplifications of this equation are obtained assuming constancy of the mass density for the liquid phase (incompressibility). However, it should be considered that the liquid density depends on temperature (as well as all the other thermo-physical properties) and that during the heating phase of a drop injected into a hot gaseous environment, the variation of temperature yields a variation of the density that is the cause of the observed drop swelling during this initial phase [7].

It may be now observed that for a drop heating and evaporating in a gaseous environment, the temperature field inside the drop is often relatively homogeneous. Also when the drop is positioned inside a hot environment, where the temperature field can be highly dishomogeneous, the difference of temperature from drop center to the surface is limited, since the evaporation cooling keeps the drop surface temperature quite below its boiling temperature. This observation is used to support the assumption that the liquid thermo-physical properties, which depend on temperature as above mentioned, may not experience large variations inside the drop and then taking their value constant is a reasonable approximation. The constancy of the thermo-pysical properties transform equation (9.46) into the so called *heat equation*, sometimes called *Fourier equation*, not to be confused with the Fourier' law (Eq. 7.62)

$$\frac{\partial T}{\partial t} = \alpha \nabla^2 T \tag{9.47}$$

This equation, first studied by Fourier, is the prototype of the parabolic PDEs, and it appears in many different fields of science (see [8] for a survey).

In the field of drop evaporation, this equation has been widely used to study the unsteady heat transport inside the drop and in Chap. 11 a deeper analysis of the most advanced analytical approaches to this problem will be presented.

## 9.2.2  The Quasi-steady Single-Component Drop Evaporation

As for the mass transfer case, quasi-steadiness is often assumed when modelling heating of an evaporating drop, then Eq. (9.45) for a single-component drop becomes

$$\nabla_k \left( n_k^{(0)} \hat{c}_p^{(0)} T \right) + \nabla_k \left( n_k^{(1)} \hat{c}_p^{(1)} T \right) = k \nabla^2 T \tag{9.48}$$

where the RHS has been transformed assuming constancy of the gas mixture thermal conductivity.

The usual assumption of negligible diffusion of component $p = 0$ (air) into the liquid drop lead to consider the effect of mass transfer only in the gas phase, and for this special case the gas mixture can be assumed as binary (as above discussed, since air does not diffuse into the liquid, its composition can be assumed constant and treated as a single component ideal gas). Quasi-steadiness assumption implies that the drop surface can be considered still, and this further implies (see Sect. 9.1) that the flux $n^{(0)}$ is nil everywhere, then

$$\rho v_k = n_k^{(T)} = n_k^{(1)} \tag{9.49}$$

To notice that this assumption has other implications, since the condition $n_k^{(0)} = \rho v_k \chi^{(0)} + j_k^{(0)} = 0$ yields

$$j_k^{(0)} = -\rho v_k \chi^{(0)} \tag{9.50}$$

and since $j_k^{(1)} + j_k^{(0)} = 0$, then

$$j_k^{(1)} = \rho v_k \chi^{(0)} \tag{9.51}$$

With these assumptions, Eq. (9.48) assumes the final form

$$\nabla_k \left( \rho v_k \hat{c}_p^{(1)} T \right) = k \nabla^2 T \tag{9.52}$$

which is the form usually adopted to model the heat transfer for single drop evaporation [7, 9].

### 9.2.2.1  Example: Solution for Potential Flow

As an example, let us consider the case of a drop evaporating in a gaseous environment, assuming constant thermo-physical properties and steady-state. In this case, Eq. (9.52) can be written making use of the continuity equation, as

$$v_k \nabla_k T = \alpha \nabla^2 T \tag{9.53}$$

Let us now consider the case of *potential flow*, i.e. when the velocity field is the gradient of a scalar function, called the *velocity potential* $\phi$

$$v_k = \nabla_k \phi \tag{9.54}$$

Potential flow is an oversimplification of real flows and this assumption should always be taken with some care since it may yield unphysical results (like the d'Alembert's paradox, see [10]). Generally, it cannot account for the behaviour of internal and boundary layer flows [10]. Despite of the many criticisms that can be raised about this strong assumption, it is widely used in fluid mechanics since it may yield analytical solutions that well approximate real-world flows, and it is then encountered also in many engineering problems such as aircraft design for example.

In drop heating and evaporation, under the assumption of constant thermo-physical properties, the Stefan flow caused by evaporation is a potential flow (see Sect. 10.1.1, Example 1). In an incompressible flow, where $\rho$ is assumed to be constant, the function $\phi$ is harmonic, i.e. $\nabla^2\phi = 0$, since the velocity field is solenoidal ($\nabla_k v_k = 0$).

Let now substitute Eq. (9.54) into Eq. (9.53), writing $\phi'$ for $\phi/\alpha$, one obtains

$$\nabla_k\phi' \, \nabla_k T = \nabla^2 T \tag{9.55}$$

This equation has a simple general solution:

$$T = Ae^{\phi'} + B \tag{9.56}$$

in fact

$$\nabla_k T = A\nabla_k\phi' e^{\phi'} \tag{9.57}$$

$$\nabla^2 T = \nabla_k\left(A\nabla_k\phi' e^{\phi'}\right) = Ae^{\phi'}\left(\nabla^2\phi' + \nabla_k\phi'\nabla_k\phi'\right) = Ae^{\phi'}\nabla_k\phi'\nabla_k\phi' \tag{9.58}$$

where the last equality is obtained observing that $\phi'$ is harmonic. Substitution of Eqs. (9.57) and (9.58) into Eq. (9.55) yields an identity. This solution may be used for example when the boundary conditions on the drop surface are uniform, since in such a case the degrees of freedom given by the two constants $A$ and $B$ in the general solution (9.56) are enough to satisfy the prescribed boundary conditions. When the boundary conditions are non uniform, the relation between the value of $\phi'$ and $T$ given by Eq. (9.56) does not allow, in general, to satisfy the boundary conditions.

### 9.2.3 Time-Dependent Integral Balances on a Single Component Drop

In some applications the effect of temperature gradients *inside* the evaporating drop is not modelled and an integral balance of mass and energy on the whole drop is used. We will assume here that the drop is spherical and liquid and gas thermo-physical properties are constant. The integral mass balance is simply expressed by the equation

$$\frac{dM_d}{dt} = -m_{ev} \tag{9.59}$$

where $M_d$ is the drop mass and $m_{ev}$ is the evaporation rate, assumed positive. The assumption of spherical shape allows to write

$$4\pi R_d^2\rho_L \dot{R}_d = -m_{ev} \tag{9.60}$$

where $R_d(t)$ is the time-dependent drop radius.

The species balances at interface (Eq. 6.101) are

$$\rho_G^{(0)} \left( v_{Gn}^{(0)} - V_n \right) = \rho_L^{(0)} \left( v_{Ln}^{(0)} - V_n \right) \tag{9.61}$$

$$\rho_G^{(1)} \left( v_{Gn}^{(1)} - V_n \right) = \rho_L^{(1)} \left( v_{Ln}^{(1)} - V_n \right) \tag{9.62}$$

where as usual the superscript $p = 1$ stands for the evaporating species while $p = 0$ stands for the *gas*, the subscript $G$ and $L$ stand for the gaseous and liquid phases, respectively, and $V_n$ is the normal component of the surface velocity (equal to $\dot{R}_d$). In the following the subscript $n$, used to indicate the vector component normal to the interface, will be dropped since the only non-nil component is the radial one, normal to the drop interface.

Assuming negligible gas absorption $\rho_L^{(0)} = 0$ and in absence of internal recirculation $v_L^{(1)} = 0$ Eqs. (9.61) and (9.62) yield

$$v_G^{(0)} = V_n = \dot{R}_d \tag{9.63}$$

$$\rho_G^{(1)} v_G^{(1)} = \left( \rho_G^{(1)} - \rho_L \right) V_n = \left( \rho_G^{(1)} - \rho_L \right) \dot{R}_d \tag{9.64}$$

and

$$\rho_G v_G = \rho_G^{(0)} v_G^{(0)} + \rho_G^{(1)} v_G^{(1)} = \rho_G^{(0)} \dot{R}_d + \left( \rho_G^{(1)} - \rho_L \right) \dot{R}_d = (\rho_G - \rho_L) \dot{R}_d \tag{9.65}$$

The energy conservation equation inside the drop can be written as (Eq. 6.77)

$$\frac{\partial \rho_L \hat{h}_L}{\partial t} + \nabla_j \left( \rho_L v_{Lj} \hat{h}_L \right) = -\nabla_j q_{Lj} + \tau_{jk} \nabla_k v_{Lj} + v_{Lj} \nabla_j P_L + \frac{\partial P_L}{\partial t} \tag{9.66}$$

Since $v_L = 0$ and neglecting pressure terms ($P$ is assumed constant) the equation reduces to the simple form:

$$\frac{\partial \rho_L \hat{h}_L}{\partial t} = -\nabla_j q_{Lj} \tag{9.67}$$

The integration of both sides of Eq. (9.67) yields for the LHS

$$\int_V \frac{\partial \rho_L \hat{h}_L}{\partial t} = \frac{d}{dt} \int_V \rho_L \hat{h}_L dV - \int_A \rho_L \hat{h}_L V_n dA = \frac{d \left( M_d \, \bar{h}_{L,d} \right)}{dt} - 4\pi R_d^2 \rho_L \hat{h}_{Ls} \dot{R}_d \tag{9.68}$$

where $\bar{h}_{L,d}$ is the average enthalpy of the liquid per unit of mass, defined as

$$\bar{h}_{L,d} = \frac{\int_V \rho_L \hat{h}_L dV}{M_d} \tag{9.69}$$

and $\hat{h}_{Ls}$ is the enthalpy of the liquid for unit of mass, evaluated at surface conditions.

Integrating the RHS of (9.67) yields

$$- \int \nabla_j q_j dV = - \int q_{Ln} dA = -q_L 4\pi R_d^2 \tag{9.70}$$

and equating (9.68) and (9.70) yields

$$M_d \frac{d \bar{h}_{L,d}}{dt} - m_{ev} \left( \bar{h}_{L,d} - \hat{h}_{Ls} \right) = -q_L 4\pi R_d^2 \tag{9.71}$$

where Eqs. (9.59) and (9.60) have been used.

The energy balance at interface (Eq. 6.117)

$$0 = \left\{ \rho \left( \hat{h} + \frac{1}{2} |v|^2 \right) (v - V_n) + P V_n + q_n - \tau_{jk} v_j \check{N}_k \right\} \tag{9.72}$$

can be simplified neglecting the work of viscous forces (and recalling that $v_L = 0$) yielding

$$- \rho_L \left( \hat{h}_L + \frac{1}{2} |v_L|^2 \right) V_n + P_L V_n + q_L = \rho_G \left( \hat{h}_G + \frac{1}{2} |v_G|^2 \right) (v_G - V_n) + P_G V_n + q_G \tag{9.73}$$

where $\hat{h}_G$ is the gas mixture enthalpy

$$\rho_G \hat{h}_G = \hat{h}_G^{(0)} \rho^{(0)} + \hat{h}_G^{(1)} \rho^{(1)} \tag{9.74}$$

while $\hat{h}_L$ is that of the liquid (and $\hat{h}_L = \hat{h}_L^{(1)}$ since the liquid is made only by component $p - 1$).

Assuming equal pressure of gas and liquid, which is equivalent to neglecting the effect of surface tension, and neglectful kinetic energy terms when compared to enthalpy, yields

$$- \rho_L \hat{h}_L^{(1)} \dot{R}_d + q_L = \rho_G \hat{h}_G v_G - \rho_G \hat{h}_G \dot{R}_d + q_G \tag{9.75}$$

and using (9.65)

$$q_L = -\rho_L \left( \hat{h}_G - \hat{h}_L^{(1)} \right) \dot{R}_d + q_G \tag{9.76}$$

The heat flux in the gas mixture, considering the diffusive terms, can be written as (Eq. 7.65, the Dufour effect is assumed neglectful)

$$q_G = -k_G \nabla_n T_G + \sum_{p=0}^{n} j_{Gn}^{(p)} \hat{h}_G^{(p)} = -k_G \nabla_n T_G + j_G^{(0)} \hat{h}_G^{(0)} + j_G^{(1)} \hat{h}_G^{(1)} \tag{9.77}$$

where $\nabla_n T_G$ is the component of the gas temperature gradient normal to the drop surface and $j_G^{(p)}$ are the components of the diffusive fluxes normal to the surface. Since the diffusive fluxes can be written as (Eq. 6.35a)

$$j_G^{(p)} = \rho_G^{(p)} v_G^{(p)} - \rho_G v_G \chi^{(p)} \tag{9.78}$$

then using Eqs. (9.63) and (9.64) yields

$$j_G^{(0)} = \rho_G^{(0)} \dot{R}_d - (\rho_G - \rho_L) \chi^{(0)} \dot{R}_d = \rho_L \chi^{(0)} \dot{R}_d = -j_G^{(1)} \tag{9.79}$$

and

$$j_G^{(0)} \hat{h}_G^{(0)} + j_n^{(1)} \hat{h}_G^{(1)} = \rho_L \chi^{(0)} \dot{R}_d \left( \hat{h}_G^{(0)} - \hat{h}_G^{(1)} \right) \tag{9.80}$$

Substitution of Eqs. (9.80) and (9.77) into Eq. (9.76) yields:

$$q_L = -\rho_L \left( \widehat{h}_G - \widehat{h}_L^{(1)} \right) \dot{R}_d - k_G \nabla_n T_G + \rho_L \chi^{(0)} \dot{R}_d \left( \hat{h}_G^{(0)} - \hat{h}_G^{(1)} \right) \tag{9.81}$$
$$= \rho_L \dot{R}_d \left( \widehat{h}_L^{(1)} - \widehat{h}_G^{(1)} \right) - k_G \nabla_n T_G = -\rho_L \dot{R}_d \widehat{h}_{LV}^{(1)} - k_G \nabla_n T_G$$

where $\widehat{h}_{LV}^{(1)} = \widehat{h}_G^{(1)} - \widehat{h}_L^{(1)}$ is the latent heat of vaporisation of component $p = 1$.
Equation (9.71) can then be written as

$$M_d \frac{d \bar{h}_{L,d}}{dt} - m_{ev} \left( \bar{h}_{L,d} - \hat{h}_{Ls} \right) = -m_{ev} \widehat{h}_{LV}^{(1)} + Q_s \tag{9.82}$$

where Eq. (9.60) has been used to eliminate $\dot{R}_d$ in favour of $m_{ev}$ and

$$Q_s = 4\pi R_r^2 k_G \nabla_n T \tag{9.83}$$

is the sensible heat rate, assumed positive when it enters the drop. Equation (9.82) can be further simplified when the difference between the drop average temperature and the surface temperature can be assumed to be neglectful, yielding the widely used equation

$$M_d \frac{d \bar{h}_{L,d}}{dt} = -m_{ev} \widehat{h}_{LV}^{(1)} + Q_s \tag{9.84}$$

### 9.2.4   The Effective Temperature

When modelling the energy transfer from evaporating spherical drops, the effect of evaporation is sometimes taken into account introducing the so-called *effective temperature* [7]

$$T_{eff} = T_\infty + \frac{\rho_L}{h_c} h_{lv} \dot{R}_d \tag{9.85}$$

where $R_d(t)$ is the time-varying drop radius. This equation can be found by an energy balance at the drop surface.

Considering again the Eq. (9.81), we introduce the heat transfer coefficient $h_c$ through the relation $-k_G \nabla_n T_G = h_c (T_s - T_\infty)$, where $T_s$ is the drop surface temperature and, recalling that $q_{Ln} = -k_L \left(\frac{\partial T}{\partial R}\right)_{R=R_d}$, Eq. (9.81) yields

$$h_c (T_s - T_\infty) - \rho_L \dot{R}_d (t) h_{LV} = -k_L \left(\frac{\partial T}{\partial R}\right)_{R=R_d} \tag{9.86}$$

which can be written as

$$h_c \left[T_s - T_{eff}(t)\right] = -k_L \left(\frac{\partial T}{\partial R}\right)_{R=R_d} \tag{9.87}$$

where

$$T_{eff} = T_\infty + \frac{\rho_L}{h_c} h_{LV} \dot{R}_d (t) \tag{9.88}$$

is the *effective* temperature.

# References

1. Brenn, G., Deviprasath, L.J., Durst, F., Fink, C.: Evaporation of acoustically levitated multi-component liquid droplets. Int. J. Heat Mass Transf. **50**, 5073–5086 (2007)
2. Wilms, J.: Evaporation of multicomponent droplets. Ph.D. Thesis Universität Stuttgart (2005)
3. Tonini, S., Cossali, G.E.: A novel formulation of multi-component drop evaporation models for spray applications. Int. J. Therm. Sci. **89**, 245–253 (2015)
4. Tonini, S., Cossali, G.E.: On molar- and mass-based approaches to single component drop evaporation modelling. Int. Commun. Heat Mass **77**, 87–93 (2016)
5. Incropera, F.P., DeWitt, D.P.: Fundamentals of Heat and Mass Transfer, 2nd edn. Wiley (1985)
6. Gyftopoulos, E.P., Beretta, G.P.: Thermodynamics: Foundations and Applications. Macmillan, New York (1991)
7. Sazhin, S.: Droplet and Sprays. Springer (2014)
8. Narasimhan, T.N.: Fourier's heat conduction equation: history, influence and connections. Rev. Geophys. **37**(1), 151–172 (1999)
9. Abramzon, B., Sirignano, W.A.: Droplet vaporization model for spray combustion calculations. Int. J. Heat Mass Transf. **32**(9), 1605–1618 (1989)
10. Batchelor, G.K.: An Introduction to Fluid Dynamics. Cambridge University Press (1973)

# Part III
# Analytical Modelling of Drop Heating and Evaporation

The modelling of liquid drop evaporation in gaseous environment has been the topic of extensive research since the nineteenth century, when Maxwell [87] proposed the first model on this subject. Since then the interest on this phenomenon has grown, driven by its importance in a wide range of applicative fields, like spray combustion, spray painting, fire control, medical applications, etc.

In the following chapters we will use the methods described in the first part of the book to analytically model the heat and mass transfer processes that take place when a drop is evaporating in a gaseous environment. Conservation and constitutive equations described in Part II will be used to describe those processes and analytical solutions will be sought, under different simplifying assumptions, in coordinate systems where the problem description simplifies.

The first two chapters will treat steady problems that admit, in a proper coordinate system, one dimensional (Chap. 10) and two or three dimensional (Chap. 11) solutions, for drops of different shapes. The third chapter describes time-dependent drop evaporation considering the effect of moving boundaries caused by drop shrinking or oscillation, reporting recent advances in this field and a classical analytical solution to the time-dependent heat transfer problem in spheroidal bodies. The last chapter is devoted to the modelling of multi-component drop evaporation, for spherical and ellipsoidal drops.

# Chapter 10
# One-Dimensional Modelling of Drop Heating and Evaporation Under Steady Conditions

We have seen how modelling of drop evaporation implies the solution of a set of PDEs (momentum, energy and species conservation), which may represent a remarkable challenge, particularly when an analytical approach is chosen. The problem can be greatly simplified introducing some assumptions, like sphericity of the drop, constancy of the thermo-physical properties and steadiness. The classical Fuchs model [1] is based on all these assumptions and yields a very simple result, which is the basis of the most widely used evaporation models (like [2]), nowadays implemented in most of CFD codes for dispersed flow applications. However some of these assumptions can be relieved and still an analytical approach can be used to solve the problem. In this chapter we will see an approach that allows to solve the energy and species transport equations for the gaseous phase, considering the effect of variable thermo-physical properties and general drop shape. The drop is assumed to be made of a pure substance evaporating under steady-state conditions, and gas absorption and diffusion through the liquid phase is neglected.

## 10.1 Analytical Modelling of the Heat and Mass Transfer from a Single-Component Drop in a Gaseous Mixture

When a finite bulk of a single-component liquid is injected into a gaseous environment, simultaneous heat, mass and momentum transfer through the gas phase take place. Figure 10.1 shows a schematic of the problem. In the following the apex $p = 1$ stands for the evaporating component, while $p = 0$ stands for the mixture of all the other components, treated as a single gaseous component; $y_s^{(0)}$ and $y_\infty^{(0)}$ are the molar fractions of the component $p = 0$ at drop surface and free stream conditions, respectively, $y_s^{(1)}$ and $y_\infty^{(1)}$ are the corresponding values of the evaporating component, $p = 1$; $T_L$ is the liquid temperature (assumed to be uniform and equal to the surface temperature $T_s$), $T_\infty$ is the gas temperature at free stream conditions; $\mathbf{n}^{(p)}$ are the mass fluxes ($p = 0, 1$) and $\mathbf{q}$ is the heat flux. The temperature of the liquid phase

© Springer Nature Switzerland AG 2021

G. E. Cossali and S. Tonini, *Drop Heating and Evaporation: Analytical Solutions in Curvilinear Coordinate Systems*, Mathematical Engineering, https://doi.org/10.1007/978-3-030-49274-8_10

**Fig. 10.1** Sketch of a finite
bulk of liquid in a gaseous
environment

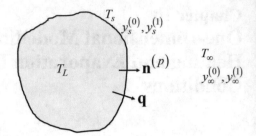

is assumed homogeneously distributed, then the thermal diffusion in the liquid phase
does not need to be modelled; in Chap. 12 we will see how this assumption can be
relieved.

In the gaseous mixture the evaporating component, $p = 1$, will be sometimes
referred as *vapour*, while the mixture of the non-evaporating components will be
sometimes referred as *gas*.

It is worth to notice that a steady problem with steady boundaries of the liquid
bulk can be set only assuming a mass source inside the drop to balance the evaporated
mass (cfr Sect. 9.1).

Considering the transport phenomena in the gas phase, the quasi-steady assumption reduces the species conservation equations (9.1) to:

$$\nabla_j N_j^{(p)} = 0 \qquad\qquad (10.1a)$$
$$\nabla_j n_j^{(p)} = 0 \qquad\qquad (10.1b)$$

(in molar 10.1a and mass 10.1b forms, respectively), where the vectors of molar and
mass fluxes for the species $p$ can be expressed (see Part II) as the sum of convective
and diffusive fluxes

$$\mathbf{N}^{(p)} = \mathbf{N}^{(T)} y^{(p)} + \mathbf{J}^{(p)} \qquad\qquad (10.2a)$$
$$\mathbf{n}^{(p)} = \mathbf{n}^{(T)} \chi^{(p)} + \mathbf{j}^{(p)} \qquad\qquad (10.2b)$$

with $\chi^{(p)}$ being the mass fraction of the species $p$

$$\chi^{(p)} = \frac{cMm^{(p)}}{\rho^{(p)}} y^{(p)} \qquad\qquad (10.3)$$

and being $J^{(p)}$ and $j^{(p)}$ the molar and mass diffusive fluxes of the species $p$, respectively.

In Chap. 7 it was shown that the Maxwell-Stefan equations for a binary mixture
yield the well known Fick's law, and the diffusive fluxes, in molar and mass forms,
can be modelled as

$$\mathbf{J}^{(p)} = -cD_{10}\nabla_k y^{(p)} \tag{10.4a}$$

$$\mathbf{j}^{(p)} = -\rho D_{10}\nabla_k \chi^{(p)} \tag{10.4b}$$

Inserting Eq. (10.4) in Eq. (10.2) and then using the steady-state species conservation equations (10.1a, 10.1b) yields

$$\nabla_k \left( N_k^{(T)} y^{(p)} - cD_{10}\nabla_k y^{(p)} \right) = 0 \tag{10.5a}$$

$$\nabla_k \left( n_k^{(T)} \chi^{(p)} - \rho D_{10}\nabla_k \chi^{(p)} \right) = 0 \tag{10.5b}$$

We can notice that for a binary mixture only one of the two Eqs. (10.5a) (or Eqs 10.5b) is needed since the molar fractions $y^{(0)}$, $y^{(1)}$ (and the mass fractions $\chi^{(0)}$, $\chi^{(1)}$) are related to each other by the equation $y^{(0)} + y^{(1)} = 1$ ( $\chi^{(0)} + \chi^{(1)} = 1$ for the mass fractions).

Classical models (refer to [2] as the most commonly used for dispersed phase applications) assume that the thermo-physical properties of the gas phase are constant. Then, the density (either molar or mass density) and the binary diffusion coefficients in Eq. (10.5a) and (10.5b) are assumed constant and the constancy of the molar density $c$ does not imply the constancy of the mass density $\rho$ (and vice-versa).

As an example, writing Eqs. (10.5a) and (10.5b) for $p = 1$ and for *constant* molar and mass densities, respectively, yields

$$N_j^{(T)}\nabla_j y^{(1)} - cD_{10}\nabla^2 y^{(1)} = 0 \tag{10.6a}$$

$$n_j^{(T)}\nabla_j \chi^{(1)} - \rho D_{10}\nabla^2 \chi^{(1)} = 0 \tag{10.6b}$$

These two equations are not anymore equivalent, since they are derived under different assumptions that may yield substantial differences in the solutions (see [3] for a detailed analysis). We will now see how this strong assumption (constancy of thermo-physical properties) can be relieved and still analytical solutions to the conservation equations can be found.

In the following analysis we will allow density, diffusion coefficient, thermal conductivity and heat capacity to depend on temperature.

As already discussed in PartII, when modelling a two-phase flow problem, as for the case of drop heating and evaporation, balances of mass, species, momentum and energy through the interface must be considered, and the parameters describing the problem may experience discontinuous (jump) variations at the interface.

The species jump balances for the case of a binary mixture are

$$\rho_L^{(0)} \left( v_{Ln}^{(0)} - V_{In} \right) = \rho_G^{(0)} \left( v_{Gn}^{(0)} - V_{In} \right) \tag{10.7}$$

$$\rho_L^{(1)} \left( v_{Ln}^{(1)} - V_{In} \right) = \rho_G^{(1)} \left( v_{Gn}^{(1)} - V_{In} \right) \tag{10.8}$$

We have seen in PartII that when modelling drop evaporation the assumption of negligible diffusion of the gas species in the liquid is considered acceptable, in particular under not too high pressure conditions. Then, both the velocity and mass density of the species $p = 0$ in the liquid phase are set equal to 0 and the first jump condition, Eq. (10.7), leads to

$$V_{In} = v_{Gn}^{(0)} \tag{10.9}$$

i.e. the gas velocity and surface velocity components normal to the surface are equal.

Since the quasi-steadiness implies that $V_{In} = 0$, then $v_{Gn}^{(0)} = 0$ and the other components of $\mathbf{v}_G^{(0)}$ are nil for the no-slip conditions. From the second jump condition (10.8), it stems

$$\rho_L v_{Ln} = \rho_G^{(1)} v_{Gn}^{(1)} = n_n^{(1)} \tag{10.10}$$

where $\rho_L v_{Ln} = \rho_L^{(1)} v_{Ln}^{(1)}$ since the drop is made only of component $p = 1$, and the second equality in Eq. (10.10) is obtained from Eq. (6.29).

The mass average velocity in the gas phase is defined as

$$\bar{v}_j = v_j = \frac{\rho_G^{(0)} v_{Gj}^{(0)} + \rho_G^{(1)} v_{Gj}^{(1)}}{\rho} = \frac{\rho_G^{(1)} v_{Gj}^{(1)}}{\rho} \tag{10.11}$$

where $\rho = \rho_G^{(0)} + \rho_G^{(1)}$ is the gas mixture mass density, and $v_{Gj}^{(0)} = 0$ has been used; the vapour flux leaving the interface (the so-called *Stefan flow*) can then be written as

$$n_j^{(1)} = Mm^{(1)} N_j^{(1)} = \rho v_j \tag{10.12}$$

Since the drop is assumed to float in a steady gas, i.e. convection is negligible, and the *molar* (and *mass*) flux of the species *gas* ($p = 0$) is nil on the drop surface, then it is nil everywhere

$$N_j^{(0)} = N_j^{(T)} y^{(0)} - c D_{10} \nabla_j y^{(0)} = 0 \tag{10.13}$$

or

$$N_j^{(T)} = c D_{10} \frac{1}{y^{(0)}} \nabla_j y^{(0)} = c D_{10} \nabla_j H \tag{10.14}$$

where

$$H = \ln \left( y^{(0)} \right) = \ln \left( 1 - y^{(1)} \right)$$

Equation (10.12) then becomes

$$n_j^{(1)} = Mm^{(1)} c D_{10} \nabla_j H \tag{10.15}$$

and the steady-state species conservation equation (10.1b) yields

$$\nabla_j \left( M m^{(1)} c D_{10} \nabla_j H \right) = 0 \tag{10.16}$$

The energy conservation equation (9.52), under steady-state conditions can be written, using the mass conservation equation, $\nabla_k (\rho v_k) = 0$, as

$$\rho v_k \nabla_k \left( \hat{c}_p^{(1)} T \right) = \nabla_k \left( k_{mix} \nabla_k T \right) \tag{10.17}$$

and using Eqs. (10.12) and (10.15) Eq. (10.17) becomes

$$M m^{(1)} c D_{10} \nabla_j H \nabla_j \left( \hat{c}_p^{(1)} T \right) = \nabla_j \left( k_{mix} \nabla_j T \right) \tag{10.18}$$

The momentum conservation equation (6.61) can be written, for steady-state conditions, Newtonian fluids and no external forces, as

$$\rho v_k \nabla_k \left( v_j \right) = -\nabla_k P + \mu_{mix} \nabla^2 v_k \tag{10.19}$$

where again the mass conservation equation has been used and the viscosity is assumed constant.

Then using again Eqs. (10.12) and (10.15) Eq. (10.19) becomes

$$M m^{(1)} c D_{10} \nabla_k H \nabla_k \left( \bar{v}_j \right) = -\nabla_j P + \mu_{mix} \nabla^2 \bar{v}_j \tag{10.20}$$

The thermo-physical properties appearing in Eqs. (10.16), (10.18) and (10.20) are in general function of temperature, pressure and composition, i.e.

$$c = c \left( T, P \right) \tag{10.21a}$$

$$D_{10} = D_{10} \left( T, P \right) \tag{10.21b}$$

$$\hat{c}_p^{(1)} = \hat{c}_p^{(1)} \left( T, P \right) \tag{10.21c}$$

$$k_{mix} = k_{mix} \left( T, P, y^{(p)} \right) \tag{10.21d}$$

$$\mu_{mix} = \mu_{mix} \left( T, P, y^{(p)} \right) \tag{10.21e}$$

As above stated, in the following the mixture viscosity is assumed constant, although, as it will be shown later, this assumption does not influence the results of the model. The gas mixture is assumed to behave like an ideal gas, then the ideal gas law is used to calculate the molar gas density

$$c = P \bar{R}^{-1} T^{-1} \tag{10.22}$$

where $\bar{R}$ is the universal gas constant.

The pressure and temperature dependence of the binary diffusion coefficient $D_{10}$ is modelled by the equation

$$D_{10} = d_0 T^m P^{-1} \tag{10.23}$$

where $d_0$ is a constant, which depends on the species, while the index $m$ is usually assumed to be equal to $7/4$, as proposed by the widely used Fuller-Schettler-Giddings (FSG) correlation [4], (a value of $m = 3/2$ is instead predicted by the classical statistical thermodynamics [5]). It is worth to notice that any reasonable choice of $m$ can be used in the present modelling and that in the conservation equations (10.16), (10.18) and (10.20) the product $c D_{10}$ appears, then the dependence on pressure is cancelled and $c D_{10}$ is actually independent of the pressure.

The dependence of vapour heat capacity on temperature at low pressure can be described as polynomials or power series [6]. However, for the purposes of the present analytical approach, the following simpler relation is used

$$\hat{c}_p^{(1)} = c_h T^b \tag{10.24}$$

where the constants $c_h$ and $b$ can be found by data fitting [6]; the interested reader can refer to [7] for a detailed analysis of the error introduced by this assumption.

The thermal conductivity of a pure monatomic gaseous substance $p$ can be modelled through the use of power laws

$$k^{(p)} = r_p T^{q_p} \tag{10.25}$$

in accordance with the Chapman-Enskog treatment for monatomic gases at low density [8], where $q_p = 1/2$. For polyatomic gases a more complex temperature dependence, the Eucken formula [9], can be used; however, the power law (10.25) can still be used, where $q_p$ and $r_p$ can be found by best fitting of available data.

We may notice that the energy conservation equation (10.18) contains the thermal conductivity of the gaseous mixture, $k_{mix}$, which can be calculated as a combination of the thermal conductivities of each species, according to the Wassiljewa relation [10], as:

$$k_{mix} = \frac{k^{(0)} y_{ref}^{(0)}}{y_{ref}^{(0)} + A_{01} y_{ref}^{(1)}} + \frac{k^{(1)} y_{ref}^{(1)}}{y_{ref}^{(1)} + A_{10} y_{ref}^{(0)}} \tag{10.26}$$

where the coefficients $A_{jk}$ are obtained from the Lindsay and Bromley relationship [11] (in the present approach the temperature dependence of these coefficients will be neglected). This relation is often used to predict the thermal conductivity of gas mixtures at low pressure, due to its simple linear dependence on the conductivities of the pure species. The dynamic viscosity of the gaseous mixture, $\mu_{mix}$, appearing in the momentum conservation equations (10.20) can be calculated with the same procedure [10], and since it will be assumed constant, it can be calculated at a proper reference temperature.

### 10.1.1 The Analytical Solution

The search for an analytic solution of the system of the non linear PDEs (10.16), (10.18) and (10.20), considering the nonlinear dependence of the properties on the temperature field, appears to be an arduous task, and this observation has forced in the past the introduction of strong simplifications.

Classical evaporation models (refer to [1, 2] for the most common ones) assume uniform thermo-physical properties within the gas field, neglecting the effect of temperature gradients on their values. The thermo-physical properties are then calculated at reference temperature and composition, which are defined as a weighted average between the corresponding values at drop surface and at infinite distance from the drop

$$T_{ref} = (1 - \alpha_{ref}) T_s + \alpha_{ref} T_\infty \tag{10.27}$$

$$\chi_{ref} = (1 - \alpha_{ref}) \chi_s + \alpha_{ref} \chi_\infty \tag{10.28}$$

The averaging parameter $\alpha_{ref}$ is usually taken equal to $1/2$ at low evaporating conditions and equal to $1/3$ at high evaporating conditions [2, 12]. In this case the species conservation equation (10.16) simplifies to

$$\nabla^2 H = 0 \tag{10.29}$$

which may have relatively simple solutions, particularly for uniform boundary conditions and for a spherical drop shape.

The energy equation (10.18), when thermo-physical properties are assumed constant, simplifies to

$$\nabla_j \hat{H} \nabla_j T = \nabla^2 T \tag{10.30}$$

where $\hat{H} = H \frac{Mm^{(1)} c D_{10} c_p^{(1)}}{k_{mix}}$ and a possible general solution is obtained in terms of the function $\hat{H}$ as

$$T = Ae^{\hat{H}} + B \tag{10.31}$$

where $A$ and $B$ are unknown constants (see Sect. 9.2.2).

Despite of the apparent difficulty in approaching the problem with varying thermo-physical properties, we will see that an analytical solution can be found. To this end, we will first non-dimensionalise the problem to get an easier to handle set of equations.

The non-dimensional temperature is defined, using the value at free-stream conditions $T_\infty$, as

$$\hat{T} = \frac{T}{T_\infty} \tag{10.32}$$

Since the aim of the present approach is to obtain a general solution that may comprise also the case with constant properties, a reference temperature, defined as in (10.27), will be introduced below; but when the full dependence of the thermo-physical properties on the temperature is considered, the actual value of this reference temperature becomes uninfluential, i.e. the model predictions will be independent of it.

The physical properties can now be written (see Eqs. 10.22, 10.23, 10.24, 10.25) as

$$D_{10} = \left( D_{10,ref} \hat{T}_{ref}^{-m} \right) \hat{T}^m \tag{10.33a}$$

$$c = \left( c_{ref} \hat{T}_{ref} \right) \hat{T}^{-1} \tag{10.33b}$$

$$k^{(p)} = \left( k_{ref}^{(p)} \hat{T}_{ref}^{-q_p} \right) \hat{T}^{q_p} \tag{10.33c}$$

$$\hat{c}_p^{(1)} = \left( \hat{c}_{p,ref}^{(1)} \hat{T}_{ref}^{-b} \right) \hat{T}^b \tag{10.33d}$$

where $D_{10,ref}$, $c_{ref}$, $k_{ref}^{(p)}$ and $\hat{c}_{p,ref}^{(1)}$ are the binary diffusion coefficient, the molar density, the thermal conductivity of the species $p$ and the vapour specific heat capacity, respectively, all calculated at the selected reference temperature conditions $T_{ref}$ (Eq. 10.27); and, as above remarked, the products in brackets are independent of $T_{ref}$. Since the conductivity of the gas mixture ($k_{mix}$) is a function also of the mixture composition (Eq. 10.26), a reference value for $y^{(p)}$ must be chosen when using Eq. (10.26). The choice of the averaging parameter $\alpha_{ref}$ in Eq. (10.28) is then expected to (slightly) influence the predictions of the model. In fact, it is shown that the dependence of the model predictions on $\alpha_{ref}$ is weak under the conditions typical of drop evaporation at temperature below the boiling conditions (refer to [7] for details). This motivates the choice to calculate the gas mixture thermal conductivity by imposing a reference composition with a constant value of the averaging parameter $\alpha_{ref}$.

The conductivity of the gas mixture is then modelled as

$$k_{mix} = k_{mix,ref} \frac{a_0 \hat{T}^{q_0} + a_1 \hat{T}^{q_1}}{a_0 \hat{T}_{ref}^{q_0} + a_1 \hat{T}_{ref}^{q_1}} \tag{10.34}$$

where the thermal conductivities of each species (Eq. 10.33c) are substituted in Eq. (10.26) and the parameters $a_0 = \frac{y_{ref}^{(0)} k_{ref}^{(0)} \hat{T}_{ref}^{-q_0}}{y_{ref}^{(0)} + A_{01} y_{ref}^{(1)}}$ and $a_1 = \frac{y_{ref}^{(1)} k_{ref}^{(1)} \hat{T}_{ref}^{-q_1}}{y_{ref}^{(1)} + A_{10} y_{ref}^{(0)}}$ are introduced.

Substituting the physical properties given by Eqs. (10.33a), (10.33b), (10.34) and (10.33d) into the energy and species conservation equations (10.18) and (10.16), yields

$$\hat{T}_j^{m-1+b} \hat{\nabla} \hat{H} \hat{\nabla}_j \hat{T} = \hat{\nabla}_j \left[ \left( a_0 \hat{T}^{q_0} + a_1 \hat{T}^{q_1} \right) \hat{\nabla}_j \hat{T} \right] \tag{10.35}$$

$$\hat{\nabla}_j \left( \hat{T}_j^{m-1} \hat{\nabla}_j \hat{H} \right) = 0 \tag{10.36}$$

where $\hat{H} = \dfrac{H\left(a_0 \hat{T}_{ref}^{q0} + a_1 \hat{T}_{ref}^{q1}\right) \hat{T}_{ref}^{1-m-b}}{Le_M}$ and $Le_M = \dfrac{k_{mix,ref}}{(b+1)Mm^{(1)} \hat{c}_{p,ref}^{(1)} D_{10,ref} c_{ref}}$ is a *modified*

*Lewis* number (named after the American Chemical Engineer Warren K. Lewis, 1882–1975). The non-dimensional *nabla* operator is defined as $\hat{\nabla}_j = R_d \nabla_j$, with $R_d$ being the equivalent drop radius, i.e. the radius of a spherical drop having the same volume of the actual drop.

The momentum conservation equation (10.20) can written in non-dimensional form as

$$\hat{\nabla}_k \tilde{P} = \frac{1}{\Lambda}\left(Sc^M \hat{\nabla}^2 \tilde{v}_k - \hat{T}_{ref}^{1-m}\hat{T}^{m-1}\hat{\nabla}_j H \hat{\nabla}_j \tilde{v}_k\right) \tag{10.37}$$

where the non-dimensional velocity ($\tilde{v}_k = v_k \frac{R_d}{D_{10}}$) and pressure ($\tilde{P} = \frac{P}{c_{ref} \bar{R} T_\infty}$) are

introduced, $Sc_M = \dfrac{\mu_{ref}}{D_{10,ref}\, c_{ref}\, M_m^{(1)}}$ is a *modified Schmidt* number (named after the

German engineer Ernst Heinrich Wilhelm Schmidt, 1892–1975), and $\Lambda = \dfrac{\bar{R} T_\infty R_d^2}{M_m^{(1)} D_{10,ref}^2}$

is a non-dimensional parameter, which, for a wide variety of conditions of interest for applications, assumes quite large values ($\Lambda \sim 10^5$), generally orders of magnitudes larger than the term $\left(Sc_M \hat{\nabla}^2 \tilde{v}_k - \tilde{P}\tilde{T}^{-1}\hat{\nabla}_j H \hat{\nabla}_j \tilde{v}_k\right)$. When this is verified, the asymptotic form (for $\Lambda \to \infty$) can be used, yielding $\hat{\nabla}_k \tilde{P} = 0$ and then $\tilde{P} = const.$ The constancy of the pressure $P$ allows then to eliminate the momentum conservation equation from the problem.

The energy and species conservation equations (10.35) and (10.36) define a system of non-linear partial differential equations (PDE) in the unknown fields $\hat{H}$ and $\hat{T}$.

We will see that this system can be analytically solved once uniform Dirichlet boundary conditions are imposed at the drop surface and at infinite distance from the drop. The temperature at drop surface is then assumed uniform and equal to $T_s$ and that at infinity is $T_\infty$. The species boundary conditions are defined by the value of the molar fractions at infinity, $y_\infty^{(p)}$, which may be arbitrarily chosen, and that on the surface, $y_s^{(p)}$, which is defined by the surface temperature since thermodynamic equilibrium conditions are assumed.

The analytic solution to the non-linear PDE system is found following an approach introduced by Laboswky [13], (for details refer to [7]). An auxiliary harmonic function $\Phi$ (i.e. $\nabla^2 \Phi = 0$) equal to 1 at the drop surface and nil at infinity is introduced and assumed to be known. In the next sections of this chapter and in Chap. 11, methods to find $\Phi$ for different drop shapes will be analysed, but it is worth to observe since now that the function $\Phi$ can only depend on the geometry of the problem.

The analytic solution of the system of non-linear PDE equations (10.35) and (10.36) in the unknowns $\hat{T}$ and $\hat{H}$ can be written in implicit form as

$$\hat{H} = -\frac{1}{K_0}\left[\frac{a_0}{h_0}\hat{T}^{2+q_0-m}W_{h_0}\left(\frac{\hat{T}^{1+b}}{K_0}\right) + \frac{a_1}{h_1}\hat{T}^{2+q_1-m}W_{h_1}\left(\frac{\hat{T}^{1+b}}{K_0}\right)\right] + \hat{H}_0$$

(10.38a)

$$\Phi = \frac{\Phi_1}{K_0}\left[\frac{a_0}{1+q_0}\hat{T}^{1+q_0}W_{g_0}\left(\frac{\hat{T}^{1+b}}{K_0}\right) + \frac{a_1}{1+q_1}\hat{T}^{1+q_1}W_{g_1}\left(\frac{\hat{T}^{1+b}}{K_0}\right)\right] + \Phi_0$$

(10.38b)

where $K_0$, $\hat{H}_0$, $\Phi_0$ and $\Phi_1$ are constants to be obtained from the boundary conditions, $h_p$ and $g_p$ are constants depending on the parameters that define the species thermophysical properties (refer to Eqs. 10.33 and 10.34), explicitly

$$h_p = \frac{2+q_p-m}{1+b}$$

(10.39)

$$g_p = \frac{1+q_p}{1+b}$$

(10.40)

and $W_n(x)$ is a particular case of the *hypergeometric function*, defined as (see Appendix)

$$W_n(x) = {_2F_1}(n, 1, 1+n, x)$$

(10.41)

We shall now prove that Eqs. (10.38a) (10.38b) are actually solutions to the Eqs. (10.35) and (10.36). Consider first the function $W_n(x)$. By the definition (10.41), it is a solution of the hypergeometric differential equation

$$x(1-x)W_n'' + [1+n-(n+2)x]W_n' - nW_n = 0$$

(10.42)

and it satisfies the identity

$$(1-x)\left[xW_n' + nW_n\right] = n$$

(10.43)

In fact, setting $xW_n' + nW_n = A(x)$, Eq. (10.42) can be written as

$$(1-x)A'(x) - A(x) = 0$$

(10.44)

which admits the general solution

$$A(x) = \frac{C}{1-x}$$

(10.45)

The constant $C$ in Eq. (10.45) can be calculated using the properties of the hypergeometric function [14]: $W_n(0) = 1$, then $A(0) = n$, yielding $C = n$ and Eq. (10.43) is proven.

From the two Eqs. (10.38a) (10.38b) the gradients of $\hat{H}$ and $\Phi$ are found as

$$\nabla_j \hat{H} = - \left\{ a_0 \nabla_j \left[ \frac{\hat{T}^{2+q_0-m}}{K_0 h_0} W_{h_0} \left( \frac{\hat{T}^{1+b}}{K_0} \right) \right] + a_1 \nabla_j \left[ \frac{\hat{T}^{2+q_1-m}}{K_0 h_1} W_{h_1} \left( \frac{\hat{T}^{1+b}}{K_0} \right) \right] \right\}$$

(10.46)

$$\nabla_j \Phi = \Phi_1 \left\{ a_0 \nabla_j \left[ \frac{\hat{T}^{1+q_0}}{K_0 (1+q_0)} W_{g_0} \left( \frac{\hat{T}^{1+b}}{K_0} \right) \right] + a_1 \nabla_j \left[ \frac{\hat{T}^{1+q_1}}{K_0 (1+q_1)} W_{g_1} \left( \frac{\hat{T}^{1+b}}{K_0} \right) \right] \right\}$$

(10.47)

The terms inside the curly brackets can be simplified observing that

$$\nabla_j \left[ \frac{\hat{T}^{2+q_p-m}}{K_0 h_p} W_{h_p} \left( \frac{\hat{T}^{1+b}}{K_0} \right) \right] = \frac{\hat{T}^{1+q_p-m} (1+b)}{K_0 h_p} \left[ h_p W_{h_p} \left( \frac{\hat{T}^{1+b}}{K_0} \right) + \frac{\hat{T}^{1+b}}{K_0} W'_{h_p} \left( \frac{\hat{T}^{1+b}}{K_0} \right) \right] \nabla_j \hat{T} =$$

(10.48)

$$= \frac{\hat{T}^{q_p-m+1} (b+1)}{K_0 - \hat{T}^{1+b}} \nabla_j \hat{T}$$

where the first equality comes from the application of the product and chain rules and recalling Eq. (10.39), and the second equality derives from the identity (10.43). The same procedure is applied to the curly bracket terms in Eq. (10.47), using Eq. (10.40)

$$\nabla_j \left[ \frac{\hat{T}^{1+q_p} W_{g_p} \left( \frac{\hat{T}^{1+b}}{K_0} \right)}{K_0 (1+q_p)} \right] = \frac{\hat{T}^{q_p} (1+b)}{K_0 (1+q_p)} \left[ g_p W_{g_p} \left( \frac{\hat{T}^{1+b}}{K_0} \right) + \frac{\hat{T}^{1+b}}{K_0} W'_{g_p} \left( \frac{\hat{T}^{1+b}}{K_0} \right) \right] \nabla_j \hat{T} =$$

(10.49)

$$= \frac{\hat{T}^{q_p}}{K_0 - \hat{T}^{1+b}} \nabla_j \hat{T}$$

Then Eqs. (10.46) and (10.47) become

$$\nabla_j \hat{H} = - \left( a_0 \hat{T}^{q_0} + a_1 \hat{T}^{q_1} \right) \frac{\hat{T}^{1-m} (1+b) \nabla_j \hat{T}}{K_0 - \hat{T}^{1+b}}$$

(10.50)

$$\nabla_j \Phi = \Phi_1 \left( a_0 \hat{T}^{q_0} + a_1 \hat{T}^{q_1} \right) \frac{\nabla_j \hat{T}}{K_0 - \hat{T}^{1+b}}$$

(10.51)

and consequently

$$\hat{T}^{m-1} \nabla_j \hat{H} = - \frac{(1+b)}{\Phi_1} \nabla_j \Phi$$

(10.52)

which shows that Eq. (10.36) is satisfied since $\nabla^2 \Phi = 0$.

Using Eq. (10.52), the LHS of Eq. (10.35) can be written as

$$\text{LHS} = \left( \hat{T}^{m-1+b} \nabla_j \hat{H} \right) \nabla_j T = - \hat{T}^b \frac{(1+b)}{\Phi_1} \nabla_j \Phi \nabla_j T$$

(10.53)

whereas, using Eqs. (10.50) and (10.52), the RHS of Eq. (10.35) becomes

$$\text{RHS} = \nabla_j \left[ \left( a_0 \hat{T}^{q_0} + a_1 \hat{T}^{q_1} \right) \nabla_j T \right] = \frac{1}{\Phi_1} \nabla_j \left[ \left( K_0 - \hat{T}^{1+b} \right) \nabla_j \Phi \right] = -\hat{T}^b \frac{(1+b)}{\Phi_1} \nabla_j \hat{T} \nabla_j \Phi$$

(10.54)

proving that also Eq. (10.35) is satisfied.

As above stated, the constants $K_0$, $\hat{H}_0$, $\Phi_0$ and $\Phi_1$ in Eqs. (10.38a, 10.38b) are calculated imposing the boundary conditions on the surface and at infinity, i.e.

$$\text{on the surface: } \hat{T} = \hat{T}_s; \ \hat{H} = \hat{H}_s; \ \Phi = 1 \qquad (10.55a)$$

$$\text{at infinity: } \hat{T} = 1; \ \hat{H} = \hat{H}_\infty; \ \Phi = 0 \qquad (10.55b)$$

Then, substituting the two boundary conditions into Eq. (10.38a) yields

$$\hat{H}_\infty = -\frac{1}{K_0} \left[ \frac{a_0}{h_0} W_{h_0} \left( \frac{1}{K_0} \right) + \frac{a_1}{h_1} W_{h_1} \left( \frac{1}{K_0} \right) \right] + \hat{H}_0 \qquad (10.56)$$

$$\hat{H}_s = -\frac{1}{K_0} \left[ \frac{a_0}{h_0} \hat{T}_s^{2+q_0-m} W_{h_0} \left( \frac{\hat{T}_s^{1+b}}{K_0} \right) + \frac{a_1}{h_1} \hat{T}_s^{2+q_1-m} W_{h_1} \left( \frac{\hat{T}_s^{1+b}}{K_0} \right) \right] + \hat{H}_0$$

(10.57)

The constant $K_0$ can be calculated by solving the transcendental equation obtained subtracting Eq. (10.57) from Eq. (10.56)

$$\hat{H}_\infty - \hat{H}_s = a_0 X_0 + a_1 X_1 \qquad (10.58)$$

where the functions $X_j$ are defined as

$$X_j = \frac{1}{K_0 h_j} \left[ \hat{T}_s^{2+q_0-m} W_{h_j} \left( \frac{\hat{T}_s^{1+b}}{K_0} \right) - W_{h_j} \left( \frac{1}{K_0} \right) \right] \qquad (10.59)$$

Once $K_0$ is known, the other constant $\hat{H}_0$ can be calculated from one of the two Eqs. (10.56) or (10.57).

Substituting the B.C.s into Eq. (10.38b) yields the values of $\Phi_0$ and $\Phi_1$, in fact

$$1 = \frac{\Phi_1}{K_0} \left[ \frac{a_0}{1+q_0} \hat{T}_s^{1+q_0} W_{g_0} \left( \frac{\hat{T}_s^{1+b}}{K_0} \right) + \frac{a_1}{1+q_1} \hat{T}_s^{1+q_1} W_{g_1} \left( \frac{\hat{T}_s^{1+b}}{K_0} \right) \right] + \Phi_0$$

(10.60)

$$0 = \frac{\Phi_1}{K_0} \left[ \frac{a_0}{1+q_0} W_{g_0} \left( \frac{1}{K_0} \right) + \frac{a_1}{1+q_1} W_{g_1} \left( \frac{1}{K_0} \right) \right] + \Phi_0 \qquad (10.61)$$

subtracting Eq. (10.60) from Eq. (10.61), the constant $\Phi_1$ is calculated as

$$\Phi_1 = \frac{K_0}{\frac{a_0}{1+q_0}\left[\hat{T}_s^{1+q_0}\, W_{g0}\left(\frac{\hat{T}_s^{1+b}}{K_0}\right) - W_{g0}\left(\frac{1}{K_0}\right)\right] + \frac{a_1}{1+q_1}\left[\hat{T}_s^{1+q_1}\, W_{g1}\left(\frac{\hat{T}_s^{1+b}}{K_0}\right) - W_{g1}\left(\frac{1}{K_0}\right)\right]}$$

$$(10.62)$$

Once the system of linear algebraic equations (10.60) and (10.61) is solved, the temperature $\hat{T}$ and composition $\hat{H}$ fields can be calculated from Eqs. (10.38b) and (10.38a).

We can now observe that the solution of this problem is defined by two arrays of constant parameters

$$\Pi^{BC} = \left[K_0,\, \hat{H}_0,\, \Phi_0,\, \Phi_1\right] \tag{10.63}$$

$$\Pi^{P} = [m,\, b,\, q_0,\, q_1] \tag{10.64}$$

where the array $\Pi^{BC}$ is a function of the boundary conditions on the drop surface and at infinity (see Fig. 10.1): $T_s$, $T_\infty$, $y_s^{(1)}$, $y_\infty^{(1)}$, while the array $\Pi^{P}$ depends on the thermo-physical property constants (see Eq. 10.21).

The general approach reported above accounts for the temperature dependence of the thermo-physical properties; the solution appears a little cumbersome, due also to the implicit form of the Eqs. (10.38a) and (10.38b), however we will see now that when the main aim is the evaluation of the heat and vapour fluxes from the drop surface, the approach yields a relatively simple model.

Moreover, this result comprises many of the simplified models that can be found in literature, yielding simpler forms when some of the thermo-physical properties are assumed constant. By way of example, we will consider first the case when all thermo-physical properties are assumed constant and then the case of constant vapour heat capacity and thermal conductivities and variable density and diffusion coefficient.

#### 10.1.1.1  Example 1: Constant Thermo-Physical Properties

The case of constant thermo-physical properties corresponds to the array $\Pi^{P} = [m, b, q_0, q_1] = [1, 0, 0, 0]$ (refer to Eq. 10.21), then the parameters $h_p$ and $g_p$ are both equal to 1 (Eqs. 10.39 and 10.40, respectively). Under these simplifications the system of non-linear PDEs (Eqs. 10.35 and 10.36) becomes

$$\hat{\nabla}\hat{H}\hat{\nabla}_j\hat{T} = (a_0 + a_1)\,\hat{\nabla}^2\hat{T} \tag{10.65a}$$

$$\hat{\nabla}^2\hat{H} = 0 \tag{10.65b}$$

which, apart from some uninfluential multiplicative constants, are equivalent to Eqs. (10.30) and (10.29). The general solution of the system (10.65) is

$$\hat{H} = A\Phi + B \tag{10.66a}$$

$$\hat{T} = Ce^{(a_0+a_1)\hat{H}} + D \tag{10.66b}$$

where the constants $A, B, C$ and $D$ are calculated imposing the boundary conditions (10.55) and the solution to the problem is

$$\hat{H} = \left(\hat{H}_s - \hat{H}_\infty\right)\Phi + \hat{H}_\infty \tag{10.67a}$$

$$\hat{T} = \frac{1 - \hat{T}_s}{e^{(a_0+a_1)\hat{H}_\infty} - e^{(a_0+a_1)\hat{H}_s}}e^{(a_0+a_1)\hat{H}} + \frac{\hat{T}_s e^{(a_0+a_1)\hat{H}_\infty} - e^{(a_0+a_1)\hat{H}_s}}{e^{(a_0+a_1)\hat{H}_\infty} - e^{(a_0+a_1)\hat{H}_s}} \tag{10.67b}$$

An equivalent form of the solution can be derived starting from Eqs. (10.38a, 10.38b), substituting the vector $\Pi^P = [m, b, q_0, q_1] = [1, 0, 0, 0]$

$$\hat{H} = -(a_0 + a_1)\frac{\hat{T}}{K_0}W_1\left(\frac{\hat{T}}{K_0}\right) + \hat{H}_0 \tag{10.68a}$$

$$\Phi = \Phi_1(a_0 + a_1)\frac{\hat{T}}{K_0}W_1\left(\frac{\hat{T}}{K_0}\right) + \Phi_0 \tag{10.68b}$$

and in this case the hypergeometric function simplifies to $W_1(x) = {}_2F_1(1, 1, 2, x) = -\frac{\ln(1-x)}{x}$; the general solution of the problem is then

$$\hat{H} = (a_0 + a_1)\ln\frac{K_0 - \hat{T}}{K_0} + \hat{H}_0 \tag{10.69a}$$

$$\Phi = \Phi_1(a_0 + a_1)\frac{\hat{T}}{K_0}\ln\frac{K_0 - \hat{T}}{K_0} + \Phi_0 \tag{10.69b}$$

which is equivalent to the solution given by Eq. (10.67).

### 10.1.1.2   Example 2: Constant Vapour Heat Capacity and Thermal Conductivities

The case of constant vapour heat capacity and thermal conductivities was treated in [15] and it corresponds to the array $\Pi^P = [m, b, q_0, q_1] = [m, 0, 0, 0]$ (refer to Eq. 10.21), then the parameters $h_p$ and $g_p$ are equal to $h_p = 2 - m$ and $g_p = 1$ (Eqs. 10.39 and 10.40, respectively). In this case the system of non-linear PDEs (Eqs. 10.35 and 10.36) simplifies to

$$\hat{T}_j^{m-1}\hat{\nabla}\hat{H}\hat{\nabla}_j\hat{T} = (a_0 + a_1)\hat{\nabla}^2\hat{T} \tag{10.70a}$$

$$\hat{\nabla}_j\left(\hat{T}_j^{m-1}\hat{\nabla}_j\hat{H}\right) = 0 \tag{10.70b}$$

and, using Eqs. (10.38a, 10.38b), the solution of the problem reduces to

$$\hat{H} = -(a_0 + a_1) \frac{\hat{T}^{2-m}}{(2-m) K_0} W_{2-m} \left( \frac{\hat{T}}{K_0} \right) + \hat{H}_0 \qquad (10.71a)$$

$$\Phi = \Phi_1 (a_0 + a_1) \frac{\hat{T}}{K_0} W_1 \left( \frac{\hat{T}}{K_0} \right) + \Phi_0 \qquad (10.71b)$$

It is interesting now to observe that the function $W_n(x)$ can be written in terms of the incomplete *Beta function* [14]

$$W_n(x) = nx^{-n} B(x, n, 0) = nx^{-n} \int_0^x \frac{t^{n-1}}{1-t} dt \qquad (10.72)$$

Then, the general solution of the problem can be also written in a more expressive integral form

$$\hat{H} = \int \frac{\hat{T}^{1-m}}{\hat{T} - K_0} d\hat{T} \qquad (10.73a)$$

$$\Phi = \int \frac{1}{\hat{T} - K_0} d\hat{T} \qquad (10.73b)$$

The proof that the two functions (10.73a) and (10.73b) satisfy the conservation equations (10.70) is now straightforward: the gradients of the two functions $\hat{H}$ and $\Phi$ are

$$\hat{\nabla}_j \hat{H} - \frac{\hat{T}^{1-m}}{\hat{T} - K_0} \hat{\nabla}_j \hat{T} \qquad (10.74)$$

$$\hat{\nabla}_j \Phi = \frac{1}{\hat{T} - K_0} \hat{\nabla}_j \hat{T} \qquad (10.75)$$

and substituted into Eq. (10.70b) yields

$$\hat{\nabla}_j \left( \hat{T}^{m-1} \hat{\nabla}_j \hat{H} \right) = \hat{\nabla}_j \left( \hat{T}^{m-1} \frac{\hat{T}^{1-m}}{\hat{T} - K_0} \hat{\nabla}_j \hat{T} \right) = \hat{\nabla}_j \left( \hat{\nabla}_j \Phi \right) = \hat{\nabla}^2 \Phi = 0 \quad (10.76)$$

which is satisfied since the function $\Phi$ is an harmonic function. Substituting Eqs. (10.74) and (10.75) in the LHS of the energy conservation equation (10.70a) yields

$$\left( \hat{T}^{m-1} \hat{\nabla}_j \hat{H} \right) \hat{\nabla}_j \hat{T} = \left( \hat{T}^{m-1} \frac{\hat{T}^{1-m}}{\hat{T} - K_0} \hat{\nabla}_j \hat{T} \right) \hat{\nabla}_j T = \hat{\nabla}_j \Phi \hat{\nabla}_j T \qquad (10.77)$$

while calculating the RHS of the energy conservation equation (10.70a) using Eq. (10.75) yields

$$\hat{\nabla}^2 \hat{T} = \hat{\nabla}_j \left( \hat{\nabla}_j T \right) = \hat{\nabla}_j \left[ \hat{\nabla}_j \Phi \left( \hat{T} - K_0 \right) \right] = \hat{\nabla}^2 \Phi \left( \hat{T} - K_0 \right) + \hat{\nabla}_j \Phi \hat{\nabla}_j T = \hat{\nabla}_j \Phi \hat{\nabla}_j T \quad (10.78)$$

which proves the statement.

## 10.1.2   The Species and Energy Fluxes

Having obtained the analytical form of the functions $\hat{H}$ and implicitly that of $\hat{T}$ from Eqs. (10.38a, 10.38b), we can now use them to evaluate the mass and heat fluxes at the drop surface. To evaluate the mass flux we can use Eq. (10.15), and for the sensible heat flux we can use Fourier's law

$$q_j = -k_{mix} \nabla_j T \quad (10.79)$$

The gradients of $\hat{H}$ and $\hat{T}$ that enter in these equations can be calculated from the gradients

$$\nabla_j \hat{H} = -\frac{1}{\Phi_1} (1 + b) \, \hat{T}^{1-m} \nabla_j \Phi \quad (10.80)$$

$$\nabla_j \hat{T} = -\frac{\left( K_0 - \hat{T}^{1+b} \right)}{\Phi_1 \left( a_0 \hat{T}^{q_0} + a_1 \hat{T}^{q_1} \right)} \nabla_j \Phi \quad (10.81)$$

recalling the definition of $\hat{H} = \frac{H \left( a_0 \hat{T}_{ref}^{q_0} + a_1 \hat{T}_{ref}^{q_1} \right) \hat{T}_{ref}^{1-m-b}}{Le_M}$ and $\hat{T} = \frac{T}{T_\infty}$ and then

$$n_j^{(1)} = f_n^{\ (1)} \nabla_j \Phi \quad (10.82)$$

$$q_j = f_q \nabla_j \Phi \quad (10.83)$$

where

$$f_n^{\ (1)} = -\frac{k_{mix,ref} \hat{T}_{ref}^b}{c_{p,ref}^{(1)} \left( a_0 \hat{T}_{ref}^{q_0} + a_1 \hat{T}_{ref}^{q_1} \right) \Phi_1} \quad (10.84)$$

$$f_q = -T_\infty \frac{k_{mix,ref} \left( K_0 - \hat{T}_{ref}^{1+b} \right)}{\left( a_0 \hat{T}_{ref}^{q_0} + a_1 \hat{T}_{ref}^{q_1} \right) \Phi_1} \quad (10.85)$$

Equations (10.82) and (10.83) explicitly show that the mass and heat fluxes can be written as the product of two functions: the gradient of the auxiliary function $\Phi$, which

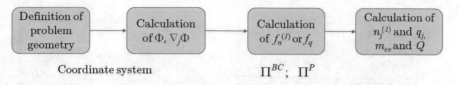

Coordinate system                                  $\Pi^{BC}$;   $\Pi^{P}$

**Fig. 10.2**  Sketch of the procedure to find heat and vapour fluxes and rates

is independent of the operating conditions and species thermo-physical properties and only depends on the geometry of the problem, and a function ($f_n^{(1)}$ and $f_q$ for the mass and heat flux, respectively) that depends on the operating conditions and the values of the thermo-physical properties coefficients (the array $\Pi^P$), while it is independent of the geometry of the problem.

The evaporation rate and the heat rate can be calculated by integrating the local fluxes over the drop surface

$$m_{ev} = \int_A n_j^{(1)} dA = f_n^{(1)} \int_A \nabla_j \Phi \, dA \tag{10.86a}$$

$$Q = \int_A q_j \, dA = f_q \int_A \nabla_j \Phi \, dA \tag{10.86b}$$

As a result of this analysis, we can say that the modelling of heating and evaporation of a single-component drop can be addressed following the procedure schematised in Fig. 10.2. As we will see in the following sections, the calculation of the auxiliary function $\Phi$ can be facilitated by a proper choice of the coordinate system that better fits the problem geometry, and the evaluation of the influence of the thermo-physical properties is then obtained from the calculation of the two function $f_n^{(1)}$ and $f_q$.

### 10.1.3   The Effect of Temperature Dependence of the Thermo-Physical Properties

To study the effect of modelling the thermo-physical properties, let first start with the *classical* model, assuming all properties as constants. This corresponds to the array $\Pi^P = [m, b, q_0, q_1] = [1, 0, 0, 0]$, then the parameter $f_n^{(1)}$ in Eq. (10.84) reduces to

$$f_n^{(1),c} = M m^{(1)} c_{ref} D_{10,ref} (H_s - H_\infty) \tag{10.87}$$

where the superscript $c$ stands for '*constant properties*'. The parameter $f_n^{(1),c}$ is a function of the two arrays $\Pi^{BC}$ and $\Pi^P$ (Eqs. 10.63 and 10.64) and of the averaging parameter $\alpha_{ref}$ (refer to Eq. 10.27), which enters in the calculation of the molar density $c_{ref}$ and the diffusion coefficient $D_{10,ref}$ at reference conditions. Studies

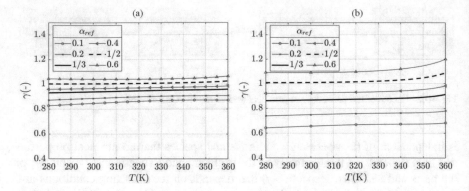

**Fig. 10.3** Evaporation rate ratio $\gamma$ as a function of the liquid temperature, for different averaging parameter $\alpha_{ref}$ in the constant gas density model, for water drops at **a** 500 K and **b** 1000 K gas temperature

from the literature [2, 12] have shown that a value of $\alpha_{ref}$ equal to $1/2$ gives better results at moderately low evaporating conditions, while a value of $\alpha_{ref}$ equal to $1/3$ should be preferred at moderately high evaporating conditions. To better quantify the effect of the choice of a value for $\alpha_{ref}$ let us define the non-dimensional parameter $\bar{\gamma}$

$$\bar{\gamma} = \frac{m_{ev}^c \left( \alpha_{ref} \right)}{m_{ev}} \tag{10.88}$$

as the ratio between the evaporation rate predicted using the constant properties model (Eq. 10.86a, with the parameter parameter $f_n^{(1)}$ given by Eq. 10.87) and the evaporation rate predicted using the variable properties model (Eq. 10.86a, with the parameter $f_n^{(1)}$ from Eq. 10.84); the latter is taken as a reference since it is expected to yield a more accurate prediction.

Figure 10.3 shows the effect of this choice, the cases reported correspond to a water drop vaporising in stagnant air, with a free stream temperature equal to 500 and 1000 K. The drop temperature has been varied from 280 K up to the boiling point. Six values of the averaging parameter $\alpha_{ref}$ have been selected, from 0.1 up to 0.6. The graphs clearly show that the choice of the averaging parameter $\alpha_{ref}$ does affect the predictions of the *classical* model and, compared to the more accurate modelling, the results confirm that at low drop temperature (i.e. low evaporating conditions) the classical constant property model with $\alpha_{ref}$ equal to $1/2$ better captures the phenomenon (with relative difference from the variable property model lower than 1%), while a lower value of $\alpha_{ref}$ should be used as the drop temperature increases. This is more evident at higher temperatures of the gaseous mixture at free stream conditions ($T_\infty = 1000$ K) since the effect of flow field temperature gradients on the thermo-physical properties are more significant.

These results also suggest that the choice of the best value for the averaging parameter $\alpha_{ref}$, when using the *classical* model, may vary with evaporating conditions, which are described by the Spalding mass transfer number, defined as

$$B_M = \frac{\chi_s - \chi_\infty}{1 - \chi_s} \tag{10.89}$$

A way to improve the performance of the *classical* constant properties model was suggested in [3, 16]: the value of $\alpha_{ref}$ could be chosen as a function of the evaporating conditions, and the following relation between $\alpha_{ref}$ and $B_M$ was proposed

$$\alpha_{ref} = \frac{A_{\alpha_{ref}}}{\log(1 + B_M)} + \frac{1}{1 - (1 + B_M)^{\frac{1}{A_{\alpha_{ref}}}}} \tag{10.90}$$

where the coefficient $A_{\alpha_{ref}}$ depends on the drop species and the gas temperature, and it is calculated as the optimal value that minimises the integral (see [16] for details)

$$\int_0^{B_{M,max}} \left| \bar{\gamma}(B_M, A_{\alpha_{ref}}) - 1 \right|^2 dB_M \tag{10.91}$$

To assess the validity of this approach, the evaporation ratio $\bar{\gamma}$ as a function of the Spalding mass transfer number $B_M$ is reported in Fig. 10.4, for three cases: $\alpha_{ref} = 1/2$, $\alpha_{ref} = 1/3$ and $\alpha_{ref}$ given by Eq. (10.90). The test case corresponds to a water drop vaporising in stagnant air at 1000 $K$. The results evidence that the estimation of $\alpha_{ref}$, given by Eq. (10.90), yields a satisfactory agreement over all the

**Fig. 10.4  a**
Non-dimensional
evaporation rate as a function
of the Spalding mass transfer
number $B_M$, with the
averaging parameter $\alpha_{ref}$
reported in figure (**b**) and
equal to 1/3 and 1/2; water
drop in air at 1000 K

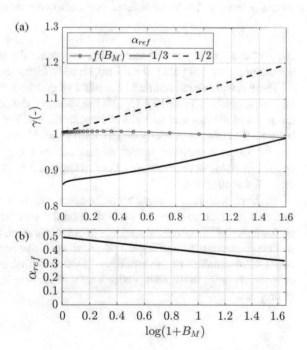

**Table 10.1** Values of the averaging constant $A_{\alpha_{ref}}$ (Eq. 10.90), for six species and two operating conditions

| Species | $T_\infty = 500\,\mathrm{K}$ | $T_\infty = 1000\,\mathrm{K}$ |
|---|---|---|
| Water | 0.752 | 0.752 |
| Acetone | 0.753 | 1.24 |
| Ethanol | 0.85 | 0.68 |
| n-hexane | 0.66 | 0.75 |
| n-octane | 0.6 | 0.65 |
| n-dodecane | 0.545 | 0.54 |

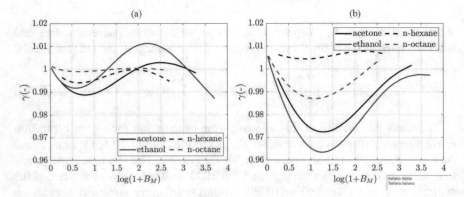

**Fig. 10.5** Evaporation rate ratio $\gamma$, as function of the Spalding mass transfer number, calculated using $\alpha_{ref}$ from Eq. (10.90), for different drop species evaporating in air at **a** 500 K and **b** 1000 K

range of evaporating conditions, with a maximum discrepancy less than 4%, while the use of a constant value of $\alpha_{ref}$ may lead to discrepancies larger than 20%.

These results were confirmed for other evaporating species (acetone, ethanol, n-hexane, n-octane). Table 10.1 reports the values of the constant $A_{\alpha_{ref}}$ (Eq. 10.90) for the selected species and two gas temperatures, and Fig. 10.5 show the discrepancy between the constant property model, with a variable average coefficient (Eq. 10.90), and the variable property model as function of the operating conditions for four species and two gas temperatures, confirming that the maximum discrepancy is lower than 4% for all test cases.

We have seen how a more accurate approach, in term of modelling the dependence of gas thermo-physical properties on the temperature, allows improving the performance of the *classical* constant properties model, undoubtedly simpler to use. But the final evaluation of important parameters like heat and mass fluxes and rates needs the evaluation of the auxiliary function $\Phi$ that depends on the shape of the evaporating drop. In the next sections we will see how to take into account the effect of drop shape.

## *10.1.4   The Effect of Drop Shape*

The early models of drop heating and evaporation were developed under the hypothesis that the evaporating drop is spherical. The spherical shape is the shape of smallest surface for a given volume, and it is then assumed by any lump of liquid subject to the sole action of the surface tension. However, when a drop is interacting with the environment other forces may act on it. A drop kept still in a gravity field will be deformed by the action of gravity, as for a drop suspended on a wire or for a drop laying on a surface (sessile drop). If the drop is freely falling, gravitation does not alter its form, but aerodynamic interactions generate surface forces that modify its shape [17]. The gas flow around the drop causes a peak pressure at the leading edge and a minimum pressure at the equator, that leads to an oblate shape. On the other hand, the liquid circulation inside the drop, caused by the shear stresses acting on the surface, induces a deformation towards the prolate shape.

Moreover, once that the shape of a drop has been modified from the spherical one, the action of the surface tension will tend to restore the initial condition of minimum potential (surface) energy and the shape will begin to modify with time, generating drop oscillation, a phenomenon that has been largely studied since the nineteenth century [18].

Also the practical generation of drops influences their shapes. A drop detached from a tiny tube by gravity has an initial shape far from spherical; drops generated by injecting a liquid at high pressure through an orifice have the most different shapes due to the jet break-up mechanisms [19]; drops interacting with a gas stream at high relative velocity may deform and break-up, generating other drops that are non-spherical, etc. To summarise, spherical shape is likely to be the less common shape for a drop in the real world and this justifies the growing research activities on non-spherical drop evaporation in the recent years [20]. It should also be mentioned that the theory above developed does not assume that the evaporating liquid is made by a single lump of fluid. In fact the theory applies to any configuration of the liquid surface, with the only constraint that the total mass of liquid is finite. This allows its application to another interesting (and quite important from a practical point of view) case: the evaporation of more than one drop in the same environment, which is what happens for example in sprays and aerosols.

The effect of the drop shape is given by the function $\Phi$ that satisfies the Laplace equation $\nabla^2 \Phi = 0$ and it is equal to 1 on the interface, while it is nil at infinity. The solution of this problem is usually greatly facilitated when the shape of the interface fits an iso-surface of a curvilinear coordinate system. We will see in the next sections examples of such method to find $\Phi$, making use of the mathematical tools developed in Part I, to which the reader will be redirected for more details.

## 10.2   The Spherical Drop

As already said, the majority of the analytical models predicting drop evaporation and implemented in CFD codes for dispersed phase applications assume that the evaporating drop is perfectly spherical, and that the problem can be analysed assuming spherical symmetry. The *natural* coordinate system is then the spherical one, which is defined as follows

$$x = R\sqrt{1 - \eta^2} \cos(\varphi) \tag{10.92a}$$
$$y = R\sqrt{1 - \eta^2} \sin(\varphi) \tag{10.92b}$$
$$z = R\eta \tag{10.92c}$$

where $\eta = \cos(\theta)$ is used as angular coordinate instead of $\theta$, since the mathematical formulation of the problem assumes a more compact from (see Chap. 4 for a comparison of the two equivalent systems).

The coordinate space $(R, \eta, \varphi)$ is limited *by*

$$0 \leq R < \infty; \quad -1 \leq \eta \leq 1; \quad 0 \leq \varphi < 2\pi \tag{10.93}$$

The iso-surfaces $R = const.$ correspond to concentric spheres centred in the origin, the iso-surfaces $\eta = const.$ correspond to cones with vertices in the origin and the iso-surfaces $\varphi = const.$ correspond to half-planes intersecting on the $z$-axis. Figure 4.4 in Chap. 4 shows a sketch of the spherical coordinate system where the interested reader can find all the details about this coordinate system and the corresponding differential operators used below.

In this coordinate system the interface fits the iso-coordinate surface $R = R_d$, where $R_d$ is the drop radius and we can see how this observation will simplify the solution. The Laplace equation in this coordinate system is

$$\frac{\partial^2 \Phi}{\partial R^2} + \frac{2}{R}\frac{\partial \Phi}{\partial R} + \frac{1 - \eta^2}{R^2}\frac{\partial^2 \Phi}{\partial \eta^2} - \frac{2\eta}{R^2}\frac{\partial \Phi}{\partial \eta} + \frac{1}{R^2\left(1 - \eta^2\right)}\frac{\partial^2 \Phi}{\partial \varphi^2} = 0 \tag{10.94}$$

When uniform boundary conditions are imposed on the iso-surface, the problem becomes symmetric around the origin and the auxiliary function $\Phi$ only depends on the $R$-coordinate, then the Laplace equation (10.94) reduces to

$$\frac{d^2 \Phi}{dR^2} + \frac{2}{R}\frac{d\Phi}{dR} = 0 \tag{10.95}$$

which admits as general solution

$$\Phi(R) = A + \frac{B}{R} \tag{10.96}$$

**Fig. 10.6** Schematic of
B.C.s imposed using the
spherical coordinate system

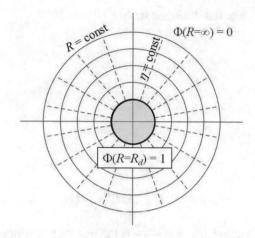

The boundary conditions at the iso-surface corresponding to the drop surface,
$R = R_d$, and the infinite, $R \to \infty$, according to the schematic of Fig. 10.6, are

$$\Phi \left( R = R_d \right) = 1 \tag{10.97}$$
$$\Phi \left( R \to \infty \right) = 0 \tag{10.98}$$

and the evaluation of the constants $A$ and $B$ yields

$$\Phi \left( R \right) = \frac{R_d}{R} \tag{10.99}$$

To evaluate the fluxes, the gradient of $\Phi$ along the direction normal to the surface
is needed (Eqs. 10.82 and 10.83), and for the spherical case it is

$$\nabla_R \Phi = \frac{d\Phi}{dR} = -\frac{R_d}{R^2} \tag{10.100}$$

The vapour fluxes along the $R$-direction on the drop surface ($R = R_d$) are calculated using Eqs. (10.82) and (10.100)

$$n_R^{(1)} \left( R_d \right) = f_n^{(1)} \left( \nabla_R \Phi \right)_{R=R_d} = -\frac{f_n^{(1)}}{R_d} \tag{10.101}$$

The evaporation rate can then be calculated as the integral of the vapour flux along
the $R$-direction at the drop surface

$$m_{ev} = \int_A n_R^{(1)} \left( R_d \right) dA = f_n^{(1)} \int_0^{2\pi} \int_{-1}^1 \left( \nabla_R \Phi \right)_{R=R_d} h_\eta h_\varphi d\eta d\varphi = -4\pi R_d f_n^{(1)} \tag{10.102}$$

Assuming constant values for the thermo-physical properties, Eq. (10.87) yields

**Fig. 10.7** Schematic for the film theory model

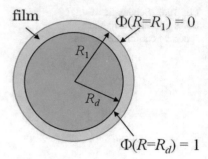

$$m_{ev}^c = 4\pi R_d D_{10,ref} M_m^{(1)} c_{ref} (H_s - H_\infty) = 4\pi R_d D_{10,ref} M_m^{(1)} c_{ref} \ln \left(1 + B_M^*\right)$$
(10.103)

where $B_M^* = \frac{y_s^{(1)} - y_\infty^{(1)}}{1 - y_s^{(1)}}$ is the modified Spalding mass transfer number, as predicted by the classical Stefan-Fuchs evaporation model [1], in molar form, which holds for stationary droplet evaporating in stagnant environment (Re = 0). The case of variable gas density and constancy of all the other thermo-physical properties was first studied in [21].

Analogously, the heat flux and the heat rate can be calculated as

$$q_R (R_d) = f_q (\nabla_R \Phi)_{R=R_d} = -\frac{f_q}{R_d}$$
(10.104)

$$\dot{Q} = \int_A q_R (R_d) \, dA = -4\pi R_d f_q$$
(10.105)

## 10.2.1 Effect of Convection

To take into account the effect of convection caused by the drop-gas relative motion (Re > 0), a model based on the so-called *film theory* was proposed in [2]. The model assumes that the evaporating drop is surrounded by a mass diffusional region, which thickness depends on the Reynolds number and the physical properties of the species, as schematically illustrated in Fig. 10.7, where $R = R_d$ is the iso-surface corresponding to the drop surface and $R = R_1$ is the iso-surface corresponding to the film edge.

The boundary conditions at infinity are now replaced by the boundary conditions at the film edge

$$\Phi (R = R_d) = 1 \rightarrow A + \frac{B}{R_d} = 1$$
(10.106)

$$\Phi (R = R_1) = 0 \rightarrow A + \frac{B}{R_1} = 0$$
(10.107)

Then the two constants $A$ and $B$ are

$$A = -\frac{R_d}{R_1 - R_d}; \quad B = \frac{R_d R_1}{R_1 - R_d} \tag{10.108}$$

and the solution (10.96) assumes the form

$$\Phi^{(Re>0)}(R) = -\frac{R_d}{R_1 - R_d}\left(1 + \frac{R_1}{R}\right) \tag{10.109}$$

while the gradient of $\Phi$ along the $R$-coordinate is

$$\nabla_R \Phi^{(Re>0)} = -\frac{R_d R_1}{R_1 - R_d}\frac{1}{R^2} \tag{10.110}$$

The evaporation rate under convective conditions can be calculated as

$$m_{ev}^{(Re>0)} = \int_A n_{R=R_d}^{(1)} dA = f_n^{(1)} \int_0^{2\pi} \int_{-1}^1 \left(\nabla_R \Phi^{(Re>0)}\right)_{R=R_d} h_\eta h_\varphi d\eta d\varphi = -4\pi \frac{R_d R_1}{R_1 - R_d} f_n^{(1)} \tag{10.111}$$

The main problem with the film theory approach is that the film thickness $(R_1 - R_d)$ is unknown and it cannot be defined in general from basic principles. The introduction of the Sherwood number $Sh$ (named after the American chemical engineer Thomas Kilgore Sherwood, 1903–1976) allows to relate the evaporation rate under convective conditions ($Re > 0$) to that under non-convective conditions ($Re = 0$), as

$$m_{ev}^{(Re>0)} = \frac{Sh}{2} m_{ev} \tag{10.112}$$

Following [2], $Sh$ can be defined by a semi-empirical correlation

$$Sh = 2 + \frac{Sh_0 - 2}{F_M(B_M)} \tag{10.113}$$

where

$$Sh_0 = 2 + 0.552\sqrt{Re}\sqrt[3]{Sc} \tag{10.114}$$

$$F_M(B_M) = \frac{(1 + B_M)^{0.7} \log(1 + B_M)}{B_M} \tag{10.115}$$

being $Sc$ the Schmidt number

$$Sc = \frac{\nu}{D_{10}} \tag{10.116}$$

with $\nu = \frac{\mu}{\rho}$ being the kinematic viscosity and $B_M$ the Spalding mass transfer, defined by Eq. (10.89).

The effect of convection is then introduced in a semi-empirical way, starting from an exact solution of the Laplace problem. A similar approach can be used to evaluate the heat rate (see [22]).

## 10.3   The Spheroidal Drop

Experimental observations have reported relatively stable oblate spheroidal shape of rain drops freely falling due to gravity [17]. Moreover, during a drop oscillation caused by an initial deformation from the spherical shape, different modes characterise the unsteady surface shape [18], but, due to the action of viscous forces that damp the oscillations, the one that survives longer is the mode $n = 2$ that can be represented as an oscillation between oblate and the prolate spheroidal shapes. Spheroids are then a good representation of the shape of liquid drops in many realistic conditions and in this section we will analyse the evaporation characteristics of this kind of non-spherical drops.

The spheroid surface is obtained by the rotation of an ellipse around one of the axis and it is defined by the equation

$$\frac{x^2 + y^2}{a_r^2} - \frac{z^2}{a_z^2} = 1 \tag{10.117}$$

where $z$ is the symmetry axis. The drop shape is completely characterised by the two parameters $a_r$ and $a_z$ (see Fig. 10.8).

For the present purpose, the interface shape can be more conveniently defined by two other parameters, $\varepsilon$ and $R_d$. The deformation parameter $\varepsilon$, also called *aspect ratio*, is defined as

$$\varepsilon = \frac{a_z}{a_r} \tag{10.118}$$

and it is smaller than one ($\varepsilon < 1$) for the oblate shape and larger than one ($\varepsilon > 1$) for the prolate shape. The equivalent radius $R_d$ is the radius of a spherical drop having the same volume of the spheroid, and since the volume of the spheroid is given by $V = \frac{4}{3}\pi a_r^2 a_z$, the equivalent drop radius is defined as

$$R_d = \left(a_r^2 a_z\right)^{1/3} \tag{10.119}$$

The *natural* coordinate systems to solve the problem of heating and evaporation of a spheroidal (either prolate or oblate) drop are the corresponding spheroidal coordinate systems. The description of these two coordinate systems was given in Chap. 4 and here we just recall the main features, which are useful for the present problem.

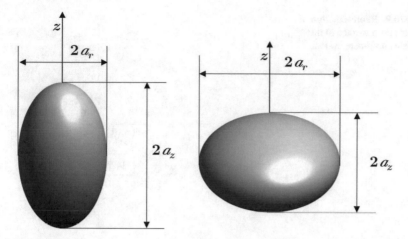

**Fig. 10.8** Definition of axial ($a_z$) and radial ($a_r$) spheroid semi-axes for prolate (left) and oblate (right) spheroids

## 10.3.1 The Prolate Spheroidal Drop

The prolate spheroidal coordinate system is defined by the following equations

$$x = a \sqrt{\zeta^2 - 1}\sqrt{1 - \eta^2}\cos(\varphi) \tag{10.120a}$$

$$y = a \sqrt{\zeta^2 - 1}\sqrt{1 - \eta^2}\sin(\varphi) \tag{10.120b}$$

$$z = a \zeta\eta \tag{10.120c}$$

where $a$ is a scaling parameter and the coordinate space ($\zeta, \eta, \varphi$) is limited by

$$1 \leq \zeta < \infty; \quad -1 \leq \eta \leq 1; \quad 0 \leq \varphi < 2\pi \tag{10.121}$$

The iso-surfaces $\zeta = const.$ correspond to prolate spheroids, with semi-axes $a_r = a\sqrt{\zeta^2 - 1}$ and $a_z = a\zeta$, the iso-surfaces $\eta = const.$ correspond to hyperboloids and the iso-surfaces $\varphi = const.$ correspond to half-planes passing through the $z$-axis.

If the drop has a prolate spheroidal shape, the surface can be defined in this curvilinear coordinate system by the simple equation: $\zeta = \zeta_0$, as in Fig. 10.9.

From Eqs. (10.117) and (10.120a), the aspect ratio can be evaluated as

$$\varepsilon = \frac{a_z}{a_r} = \frac{\zeta_0}{\sqrt{\zeta_0^2 - 1}} \Rightarrow \zeta_0 = \frac{\varepsilon}{\sqrt{\varepsilon^2 - 1}} \tag{10.122}$$

The scale parameter $a$ in Eq. (10.120) can be related to $\varepsilon$ and the equivalent radius, $R_d$, using Eq. (10.119) and the definition of the semi-axes

**Fig. 10.9** Representation of
prolate drop surface in the
prolate coordinate system

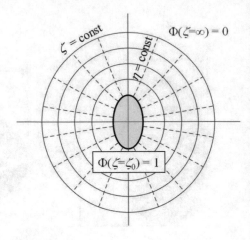

$$a = R_d \frac{\left|1 - \varepsilon^2\right|^{1/2}}{\varepsilon^{1/3}} \tag{10.123}$$

The Laplace equation in this coordinate system is (cfr Sect. 4.4.5)

$$\frac{1}{a^2 \left(\zeta^2 - \eta^2\right)} \left[ \left(\zeta^2 - 1\right) \frac{\partial^2 \Phi}{\partial \zeta^2} + 2\zeta \frac{\partial \Phi}{\partial \zeta} + \frac{\partial^2 \Phi}{\partial \eta^2} - 2\eta \frac{\partial \Phi}{\partial \eta} \right] + \frac{1}{a^2 \left(\zeta^2 - 1\right) \left(1 - \eta^2\right)} \frac{\partial^2 \Phi}{\partial \varphi^2} = 0 \tag{10.124}$$

When uniform boundary conditions are imposed on the iso-surface corresponding to the drop surface ($\zeta = \zeta_0$), the auxiliary function $\Phi$ only depends on the $\zeta$-coordinate. Then the Laplace equation (10.124) reduces to

$$2\zeta \frac{d\Phi}{d\zeta} + \left(\zeta^2 - 1\right) \frac{d^2\Phi}{d\zeta^2} = 0 \tag{10.125}$$

which has the general solution

$$\Phi\left(\zeta\right) = A + B \ln \frac{\zeta - 1}{\zeta + 1} \tag{10.126}$$

The boundary conditions at the iso-surface corresponding to the drop surface, $\zeta = \zeta_0$, and at infinity, $\zeta = \infty$, are (see Fig. 10.9)

$$\Phi\left(\zeta = \zeta_0\right) = 1 \tag{10.127}$$

$$\Phi\left(\zeta = \infty\right) = 0 \tag{10.128}$$

and they allow calculating the constants $A$ and $B$ yielding

$$\Phi\left(\zeta\right) = \frac{\ln\frac{\zeta-1}{\zeta+1}}{\ln\frac{\zeta_0-1}{\zeta_0+1}} \tag{10.129}$$

Again, to evaluate the fluxes, the gradient of $\Phi$ along the direction normal to the surface is needed (Eqs. 10.82 and 10.83), and for the present case this is the gradient along the $\zeta$-direction, i.e.

$$\nabla_\zeta \Phi = \frac{1}{h_\zeta}\frac{d\Phi}{d\zeta} = \frac{2}{a\sqrt{\left(\zeta^2 - \eta^2\right)\left(\zeta^2 - 1\right)}}\frac{1}{\ln\frac{\zeta_0-1}{\zeta_0+1}} \tag{10.130}$$

## 10.3.2 The Oblate Spheroidal Drop

Similarly to the prolate case, the oblate spheroidal coordinate system is defined by the following equations

$$x = a\sqrt{\zeta^2 + 1}\sqrt{1 - \eta^2}\cos\left(\varphi\right) \tag{10.131a}$$
$$y = a\sqrt{\zeta^2 + 1}\sqrt{1 - \eta^2}\sin\left(\varphi\right) \tag{10.131b}$$
$$z = a\,\zeta\eta \tag{10.131c}$$

where again $a$ is a scale parameter and the coordinate space $(\zeta, \eta, \varphi)$ is limited as:

$$0 \leq \zeta < \infty; \quad -1 \leq \eta \leq 1; \quad 0 \leq \varphi < 2\pi \tag{10.132}$$

The iso-surfaces $\zeta = const.$ correspond to oblate spheroids, the iso-surfaces $\eta = const.$ correspond to hyperboloids and the iso-surfaces $\varphi = const$ correspond to half-planes passing through the $z$-axis.

For a drop with an oblate spheroidal shape, the surface can be defied in this curvilinear coordinate system by the simple equation: $\zeta = \zeta_0$, as in Fig. 10.10.

Following the same steps already seen for the prolate case, the aspect ratio can be evaluated as

$$\varepsilon = \frac{a_z}{a_r} = \frac{\zeta_0}{\sqrt{\zeta_0^2 + 1}} \Rightarrow \zeta_0 = \frac{\varepsilon}{\sqrt{1 - \varepsilon^2}} \tag{10.133}$$

The scale parameter $a$ in Eq. (10.131a) is related to $\varepsilon$ and the equivalent radius, $R_d$, as for the prolate case, i.e. by Eq. (10.123).

The Laplace equation is now (cfr Sect. 4.4.6)

$$\frac{1}{a^2\left(\zeta^2 + \eta^2\right)}\left[\left(\zeta^2 + 1\right)\frac{\partial^2\Phi}{\partial\zeta^2} + 2\zeta\frac{\partial\Phi}{\partial\zeta} + \left(1 - \eta^2\right)\frac{\partial^2\Phi}{\partial\eta^2} - 2\eta\frac{\partial\Phi}{\partial\eta}\right] + \frac{1}{a^2\left(\zeta^2 + 1\right)\left(1 - \eta^2\right)}\frac{\partial^2\Phi}{\partial\varphi^2} = 0 \tag{10.134}$$

**Fig. 10.10** Representation of oblate drop surface in the oblate coordinate system

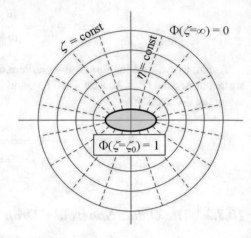

and again, when uniform boundary conditions are imposed on the iso-surface corresponding to the drop surface ($\zeta = \zeta_0$), the auxiliary function $\Phi$ depends only on the $\zeta$-coordinate and the Laplace equation simplifies to

$$2\zeta \frac{d\Phi}{d\zeta} + (\zeta + 1)\frac{d^2\Phi}{d\zeta^2} = 0 \tag{10.135}$$

The general solution for this equation is

$$\Phi\left(\zeta\right) = A + B \arctan\left(\zeta\right) \tag{10.136}$$

Using the boundary conditions

$$\Phi\left(\zeta = \zeta_0\right) = 1 \tag{10.137}$$
$$\Phi\left(\zeta = \infty\right) = 0 \tag{10.138}$$

to determine the constants $A$ and $B$ yields the final form

$$\Phi\left(\zeta\right) = \frac{\arctan\left(\zeta\right) - \frac{\pi}{2}}{\arctan\left(\zeta_0\right) - \frac{\pi}{2}} \tag{10.139}$$

It is worth to notice that an apparently different expression for the solution of the 1D Laplace equation (10.136) can be written as

$$\Phi\left(\xi\right) = \frac{\arctan\left(e^{\xi}\right) - \frac{\pi}{2}}{\arctan\left(e^{\xi_0}\right) - \frac{\pi}{2}} \tag{10.140}$$

**Fig. 10.11** Vapour distribution around an oblate drop (left) with and a prolate drop (right)

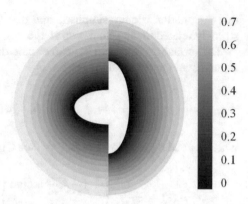

where $\sinh(\xi) = \zeta$. The two expressions are identical and this can be shown making use of the identity

$$2 \arctan^{-1}\left(e^{\xi}\right) - \arctan\left[\sinh\left(\xi\right)\right] = \frac{\pi}{2} \tag{10.141}$$

As for the prolate spheroid case, to evaluate the fluxes, the gradient of $\Phi$ along the direction normal to the surface, i.e. $\zeta$, is needed and in this case

$$\nabla_{\zeta}\Phi = \frac{1}{h_{\zeta}}\frac{d\Phi}{d\zeta} = \frac{1}{a\sqrt{\zeta^2 + \eta^2}\sqrt{\zeta^2 + 1}}\frac{1}{\arctan(\zeta) - \frac{\pi}{2}} \tag{10.142}$$

Figure 10.11 shows an example of the distribution of $\Phi$ around an oblate (left) and a prolate (right) liquid drop, with $\varepsilon = 0.5$ and $\varepsilon = 2.18$, respectively. The two drops have the same volume, equal to $\frac{4}{3}\pi R_d^3$, where $R_d$ is the radius of the equivalent volume spherical drop

### 10.3.3 The Vapour Fluxes from Spheroidal Drops

Once the analytical expression for the gradient of $\Phi$ is known, the vapour fluxes along the $\zeta$-direction at the drop surface ($\zeta = \zeta_0$) can be analytically calculated using Eqs. (10.82), (10.130) and (10.142)

$$n_{\zeta}^{(1)}\left(\zeta_0\right) = f_n^{(1)}\left(\nabla_{\zeta}\Phi\right)_{\zeta=\zeta_0} = \frac{f_n^{(1)}}{a}\begin{cases}\dfrac{2}{\sqrt{\zeta_0^2-\eta^2}\sqrt{\zeta_0^2-1}}\dfrac{1}{\ln\frac{\zeta_0-1}{\zeta_0+1}} & \text{prolate}\\[2ex]\dfrac{1}{\sqrt{\zeta_0^2+\eta^2}\sqrt{\zeta_0^2+1}}\dfrac{1}{\arctan(\zeta_0)-\pi/2} & \text{oblate}\end{cases} \tag{10.143}$$

The expected increase of the evaporation rates in deformed drops is often ascribed to the increase of surface with respect to the spherical shape. We will see now that this explanation is quite simplistic and that the problem is more involved and it is strongly related to the actual drop shape.

The surface of a spheroid $A_{sd}$ can be calculated as a function of the two parameters $R_d$ and $\varepsilon$ and the general expression is

$$A_{sd} = \int_0^{2\pi} \int_{-1}^{+1} h_\eta h_\varphi d\eta d\varphi = 4\pi R_d^2 \beta(\varepsilon) \tag{10.144}$$

where $h_n$ and $h_\varphi$ are the scale factors (see Chap. 4) and the function $\beta(\varepsilon)$

$$\beta(\varepsilon) = \frac{1}{2\varepsilon^{2/3}} \begin{cases} 1 + \frac{\varepsilon^2}{\sqrt{\varepsilon^2-1}} \arctan\left(\sqrt{\varepsilon^2-1}\right) & \text{prolate} \\ 1 + \frac{\varepsilon^2}{\sqrt{1-\varepsilon^2}} \ln\left(\sqrt{\frac{1+\sqrt{1-\varepsilon^2}}{1-\sqrt{1-\varepsilon^2}}}\right) & \text{oblate} \end{cases} \tag{10.145}$$

can be interpreted as the non-dimensional surface $\beta = \frac{A_{sd}}{4\pi R_d^2}$.

Integrating Eq. (10.143) over the drop surface yields the evaporation rate

$$m_{ev,sd} = \int_A n_\zeta^{(1)}(\zeta_0) \, dA = 2\pi \int_{-1}^1 n_\zeta^{(1)}(\zeta_0) h_\eta h_\varphi d\eta \tag{10.146}$$

where the last equation is found using the axial-symmetry of the problem, and the scale factors are given by Eqs. (4.234) and (4.264) for the prolate and oblate case, respectively. The integration is straightforward and yields

$$m_{ev,sd} = -4\pi R_d f_n^{(1)} \frac{\left|1-\varepsilon^2\right|^{1/2}}{\varepsilon^{1/3}} \begin{cases} \frac{1}{\ln\left(\sqrt{\frac{\varepsilon+1}{\varepsilon-1}}+1\right)-\ln\left(\sqrt{\frac{\varepsilon+1}{\varepsilon-1}}-1\right)} & \text{prolate} \\ \frac{1}{\pi-2\arctan\left(\sqrt{\frac{1+\varepsilon}{1-\varepsilon}}\right)} & \text{oblate} \end{cases} \tag{10.147}$$

where Eqs. (10.122) and (10.133) have been used to eliminate $\zeta_0$ in favour of $\varepsilon$.

It is interesting to observe that for a spherical drop the evaporation rate is simply $-4\pi R_d f_n^{(1)}$, it is then of certain interest to compare the evaporation rates of spherical and spheroidal drops having the same volume (i.e. the same equivalent radius $R_d$). Defining the non-dimensional parameter

$$\hat{\Gamma} = \frac{m_{ev,sd}}{m_{ev,sph}} \tag{10.148}$$

as the ratio of the actual evaporation rate $m_{ev,sd}$ and that of a spherical drop having the same volume, $m_{ev,sph}$, the expression of $\hat{\Gamma}$ is

$$\hat{\Gamma}(\varepsilon) = \frac{\left|1-\varepsilon^2\right|^{1/2}}{\varepsilon^{1/3}} \begin{cases} \frac{1}{\ln\left(\sqrt{\frac{\varepsilon+1}{\varepsilon-1}}+1\right)-\ln\left(\sqrt{\frac{\varepsilon+1}{\varepsilon-1}}-1\right)} & \text{prolate} \\ \frac{1}{\pi-2\arctan\left(\sqrt{\frac{1+\varepsilon}{1-\varepsilon}}\right)} & \text{oblate} \end{cases} \tag{10.149a}$$

**Fig. 10.12** Non-dimensional evaporation rate as a function of the non-dimensional drop surface for prolate and oblate drops having the same volume

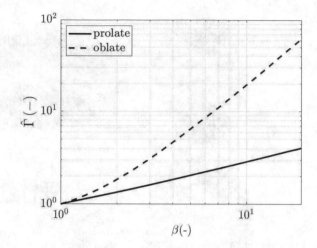

Figure 10.12 reports the profiles of the non-dimensional parameter $\hat{\Gamma}$ as a function of the non-dimensional spheroid surface $\beta$. It can be seen that the increase of evaporation rate is not simply due to the increase of the drop surface, the shape of the drop plays an important role. A spherical drop of a given volume can be deformed towards the oblate or the prolate shape, then increasing the surface, obtaining two non-spherical drops having the same volume and surface, but with different shape (oblate or prolate). For spheroidal drops having the same volume and surface area (then the same $\beta$), the prolate shape is that having the highest evaporation rate, and the increases in $m_{ev}$ is not proportional to the increase in surface area.

Let now turn to the local flux; both spheroidal drops plotted in Fig. 10.11 have the same surface ($\beta = 1.1$) and the figure shows that the vapour concentration gradients are higher in the region of highest surface curvature, corresponding to the equator for the oblate drop and the two poles for the prolate one. This suggests that the corresponding vapour fluxes are higher on the poles for the prolate drop, and on the equator for the oblate one. This is confirmed in Fig. 10.13, which shows the distribution of the mass flux vectors on the surface of a prolate and an oblate drop.

This observation suggests that a relation may exist between the vapour fluxes, which are proportional to the vapour gradients, and the surface curvature, and conjectures in this sense can be found in literature, e.g. [23, 24], where a proportionality between flux and mean curvature was proposed. However, the analytic expression given in Eq. (10.143) allows finding the exact relationship. To this end consider the Gaussian curvature of a spheroid (see Chap. 2), which can be calculated as

$$K_G = \frac{\varepsilon^{8/3}}{R_d^2 \left|1 - \varepsilon^2\right|^2} \begin{cases} \dfrac{1}{\left(\frac{1}{\varepsilon^2-1}+1-\eta^2\right)^2} & \text{prolate} \\[4mm] \dfrac{1}{\left(\frac{1}{1-\varepsilon^2}-1+\eta^2\right)^2} & \text{oblate} \end{cases} \tag{10.150}$$

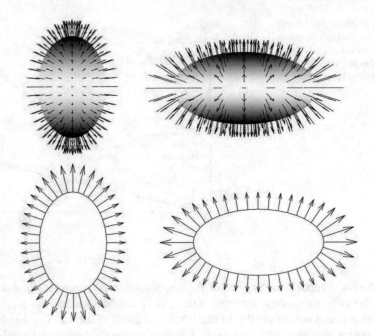

**Fig. 10.13** Distribution of the mass flux vectors on the surface of a prolate (left) and an oblate (right) drop

It can be seen that the evaporation mass fluxes can be expressed as a single function of the Gaussian curvatures

$$n_\zeta^{(1)}(\zeta_0) = \left(R_d^2 K_G\right)^{1/4} \frac{m_{ev}}{4\pi R_d^2} \qquad (10.151)$$

where Eq. (10.147) has been used.

A non-dimensional vapour flux can be defined as

$$\hat{n}_\zeta^{(1)}(\zeta_0) = \frac{n_\zeta^{(1)}(\zeta_0)\, 4\pi R_d^2}{m_{ev}} = \left(K_G^*\right)^{1/4} \qquad (10.152)$$

where $K_G^* = R_d^2 K_G$ is the non-dimensional Gaussian curvature and Fig. 10.14 shows a sample of the profiles of $\hat{n}_\zeta^{(1)}(\zeta_0)$ for oblate and prolate liquid drops with same surface area and volume and three different deformations in terms of increased surface area with respect to the spherical case: a) $\beta = 1.002$ ($\varepsilon_{oblate} = 0.9$; $\varepsilon_{prolate} = 1.11$); (b) $\beta = 1.1$ ($\varepsilon_{oblate} = 0.49$; $\varepsilon_{prolate} = 2.22$); (c) $\beta = 1.4$ ($\varepsilon_{oblate} = 0.26$; $\varepsilon_{prolate} = 5.42$). In these plots the more geometrical meaningful polar angle $\theta$ is used instead of $\eta$; the relation between $\theta$ and $\eta$ can be easily found from the definition

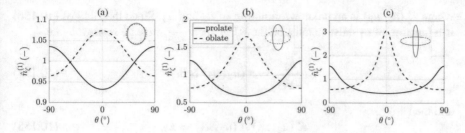

**Fig. 10.14** Vapour profile along the surface of prolate and oblate spheroidal drops

$$\theta = \arctan\left(\frac{\sqrt{x^2 + y^2}}{z}\right) = \begin{cases} \arctan\left(\frac{\sqrt{\zeta_0^2-1}\sqrt{1-\eta^2}}{\zeta_0\eta}\right) & \text{prolate} \\ \arctan\left(\frac{\sqrt{\zeta_0^2+1}\sqrt{1-\eta^2}}{\zeta_0\eta}\right) & \text{oblate} \end{cases} \tag{10.153}$$

The heat flux of the three drop surfaces is

$$q_\zeta^{(1)}(\zeta_0) = f_q \left(\nabla_\zeta \Phi\right)_{\zeta=\zeta_0} \tag{10.154}$$

where $\left(\nabla_\zeta \Phi\right)_{\zeta=\zeta_0}$ is reported in Eq. (10.143). Following the same procedure above described, it is easy to show that also the heat flux is proportional to $K_G^{1/4}$ and that $\hat{\Gamma}$ gives the ratio between the heat rate from a spheroidal drop and that from a spherical one having the same volume.

## 10.4  The Triaxial Ellipsoidal Drop

One dimensional problems are such that the auxiliary function $\Phi$ (see Sect. 10.1) depends only on one coordinate. If the drop surface in a curvilinear coordinate system $(u_1, u_2, u_3)$ is defined by the equation $u_1 = const.$, i.e. it is a coordinate surface, and $\Phi$ is assumed to depend only on $u_1$, then the Laplace equation $\nabla^2 \Phi = 0$ can be reduced to the simpler form (see Chap. 4)

$$\frac{\partial}{\partial u_1}\left[\frac{h_2 h_3}{h_1}\frac{\partial \Phi}{\partial u_1}\right] = 0 \tag{10.155}$$

where $h_j$ are the scale factors. A first integration over $u_1$ yields

$$\frac{\partial \Phi}{\partial u_1} = K(u_2, u_3)\frac{h_1}{h_2 h_3} \tag{10.156}$$

where $K\,(u_2, u_3)$ is an unknown function of $u_2$ and $u_3$. Since the LHS of (10.156) is a function of $u_1$ only, so must be the RHS, then $\frac{h_1}{h_2 h_3}$ must be *separable*, i.e.

$$\frac{h_1}{h_2 h_3} = A\,(u_2, u_3)\,B\,(u_1) \tag{10.157}$$

and then

$$K\,(u_2, u_3)\,A\,(u_2, u_3) = K_0 \tag{10.158}$$

where $K_0$ is a constant. Then the unknown function $K\,(u_2, u_3)$ must be equal to

$$K\,(u_2, u_3) = \frac{K_0}{A\,(u_2, u_3)} \tag{10.159}$$

Among the curvilinear systems reported in Chap. 4 such that a surface $u_j = const.$ can be taken to represent a drop surface, only few satisfy Eq. (10.157). We have seen in the previous section that 1-D solutions can be found in spherical and in spheroidal coordinate systems, but there is another system that allows 1-D solutions, precisely the ellipsoidal coordinate system (see Chap. 4), which is defined by the equations

$$x^2 = \frac{\xi^2 u^2 v^2}{\phi^2 \lambda^2} \tag{10.160a}$$

$$y^2 = \frac{\left(\xi^2 - \lambda^2\right)\left(u^2 - \lambda^2\right)\left(\lambda^2 - v^2\right)}{\vartheta^2 \lambda^2} \tag{10.160b}$$

$$z^2 = \frac{\left(\xi^2 - \phi^2\right)\left(\phi^2 - u^2\right)\left(\phi^2 - v^2\right)}{\vartheta^2 \phi^2} \tag{10.160c}$$

with

$$0 \le v^2 \le \lambda^2 \le u^2 \le \phi^2 \le \xi^2 < +\infty; \quad \vartheta^2 = \phi^2 - \lambda^2 \tag{10.161}$$

In this system the surface $\xi = \xi_0$ is a triaxial ellipsoid; in fact, defining

$$a_x^2 = \xi_0^2; \quad a_y^2 = \xi_0^2 - \lambda^2; \quad a_z^2 = \xi_0^2 - \phi^2 \tag{10.162}$$

the following equation holds

$$\frac{x^2}{a_x^2} + \frac{y^2}{a_y^2} + \frac{z^2}{a_z^2} = 1 \tag{10.163}$$

which shows that $a_x$, $a_y$, and $a_z$ are the half-axes of the ellipsoid.

In the following we will always assume, without limiting the generality, that $a_x \ge a_y \ge a_z$ (see Fig. 10.15).

Since the volume of the ellipsoid is $V = \frac{4}{3}\pi a_x a_y a_z$, the radius $R_d$ of a spherical drop having the same volume of the ellipsoid is $R_d = \left(a_x a_y a_z\right)^{1/3}$, and introducing

**Fig. 10.15** Schematic of
triaxial ellipsoid

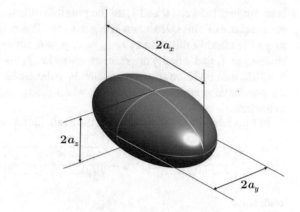

**Fig. 10.16** Representation
of the spheroidal family on
**a** the $\varepsilon_y - \varepsilon_z$ plane and
**b** $p - q$ plane

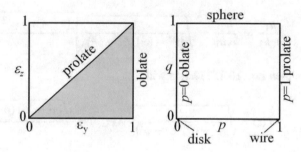

the aspect ratios $\varepsilon_y = \frac{a_y}{a_x}$ and $\varepsilon_z = \frac{a_z}{a_x}$ (to notice that $1 \geq \varepsilon_y \geq \varepsilon_z$), the following
relations can be obtained

$$a_x = \frac{R_d}{\left(\varepsilon_y \varepsilon_z\right)^{1/3}}; \quad a_y = \varepsilon_y \frac{R_d}{\left(\varepsilon_y \varepsilon_z\right)^{1/3}}; \quad a_z = \varepsilon_z \frac{R_d}{\left(\varepsilon_y \varepsilon_z\right)^{1/3}} \qquad (10.164)$$

$$\lambda^2 = \frac{R_d^2 \left(1 - \varepsilon_y^2\right)}{\left(\varepsilon_y \varepsilon_z\right)^{2/3}}; \quad \phi^2 = \frac{R_d^2 \left(1 - \varepsilon_z^2\right)}{\left(\varepsilon_y \varepsilon_z\right)^{2/3}}; \quad \vartheta^2 = \frac{R_d^2 \left(\varepsilon_y^2 - \varepsilon_z^2\right)}{\left(\varepsilon_y \varepsilon_z\right)^{2/3}} \qquad (10.165)$$

The triaxial ellipsoid is clearly the most general case of ellipsoid that comprises the
oblate spheroid ($a_x = a_y > a_z$), the prolate spheroid ($a_x > a_y = a_z$) and the sphere
($a_x = a_y = a_z$). Using the aspect ratios $\varepsilon_y$ and $\varepsilon_z$ previously defined, the *ellipsoid
family* can be represented in a $\varepsilon_y - \varepsilon_z$ plane as in Fig. 10.16a where, since $\varepsilon_z \leq \varepsilon_y$,
only the gray region represents possible ellipsoids.

In [25] a simpler classification of the *ellipsoid family* was proposed. Using two
*shape* parameters defined as

$$p^2 = \frac{1 - \varepsilon_y^2}{1 - \varepsilon_z^2}; \quad q = \varepsilon_y \varepsilon_z = \left(\frac{R_0}{a_x}\right)^3 \qquad (10.166)$$

both ranging between 0 and 1, all the possible ellipsoids can be described by points in the square of unit side shown in Fig. 10.16b. There, the spherical drop corresponds to $q = 1$, since in this case $\varepsilon_y = \varepsilon_z = 1$; $p = 0$ corresponds to the oblate spheroid, since $\varepsilon_y = 1$, and when $q$ approaches zero, (i.e. $\varepsilon_z \to 0$) the shape becomes similar to a disk; the condition $p = 1$ corresponds to the prolate spheroid ($\varepsilon_y = \varepsilon_z$) and when the parameter $q$ approaches zero (i.e. when $\varepsilon_y, \varepsilon_z \to 0$) the shape becomes similar to a wire.

In triaxial ellipsoidal coordinates the scale factors are (see again Chap. 4)

$$h_\xi = \sqrt{\frac{(\xi^2 - u^2)(\xi^2 - v^2)}{(\xi^2 - \lambda^2)(\xi^2 - \phi^2)}}; \quad h_u = \sqrt{\frac{(u^2 - v^2)(\xi^2 - u^2)}{(u^2 - \lambda^2)(\phi^2 - u^2)}}; \quad h_v = \sqrt{\frac{(\xi^2 - v^2)(u^2 - v^2)}{(\lambda^2 - v^2)(\phi^2 - v^2)}}$$

(10.167)

and, since

$$\frac{h_1}{h_2 h_3} = \frac{h_\xi}{h_u h_v} = \sqrt{\frac{1}{(\xi^2 - \lambda^2)(\xi^2 - \phi^2)}} \sqrt{\frac{(u^2 - \lambda^2)(\phi^2 - u^2)(\lambda^2 - v^2)(\phi^2 - v^2)}{(u^2 - v^2)^2}}$$

(10.168)

then Eq. (10.157) holds with

$$B(\xi) = \sqrt{\frac{1}{(\xi^2 - \lambda^2)(\xi^2 - \phi^2)}}; \quad A(u, v) = \frac{\sqrt{(u^2 - \lambda^2)(\phi^2 - u^2)(\lambda^2 - v^2)(\phi^2 - v^2)}}{(u^2 - v^2)}$$

(10.169)

From Eq. (10.158), the function $K(u, v)$ is now

$$K(u, v) = K_0 \frac{(u^2 - v^2)}{\sqrt{(u^2 - \lambda^2)(\phi^2 - u^2)(\lambda^2 - v^2)(\phi^2 - v^2)}}$$

(10.170)

and the integration of (10.156) yields

$$\Phi = K_0 \int B(\xi)\, d\xi + K_1$$

(10.171)

The constants $K_0$ and $K_1$ can be found from the boundary conditions ($\Phi(\xi_0) = 1$; $\Phi(\infty) = 0$) and the solution is then

$$\Phi = 1 - \frac{\int_{\xi_0}^{\xi} B(\xi)\, d\xi}{\int_{\xi_0}^{\infty} B(\xi)\, d\xi}$$

(10.172)

The integral $\int_{\xi_0}^{\xi} B(\xi)\, d\xi$ can be written in terms of incomplete elliptic integral of first kind (see Appendix)

$$F(x, k) = \int_0^x \frac{dt}{\sqrt{\left(1 - k^2 t^2\right)\left(1 - t^2\right)}} \tag{10.173}$$

in fact, the substitution $k = \frac{\gamma}{\phi}, t = \frac{\phi}{\xi}$ yields

$$\int_{\xi_0}^{\xi} B(\xi)\, d\xi = \int_{\xi_0}^{\xi} \frac{d\xi}{\sqrt{\left(\lambda^2 - \xi^2\right)\left(\phi^2 - \xi^2\right)}} = \int_{\xi_0}^{\xi} \frac{d\xi}{\xi^2 \sqrt{\left(\left(\frac{\phi}{\xi}\right)^2 \left(\frac{\lambda}{\phi}\right)^2 - 1\right)\left(\left(\frac{\phi}{\xi}\right)^2 - 1\right)}} \tag{10.174}$$

$$= -\frac{1}{\phi} \int_{\phi/\xi_0}^{\phi/\xi} \frac{dt}{\sqrt{\left(1 - t^2 k^2\right)\left(1 - t^2\right)}} = \frac{1}{\phi}\left[F\left(\frac{\phi}{\xi_0}, k\right) - F\left(\frac{\phi}{\xi}, k\right)\right]$$

and since

$$\int_{\xi_0}^{\infty} B(\xi)\, d\xi = \frac{1}{\phi}\left[F\left(\frac{\phi}{\xi_0}, k\right)\right] \tag{10.175}$$

the solution (10.172) can be written as

$$\Phi = \frac{F\left(\frac{\phi}{\xi}, k\right)}{F\left(\frac{\phi}{\xi_0}, k\right)} \tag{10.176}$$

The component of the gradient of $\Phi$ normal to the surface can then be calculated

$$\nabla_\xi \Phi = \frac{1}{h_\xi} \frac{d\Phi}{d\xi} = K(u, v) \frac{1}{h_u h_v} = K_0 \frac{1}{\sqrt{\left(\xi^2 - u^2\right)\left(\xi^2 - v^2\right)}} \tag{10.177}$$

### 10.4.1  The Fluxes and Evaporation Rates

The mass and heat fluxes can be calculated from Eqs. (10.82) and (10.83) and the evaporation rate can be calculated integrating the mass flux

$$n_\xi^{(1)} = f_n^{(1)} \nabla_\xi \Phi \tag{10.178}$$

over the drop surface as

$$m_{ev,tx} = f_n^{(1)} \int_A \nabla_\xi \Phi\, dA = f_n^{(1)} \int_0^\lambda \int_\lambda^\phi \frac{h_u h_v}{h_\xi} \frac{\partial \Phi}{\partial \xi}\, du\, dv = f_n^{(1)} \int_0^\lambda \int_\lambda^\phi K(u, v)\, du\, dv \tag{10.179}$$

where the last equality comes from Eq. (10.156); consequently the heat rate is

$$Q_{tx} = f_q \int_0^\lambda \int_\lambda^\phi K(u, v)\, du\, dv \tag{10.180}$$

From the boundary conditions it was found that $K_0 = \left[ \int_{\xi_0}^{\infty} B(\xi) \, d\xi \right]^{-1} = \phi \left[ F\left( \frac{\phi}{\xi_0}, k \right) \right]^{-1}$ and then

$$m_{ev,tx} = f_n^{(1)} \phi \left[ F\left( \frac{\phi}{\xi_0}, k \right) \right]^{-1} \int_0^\lambda \int_\lambda^\phi \frac{1}{A(u,v)} du \, dv \qquad (10.181)$$

$$Q_{tx} = f_q \phi \left[ F\left( \frac{\phi}{\xi_0}, k \right) \right]^{-1} \int_0^\lambda \int_\lambda^\phi \frac{1}{A(u,v)} du \, dv \qquad (10.182)$$

where the integral $\int_0^\lambda \int_\lambda^\phi \frac{1}{A(u,v)} du \, dv$ can be shown to be equal to $4\pi$ (see [25] for a demonstration).

It is now possible to explore how the local fluxes may depend on the local curvature of the surface. For spheroidal drops it was shown (cfr Sect. 10.3) that the non-dimensional vapour flux

$$\hat{n}_\xi^{(1)} = \frac{n_\xi^{(1)} 4\pi R_d^2}{m_{ev,tx}} \qquad (10.183)$$

is equal to the fourth root of the non-dimensional Gaussian curvature, $K_G^* = R_d^2 K_G$. The Gaussian curvature can be calculated for the triaxial ellipsoid using Eq. (2.143), that in the present case yields

$$K_G = -\frac{1}{2g^{1/2}} \left[ \frac{\partial}{\partial u} \left( \frac{1}{g^{1/2}} \frac{\partial h_v^2}{\partial u} \right) + \frac{\partial}{\partial v} \left( \frac{1}{g^{1/2}} \frac{\partial h_u^2}{\partial u} \right) \right] = \frac{1}{R_d^2} \frac{(\varepsilon_y \varepsilon_z)^{8/3}}{\left( 1 - \frac{v^2}{\xi_0^2} \right)^2 \left( 1 - \frac{u^2}{\xi_0^2} \right)^2} \qquad (10.184)$$

where $g^{1/2} = h_\xi h_u h_v$.

Substituting (10.178) and (10.181) into (10.183) yields:

$$\hat{n}_\xi^{(1)} = f_n^{(1)} K_0 \frac{1}{\sqrt{(\xi^2 - v^2)(\xi^2 - u^2)}} \frac{4\pi R_d^2}{f_n^{(1)} \phi \left[ F\left( \frac{\phi}{\xi_0}, k \right) \right]^{-1} \int\int \frac{1}{A(u,v)} du \, dv} = \qquad (10.185)$$

$$= \frac{R_d^2}{\xi_0^2 \sqrt{\left( 1 - \frac{v^2}{\xi_0^2} \right) \left( 1 - \frac{u^2}{\xi_0^2} \right)}} = \frac{(\varepsilon_y \varepsilon_z)^{2/3}}{\sqrt{\left( 1 - \frac{v^2}{\xi_0^2} \right) \left( 1 - \frac{u^2}{\xi_0^2} \right)}} = (K_G^*)^{1/4}$$

where $\xi_0^2 = a_x^2 = \frac{R_d^2}{(\varepsilon_y \varepsilon_z)^{2/3}}$ (see Eq. 10.164) was used. This equation is exactly that found for the case of spheroidal drops (see Eq. 10.152).

The dependence of the evaporation rate on drop surface area can be found as follows. The surface area of a general triaxial ellipsoid, having the same volume of a sphere of radius $R_d$, can be calculated as (see [25])

$$S\left(R_d, \varepsilon_y, \varepsilon_z\right) = 2\pi \frac{R_d^2}{\left(\varepsilon_y\varepsilon_z\right)^{2/3}} \left\{ \varepsilon_z^2 + E\left(\varphi, m\right)\varepsilon_y\sqrt{\left(1 - \varepsilon_z^2\right)} + F\left(\varphi, m\right)\frac{\varepsilon_y\varepsilon_z^2}{\sqrt{\left(1 - \varepsilon_z^2\right)}} \right\}$$

(10.186)

where $F\left(\varphi, k\right)$ and $E\left(\varphi, k\right)$ are the *incomplete elliptical integrals* of first and second second kind, respectively, defined as (see Appendix)

$$F\left(x, k\right) = \int_0^x \frac{dt}{\sqrt{1 - t^2}\sqrt{1 - k^2t^2}}; \quad E\left(x, k\right) = \int_0^x \frac{\sqrt{1 - k^2t^2}}{\sqrt{1 - t^2}}dt$$

(10.187)

and

$$\varphi = \arcsin\sqrt{1 - \varepsilon_z^2}; \quad m^2 = \frac{\left(\varepsilon_y^2 - \varepsilon_z^2\right)}{\varepsilon_y^2\left(1 - \varepsilon_z^2\right)}$$

(10.188)

We have seen that the effect of the shape on the evaporation and heat rate of a spheroidal drop can be characterised by the ratio between the actual evaporation rate and that of a spherical drop having the same volume, i.e.

$$\hat{\Gamma} = \frac{m_{ev,tx}}{m_{ev,sph}}$$

(10.189)

which was found to be a purely geometrical parameter (see Eq. 10.149a). The evaluation of $\hat{\Gamma}$ for the case of a triaxial ellipsoidal drop, as a function of the non-dimensional drop surface $\beta = \frac{A_{tx}}{4\pi R_d^2}$ is shown in Fig. 10.17. The two limiting curves correspond to the case of a prolate spheroid (upper curve) and an oblate spheroid (lower curve). Thus, for an ellipsoidal drop of given volume and surface, the maximum evaporation rate is always obtained for a prolate drop ($p = 1$) while the minimum for the oblate drop ($p = 0$) (see [26]) and the triaxial ellipsoidal drops yield intermediate values, depending on the deformation parameter $p$.

**Fig. 10.17** Non-dimensional evaporation rate as a function of the non-dimensional drop surface for different ellipsoidal drops

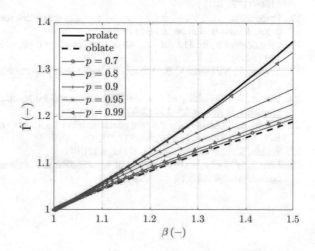

# References

1. Fuchs, N.A.: Vaporisation and Droplet Growth in Gaseous Media. Pergamon Press, London (1959)
2. Abramzon, B., Sirignano, W.A.: Droplet vaporization model for spray combustion calculations. Int. J. Heat Mass Tran. **32**(9), 1605–1618 (1989)
3. Tonini, S., Cossali, G.E.: On molar- and mass-based approaches to single component drop evaporation modelling. Int. Commun. Heat Mass **77**, 87–93 (2016)
4. Fuller, E.N., Schetter, P.D., Giddings, J.C.: New method for prediction of binary gas-phase diffusion coefficients. Ind. Eng. Chem. **58**(5), 18–27 (1966)
5. Wannier, G.H.: Statistical Physics. Wiley (1966)
6. Bernstein, H.J.: Power series for the temperature dependence of the heat capacity of gases. J. Chem. Phys. **24**, 911–912 (1956)
7. Cossali, G.E., Tonini, S.: An analytical model of heat and mass transfer from liquid drops with temperature dependence of gas thermo-physical properties. Int. J. Heat Mass Transf. **138**, 1166–1177 (2019)
8. Hirschfelder, J.O., Curtiss, C.F., Bird, R.B.: Molecular Theory of Gases and Liquids, 2nd edn. Wiley, New York (1964)
9. Ferziger, J.H., Kaper, H.G.: Mathematical Theory of Transport Processes in Gases. North-Holland, Amsterdam (1972)
10. Wassiljewa, A.: Heat conduction in gas mixtures. Physikalische Zeitschrift **5**(22), 737–742 (1904)
11. Lindsay, A.L., Bromley, L.A.: Thermal conductivity of gas mixtures. Ind. Eng. Chem. **42**, 1508–1511 (1950)
12. Ebrahimian, V., Habchi, C.: Towards a predictive evaporation model for multi-component hydrocarbon droplets at all pressure conditions. Int. J. Heat Mass Transf. **54**, 3552-3565 (15–16) (2011)
13. Labowsky, M.: A formalism for calculating the evaporation rates of rapidly evaporating inter-acting particles. Combust. Sci. Technol. **18**, 145–151 (1978)
14. Olver, F.W., Lozier, D.W., Boisvert, R.F., Clark, C.W. (eds.): NIST Handbook of Mathematical Functions. Cambridge University Press (2010)
15. Cossali, G.E., Tonini, S.: Modelling the effect of variable density and diffusion coefficient on heat and mass transfer from a single component spherical drop evaporating in high temperature air streams. Int. J. Heat Mass Transf. **118**, 628–636 (2018)
16. Cossali, G.E., Tonini, S.: An analytical model of heat and mass transfer from liquid drops with temperature dependence of gas thermophysical properties. Int. J. Heat Mass Transf. **138**, 1166–1177 (2019)
17. Pruppacher, H.R., Pitter, R.L.: A semi-empirical determination of the shape of cloud and rain drops. J. Atmos. Sci. **28**, 86–94 (1971)
18. Chandrasekhar, S.: The oscillations of a viscous liquid globe. Proc. Lond. Math. Soc. **9**, 141–149 (1959)
19. Eggers, J., Villermaux, E.: Physics of liquid jets. Reports on Progress in Physics (036601) (2008)
20. Sazhin, S.: Modelling of fuel droplet heating and evaporation: recent results and unsolved problems. Fuel **196**, 69–101 (2017)
21. Tonini, S., Cossali, G.E.: An analytical model of liquid drop evaporation in gaseous environment. Int. J. Therm. Sci. **57**, 45–53 (2012)
22. Sazhin, S.: Droplet and Sprays. Springer (2014)
23. Lian, Z.W., Reitz, R.D.: The effect of vaporization and gas compressibility on liquid jet atomization. At. Sprays **3**(3), 249–264 (1993)

24. Mashayek, F.: Dynamics of evaporating drops. Part I: formulation and evaporation model. Int. J. Heat Mass Transf. **44**(8), 1517–1526 (2001)
25. Tonini, S., Cossali, G.E.: One-dimensional analytical approach to modelling evaporation and heating of deformed drops. Int. J. Heat Mass Transf. **9**, 301–307 (2016)
26. Tonini, S., Cossali, G.E.: An exact solution of the mass transport equations for spheroidal evaporating drops. Int. J. Heat Mass Transf. **60**, 236–240 (2013)

# Chapter 11
# Two- and Three-Dimensional Modelling of Heating and Evaporation Under Steady Conditions

In Chap. 10 we have seen that when an evaporating drop has the shape of a sphere, a spheroid or an ellipsoid, and the boundary conditions are uniform over the drop surface, the whole problem simplifies when proper coordinate systems are used and one-dimensional solutions of the conservation equations can be found. When the drop assumes different shapes, or the boundary conditions are not uniform, two- or three-dimensional solutions appear, even using proper coordinate systems. In this chapter we will explore some cases of practical interest when 2-D or even 3-D solutions can be found analytically.

The case of two drops evaporating in the same environment will be analysed using the bispherical coordinate system. A drop having the shape of a cyclide surface will be analysed in inverse spheroidal coordinates, the evaporation of a sessile drop will be modelled using toroidal coordinates and finally, spherical and spheroidal drop evaporation with non-uniform boundary conditions on the surface will be considered.

## 11.1  A Pair of Interacting Spherical Drops

In Sect. 10.1 it was shown that the analytical approach to drop heating and evaporation developed there can be applied to any finite evaporating lump of liquid, and even to the case when the liquid is separated in a finite number of parts. We will now analyse the case of two spherical drops evaporating in a steady environment. The drops are assumed to be spheres of any radius, positioned at any distance from each other, larger than the sum of the two drop radii to avoid interference (see Fig. 10.17a). This problem was first analytically solved by [1] for the case of negligible Stefan flow for two identical isothermal drops, the extension to the case with the inclusion of Stefan flow can be found in [2, 3] for identical isothermal drops and in [4] for non-identical isothermal drops, under the assumption of constant gas properties. An extension to the case of non-identical and non-isothermal drops and non-constant gas density can be found in [5].

© Springer Nature Switzerland AG 2021                                    297
G. E. Cossali and S. Tonini, *Drop Heating and Evaporation: Analytical Solutions in Curvilinear Coordinate Systems*, Mathematical Engineering,
https://doi.org/10.1007/978-3-030-49274-8_11

The natural coordinate system for this geometrical configuration is the bispherical one, defined by the equations (see also Chap. 4)

$$x = a \frac{\sin(\xi) \sin(\varphi)}{\Theta(\eta, \xi)} \tag{11.1a}$$

$$y = a \frac{\sin(\xi) \cos(\varphi)}{\Theta(\eta, \xi)} \tag{11.1b}$$

$$z = a \frac{\sinh(\eta)}{\Theta(\eta, \xi)} \tag{11.1c}$$

where $\Theta(\eta, \xi) = \cosh(\eta) - \cos(\xi)$. The coordinate space $(\eta, \xi, \varphi)$ is limited to

$$-\infty \le \eta < \infty; \quad 0 \le \xi \le \pi; \quad 0 \le \varphi < 2\pi \tag{11.2}$$

The surface $\eta = \eta_0$ represents a sphere with radius $R_d = \frac{a}{|\sinh(\eta_0)|}$ centered in $z_0 = a \coth(\eta_0)$, the surface $\xi = \xi_0$ is generated by the rotation of an arc of a circumference around the $z$-axis and it can be seen as a self-intersecting torus (sometimes called *lemon* when the arc length is shorter than $\pi R$ and *apple* when the arc is longer than $\pi R$ [6]). The surface $\varphi = \varphi_0$ is a half-plane passing through the $z$-axis (refer to Chap. 4 for more details on this coordinate system).

Considering now the problem of two spherical drops evaporating in the same environment, the two spherical surfaces are defined in bispherical coordinates by the equations $\eta = \eta_1$ and $\eta = \eta_2$, the radii of the spheres $R_{d1}$, $R_{d2}$ and the drop distance, $L_{1-2}$, define the values of the parameters $a$, $\eta_1$ $\eta_2$ through the equations

$$R_{d1} = \frac{a}{\sinh(\eta_1)}; \quad R_{d2} = \frac{a}{|\sinh(\eta_2)|} \tag{11.3}$$

$$L_{1-2} = z_1 - z_2 = a [\coth(\eta_1) - \coth(\eta_2)] \tag{11.4}$$

where $\eta_1 > 0$ and $\eta_2 < 0$ and $z_1 = a \coth(\eta_1)$; $z_2 = a \coth(\eta_2)$ are the coordinates of the drop centres, see Fig. 11.1 for a sketch of the two-drops configuration.

The Laplace equation in bispherical coordinates is

$$\frac{\Theta^3}{a^3} \left[ \frac{\partial}{\partial \eta} \left( \frac{1}{\Theta} \frac{\partial \Phi}{\partial \eta} \right) + \frac{1}{\sin(\xi)} \frac{\partial}{\partial \xi} \left( \frac{\sin(\xi)}{\Theta} \frac{\partial \Phi}{\partial \xi} \right) \right] + \frac{\Theta^2}{a^2 \sin^2(\xi)} \frac{\partial^2 \Phi}{\partial \varphi^2} = 0 \tag{11.5}$$

When uniform boundary conditions are imposed on the iso-surfaces corresponding to the drop surfaces ($\eta = \eta_1$ and $\eta = \eta_2$), the auxiliary function $\Phi$ only depends on the $\eta$ and $\xi$ coordinates, due to the axial-symmetry around the $z$−axis; then the Laplace equation (11.5) reduces to

$$\frac{\partial}{\partial \eta} \left( \frac{\sin \xi}{\Theta} \frac{\partial \Phi}{\partial \eta} \right) + \frac{\partial}{\partial \xi} \left( \frac{\sin \xi}{\Theta} \frac{\partial \Phi}{\partial \xi} \right) = 0 \tag{11.6}$$

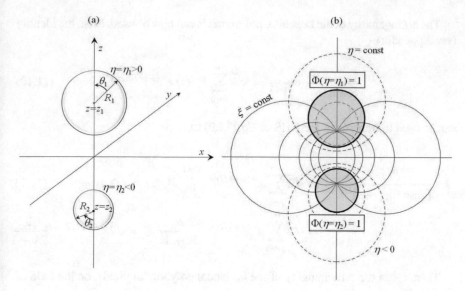

**Fig. 11.1  a** Sketch of a pair of interacting spherical drops. **b** $\eta - \xi$ bispherical coordinate system

The explicit form of the solution of Eq. (11.6) under the condition that $\Phi = 0$ at infinity is (see also [6])

$$
\Phi = \Theta^{1/2} \sum_{p=0}^{\infty} P_p(s) \left\{ m_p \cosh\left[\left(p + \frac{1}{2}\right)\eta\right] + n_p \sinh\left[\left(p + \frac{1}{2}\right)\eta\right] \right\} \quad (11.7)
$$

where $s = \cos\zeta$ and $P_p(s)$ is the Legendre polynomial of degree $p$ (see Appendix for details).

The coefficients $m_p$ and $n_p$ in Eq. (11.7) are calculated imposing the boundary conditions at the drop surfaces. Let first impose on Eq. (11.7) the boundary condition on the surface of drop 1

$$
[\cosh(\eta_1) - s]^{1/2} \sum_{p=0}^{\infty} P_p(s) \left\{ m_p \cosh\left[\left(p + \frac{1}{2}\right)\eta_1\right] + n_p \sinh\left[\left(p + \frac{1}{2}\right)\eta_1\right] \right\} = 1
$$

$$(11.8)$$

after rearranging Eq. (11.8), let multiply each side by the generic Legendre polynomial $P_k(s)$ and integrate both sides on $s \in [-1, 1]$ to obtain:

$$
\int_{-1}^{1} \sum_{p=0}^{\infty} P_p(s) P_k(s) \left\{ m_p \cosh\left[\left(p + \frac{1}{2}\right)\eta_1\right] + n_p \sinh\left[\left(p + \frac{1}{2}\right)\eta_1\right] \right\} ds
$$

$$
= \int_{-1}^{1} \frac{P_k(s)}{[\cosh(\eta_1) - s]^{1/2}} ds
$$

$$(11.9)$$

The orthogonality of the Legendre polynomials can now be used. First, the identity (see Appendix)

$$\frac{1}{[\cosh(\eta) - s]^{1/2}} = \sqrt{2} \sum_{p=0}^{\infty} P_p(s) e^{-(p+1/2)|\eta|} \tag{11.10}$$

can be used to transforms the RHS of Eq. (11.9) to

$$\int_{-1}^{1} \frac{P_k(s)}{[\cosh(\eta_1) - s]^{1/2}} ds = \sqrt{2} \sum_{p=0}^{\infty} e^{-(p+1/2)|\eta_1|} \int_{-1}^{1} P_p(s) P_k(s) ds = \tag{11.11}$$

$$= \sqrt{2} \sum_{p=0}^{\infty} e^{-(p+1/2)|\eta_1|} \frac{2}{2k+1} \delta_{pk} = \sqrt{2} e^{-(k+1/2)|\eta_1|} \frac{2}{2k+1}$$

Then, again the orthogonality of the Legendre polynomial yields, on the LHS of Eq. (11.9)

$$\int_{-1}^{1} \sum_{p=0}^{\infty} P_p(s) P_k(s) ds = \delta_{pk} \frac{2}{2k+1} \tag{11.12}$$

and finally

$$m_p \cosh\left[\left(p + \frac{1}{2}\right)\eta_1\right] + n_p \sinh\left[\left(p + \frac{1}{2}\right)\eta_1\right] = \sqrt{2} e^{-(p+1/2)|\eta_1|} \tag{11.13}$$

That is an equations (precisely one equation for each value of $p$) for the unknowns $m_p$ and $n_p$.

The same procedure can now be repeated imposing the B.C. on the surface $\eta = \eta_2$, obtaining

$$m_p \cosh\left[\left(p + \frac{1}{2}\right)\eta_2\right] + n_p \left[\sinh\left(p + \frac{1}{2}\right)\eta_2\right] = \sqrt{2} e^{-(p+1/2)|\eta_2|} \tag{11.14}$$

The linear system made by the two Eqs. (11.13) and (11.14) can be easily solved obtaining the explicit form of the coefficients $m_p$ and $n_p$

$$m_p = \sqrt{2} \frac{\sinh\left[\left(p + \frac{1}{2}\right)\eta_2\right] e^{-\left(p+\frac{1}{2}\right)\eta_1} - \sinh\left[\left(p + \frac{1}{2}\right)\eta_1\right] e^{\left(p+\frac{1}{2}\right)\eta_2}}{\sinh\left[\left(p + \frac{1}{2}\right)(\eta_2 - \eta_1)\right]} \tag{11.15a}$$

$$n_p = \sqrt{2} \frac{\cosh\left[\left(p + \frac{1}{2}\right)\eta_2\right] e^{-\left(p+\frac{1}{2}\right)\eta_1} - \cosh\left[\left(p + \frac{1}{2}\right)\eta_1\right] e^{\left(p+\frac{1}{2}\right)\eta_2}}{\sinh\left[\left(p + \frac{1}{2}\right)(\eta_1 - \eta_2)\right]} \tag{11.15b}$$

where $\eta_2 < 0$ and $\eta_1 > 0$.

To notice that for the case of two identical drops, $\eta_2 = -\eta_1$, the coefficients $n_p$ are all nil and

$$m_p = \sqrt{2}\frac{2\sinh\left[\left(p+\frac{1}{2}\right)\eta_1\right]e^{-\left(p+\frac{1}{2}\right)\eta_1}}{\sinh\left[2\left(p+\frac{1}{2}\right)\eta_1\right]} = \sqrt{2}\frac{e^{-\left(p+\frac{1}{2}\right)\eta_1}}{\cosh\left[\left(p+\frac{1}{2}\right)\eta_1\right]} \qquad (11.16)$$

The vapour distribution can be found from the auxiliary function $\Phi$, through Eq. (10.38). The relative distance between the two drops is defined as

$$\Lambda_{1-2} = \frac{L_{1-2}}{\bar{D}_d} = \frac{z_1 - z_2}{\bar{D}_d} \qquad (11.17)$$

where $\bar{D}_d$ is a reference drop diameter $\bar{D}_d = R_{d1} + R_{d2}$; for two identical drops $\Lambda_{1-2} = \cosh(\eta_1)$.

For the two non-identical drops, the ratio of the drop radii is defined by

$$\delta_{1-2} = \frac{R_{d2}}{R_{d1}} \qquad (11.18)$$

where drop 2 is assumed to be the smaller one and centered in $z_2 < 0$, so that $\delta_{1-2} \leq 1$.

From the approach described in Sect. 10.1 it is now possible to evaluate the vapour distribution in the gaseous mixture taking into account the dependence of thermo-physical properties on temperature. In [5] the effect of different drop size and the gas density dependence on temperature was analysed; those results can be obtained from the most general approach imposing the vector of thermo-physical properties (refer to Eq. 10.64) as

$$\Pi^P = [m, b, q_0, q_1] = [0, 0, 0, 0] \qquad (11.19)$$

Figure 11.2 shows the vapour distribution around two pair of drops for $\Lambda_{1-2} = 2$ and $\delta_{1-2} = 1$ and 0.5. The effect of a varying gas density can be appreciated by comparing the vapour field with that obtained assuming constant thermo-physical properties as in [3, 4] (lower half of the figure). This second case can be obtained simply setting the vector of thermo-physical properties equal to

$$\Pi^P = [m, b, q_0, q_1] = [1, 0, 0, 0] \qquad (11.20)$$

The effect of drop proximity is evident in the vapour field distribution for both (variable density and constant properties) test cases. The vapour iso-surfaces, which in the region close to the drop surface have shapes that depend on drop radii and distance, tend to become spherical farther from the two drops.

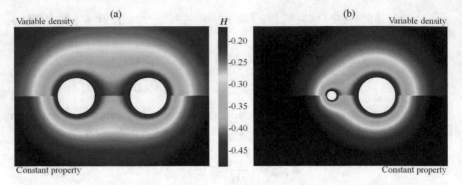

**Fig. 11.2** Effect of thermo-physical properties and drop neighbouring on the vapour molar fraction distribution around two **a** drops of the same size and **b** non-identical drops

### 11.1.1   The Vapour and Heat Fluxes

The drop proximity influences the local surface vapour and thermal fluxes, which are calculated from Eqs. (10.82) and (10.83), where the gradient of the auxiliary function $\Phi$ normal to the drop surface $\eta_j$ is now obtained from the solution (Eq. 11.7) as

$$(\nabla_\eta \Phi)_{\eta=\eta_j} = \left(\frac{1}{h_\eta} \frac{\partial \Phi}{\partial \eta}\right)_{\eta=\eta_j} = \tag{11.21}$$

$$= \frac{1}{2a} \sinh(\eta_j) + \frac{1}{a}\Theta^{3/2} \sum_{p=0}^{\infty} \left(p + \frac{1}{2}\right) P_p(y) \left\{ \begin{array}{l} m_p \cosh\left[\left(p + \frac{1}{2}\right)\eta_j\right] + \\ + n_p\left[\sinh\left(p + \frac{1}{2}\right)\eta_j\right] \end{array} \right\} \tag{11.22}$$

where $\Phi(\eta = \eta_j) = 1$ has been used and $h_\eta$ is the scale factor (cfr Sect. 4.4.13)

$$h_\eta = \frac{a}{\Theta}; \quad h_\xi = \frac{a}{\Theta}; \quad h_\varphi = \frac{a\sin(\xi)}{\Theta} \tag{11.23}$$

The mass and heat rates from each drop are calculated by integrating the corresponding fluxes over each drop surface $A_j$

$$m_{ev,j} = \int_A n_{\eta_j}^{(1)} dA = f_n^{(1)} \int_{A_j} (\nabla_\eta \Phi)_{\eta=\eta_j} dA_j \tag{11.24a}$$

$$Q_j = \int_A q_{\eta_j} dA = f_q \int_{A_j} (\nabla_\eta \Phi)_{\eta=\eta_j} dA_j \tag{11.24b}$$

where

$$\int_{A_j} (\nabla_\eta \Phi)_{\eta=\eta_j} \, dA_j = \int_0^{2\pi} \int_0^\pi \left( \frac{1}{h_\eta} \frac{\partial \Phi}{\partial \eta} \right)_{\eta=\eta_j} h_\xi h_\varphi d\xi d\varphi \tag{11.25}$$

$$= 2\pi a \int_0^\pi \left( \frac{\partial \Phi}{\partial \eta} \right)_{\eta=\eta_j} \frac{\sin \xi}{\Theta} d\xi = -2\sqrt{2}\pi a \sum_{p=0}^\infty \left( m_p + (-1)^{1+j} n_p \right)$$

obtained using Eq. (11.10) and the following integral (see Appendix)

$$\int_{-1}^1 \frac{P_p(y)}{\Theta^{3/2}} dy = \frac{2}{\sinh(\eta)} \sqrt{2} e^{-(p+\frac{1}{2})|\eta|} \tag{11.26}$$

The effect of the interaction between the two evaporating drops can be quantified by evaluating the non-dimensional ratio between the surface vapour flux over one interacting drop, $n_{\eta_j}^{(1)}$, and the corresponding flux, at the same point, over the same drop evaporating alone in the same ambient, $n_{R_j}^{(1)}$, i.e.

$$\psi_d = \frac{n_{\eta_j}^{(1)}}{n_{R_j}^{(1)}} = \frac{q_{\eta_j}}{q_{R_j}} = -R_{dj} \left( \nabla_\eta \Phi \right)_{\eta=\eta_j} \tag{11.27}$$

where the last equality is obtained using Eq. (10.101).

It is easy to see from Eqs. (11.27) that $\psi_d$ is also equal to the ratio of the heat fluxes, it is independent of the thermo-physical properties and operating conditions and it only depends on the geometrical configuration, i.e. on the parameters $\Lambda_{1-2}$, $\delta_{1-2}$.

Figure 11.3 shows the ratio $\psi_d$ as a function of the polar angle $\theta$ (see Fig. 11.1) for the case of two identical drops ($\delta_{1-2} = 1$) and different drop distances ($\Lambda_{1-2}$ varying from 2 up to 10). The graph shows that the minimum flux is located in the region closest to the other drop ($\theta = 180$ deg), corresponding to the maximum local screening effect. Moving towards the opposite side ($\theta = 0$ deg) the flux increases since the local screening effect decreases. The increase of the drop distance (higher $\Lambda_{1-2}$), as expected, reduces the interaction effect on the vapour and heat fluxes.

This screening effect can also be evaluated from a global point of view, by comparing the evaporation rate from one of the interacting drops, $m_{ev}^{(j)}$, to that from the corresponding isolated drop, $m_{ev0}^{(j)}$. A screening coefficient can be defined by the non-dimensional parameter $\Psi^{(j)}$

$$\Psi^{(j)} = \frac{m_{ev}^{(j)}}{m_{ev0}^{(j)}} = \frac{\sqrt{2} \sinh \eta_j \sum_{p=0}^\infty \left( m_p + (-1)^{1+j} n_p \right)}{2} \tag{11.28}$$

In case of two identical drops (twin-drops), $n_p = 0$ and $m_p$ are defined by Eqs. (11.16); the screening coefficient $\Psi^{(j)}$ is then equal for the two drops, i.e. $\Psi^{(j)} = \Psi^{(1)} = \Psi^{(2)}$

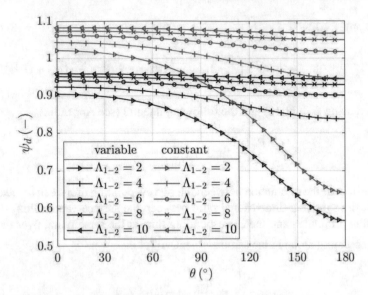

**Fig. 11.3** Effect of thermo-physical properties and drop distance on the surface non-dimensional vapour mass flux for water twin drops

$$\Psi^{(j)} = \sqrt{1 - \Lambda_{1-2}^2} \sum_{p=0}^{\infty} \left( \frac{2}{\left( \Lambda_{1-2} + \sqrt{1 - \Lambda_{1-2}^2} \right)^{2p+1} + 1} \right) \qquad (11.29)$$

Figure 11.4 shows the non-dimensional parameter $\Psi^{(j)}$ for each droplet as a function of the non-dimensional distance $\Lambda_{1-2}$ and for different values of the size ratio $\delta_{1-2}$. The screening effect increases as the drop distance decreases, as expected, and it is still not negligible for $\Lambda_{1-2}$ as large as 10 (for two identical drops). The screening effect is higher for the smaller drop and the difference increases as the size ratio $\delta_{1-2}$ decreases. An attempt of comparing these predictions with experiments is shown in [5], where it is also reported an analytical solution for the vapour field distribution in the case of two drops at different temperatures.

## 11.2   The Inverse Spheroidal Drops

The dependence of the heat and mass fluxes on local surface curvature is intuitive and supported by everyday experience on heat transfer; as a consequence the early works on non-spherical drop evaporation modelling [7, 8] assumed a direct relationship between the mean curvature and the fluxes. In [8] the author proposed a correlation based on a suggestion of Lian and Reitz [9], which postulated that the local evaporation flux from a deformed surface could be equated to that from a spherical drop

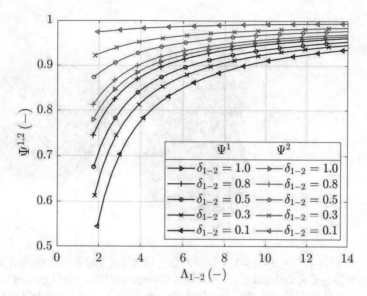

**Fig. 11.4** Effect of drop distance and size ratio on the drop screening coefficients

having a mean curvature equal to the local mean curvature of the surface under consideration. In Chap. 10 we have seen that the local heat and mass fluxes on the surface of an ellipsoidal drop are linked to the surface curvature by a direct proportionality between fluxes and the fourth root of the Gaussian curvature, at least for drops with ellipsoidal shape. One may wonder whether a direct link between local curvature properties and fluxes may exist as a general rule. In this section we will explore this possibility showing that an universal relationship of that kind cannot exist.

To this end we can evaluate the evaporation fluxes from drops having a shape different from an ellipsoid, and we can choose to study drops having the forms shown in Fig. 11.5, i.e. those of *inverse spheroids*. These shapes can be observed in case of binary drop collisions at low velocity (see for example [10]), and also when considering large amplitude drop oscillations (see Chap. 12).

The inverse-oblate and the inverse-prolate spheroidal coordinate systems are defined in a unique way by their relationship with the Cartesian coordinates

$$x = a \frac{\sqrt{\zeta^2 + \alpha}\sqrt{1 - \eta^2}}{\Theta(\zeta, \eta)} \sin(\varphi) \tag{11.30a}$$

$$y = a \frac{\sqrt{\zeta^2 + \alpha}\sqrt{1 - \eta^2}}{\Theta(\zeta, \eta)} \cos(\varphi) \tag{11.30b}$$

$$z = a \frac{\zeta \eta}{\Theta(\zeta, \eta)} \tag{11.30c}$$

**Fig. 11.5** Samples of inverse **a** oblate and **b** prolate spheroids

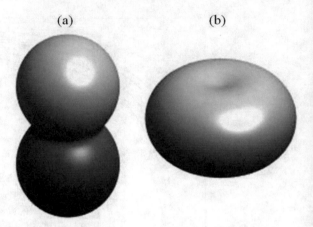

(a)                                          (b)

where $\Theta = \zeta^2 + \alpha\left(1 - \eta^2\right)$, $-1 \le \eta \le 1$, $0 \le \varphi \le 2\pi$, and $\alpha = 1$ for the inverse-oblate case $(0 \le \zeta < \infty)$ and $\alpha = -1$ for the inverse-prolate case $(1 \le \zeta < \infty)$. The equation $\zeta = \zeta_0$ describes a surface called *inverse spheroid* (either oblate or prolate) obtained rotating a quartic curve around the $z$-axis; these surfaces are also called *rotational cyclides* [6].

The two drop shapes are shown in Fig. 11.5; more details on inverse spheroidal coordinates can be found in Sects. 4.4.15 and 4.4.16.

A spherical surface in such systems is defined by the following equation

$$R_1^2 = x^2 + y^2 + z^2 = \frac{a^2}{\Theta} \tag{11.31}$$

Since $\Theta = \zeta^2 + \alpha\left(1 - \eta^2\right)$ becomes nil only for $\zeta = 0$ and $\eta = 1$ for the inverse-oblate case and for $\zeta = 1$ and $\eta = 0$ for the inverse-prolate case, the points $(\zeta, \eta) = (0, 1)$ and $(\zeta, \eta) = (1, 0)$ represent a sphere of infinite radius in the inverse-oblate and inverse-prolate systems, respectively.

Since the drop volume is an important parameter, let us calculate the volume of the space inside an inverse-oblate and an inverse-prolate spheroids (defined by the equation $\zeta = \zeta_0$) by using Pappus formula

$$V = \pi \int_{z_{min}}^{z_{max}} \left(x^2 + y^2\right) dz = \pi \int_{-1}^{1} \left(x^2 + y^2\right) \frac{dz}{d\eta} d\eta \tag{11.32}$$

$$= \pi a^3 \left(\zeta_0^2 + \alpha\right) \zeta_0 \int_{-1}^{1} \frac{\zeta_0^2\left(1 - \eta^2\right) + \alpha\left(1 - \eta^4\right)}{\Theta^4} d\eta$$

that yields, for the two cases, (using Eq. 11.30a)

$$V = \pi a^3 \zeta_0 \frac{\frac{2\alpha + 5\zeta_0^2}{\zeta_0^4} + 3\frac{M(\alpha,\zeta_0)}{\sqrt{\zeta_0^2 + \alpha}}}{6\left(\zeta^2 + \alpha\right)} \tag{11.33}$$

$$M\left(\alpha, \zeta_0\right) = \begin{cases} \frac{1}{2}\ln\left(\frac{\sqrt{\zeta_0^2 + 1} + 1}{\sqrt{\zeta_0^2 + 1} - 1}\right) & \alpha = 1 \\[2ex] \arctan\left(\frac{1}{\sqrt{\zeta_0^2 - 1}}\right) & \alpha = -1 \end{cases} \tag{11.34}$$

An equivalent radius $R_d$ can be defined as $V = \frac{4}{3}\pi R_d^3$, yielding a relation between the scale parameter $a$ in Eqs. (11.30a) and $R_d$

$$a = \frac{R_d}{f_\alpha\left(\zeta_0\right)} \tag{11.35}$$

$$f_\alpha\left(\zeta_0\right) = \left\{\frac{1}{8}\frac{\zeta_0}{\left(\zeta^2 + \alpha\right)}\left[\frac{5\zeta_0^2 + 2\alpha}{\zeta_0^4} + 3\frac{M\left(\alpha, \zeta_0\right)}{\sqrt{\zeta_0^2 + \alpha}}\right]\right\}^{1/3} \tag{11.36}$$

The method for finding an analytical solution of the energy and species conservation equations described in Chap. 10 can be applied to the case of inverse-spheroidal drops. The method needs the analytical solution of the Laplace equation $\nabla^2\Phi = 0$ with boundary conditions $\Phi = 1$ on the surface and $\Phi = 0$ at infinity. In this coordinate system the Laplace equation for axial symmetric problems is

$$\frac{\partial}{\partial\zeta}\left(\frac{\left(\zeta^2 + \alpha\right)}{\Theta}\frac{\partial\Phi}{\partial\zeta}\right) + \frac{\partial}{\partial\eta}\left(\frac{\left(1 - \eta^2\right)}{\Theta}\frac{\partial\Phi}{\partial\eta}\right) = 0 \tag{11.37}$$

and in Chap. 4 it was shown that is $R$ *separable*; then introducing $\Phi = \Theta^{1/2}H\left(\zeta\right)X\left(\eta\right)$, the PDE (11.37) splits into two ODEs. In fact, the first term of the LHS yields

$$\frac{\partial}{\partial\eta}\left(\frac{1 - \eta^2}{\Theta}\frac{\partial\Theta^{1/2}XZ}{\partial\eta}\right) = \frac{Z}{\Theta^{1/2}}\left[\left(1 - \eta^2\right)X'' - 2\eta X' + \frac{\alpha}{\Theta^2}\left(3\eta^2\zeta^2 - \Theta\right)X\right] \tag{11.38}$$

while the second one yields

$$\frac{\partial}{\partial\zeta}\left(\frac{\zeta^2 + \alpha}{\Theta}\frac{\partial\left(\Theta^{1/2}XZ\right)}{\partial\zeta}\right) = \frac{X}{\Theta^{1/2}}\left[\left(\zeta^2 + \alpha\right)Z'' + 2\zeta Z' + \frac{\alpha}{\Theta^2}\left(\Theta - 3\zeta^2\eta^2\right)Z\right] \tag{11.39}$$

then the Laplace equation (11.37) yields

$$\frac{1}{X}\left[\left(1 - \eta^2\right)X'' - 2\eta X'\right] = -\frac{1}{Z}\left[\left(\zeta^2 + \alpha\right)Z'' + 2\zeta Z'\right] = \lambda \tag{11.40}$$

i.e. one obtains the two ordinary differential equations

$$\left(1 - \eta^2\right) X'' - 2\eta X' - p\left(p + 1\right) X = 0 \tag{11.41a}$$

$$\left(\zeta^2 + \alpha\right) Z'' + 2\zeta Z' + p\left(p + 1\right) Z = 0 \tag{11.41b}$$

where $\lambda$ has been replaced by $p\left(p + 1\right)$. Both of Eqs. (11.41) are Legendre equations. The first Eq. (11.41a) has solutions, corresponding to any given value of $p$

$$X_p\left(\eta\right) = \begin{cases} C_0 + D_0 \log\left(\frac{\eta+1}{\sqrt{1-\eta^2}}\right) & \text{for } p\left(p+1\right) = 0 \\ C_p P_p\left(\eta\right) + D_p Q_p\left(\eta\right) & \text{for } p\left(p+1\right) \neq 0 \end{cases} \tag{11.42}$$

where $P_p\left(\eta\right)$ and $Q_p\left(\eta\right)$ are the Laplace functions of first and second kind, respectively.

Since $\eta$, the argument of the Legendre functions $P_p$ and $Q_p$, ranges between $-1$ and $+1$ and since it is known that $\lim_{\eta \to +1} Q_p\left(\eta\right) = \infty$ (see Appendix), then $D_p$ must be set to zero for any $p$. Moreover, since $P_p\left(\eta\right) \to \infty$ when $\eta \to 1$ unless $p$ is a non-negative integer, we are compelled to set $p$ as being a non-negative integer.

The second Eq. (11.41b) has the following general solutions for the two values of $\alpha$

$$H_p\left(\zeta\right) = \begin{cases} A_0 + B_0 \operatorname{arccot}\left(\zeta\right) & \text{for } p\left(p+1\right) = 0 \text{ inverse oblate, } \alpha = 1 \\ A_p P_p\left(i\zeta\right) + B_p Q_p\left(i\zeta\right) & \text{for } p\left(p+1\right) \neq 0 \text{ inverse oblate, } \alpha = 1 \\ A_0 + B_0 Q_0\left(\zeta\right) & \text{for } p\left(p+1\right) = 0 \text{ inverse prolate, } \alpha = -1 \\ A_p P_p\left(\zeta\right) + B_p Q_p\left(\zeta\right) & \text{for } p\left(p+1\right) \neq 0 \text{ inverse prolate, } \alpha = -1 \end{cases} \tag{11.43}$$

Thus a general solution of the Laplace equation can be written as

$$\Phi_p = \Theta^{1/2}\left(\zeta, \eta\right) H_p\left(\zeta\right) P_p\left(\eta\right) \tag{11.44}$$

When the boundary conditions on the drop surface are uniform, the problem is symmetric with respect to $\eta$, i.e. $\Phi_p\left(-\eta\right) = \Phi_p\left(\eta\right)$, i.e. the functions $\Phi_p\left(\zeta, \eta\right)$ must be even with respect to $\eta$ and this happens only when $p = 2n$.

The functions $H_p\left(\zeta\right)$ contain, for each value of $p \neq 0$, two independent functions of $\zeta$, which again are the Legendre functions of first and second kind. We can still reduce the number of solutions to a single numerable infinity, as we have done it for $X_p\left(\eta\right)$, but this needs a more sophisticated reasoning.

The variable $\zeta$ ranges between 0 and $\infty$ for $\alpha = 1$ (inverse-oblate) and from 1 and $\infty$ for $\alpha = -1$ (inverse-prolate). Let now see what the lower limiting values of $\zeta$ represent. For the inverse-oblate case, $\alpha = 1$, the condition $\zeta = 0$ describes the surface

$$r = \sqrt{x^2 + y^2} = \frac{a}{\sqrt{1 - \eta^2}}; \quad z = 0 \tag{11.45}$$

i.e. a portion of the plane $z = 0$, from $r = a$ to $r = \infty$. For the inverse-prolate case, $\alpha = -1$, the condition $\zeta = 1$ describes the line

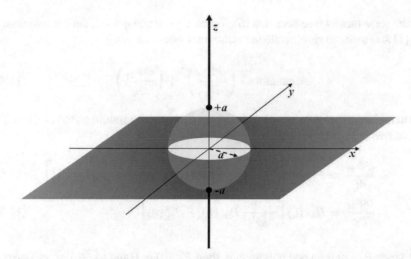

**Fig. 11.6** Schematic of the region representing $\zeta = 0$ and $\zeta = 1$ for inverse oblate and inverse prolate coordinate systems, respectively

$$r = 0; \quad z = \frac{a}{\eta} \qquad (11.46)$$

i.e. the $z$−axis with the exception of the segment between $-a$ and $+a$, see the Fig. 11.6.

Considering the inverse-prolate case, $\alpha = -1$, the functions $H_p (\zeta)$ are given by Eq. (11.43) and since $\lim_{\zeta \to 1} Q_p (\zeta) = \infty$, then all the coefficients $B_p$ must be nil if the solution is required to be bounded on the $z$−axis.

For the inverse-oblate case, $\alpha = 1$, the values that the functions $H_p (\zeta)$ attain on the inferior limit $\zeta = 0$ are bounded (since both $P_p (i\zeta)$ and $Q_p (i\zeta)$ are bounded in $\zeta = 0$), then apparently we cannot set the coefficients $B_p$ to zero as we did for the inverse-prolate case. However, we can further investigate the behaviour of the solutions $\Phi_p$ when $\zeta = 0$, following the analysis reported in [11]. For this, we evaluate the absolute value of the gradient $\nabla \Phi_{2n}$

$$
\begin{aligned}
|\nabla \Phi_{2n}|^2 &= \left( \frac{1}{h_\zeta} \frac{\partial \Phi_{2n}}{\partial \zeta} \right)^2 + \left( \frac{1}{h_\eta} \frac{\partial \Phi_{2n}}{\partial \eta} \right)^2 = \\
&= \frac{\Theta^2 \left[ (\zeta^2 + 1) \left( \frac{\partial \Phi_{2n}}{\partial \zeta} \right)^2 + (1 - \eta^2) \left( \frac{\partial \Phi_{2n}}{\partial \eta} \right)^2 \right]}{a^2 (\zeta^2 + \eta^2)}
\end{aligned}
\qquad (11.47)
$$

where

$$h_\zeta = \frac{a}{\Theta} \sqrt{\frac{\zeta^2 + \alpha \eta^2}{\zeta^2 + a}}; \quad h_\eta = \frac{a}{\Theta} \sqrt{\frac{\zeta^2 + \eta^2}{1 - \eta}} \qquad (11.48)$$

are the scale factors (see Sect. 4.4.16). When $\zeta \to 0$ and $\eta \to 0$, the denominator of Eq. (11.47) tends to zero, while the numerator becomes

$$\text{numerator} = \left(\frac{\partial \Phi_{2n}}{\partial \zeta}\right)^2 + \left(\frac{\partial \Phi_{2n}}{\partial \eta}\right)^2 \tag{11.49}$$

and to maintain the finiteness of the gradient this quantity should go to zero, too. The derivatives in Eq. (11.49) are:

$$\frac{\partial \Phi_{2n}}{\partial \zeta} = P_{2n}(\eta) \left[\frac{\zeta H_{2n}(\zeta)}{\Theta^{1/2}} + i \left[A_{2n} P'_{2n}(i\zeta) + B_{2n} Q'_{2n}(i\zeta)\right]\right] \tag{11.50}$$

$$\frac{\partial \Phi_{2n}}{\partial \eta} = H_{2n}(\zeta) \left[-\frac{\eta}{\Theta^{1/2}} P_{2n}(\eta) + P'_{2n}(\eta)\right] \tag{11.51}$$

and, since $P'_{2n}(\eta)$ is an odd polynomial, then $P'_{2n}(0) = 0$ and $\left(\frac{\partial \Phi_{2n}}{\partial \eta}\right)_{\eta=0} = 0$. About the $\zeta-$derivative $\frac{\partial \Phi_{2n}}{\partial \zeta}$, the term $A_{2n} P'_{2n}(i\zeta)$ goes to zero when $\zeta = 0$ ($P'_{2n}(i\zeta)$ is an odd polynomial) but the term $B_{2n} Q'_{2n}(i\zeta)$ does not go to zero. Then, for preserving the finiteness of the gradient on the $z-$plane all the coefficients $B_{2n}$ must be set to zero. The functions $H_p(\zeta)$ are then

$$H_p(\zeta) = \begin{cases} P_p(i\zeta) & \text{inverse oblate, } \alpha = 1 \\ P_p(\zeta) & \text{inverse prolate, } \alpha = -1 \end{cases} \tag{11.52}$$

and the most general solution of the Laplace equation can then be written as the series

$$\Phi = \Theta^{1/2}(\zeta, \eta) \sum_{n=0}^{\infty} s_{2n} H_{2n}(\zeta) P_{2n}(\eta) \tag{11.53}$$

We have already seen that the points $(\zeta, \eta) = (0, 1)$ and $(\zeta, \eta) = (1, 0)$ represent a sphere of infinite radius in the inverse-oblate and inverse-prolate systems, respectively. Since the functions $H_p(\zeta)$ are limited on that sphere, and the function $\Theta$ tends to zero on the same sphere (see Eq. 11.31), then the function $\Phi$ is nil on that sphere and one of the boundary condition is satisfied.

Let now turn to the last condition: $\Phi = 1$ on the drop surface, i.e. on $\zeta = \zeta_0$

$$\Theta^{1/2}(\zeta_0, \eta) \sum_{p=0}^{\infty} s_{2n} H_{2n}(\zeta_0) P_{2n}(\eta) = 1 \tag{11.54}$$

This condition allows to evaluate the coefficients $s_{2n}$ by using the orthogonality of the Legendre polynomials. In fact, multiplying both sides of Eq. (11.54) by $P_{2k}(\eta) \Theta^{-1/2}$ and integrating over $\eta \in [-1, 1]$ yields

$$\sum_{p=0}^{\infty} s_{2n} H_{2n}\left(\zeta_0\right) \int_{-1}^{1} P_{2n}\left(\eta\right) P_k\left(\eta\right) d\eta = \int_{-1}^{1} \frac{P_{2k}\left(\eta\right)}{\sqrt{\zeta_0^2 + \alpha\left(1 - \eta^2\right)}} d\eta \qquad (11.55)$$

and then (see Appendix)

$$s_{2k} = \frac{2}{H_{2k}\left(\zeta_0\right)\left(4k + 1\right)} \int_{-1}^{1} \frac{P_{2k}\left(\eta\right)}{\sqrt{\zeta_0^2 + \alpha\left(1 - \eta^2\right)}} d\eta \qquad (11.56)$$

The integral

$$I_{2k}^{(\alpha)} = \int_{-1}^{1} \frac{P_{2k}\left(\eta\right)}{\sqrt{\zeta_0^2 + \alpha\left(1 - \eta^2\right)}} d\eta \qquad (11.57)$$

can be analytically calculated using the identity (see Appendix)

$$\int_{-1}^{1} \frac{P_{2n}\left(\eta\right)}{\sqrt{\cosh^2\left(\xi\right) - \eta^2}} d\eta = 2i\, P_{2n}\left(0\right) Q_{2n}\left(i \sinh\left(\xi\right)\right) \qquad (11.58)$$

In fact, setting $\zeta_0 = \sinh\left(\xi\right)$, Eq. (11.58) transforms to

$$2i\, P_{2k}\left(0\right) Q_{2k}\left(i\zeta_0\right) = \int_{-1}^{1} \frac{P_{2k}\left(\eta\right)}{\sqrt{\zeta_0^2 + 1 - \eta^2}} d\eta = I_{2k}^{(1)} \qquad (11.59)$$

while setting $-\zeta_0 - i \sinh\left(\xi\right)$, Eq. (11.58) transforms to

$$- 2P_{2k}\left(0\right) Q_{2k}\left(-\zeta_0\right) = \int_{-1}^{1} \frac{P_{2k}\left(\eta\right)}{\sqrt{\zeta_0^2 - 1 + \eta^2}} d\eta - I_{2k}^{(-1)} \qquad (11.60)$$

The coefficients $s_n$ can then be calculated analytically as

$$s_{2k+1} = 0 \qquad (11.61)$$

$$s_{2k} = \frac{2}{\left(4k + 1\right)} \begin{cases} 2P_{2k}\left(0\right) \frac{iQ_{2k}\left(i\zeta_0\right)}{P_{2k}\left(i\zeta\right)} & \text{inverse oblate, } \alpha = 1 \\ 2P_{2k}\left(0\right) \frac{Q_{2k}\left(\zeta_0\right)}{P_{2k}\left(\zeta\right)} & \text{inverse prolate, } \alpha = -1 \end{cases} \qquad (11.62)$$

which are all real. Having found an analytical form of $\Phi$, the approach developed in Chap. 10 can now be applied to find the vapour fractions and temperature fields (see Eqs. 10.38) and the vapour and heat fluxes can be explicitly calculated evaluating $\nabla_j \Phi$

$$\nabla \Phi = \left[ \frac{1}{h_\zeta} \frac{\partial \Phi}{\partial \zeta}, \frac{1}{h_\eta} \frac{\partial \Phi}{\partial \eta}, 0 \right] \qquad (11.63)$$

**Table 11.1** Values of $\alpha$, definition of $\Theta$ and ranges of $\zeta$-coordinate for different coordinate systems

| $\alpha$ | $\Theta$ | $\zeta$ range | Shape of iso-$\zeta$ surfaces |
|---|---|---|---|
| 0 | 1 | $0 \leq \zeta < \infty$ | Sphere |
| +1 | 1 | $0 \leq \zeta < \infty$ | Oblate |
| −1 | 1 | $1 \leq \zeta < \infty$ | Prolate |
| +1 | $\zeta^2 + \alpha\left(1 - \eta^2\right)$ | $0 \leq \zeta < \infty$ | Inverse oblate |
| −1 | $\zeta^2 + \alpha\left(1 - \eta^2\right)$ | $1 \leq \zeta < \infty$ | Inverse prolate |

## 11.2.1   Effect of Drop Surface Curvature

We can now analyse the effect of surface curvature on vapour fluxes comparing the two families of surfaces (spheroidal and inverse-spheroidal). The strong geometrical relationship between the two families of surfaces can be used to compact the five coordinate systems (spherical, prolate and oblate, inverse-prolate and inverse-prolate) in a unique parametric one

$$x = a \frac{\sqrt{\zeta^2 + \alpha}}{\Theta} \sqrt{1 - \eta^2} \cos(\varphi) \tag{11.64a}$$

$$y = a \frac{\sqrt{\zeta^2 + \alpha}}{\Theta} \sqrt{1 - \eta^2} \sin(\varphi) \tag{11.64b}$$

$$z = a \frac{\zeta \eta}{\Theta} \tag{11.64c}$$

where $-1 \leq \eta \leq +1$ and the range of $\zeta$-coordinate, the parameter $\alpha$ and the function $\Theta$ are reported in Table 11.1 for all the coordinate systems.

It must be noticed that the meaning of each coordinate variable are different in different systems; for example, for spherical coordinates $\eta = \cos(\theta)$ where $\theta$ is the polar angle, while this is not true for the other coordinate systems, although a relationship between $\eta$ and the polar angle can be easily found. The same holds for $\zeta$; for the spherical coordinate system $a\zeta$ is simply the distance from the origin, while for the other coordinate systems $\zeta$ has no simple geometrical meaning.

The value of the scaling parameter $a$ can be related to the shape of the drop surface, i.e. the value of $\zeta_0$ in the surface equation $\zeta = \zeta_0$, and the equivalent radius $R_d$ (the radius of a spherical drop having the same volume), precisely the following relation holds

$$a = \frac{R_d}{f_\alpha(\zeta_0)} \tag{11.65}$$

where the explicit form of $f_\alpha(\zeta_0)$ is given in Table 11.2.

The scale factors for the five coordinate systems can also be written in a compact way as:

**Table 11.2** Definition of the functions $f_\alpha(\zeta)$ and $H_p(\zeta)$ and values of the coefficients $s_p$, Eq. (11.53), for different drop shapes

| $f_\alpha$ | $\alpha$ | $H_p(\zeta)$ | $s_p$ | Shape of iso-$\zeta$ surfaces |
|---|---|---|---|---|
| $\zeta_0$ | 0 | $\zeta^{-p-1}$ | $s_0 = \zeta_0$; $s_{p>0} = 0$ | Sphere |
| $\left(\zeta_0^2 + \alpha\right)^{1/3} \zeta_0^{1/3}$ | +1 | $Q_p(i\zeta)$ | $s_0 = Q_0^{-1}(i\zeta_0)$; $s_{p>0} = 0$ | Oblate |
| $\left(\zeta_0^2 + \alpha\right)^{1/3} \zeta_0^{1/3}$ | −1 | $Q_p(\zeta)$ | $s_0 = Q_0^{-1}(\zeta_0)$; $s_{p>0} = 0$ | Prolate |
| $f_1$ (Eq. 11.36) | +1 | $P_p(i\zeta)$ | $s_{2k+1} = 0$; $s_{2k} = \frac{4P_{2k}(0)iQ_{2k}(i\zeta_0)}{(4k+1)P_{2k}(i\zeta)}$ | Inverse oblate |
| $f_{-1}$ (Eq. 11.36) | −1 | $P_p(i\zeta)$ | $s_{2k+1} = 0$; $s_{2k} = \frac{4P_{2k}(0)Q_{2k}(\zeta_0)}{(4k+1)P_{2k}(\zeta)}$ | Inverse prolate |

$$h_\zeta = \frac{a}{\Theta}\sqrt{\frac{(\zeta^2 + \alpha\eta^2)}{(\zeta^2 + \alpha)}} \tag{11.66a}$$

$$h_\eta = \frac{a}{\Theta}\sqrt{\frac{(\zeta^2 + \alpha\eta^2)}{(1 - \eta^2)}} \tag{11.66b}$$

$$h_\varphi = \frac{a}{\Theta}\sqrt{(\zeta^2 + \alpha)(1 - \eta^2)} \tag{11.66c}$$

and also the solution of the Laplace equation can be written in a unique way

$$\Phi = \Theta^{1/2} \sum_{p-0}^{\infty} s_p H_p(\zeta) P_p(\eta) \tag{11.67}$$

where the functions $H_p(\zeta)$ are reported in Table 11.2, as well as the analytical expressions for the coefficients $s_p$, for the particular case of uniform Dirichlet conditions. Let now turn to the gradient $\nabla\Phi$; to evaluate the derivatives in Eq. (11.63) we can take advantage of the series solution by calculating first

$$\frac{1}{h_\eta}\frac{\partial\Phi_p}{\partial\eta} = \frac{f_\alpha(\zeta_0)}{R_d}\sqrt{\frac{1 - \eta^2}{\zeta^2 + \alpha\eta^2}} H_p(\zeta) \begin{cases} P_p'(\eta) & \text{spheroids} \\ \left[-\alpha\eta\Theta^{1/2}P_p(\eta) + \Theta^{3/2}P_p'(\eta)\right] & \text{inverse-spheroids} \end{cases} \tag{11.68}$$

$$\frac{1}{h_\zeta}\frac{\partial\Phi_p}{\partial\zeta} = \frac{f_\alpha(\zeta_0)}{R_d}\sqrt{\frac{\zeta^2 + \alpha}{\zeta^2 + \alpha\eta^2}} P_p(\eta) \begin{cases} H_p'(\zeta) & \text{spheroids} \\ \left[\zeta\Theta^{1/2}H_p(\zeta) + \Theta^{3/2}H_p'(\zeta)\right] & \text{inverse-spheroids} \end{cases} \tag{11.69}$$

Then using the series (11.67) the two components of the gradient $\nabla\Phi$ can be calculated in $\zeta = \zeta_0$.

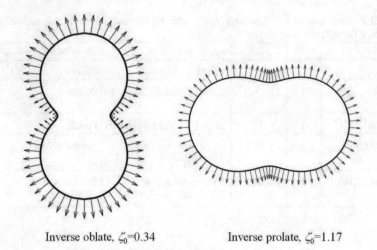

Inverse oblate, $\zeta_0$=0.34                       Inverse prolate, $\zeta_0$=1.17

**Fig. 11.7** Sample of fluxes distribution over the surface of an inverse oblate and inverse prolate drops

For the case of uniform Dirichlet boundary conditions, the component tangential to the surface is nil. Figure 11.7 shows an example of the distribution of the gradient of $\Phi$, and then of the fluxes, over the surface of two inverse spheroids.

Since the net vapour or heat flux leaving the surface is given by the component normal to the surface, we turn our attention to this one in a comparative way and, since both vapour and heat fluxes are proportional to $\nabla\Phi$, here we will analyse only one of them, and to make the analysis general we will introduce the non-dimensional flux as in Sect. 10.3.3

$$\hat{n}_\zeta^{(1)} = \frac{n_\zeta^{(1)} 4\pi R_d^2}{m_{ev}} \tag{11.70}$$

where

$$n_\zeta^{(1)} = -R_d \nabla_\zeta \Phi = R_d \frac{1}{h_\zeta} \left( \frac{\partial \Phi}{\partial \zeta} \right)_{\zeta=\zeta_0} \tag{11.71}$$

the evaporation rate $m_{ev}$ can be written as

$$m_{ev} = \int_A n_\zeta^{(1)}(\zeta_0, \eta) \, dA = f_n^{(1)} \int_A \nabla_\zeta \Phi dA \tag{11.72}$$

and from the previous results

$$\hat{n}_\zeta^{(1)} = \frac{4\pi R_d}{\int_A \nabla_\zeta \Phi dA} \frac{f_\alpha(\zeta_0)\sqrt{\zeta_0^2+\alpha}}{\sqrt{\zeta_0^2+\alpha\eta^2}} c_0 H_0'(\zeta_0) \tag{11.73}$$

for the spheroids, while for the inverse spheroids

$$\hat{n}_\zeta^{(1)} = \frac{4\pi R_d}{\int_A \nabla_\zeta \Phi dA} f_\alpha(\zeta_0) \frac{\sqrt{\zeta_0^2 + \alpha}}{\sqrt{\zeta_0^2 + \alpha\eta^2}} \left[ \zeta_0 + \Theta^{3/2} \sum_{k=1}^\infty (-1)^k \frac{(4k+1)}{4^k} \binom{2k}{k} P_{2k}(\eta) \, \Xi_k(\zeta_0) \right]$$

(11.74)

with

$$\Xi_k(\zeta_0) = \begin{cases} -\frac{Q_{2k}(i\zeta_0)P'_{2k}(i\zeta_0)}{P_{2k}(i\zeta_0)} & \alpha = 1 \quad \text{inverse oblate} \\ \frac{Q_{2k}(\zeta_0)P'_{2k}(\zeta_0)}{P_{2k}(\zeta_0)} & \alpha = -1 \,\, \text{inverse prolate} \end{cases}$$

(11.75)

where the particular values $P'_0(i\zeta_0) = 0$ and $P_{2k}(0) = \frac{(-1)^k}{4^k}\binom{2k}{k}$ have been used, and $\Theta$ is evaluated at $\zeta_0$.

Since we want to analyse possible relationships between fluxes and curvatures, let calculate the local curvatures for all these surfaces. From Chap. 2 we know that for a rotational symmetric surface defined in a curvilinear coordinate system $(\zeta, \eta, \varphi)$ by the equations

$$x = \Phi(\zeta_0, \eta)\cos(\varphi)$$ (11.76a)
$$y = \Phi(\zeta_0, \eta)\sin(\varphi)$$ (11.76b)
$$z = \Psi(\zeta_0, \eta)$$ (11.76c)

the principal curvatures on the surface $\zeta = \zeta_0$ are

$$\kappa_1 = -\frac{sgn(\Phi)\left[\Phi_{\eta\eta}\Psi_\eta - \Phi_\eta\Psi_{\eta\eta}\right]}{\left(\Phi_\eta^2 + \Psi_\eta^2\right)^{3/2}}$$ (11.77a)

$$\kappa_2 - \frac{\Psi_\eta}{|\Phi|\sqrt{\Phi_\eta^2 + \Psi_\eta^2}}$$ (11.77b)

For the present case, since

$$\Phi(\zeta, \eta) = a\frac{\sqrt{\zeta^2 + \alpha}\sqrt{1 - \eta^2}}{\Theta(\zeta, \eta)}$$ (11.78)

$$\Psi(\zeta, \eta) = a\frac{\zeta\eta}{\Theta(\zeta, \eta)}$$ (11.79)

the principal curvatures are reported (in non-dimensional form: $\kappa_j^* = R_0\kappa_j$) in Table 11.3, together with the (non-dimensional) Gaussian ($K_G^* = R_d^2 K_G = \kappa_1^*\kappa_2^*$) and mean ($C^* = R_d C = \kappa_1^* + \kappa_2^*$) curvatures.

From Eqs. (11.73) and (11.74) it can immediately be appreciated that for inverse spheroids the non-dimensional flux, Eq. (11.70), cannot be simply proportional to $K_G^{*1/4}$.

**Table 11.3** Non-dimensional principal, Gaussian and mean curvatures

| Curvatures | Spheroids | Inverse spheroids |
|---|---|---|
| $\kappa_1^*$ | $\dfrac{\zeta_0^{4/3}(\zeta_0^2+\alpha)^{5/6}}{(\zeta_0^2+\alpha\eta^2)^{3/2}}$ | $\dfrac{\zeta_0\sqrt{\zeta_0^2+\alpha}}{(\zeta_0^2+\alpha\eta^2)^{3/2}}\left[\zeta_0^2+\alpha\left(3\eta^2-1\right)\right]f_\alpha(\zeta_0)$ |
| $\kappa_2^*$ | $\dfrac{\zeta_0^{4/3}}{(\zeta_0^2+\alpha)^{1/6}\sqrt{\zeta_0^2+\alpha\eta^2}}$ | $\dfrac{\zeta_0}{\sqrt{\zeta_0^2+\alpha}\sqrt{(\zeta_0^2+\alpha\eta^2)}}\left[\zeta_0^2+\alpha\left(1+\eta^2\right)\right]f_\alpha(\zeta_0)$ |
| $K_G^*$ | $\dfrac{\zeta_0^{8/3}(\zeta_0^2+\alpha)^{2/3}}{(\zeta_0^2+\alpha\eta^2)^2}$ | $\dfrac{\zeta_0^2[\zeta_0^2+\alpha(3\eta^2-1)][\zeta_0^2+\alpha(1+\eta^2)]}{(\zeta_0^2+\alpha\eta^2)^2}f_\alpha^2(\zeta_0)$ |
| $C^*$ | $\dfrac{\zeta_0^{4/3}[2\zeta_0^2+\alpha(1+\eta^2)]}{(\zeta_0^2+\alpha\eta^2)^{3/2}(\zeta_0^2+\alpha)^{1/6}}$ | $\dfrac{2\zeta^4+\alpha\zeta^2-1+(5\alpha\zeta^2+4)\eta^2+\eta^4}{(\zeta^2+\alpha\eta^2)^{3/2}\sqrt{\zeta^2+\alpha}}\zeta_0 f_\alpha(\zeta_0)$ |

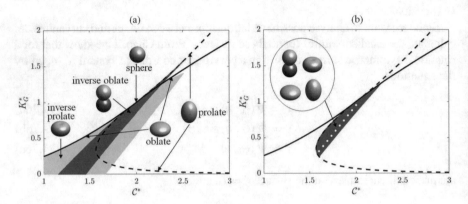

**Fig. 11.8** $K_G^* - C^*$ regions for the selected drop shapes; **b** $K_G^* - C^*$ overlapping region for selected drop shapes (gray) and selected points (white circles) for flux comparison

However, it could be argued that, as a general rule, there may exist a relationship, as complex as it could be, between the flux and the *two* curvatures, like for example $\hat{n}_\zeta^{(1)} = \hat{n}_\zeta^{(1)}\left(K_G^*, C^*\right)$ (or with the two principal curvatures, which would be equivalent), and that the rule for the spheroidal surface could just be seen as a special case. We will show now that this cannot be true, in general, since we will find a counterexample. Consider first the map reported in Fig. 11.8.

In this map, the non-dimensional Gaussian curvature, $K_G^*$, and the non-dimensional mean curvature, $C^*$, are reported on the $y-$ and $x-$ axes, respectively. The point with coordinates $(2, 1)$ represents a sphere (both curvatures have a constant value over all the surface). The areas coloured in different ways represent those values of $K_G^*$ and $C^*$ that could be found in some points of the different surfaces (spheroids and inverse spheroids) depicted on the map. It can be noticed that different surfaces may have the same values of both curvatures in some regions, and this is indicated in the map by the overlapping of the regions representing each surface. There is a region (the dark gray area in Fig. 11.6b) that represents some part of the four surfaces (two spheroids and two inverse-spheroids) where curvatures are the same for all the shapes. In [12] the non-dimensional flux, $\hat{n}_\zeta^{(1)}$, was evaluated in these

**Table 11.4** Non-dimensional fluxes for spheroids and inverse spheroids (oblate and prolate)

| $C^*$ | $K_G^*$ | $\hat{n}_{v,n}$ | | |
|---|---|---|---|---|
| | | Spheroids | Inverse oblate | Inverse prolate |
| 1.600 | 0.301 | 0.746 | 0.280 | 0.930 |
| 1.644 | 0.381 | 0.786 | 0.327 | 0.940 |
| 1.689 | 0.452 | 0.820 | 0.384 | 0.950 |
| 1.733 | 0.523 | 0.851 | 0.449 | 0.960 |
| 1.778 | 0.594 | 0.878 | 0.522 | 0.969 |
| 1.822 | 0.666 | 0.903 | 0.600 | 0.978 |
| 1.867 | 0.737 | 0.926 | 0.681 | 0.986 |
| 1.911 | 0.808 | 0.948 | 0.762 | 0.992 |
| 1.956 | 0.879 | 0.968 | 0.840 | 0.998 |
| 2.000 | 0.950 | 0.987 | 0.911 | 1.002 |

parts of the surfaces for the four cases; Table 11.4 reports the values of the fluxes in some selected points inside the dark gray region of Fig. 11.6b (the white circles) for the four shapes. The values of the non-dimensional flux is the same for the two spheroids (as expected, since it is proportional on $K_G^{*1/4}$), while for the two inverse spheroids the values are different. If there were a universal relationship between curvature and non-dimensional flux, then the values would be the same. This proves that such a relation cannot be a general rule, and that, in general, the non-dimensional flux should depend also on other parameters; the result obtained for spheroids (and incidentally also for triaxial ellipsoids) is then just a special case.

## 11.3 The Sessile Drop

A drop at rest on a surface is called *sessile* drop.

Figure 11.9 shows a sessile drop on a perfectly flat surface. We can first observe that the shape of the drop, when the effect of gravity can be neglected, is defined by three quantities: the solid-gas interfacial energy, $\gamma_{S-G}$, the solid-liquid interfacial energy, $\gamma_{S-L}$, and the liquid-gas interfacial energy (more often called *surface tension*), $\gamma_{L-G}$. The *contact angle*, $\theta_c$, i.e. the angle measured through the liquid as shown in Fig. 11.9, is defined by the *Young's equation* (named after the British scientist Thomas Young, 1773–1829)

**Fig. 11.9** Schematic of a sessile drop; $\theta_c$ is the contact angle

$$\cos{(\theta_c)} = \frac{\gamma_{S-G} - \gamma_{S-L}}{\gamma_{L-G}} \tag{11.80}$$

(for a discussion on this and other equations that are used to evaluate the contact angle see [13]). The effect of surface tension is to minimise the liquid-gas interface, and given the constraint of a fixed contact angle with the flat solid surface, the equilibrium shape of the sessile drop, in the absence of gravity, is that of a spherical cap.

Consider now the toroidal coordinate system defined by the equations

$$x = a\frac{\sinh{(\xi)}\cos{(\varphi)}}{\Theta} \tag{11.81a}$$

$$y = a\frac{\sinh{(\xi)}\sin{(\varphi)}}{\Theta} \tag{11.81b}$$

$$z = a\frac{\sin{(\theta)}}{\Theta} \tag{11.81c}$$

where $\Theta = \cosh{(\xi)} - \cos{(\theta)}$, $0 \le \xi < \infty$, $-\pi \le \theta \le +\pi$, $0 \le \varphi \le 2\pi$, (see Chap. 4 for all the details).

The surface defined by the parametric equation $\theta = \theta_0$ is a spherical cap, in fact, since

$$x^2 + y^2 + \left(z - \frac{a}{\tan{(\theta_0)}}\right)^2 = \frac{a^2}{\sin^2{(\theta_0)}} \tag{11.82}$$

the surface is a sphere with center on the $z$-axis at $z_0 = \frac{a}{\tan{(\theta_0)}}$ and with radius $\frac{a}{\sin{(\theta_0)}}$. This coordinate system is then the natural one to be used for modelling heat and mass transfer from a sessile drop since the drop surface is defined, in this system, by the equation: $\theta = \theta_0$.

Taking now the section $\varphi = 0$ (i.e. the intersection of this surface with the plane $y = 0$) the circle intersects the line $z = 0$ in $x = \pm a$, and $\theta_0 = \arctan\left(\frac{a}{z_0}\right)$ is the angle formed with the $x$-axis (see Fig. 11.10). It should be noticed that the angle $\theta_0$ is then the complement to $\pi$ of the contact angle as defined above. i.e. $\theta_0 = \pi - \theta_c$.

All the spherical caps (i.e. all surfaces $\theta = const.$) intersect the $z = 0$ plane in a circle of radius $a$, which is approached when $\xi \to \infty$, for any value of $\theta$, while the $z$-axis is approached when $\xi \to 0$.

The Laplace equation $\nabla^2 \Phi = 0$ in this coordinate system, assuming axial symmetry (i.e. $\Phi$ independent of $\varphi$) has the form:

$$\frac{\partial}{\partial \xi}\left[\frac{\sinh{(\xi)}}{\Theta}\frac{\partial \Phi}{\partial \xi}\right] + \frac{\partial}{\partial \theta}\left[\frac{\sinh{(\xi)}}{\Theta}\frac{\partial \Phi}{\partial \theta}\right] = 0 \tag{11.83}$$

and it is *R-separable* in this coordinate system (see Chap. 4) and a general solution, symmetric around the $z$-axis, is given by [6]

$$\Phi_p = \Theta^{1/2} H_p{(\xi)} N_p{(\theta)} \tag{11.84}$$

**Fig. 11.10** Section on the $(x - z)$ plane of an iso-surface $\theta = \theta_0$ in toroidal coordinates

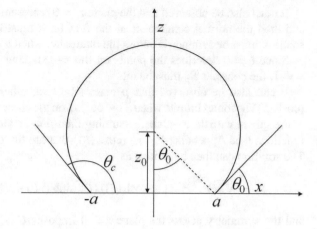

where

$$H_p (\xi) = A_p P_{p-1/2} (\cosh (\xi)) + B_p Q_{p-1/2} (\cosh (\xi)) \qquad (11.85)$$
$$N_p (\theta) = C_p \sin (p\theta) + D_p \cos (p\theta) \qquad (11.86)$$

and $P_{p-1/2}$ and $Q_{p-1/2}$ are Legendre functions of order $p - 1/2$ of the first and the second kind, respectively.

The solution of this problem is however more involved than those previously analysed, and the interested reader is referred to [11] for more details. The boundary conditions on the surface is as usual

$$\Phi (\xi, \theta_0) = 1 \qquad (11.87)$$

while that at infinity can be better described as follows. Considering a sphere centered in the origin and with radius $R_1$, its equation in toroidal coordinates is

$$R_1^2 = x^2 + y^2 + z^2 = a^2 \frac{\sinh^2 (\xi) + \sin^2 \theta}{\Theta^2} = a^2 \frac{\cosh^2 (\xi) - \cos^2 \theta}{\Theta^2} \qquad (11.88)$$

$$0 = \cos^2 \theta \left(1 + \frac{a^2}{R_1^2}\right) - 2 \cosh (\xi) \cos \theta + \left(1 - \frac{a^2}{R_1^2}\right) \cosh^2 (\xi) \qquad (11.89)$$

The sphere of infinite radius is obtained letting $R_1 \to \infty$, which corresponds to $\Theta \to 0$ that can be obtained only when $\xi \to 0$ and $\theta \to 0$ simultaneously. Then, the relative boundary condition would be

$$\Phi (0, 0) = 0 \qquad (11.90)$$

It must also be observed that the plane $z = 0$ represents an impermeable surface, and then the normal component of the flux on it must be nil, in other words the solution must be symmetric across the plane, i.e. when $\theta$ changes sign.

Since $\xi = 0$ describes the points on the $z$−axis, and $Q_v(x)$ are unbounded on $x = 1$, the constant $B_p$ must be nil.

It can also be observed that $p$ cannot be real, otherwise the function $P_{p-1/2}$ (cosh $(\xi)$) becomes infinite when $\xi \to \infty$, i.e. on the circle defined by the intersection of the sphere with the $z$−plane. Assuming then $p = i\tau$ yields the *conical* or *Mehler* functions (see Appendix) $P_{i\tau-1/2}$ (cosh $(\xi)$) that are finite when $\xi \to \infty$ for $\tau \geq 0$. The solution can then be written as

$$\Phi_\tau = \sqrt{2}\Theta^{1/2} P_{i\tau-1/2} (\cosh (\xi)) \left[ C_\tau \sinh (\tau\theta) + D_\tau \cosh (\tau\theta) \right] \tag{11.91}$$

and the symmetry across the plane $z = 0$ imposes $C(\tau) = 0$, leaving the general solution

$$\Phi_\tau = \sqrt{2}\Theta^{1/2} P_{i\tau-1/2} (\cosh (\xi)) D(\tau) \cosh (\tau\theta) \tag{11.92}$$

Since the boundary conditions cannot yield any eigenvalues for $\tau$, the most general solution can be written as

$$\Phi(\xi, \theta) = \sqrt{2}\Theta^{1/2} \int_0^\infty P_{i\tau-1/2} (\cosh (\xi)) D(\tau) \cosh (\tau\theta) d\tau \tag{11.93}$$

where the functions $D(\tau)$ must be defined through the boundary conditions.

## 11.3.1   The Boundary Conditions

It can be seen that the boundary conditions at infinity is satisfied by Eq. (11.93), since $\Theta \to 0$, while that on the drop surface becomes

$$\frac{1}{\sqrt{2}\sqrt{\cosh (\xi) - \cos (\theta_0)}} = \int_0^\infty P_{i\tau-1/2} (\cosh (\xi)) D(\tau) \cosh (\tau\theta_0) d\tau \tag{11.94}$$

To obtain an analytic form of the function $D(\tau)$, we need to rely on the so-called Mehler–Fock transform (cfr Appendix). The Mehler–Fock transform of a function $f(x)$ is defined as

$$F(\tau) = \int_1^\infty f(x) P_{i\tau-1/2}(x) dx \tag{11.95}$$

where $\tau \geq 0$, and the function $f(x)$ is piecewise continuous in the interval $[1, \infty)$. The inversion formula is given by

$$f(x) = \int_0^\infty \tau \tanh(\pi\tau) P_{i\tau-1/2}(x) F(\tau) d\tau \tag{11.96}$$

A comparison of Eqs. (11.94) and (11.96), with the substitution $x = \cosh(\xi)$, shows that

$$f(x) = \frac{1}{\sqrt{2}\sqrt{x - \cos(\theta_0)}} \tag{11.97}$$

$$F(\tau) = D(\tau) \frac{\cosh(\tau\theta_0)}{\tau \tanh(\pi\tau)} \tag{11.98}$$

and the following result (see [11] Eq. 8.12.8):

$$\frac{1}{\sqrt{2}\sqrt{\cosh(\xi) - \cos(\theta_0)}} = \int_0^\infty \frac{\cosh(\pi - \theta_0)\tau}{\cosh(\pi\tau)} P_{i\tau-1/2}(\cosh(\xi)) d\tau \tag{11.99}$$

shows that the Mehler–Fock transform of $f(x) = \frac{1}{\sqrt{2}\sqrt{x-\cos(\theta_0)}}$ is $F(\tau) = \frac{\cosh(\pi-\theta_0)}{\sinh(\pi\tau)}$, then

$$D(\tau) = \frac{\cosh(\pi - \theta_0)\tau}{\cosh(\pi\tau)\cosh(\tau\theta_0)} \tag{11.100}$$

and the solution is now

$$\Phi(\xi, \theta) = \sqrt{2}\Theta^{1/2} \int_0^\infty P_{i\tau-1/2}(\cosh(\xi)) \frac{\cosh(\pi - \theta_0)\tau}{\cosh(\pi\tau)} \frac{\cosh(\tau\theta)}{\cosh(\tau\theta_0)} d\tau \tag{11.101}$$

### 11.3.2 Explicit Solutions

An explicit solution of the problem could be obtained by solving analytically the integral appearing in Eq. (11.101). A large collection of integrals involving $P_{i\tau-1/2}(\cosh\xi)$ can be found e.g. in [14, 15].

For example the case $\theta_c = \pi/2$ admits a simple solution. This case represents a sessile drop with a contact angle exactly equal to $\pi/2$ and Eq. (11.101) becomes

$$\Phi(\xi, \theta) = \sqrt{2}\Theta^{1/2} \int_0^\infty P_{i\tau-1/2}(\cosh\xi) \frac{\cosh(\tau\theta)}{\cosh(\pi\tau)} d\tau \tag{11.102}$$

From Eq. (11.99) (setting $\theta = \pi - \theta_c \Rightarrow \theta_c = \pi - \theta$)

$$\frac{1}{\sqrt{2}\sqrt{\cosh\xi + \cos(\theta_c)}} = \int_0^\infty \frac{\cosh(\theta\tau)}{\cosh(\pi\tau)} P_{i\tau-1/2}(\cosh(\xi)) d\tau \tag{11.103}$$

and a comparison with Eq. (11.102) yields the actual solution

$$\Phi(\xi, \theta) = \frac{\Theta^{1/2}}{\sqrt{\cosh \xi + \cos(\theta_c)}} \tag{11.104}$$

This solution must be identical to that of a spherical drop floating in a gaseous atmosphere (Eq. 11.102), since in such case the boundary conditions are the same, and in fact $(a = R_d)$

$$\Phi = \frac{R_d}{R} = \frac{R_d \Theta}{a\sqrt{\cosh^2(\xi) - \cos^2 \theta}} = \frac{\Theta^{1/2}}{\sqrt{\cosh(\xi) + \cos \theta}} \tag{11.105}$$

where the last equation is found using Eq. (11.88).

The evaluation of the solution (11.101) for general values of $\theta_0$ in terms of elementary or even special functions does not appear available at present. Specialised literature on this subject is available and usually the final solution is found using numerical integration [16] or various approximations [17, 18].

## 11.4  Drop Evaporation with Non-uniform Boundary Conditions

The investigation reported in the previous sections holds when the conservation equations are solved imposing uniform Dirichlet boundary conditions at the drop surface and at infinite. In this section we will relax this hypothesis and generally non-uniform boundary conditions will be imposed at the drop surface, maintaining a constant value for the boundary condition at infinity. This is motivated by the fact that in applicative problems the conditions at the liquid/drop interface may not be perfectly uniform; for example, in a transient problem, when a drop is not spherical, even imposing initial uniform conditions at the drop surface, the heat and mass fluxes will not be uniform on the surface, then inducing a non-uniform temperature distribution within the liquid, which may lead to a non-uniform condition at the interface [19].

We will assume constant thermo-physical properties of the gaseous environment and in this case the species conservation equations in molar form, for a single-component drop, reduces to Eq. (10.29)

$$\nabla^2 H = 0 \tag{11.106}$$

where $H = \log\left(1 - y^{(1)}\right)$.

In this section the case of spherical and spheroidal (oblate and prolate) drops will be discussed. The general coordinate system can be written using the curvilinear coordinate system $(\zeta, \eta, \varphi)$

**Table 11.5** Scale factors and $g^{1/2}$ for spheroids

|  | $h_\eta$ | $h_\zeta$ | $h_\varphi$ | $g^{1/2} = h_\eta h_\zeta h_\varphi$ |
|---|---|---|---|---|
| Sphere | $a\dfrac{\zeta}{\sqrt{1-\eta^2}}$ | $a$ | $a\zeta\sqrt{1-\eta^2}$ | $a^3\zeta^2$ |
| Prolate | $a\sqrt{\dfrac{\zeta^2-\eta^2}{1-\eta^2}}$ | $a\sqrt{\dfrac{\zeta^2-\eta^2}{\zeta^2-1}}$ | $a\sqrt{(\zeta^2-1)(1-\eta^2)}$ | $a^3(\zeta^2-\eta^2)$ |
| Oblate | $a\dfrac{\sqrt{\zeta^2+\eta^2}}{\sqrt{1-\eta^2}}$ | $a\dfrac{\sqrt{\zeta^2+\eta^2}}{\sqrt{\zeta^2+1}}$ | $a\sqrt{\zeta^2+1}\sqrt{1-\eta^2}$ | $a^3(\zeta^2+\eta^2)$ |

$$x = a\sqrt{\zeta^2 + \alpha}\sqrt{1 - \eta^2}\cos\varphi \tag{11.107a}$$
$$y = a\sqrt{\zeta^2 + \alpha}\sqrt{1 - \eta^2}\sin\varphi \tag{11.107b}$$
$$z = a\zeta\eta \tag{11.107c}$$

where the parameter $\alpha$ assumes the following values according to the drop shape

$$
\begin{aligned}
&\text{sphere} \quad \alpha = 0\\
&\text{prolate} \quad \alpha = -1\\
&\text{oblate} \quad \alpha = +1
\end{aligned}
\tag{11.108}
$$

while $0 < \zeta < \infty$ for the spherical and oblate spheroidal case and $1 < \zeta < \infty$ for the prolate spheroidal case.

The drop surface is defined by the equation $\zeta = \zeta_0$. Again, it must be stressed that the coordinates $(\zeta, \eta)$ have different meanings in the different coordinate systems.

For a general shaped drop $R_d$ is assumed to be the radius of an equivalent volume spherical drop and the scale parameter $a$ in Eqs. (11.107) can be expressed as a function of $R_d$ and the deformation ratio $\varepsilon = \frac{a_z}{a_r}$, where $a_z$ and $a_r$ are the axial and radial spheroid semi-axes, respectively

$$
\begin{aligned}
&\text{sphere} \quad a = \frac{R_d}{\zeta_0}\\
&\text{spheroid} \quad a = R_d\frac{|1-\varepsilon^2|^{1/2}}{\varepsilon^{1/3}}
\end{aligned}
\tag{11.109}
$$

Equation (11.106) can be explicitly written as

$$\frac{\partial}{\partial\eta}\left(\frac{h_\zeta h_\varphi}{h_\eta}\frac{\partial H}{\partial\eta}\right) + \frac{\partial}{\partial\zeta}\left(\frac{h_\eta h_\varphi}{h_\zeta}\frac{\partial H}{\partial\zeta}\right) + \frac{\partial}{\partial\varphi}\left(\frac{h_\zeta h_\eta}{h_\varphi}\frac{\partial H}{\partial\varphi}\right) = 0 \tag{11.110}$$

where the scale factors $h_\eta$, $h_\zeta$, $h_\varphi$ are defined as in Table 11.5.

The solution of the Laplace equation (11.110) assumes the following form

$$H(\zeta, \eta, \varphi) = \sum_{n=0}^{\infty}\sum_{k=0}^{\infty} c_{nk} W_n^k(\zeta) P_n^k(\eta)\cos(k\varphi) + H_1 \tag{11.111}$$

where $P_n^m(\eta)$ are the associated Legendre functions of the first kind (see [6] for the solution and the Appendix for the definition of $P_n^k$) and the functions $W_n(\zeta)$ depend on the coordinate system as

$$
\begin{aligned}
\text{sphere} \quad & W_n^k(\zeta) = \frac{\zeta^{-(n+1)}}{\zeta_0^{-(n+1)}} \\
\text{prolate} \quad & W_n^k(\zeta) = \frac{Q_n^k(\zeta)}{Q_n^m(\zeta_0)} \\
\text{oblate} \quad & W_n^k(\zeta) = \frac{Q_n^k(i\zeta)}{Q_n^m(i\zeta_0)}
\end{aligned}
\tag{11.112}
$$

being $Q_n^k(\zeta)$ the associated Legendre functions of the second kind (see again the Appendix).

The constant $H_1$ in Eq. (11.111) is obtained by imposing the boundary condition at infinity, obtaining

$$
H_1 = H_\infty = \log\left(1 - y_\infty^{(1)}\right)
\tag{11.113}
$$

The coefficients $c_{nk}$ are calculated imposing the boundary conditions on the drop surface ($H(\zeta_0, \eta, \varphi) = H_s$), observing that $W_n(\zeta_0) = 1$,

$$
H_s(\zeta_0, \eta, \varphi) - H_\infty = \sum_{n=0}^{\infty} \sum_{k=0}^{\infty} c_{nk} P_n^k(\eta) \cos(k\varphi)
\tag{11.114}
$$

To explicitly evaluate the coefficients, the orthogonality properties of the series of functions $P_k^m(\eta)$ and $\cos(k\varphi)$ can be used.

Observing that

$$
\int_{-1}^{1} P_k^m(\eta) P_l^m(\eta)\, d\eta = \frac{2(l+m)!}{(2l+1)(l-m)!} \delta_{kl}
\tag{11.115}
$$

$$
\int_{0}^{2\pi} \cos(k\varphi) \cos(p\varphi)\, d\varphi = \pi \delta_{kp}
\tag{11.116}
$$

and multiplying both sides of Eq. (11.114) by $\cos(p\varphi)$ and $P_l^p(\eta)$ and integrating over $\eta \in [-1, 1]$ and over $\varphi \in [0, 2\pi]$, the following results are obtained

$$
\begin{aligned}
c_{00} &= \int_{-1}^{1} \int_{0}^{2\pi} \frac{1}{4\pi} H_s(\zeta_0, \eta, \varphi) P_l(\eta)\, d\varphi d\eta - H_\infty \\
c_{l0} &= \frac{2l+1}{4\pi} \int_{-1}^{1} \int_{0}^{2\pi} H_s(\zeta_0, \eta, \varphi) P_l(\eta)\, d\varphi d\eta \qquad l > 0 \\
c_{lp} &= \frac{(2l+1)(l-p)!}{2\pi(l+p)!} \int_{-1}^{1} \int_{0}^{2\pi} H_s(\zeta_0, \eta, \varphi) \cos(p\varphi) P_l^p(\eta)\, d\varphi d\eta \quad l, p > 0
\end{aligned}
\tag{11.117}
$$

**Example of Simple Non-uniform Boundary Conditions on the Drop Surface**

The solution is simplified for particular boundary conditions on the drop surface such that only a finite number of terms in Eq. (11.111) remains.

For example suppose that

$$H_s\left(\zeta_0, \eta, \varphi\right) = A\left(\eta\right) + B\left(\eta\right)\cos\left(\varphi\right) \tag{11.118}$$

where the two functions $A\left(\eta\right)$ and $B\left(\eta\right)$ can be expressed as follows

$$A\left(\eta\right) = \sum_{n=0}^{N_A} a_n P_n\left(\eta\right) \tag{11.119}$$

$$B\left(\eta\right) = \sum_{n=0}^{N_B} b_n P_n^1\left(\eta\right) \tag{11.120}$$

then $A\left(\eta\right)$ is a polynomial of degree $N_A$ while $B\left(\eta\right)$ is not a polynomial, since $P_n^1\left(\eta\right)$ are not polynomials, although they are sometimes referred as *associated Legendre polynomials*.

In this example, we analyse a simple case where the Dirichlet boundary condition is symmetric with respect to $\eta$ (i.e. $H\left(\eta, \varphi\right) = H\left(-\eta, \varphi\right)$), then only the even polynomials $P_n\left(\eta\right)$ must be used to evaluate $A\left(\eta\right)$, moreover we will assume $N_A = 4$, then

$$A\left(\eta\right) = a_0 + a_2 P_2\left(\eta\right) + a_4 P_4\left(\eta\right) \tag{11.121}$$

The function $B\left(\eta\right)$ must satisfy the same symmetry requirements on $\eta$ (i.e. $H_s\left(+\eta, \varphi\right) = H_s\left(-\eta, \varphi\right)$), then only the odd associated polynomials must be used (these are even on $\eta$) in (11.120), then choosing again $N_B = 4$,

$$B\left(\eta\right) = b_1 P_1^1\left(\eta\right) + b_3 P_3^1\left(\eta\right) \tag{11.122}$$

Assuming $y_\infty^{(1)} = 0$, then $H_\infty = 0$, Eq. (11.114) becomes

$$\sum_{n=0}^{N_A} a_n P_n^0\left(\eta\right) + \sum_{n=0}^{N_B} b_n P_n^1\left(\eta\right)\cos\left(\varphi\right) = \sum_{n=0}^{\infty}\sum_{k=0}^{\infty} c_{nk} P_n^k\left(\eta\right)\cos\left(k\varphi\right) \tag{11.123}$$

and the values of the coefficients $c_{nk}$ can be easily found comparing the two sides of this equation and the non nil coefficients $c_{nk}$ are

$$c_{00} = a_0; \quad c_{20} = a_2; \quad c_{40} = a_4; \quad c_{11} = b_1; \quad c_{31} = b_3 \tag{11.124}$$

This kind of boundary conditions can be seen as an approximation of realistic boundary conditions. The functions $A\left(\eta\right)$ and $B\left(\eta\right)$ can be evaluated by fitting equation (11.118) to the actual values of $H_s$ on a finite number of points on the surface. Since five coefficients are needed, we can choose to fit the function $H_s\left(\zeta_0, \eta, \varphi\right)$ on five points like those shown in Fig. 11.11.

Since the function $A\left(\eta\right)$ gives the values of $H_s$ along the surface at $\varphi = \pm\frac{\pi}{2}$, a choice is to select three values of $H_s$ for $\varphi = \pi/2$ and $\eta = 0, 1/2, 1$, equal to $H_{s1}$, $H_{s2}$ and $H_{s3}$, respectively, then

**Fig. 11.11** Schematic of selected points on the surface of spheroidal drops where the function $H_s\left(\zeta_0, \eta, \varphi\right)$ is fitted to given values

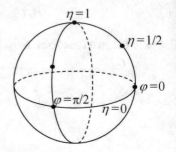

$$A\left(0\right) = a_0 + a_2 P_2\left(0\right) + a_4 P_4\left(0\right) = H_{s1} \tag{11.125a}$$

$$A\left(1/2\right) = a_0 + a_2 P_2\left(1/2\right) + a_4 P_4\left(1/2\right) = H_{s2} \tag{11.125b}$$

$$A\left(1\right) = a_0 + a_2 P_2\left(1\right) + a_4 P_4\left(1\right) = H_{s3} \tag{11.125c}$$

and this system can be solved, yielding

$$a_0 = \frac{14}{105} H_{s1} + \frac{224}{315} H_{s2} + \frac{49}{315} H_{s3}$$

$$a_2 = -\frac{22}{21} H_{s1} + \frac{32}{63} H_{s2} + \frac{34}{63} H_{s3}$$

$$a_4 = \frac{96}{105} H_{s1} - \frac{128}{105} H_{s2} + \frac{32}{105} H_{s3}$$

The function $B\left(\eta\right)$ can be defined by setting two additional values of $H_s$, corresponding to two points like those shown in Fig. 11.11, i.e. $\varphi = 0$ and $\eta = 0, 1/2$, obtaining

$$B\left(0\right) = b_1 P_1^1\left(0\right) + b_3 P_3^1\left(0\right) = H_{s4} - A\left(0\right) \tag{11.127a}$$

$$B\left(1/2\right) = b_1 P_1^1\left(1/2\right) + b_3 P_3^1\left(1/2\right) = H_{s5} - A\left(1/2\right) \tag{11.127b}$$

and explicitly

$$b_1 = -\frac{1}{3} H_{s4} - \frac{4}{3\sqrt{3}} H_{s5} + \frac{1}{3}\left[\frac{4+\sqrt{3}}{\sqrt{3}} a_0 - \frac{1+\sqrt{3}}{2\sqrt{3}} a_2 + \frac{12\sqrt{3}-37}{8*4} a_4\right] \tag{11.128a}$$

$$b_3 = \frac{4}{9} H_{s4} - \frac{4}{9}\frac{2}{\sqrt{3}} H_{s5} + \frac{4}{9}\left[\frac{2-\sqrt{3}}{\sqrt{3}} a_0 + \frac{2\sqrt{3}-1}{4\sqrt{3}} a_2 - \frac{37+24\sqrt{3}}{\sqrt{364}} a_4\right] \tag{11.128b}$$

Finally the solution (11.111) for the selected B.C. is

$$H(\zeta, \eta, \varphi) = a_0 W_0^0(\zeta) P_0(\eta) + a_2 W_2^0(\zeta) P_2(\eta) + a_4 W_4^0(\zeta) P_4(\eta) +$$
$$+ \left[ b_1 W_1^1(\zeta) P_1^1(\eta) + b_3 W_3^1(\zeta) P_3^1(\eta) \right] \cos(\varphi) \qquad (11.129)$$

where the functions $W_n^k(\zeta)$ are given in Eqs. (11.112) for each drop shape.

### 11.4.1 The Vapour Fluxes

Once the solution of the species conservation equation is obtained, the vapour mass fluxes at the drop surface along each coordinate direction are defined as

$$n_j(\zeta_0, \eta, \varphi) = Mm^{(1)}cD_{10}\left(\frac{1}{h_j}\frac{\partial H}{\partial u^j}\right)_{\zeta=\zeta_0} \qquad (11.130)$$

then

$$n_\zeta(\zeta_0, \eta, \varphi) = Mm^{(1)}cD_{10}\frac{1}{h_\zeta(\zeta_0, \eta)}\sum_{n=0}^{\infty}\sum_{k=0}^{\infty}c_{nk}W_n'^k(\zeta_0)P_n^k(\eta)\cos(k\varphi)$$
$$(11.131\text{a})$$

$$n_\eta(\zeta_0, \eta, \varphi) = Mm^{(1)}cD_{10}\frac{1}{h_\eta(\zeta_0, \eta)}\sum_{n=0}^{\infty}\sum_{k=0}^{\infty}c_{nk}W_n^k(\zeta_0)P_n'^k(\eta)\cos(k\varphi)$$
$$(11.131\text{b})$$

$$n_\varphi(\zeta_0, \eta, \varphi) = -Mm^{(1)}cD_{10}\frac{1}{h_\varphi(\zeta_0, \eta)}\sum_{n=0}^{\infty}\sum_{k=0}^{\infty}c_{nk}W_n^k(\zeta_0)P_n^k(\eta)k\sin(k\varphi)$$
$$(11.131\text{c})$$

where $W_n'^k$ and $P_n'^k(\eta)$ are the derivatives of the functions with respect to their variables (see also Appendix).

Figure 11.12 shows a sample of solutions described in this section, where the non-uniform boundary condition selected above (Eq. 11.118) are imposed at the

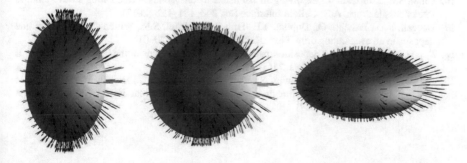

**Fig. 11.12** Fluxes on the surface of spherical, prolate and oblate drops with non-uniform Dirichlet boundary condition $\Phi_{s1} = 1$; $\Phi_{s2} = 0.9$; $\Phi_{s3} = 0.8$; $\Phi_{s4} = 0.7$; $\Phi_{s5} = 0.7$

drop surface of three drops having the same volume: a sphere, an oblate drop with $\varepsilon = 0.5$ and a prolate drop with $\varepsilon = 1.5$. The distribution of the non-dimensional quantity $\Phi = \frac{H}{H_{sl}}$ is qualitatively shown by the colours on the drop surface, while the vectors represent the $\nabla \Phi$, which is proportional to the flux vectors (Eqs. 11.131).

# References

1. Brzustowski, T.A., Twardus, E.M., Wojcicki, S., Sobiesiak, A.: Interaction of two burning fuel droplets of arbitrary size. AIAA J. **17**, 1234–1242 (1979)
2. Labowsky, M.: Calculation of the burning rates of interacting fuel droplets. Combust. Sci. Technol. **22**, 217–226 (1980)
3. Umemura, A., Ogawa, S., Oshima, N.: Analysis of the interaction between two burning droplets. Combust. Flame **41**, 45–55 (1981)
4. Umemura, A., Ogawa, S., Oshima, N.: Analysis of the interaction between two burning fuel droplets with different sizes. Combust. Flame **43**, 111–119 (1981)
5. Cossali, G.E., Tonini, S.: Variable gas density effects on transport from interacting evaporating spherical drops. Int. J. Heat Mass Transf. **127**, 485–496 (2018)
6. Moon, P., Spencer, D.E.: Field Theory Handbook. Springer, Heidelberg (1988)
7. Deng, Z.T., Litchford, R.J., Jeng, S.M.: Two-dimensional simulation of droplet evaporation at high pressure. AIAA Pap. 29–3122 (1992)
8. Mashayek, F.: Dynamics of evaporating drops. Part I: formulation and evaporation model. Int. J. Heat Mass Transf. **44**(8), 1517–1526 (2001)
9. Lian, Z.W., Reitz, R.D.: The effect of vaporization and gas compressibility on liquid jet atomization. At. Sprays **3**(3), 249–264 (1993)
10. Quan, S., Lou, J., Schmidt, D.P.: Modeling merging and breakup in the moving mesh interface tracking method for multiphase flow simulations. J. Comput. Phys. **228**(7), 2660–2675 (2009)
11. Lebedev, N.N.: Special Functions and Their Applications. Dover, Mineola (1972)
12. Lamanna, G., Cossali, G.E., Tonini, S.: The influence of curvature on the modelling of droplet evaporation at different scales. In: Lamanna, G., Tonini, S., Cossali, G.E., Weigand, B. (eds.) Droplet Interactions and Spray Processes. Springer, Berlin (2020)
13. Jasper, W.J., Anand, N.: A generalized variational approach for predicting contact angles of sessile nano-droplets on both flat and curved surfaces. J. Mol. Liq. **281**, 196–203 (2019)
14. Ditkin, V.A., Prudnikov, A.P.: Integral Transforms and Operational Calculus. Pergamon Press, Oxford (1965)
15. Mandal, B.N., Mandal, N.: Integral Expansions Related to Mehler-Fock Type Transforms. Chapman and Hall/CRC, Boca Raton (1997)
16. Chini, S.F., Amirfazli, A.: Resolving an ostensible inconsistency in calculating the evaporation rate of sessile drops. Adv. Colloid Interface Sci. **243**, 121–128 (2017)
17. Deegan, R.D., Bakajin, O., Dupont, T.F., Huber, G., Nagel, S.R., Witten, T.A.: Contact line deposits in an evaporating drop. Phys. Rev. E **62**, 756–765 (2000)
18. Popov, Y.O.: Evaporative deposition patterns: spatial dimensions of the deposit. Phys. Rev. E Stat. Nonlinear Soft Matter Phys. **71**(3), 036313 (2005)
19. Zubkov, V.S., Cossali, G.E., Tonini, S., Rybdylova, O., Crua, C., Heikal, M., Sazhin, S.S.: Mathematical modelling of heating and evaporation of a spheroidal droplet. Int. J. Heat Mass Transf. **108**, 2181–2190 (2017)

# Chapter 12
# Drop Evaporation Under Unsteady Conditions

In the previous two chapters we have investigated models of drop evaporation under steady-state conditions, an assumption widely used, although clearly unphysical: a mass source inside the drop is needed to maintain the drop shape unchanged during evaporation. To relieve this assumption a time dependent problem must be set and solved, increasing the complexity of analytical approaches. In particular, even for a spherical drop shrinking by evaporation, a moving boundary problem must be solved, which is known to be a challenging task, even for the simplest geometries. In this chapter we will see how it is possible to account for unsteadiness of the heat and mass transfer processes and still approach the modelling by analytical methods. The unsteady problem for a spherical drop will be analytically studied, from the viewpoint of an internal space-time dependent temperature field and of external varying temperature and concentration fields. The problem becomes even more complex if non-spherical drops are considered, since the effect of surface tension and inertia causes a complex time dependent modification of the drop shape, and a simple analytical model for dealing with evaporation of oscillating drops is discussed, as well as an analytical solution of the time-dependent heat transfer in spheroidal bodies.

## 12.1 Oscillating Drops

When a drop is released in a gas flow, its shape is usually far from being spherical, then the action of the surface tension will force a deformation towards the spherical shape, and the drop shape may begin to oscillate under the opposing effects of inertia and surface tension and the damping effect of viscous forces. Lord Rayleigh, in an appendix of its work on stability of jets [1], derived an equation for the angular frequencies of a linear oscillation of an inviscid drop, and since that first result a huge amount of work was published on this subject (see for example [2]). The shape of the oscillating drop surface is often evaluated through a decomposition in terms of Legendre polynomials, in a spherical coordinate system, as

© Springer Nature Switzerland AG 2021
G. E. Cossali and S. Tonini, *Drop Heating and Evaporation: Analytical Solutions in Curvilinear Coordinate Systems*, Mathematical Engineering,
https://doi.org/10.1007/978-3-030-49274-8_12

$$R(\theta, t) = R_0 + \sum_{n=1}^{\infty} a_n(t) P_n(\cos(\theta)) \tag{12.1}$$

where $a_n(t)$ represents the time-varying contribution of the mode $n$ to the departure from the spherical shape.

This particular decomposition allows separating the oscillating modes in such a way that, in the limit of small oscillation amplitudes, there is no *cross-talking* with each other. In particular, it was shown [3, 4] that when the action of liquid viscosity is taken into account, the mode $n = 2$ is the one that dissipates energy more slowly, i.e. the amplitude of the oscillations of $a_2(t)$ decreases at a smaller rate than that of all the other modes, and then, after a certain time, the drop shape may be approximately represented by

$$R(\theta, t) \simeq R_0(t) + a_2(t) P_2(\cos(\theta)) \tag{12.2}$$

It should be noticed that, to preserve the volume of the drop, $R_0$ must depend on time too, since a relationship with $a_2$ can be found from

$$\frac{2a_2^3(t)}{35} + \frac{3a_2^2(t) R_0(t)}{5} + R_0^3(t) = R_d^3 \tag{12.3}$$

which is obtained by setting the volume of the non-oscillating drop (when $a_2 = 0$ and $R_0 = R_d$ i.e. $V_d = \frac{4}{3}\pi R_d^3$) equal to that contained by the surface defined by Eq. (12.2), which can be obtained by the Pappus rule

$$V = \pi \int_{-1}^{1} r^2 \frac{dz}{d\eta} d\eta \tag{12.4}$$

where $\eta = \cos(\theta)$ and

$$r = R(\eta) \sqrt{1 - \eta^2}; \quad z = R(\eta)\, \eta \tag{12.5}$$

as

$$V = \pi \left( \frac{8a_2^3}{105} + \frac{4a_2^2 R_0}{5} + \frac{4}{3} R_0^3 \right) \tag{12.6}$$

and, introducing the ratio $\hat{a}_2 = \frac{a_2}{R_0}$,

$$R_0^3(t) = \frac{R_d^3}{\frac{2}{35}\hat{a}_2^3(t) + \frac{3}{5}\hat{a}_2^2(t) + 1} \tag{12.7}$$

Figure 12.1 shows the shape described by Eq. (12.2) for different values of $\hat{a}_2$.

It is often said that in mode $n = 2$ the drop oscillates between prolate and oblate shapes, however the shape described by Eq. (12.2) resembles that of a spheroid only

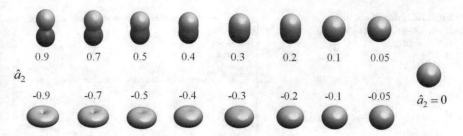

**Fig. 12.1** Drop shapes described by Eq. (12.2) as function of $\hat{a}_2$

for relatively small values of $\hat{a}_2$. The shapes described by larger values of $\hat{a}_2$ are observed in oscillating drops generated by strong initial deformation (see [5]).

It is clear that also for small values of $\hat{a}_2$ the shape is not exactly that of a spheroid, and in fact a spheroidal surface can be decomposed according to Eq. (12.1) where $a_n \neq 0$ for any $n = 2m$ (the function $R(\eta)$ is an even function). If we consider the spheroidal surface in cylindrical coordinates

$$\frac{r^2}{a_r^2} + \frac{z^2}{a_z^2} = 1 \tag{12.8}$$

the preservation of the volume imposes the conditions

$$a_r^2(t) \, a_z(t) = R_d^3 \tag{12.9}$$

Defining the aspect ratio $\varepsilon = \frac{a_z}{a_r}$ the equation transforms to

$$r^2 \varepsilon^2 + z^2 = R_d^2 \varepsilon^{4/3} \tag{12.10}$$

and in spherical coordinates, where

$$r = R \sin(\theta) ; \quad z = R \cos(\theta) \tag{12.11}$$

the same equation becomes

$$R(\theta) = \frac{R_d \varepsilon^{-1/3}}{\sqrt{1 + \cos^2(\theta) \left( \frac{1 - \varepsilon^2}{\varepsilon^2} \right)}} \tag{12.12}$$

The coefficients of the modal decomposition $R(\theta) = \sum_{k=0}^{\infty} a_k P_k(\cos(\theta))$ can be calculated exploiting the orthogonality of the Legendre polynomials; substituting $\eta = \cos \theta$

**Fig. 12.2** Map of the first three non-nil coefficients as a function of $\varepsilon$

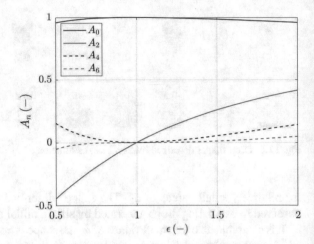

$$A_k = \frac{\int_{-1}^{1} R\left(\eta\right) P_k\left(\eta\right) d\eta}{\int_{-1}^{1} P_k^2\left(\eta\right) d\eta} = \frac{2k+1}{2} R_d \varepsilon^{-1/3} \int_{-1}^{1} \frac{P_k\left(\eta\right)}{\sqrt{1+\eta^2 \frac{1-\varepsilon^2}{\varepsilon^2}}} d\eta \qquad (12.13)$$

since $\int_{-1}^{1} P_k^2\left(\eta\right) d\eta = \frac{2}{2k+1}$. The integrals

$$I_k\left(\varepsilon\right) = \int_{-1}^{1} \frac{P_k\left(\eta\right)}{\sqrt{1+\eta^2 \frac{1-\varepsilon^2}{\varepsilon^2}}} d\eta \qquad (12.14)$$

can be analytically evaluated, defining $s = \frac{1-\varepsilon^2}{\varepsilon^2}$ the first terms are

$$I_0 = 2 \begin{cases} \frac{\sinh^{-1}\left(\sqrt{s}\right)}{\sqrt{s}} & \text{oblate} \\ \frac{\arcsin\left(\sqrt{-s}\right)}{\sqrt{-s}} & \text{prolate} \end{cases} \qquad (12.15a)$$

$$I_2 = \frac{3\sqrt{s+1}}{2s} - \frac{(3+2s)}{4s} I_0 \qquad (12.15b)$$

$$I_4 = -\frac{5\sqrt{s+1}\left(21+10s\right)}{32s^2} + \frac{3\left(35+40s+8s^2\right)}{64s^2} I_0 \qquad (12.15c)$$

$$I_6 = \frac{7\sqrt{s+1}\left(165+160s+28s^2\right)}{128s^3} - \frac{5\left(231+378s+168s^2+16s^3\right)}{256s^3} I_0 \qquad (12.15d)$$

Figure 12.2 shows the relative weight of each mode, as a function of the aspect ratio $\varepsilon$.

It can be appreciated that the weight of the modes $n > 2$ becomes negligible in comparison with that of mode $n = 2$ when $\varepsilon \to 1$, but for larger deviations from the spherical shape all the other even modes become influent; this is the reason why, for not too large deformations, the spheroidal shape may be considered a good representative of the $n = 2$ oscillation mode.

## 12.1.1 Evaporation of an Oscillating Spheroidal Drop

When a liquid drop oscillates in a gaseous environment, evaporation and possibly heat transfer should be considered since they affect the drop dynamic behaviour. A simple observation comes from the Rayleigh formula [1] to evaluate the oscillating frequency

$$\omega_n^2 = \frac{\sigma}{R_d^3 \rho^L} n (n - 1) (n + 2) \tag{12.16}$$

where $n$ is the oscillating mode; since evaporation will reduce the drop size $R_d$ the oscillation frequency of each mode must increase. But clearly the evaporation characteristics of the drop are influenced by the oscillation, since we know that the drop shape influences the vapour and heat fluxes exchanged by the drop. As we have already noticed above, there exists a huge amount of research work on drop oscillation, but only few papers can be found in the open literature specifically addressing drop oscillation under evaporating conditions (like [6–8]). The problem has been addressed mainly by numerical simulations, see for example [9–11]. A simplified analytical approach to evaluate the influence of evaporation on the oscillation frequency can be found in [11], while a way to analytically evaluate the effect of an oscillating shape on the evaporation characteristics was reported in [12]. This last approach is based on some simplifying hypotheses, like the absence of external convection (apart from the Stefan flow), constancy of thermo-physical properties, uniformity of the drop surface temperature, which are quite common when developing analytical models of drop evaporation, as we have seen in previous chapters. Besides, there are two additional assumptions: the first one is that a spheroidal surface can be considered representative of the actual drop shape during oscillation, and we have seen above the conditions under which this may be acceptable; the second one is that the evaporation flux can be calculated as the one from a deformed drop under steady-state conditions, i.e. the instantaneous values of the fluxes can be calculated from Eq. (10.143), which holds for spheroidal drops.

Under these assumptions the instantaneous evaporation rate is calculated as a function of the instantaneous deformation, as measured by the aspect ratio $\varepsilon$

$$m_{ev} = 4\pi R_d \rho D_{10} \ln (1 + B_M) \hat{\Gamma} (\varepsilon) \tag{12.17}$$

where $\hat{\Gamma} (\varepsilon)$ is given by Eq. (10.149a).

The fact that the drop is continuously changing its shape is described by assuming that $\varepsilon$ depends on time, however it is more convenient to use as main parameter the non-dimensional drop surface, defined as the ratio between the actual drop surface and that of a spherical drop having the same volume, as in Sect. 12.3, i.e.

$$\beta = \frac{A_{sd}}{A_{sh}} = \frac{1}{2\varepsilon^{2/3}} \begin{cases} 1 + \dfrac{\varepsilon^2 \ln\left(\frac{1+\sqrt{1-\varepsilon^2}}{\varepsilon}\right)}{\sqrt{1-\varepsilon^2}} & \text{Oblate} \\ 1 + \dfrac{\varepsilon^2 \arctan\left(\sqrt{\varepsilon^2-1}\right)}{\sqrt{\varepsilon^2-1}} & \text{Prolate} \end{cases} \qquad (12.18)$$

which is a univocal relation between $\varepsilon$ and $\beta$. The advantage of using $\beta$ as driving parameter consists on the fact that $\beta$ is a direct measure of the surface energy, since $E_\sigma = \sigma 4\pi R_d^2 \beta$, and that, for a low evaporating drop, the maximum value of $\beta$ in a prolate and an oblate shape should be equal when the total energy is preserved. This allows to set, as a first approximation, the following form for the time dependence of $\beta$ (see also [11])

$$\beta(t) = 1 + \Delta\beta \sin^2(\omega_2 t) \qquad (12.19)$$

where $\Delta\beta$ is the non-dimensional maximum excess area from the spherical state. The drop oscillates between a prolate shape and an oblate shape and for non-infinitesimal oscillation amplitudes the time spent in the prolate shape is larger than that spent in the oblate shape, [13], a peculiarity that is not taken into account here.

An example of the time variation of the evaporation rate, in terms of the non-dimensional parameter $\Theta$, defined as the ratio between the actual evaporation rate and that of a spherical drop having the same volume and under the same conditions, is reported in Fig. 12.3 along an oscillation period ($T_{osc} = \frac{2\pi}{\omega_2}$). The maximum value of the parameter $\beta$ during an oscillation period is taken equal to $\beta_{max} = 1.1$, which corresponds to a maximum value of $\varepsilon = 2.2$ for the prolate state and a minimum value of $\varepsilon = 0.49$ for the oblate state. In this approximate evaluation it is assumed that the loss of mass due to evaporation over an oscillating period is small enough to assume the same value of $\beta_{max}$ for both configurations. The figure shows that for these conditions the effect of oscillation leads to maximum evaporation rates that exceed that of an iso-volumic spherical drop of about 6 and 4% when the drop is in the prolate and oblate states, respectively. The values are also compared with those calculated using the correlations proposed in [11] (derived from numerical simulations) and in [9] (based on the assumed proportionality of the local vapour flux and the local mean curvature). The first correlation [11] does not account for the already observed (Sect. 10.3) different evaporation rates from oblate and prolate drops and it always overestimates the rate with a maximum deviation from the results obtained from Eq. (12.17) of about 10%. The second correlation [9] distinguishes between the two drop shapes but again it overestimates the evaporation rate along the whole oscillation period, with a maximum deviation of about 7% for the prolate shape and about 4% for the oblate one.

It is interesting to evaluate the applicability of the quasi-steady assumption, which is widely used to model the heat and mass transfer through the gaseous phase and it has also been adopted in numerical approaches.

The quasi-steady-state assumption may hold when the characteristic time scale of the evaporation process is much smaller than that of the oscillation process; the latter can be estimated (see Eq. (12.16)) as

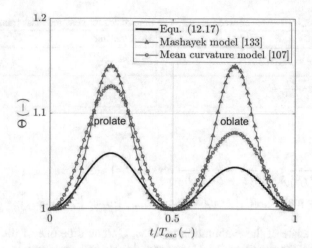

**Fig. 12.3** Transient profiles of non-dimensional evaporation rate during the oscillation period as predicted by three models

**Table 12.1** Convective, diffusive, convective-diffusive time scales for heat and mass transfer

| | Mass transfer | Heat transfer |
|---|---|---|
| Convective | $t_{conv} \approx \frac{R_d}{U_{St}}$ | $t_{conv}^{H} \approx \frac{R_d}{U_{St}}$ |
| Diffusive | $t_{diff} \approx \frac{R_d^2}{D_{10}}$ | $t_{diff}^{H} \approx \frac{R_d^2}{\alpha_{th}} = \frac{t_{diff}}{Le}$ |
| Convective-diffusive | $t_{cd} \approx \frac{D_{10}}{U_{St}^2} = \frac{t_{conv}^2}{t_{diff}}$ | $t_{cd}^{H} \approx \frac{\alpha_{th}}{U_{St}^2} = t_{cd} Le$ |

$$t_{osc} \approx \sqrt{\frac{\rho_L R_d^3}{\sigma}} \qquad (12.20)$$

The validity of the quasi-steady-state assumption can be considered acceptable when

$$t_{ev} \ll t_{osc} \qquad (12.21)$$

where $t_{ev}$ is a characteristic time-scale of the evaporation process. Since the evaporation process is a complex interaction among mass and thermal diffusion (driven by the diffusion coefficients $D_{10}$ and the thermal diffusivity $\alpha_{th}$, respectively) and Stefan flow convection (driven by a characteristic velocity $U_{St} = \frac{m_{ev}}{\beta 4\pi R_d^2 \rho_G}$), three different (but not independent) time scales can be defined from the three parameters $D_{10}$, $\alpha_{th}$ and $U_{St}$, to describe the mass and heat transfer processes (see Table 12.1). where $Le$ is the Lewis number. These time scales can be rearranged as in Table 12.2. where $\hat{m}_{ev} = \frac{m_{ev}}{4\pi R_d \rho_G D_{10}}$. From Eqs. (12.17), (12.18) and (10.149a) it can be found that, for any reasonable value of $\varepsilon$, the ratio $\frac{\beta}{\hat{m}_{ev}}$ is always larger than 1, then

**Table 12.2** Convective and convective-diffusive time scales for heat and mass transfer

| Mass transfer | Heat transfer |
|---|---|
| $t_{conv} \approx \frac{\beta R_d^2}{\hat{m}_{ev} D_{10}} = \frac{\beta t_{diff}}{\hat{m}_{ev}}$ | $t_{conv}^H \approx \frac{\beta t_{diff}^H}{\frac{\hat{m}_{ev}}{Le}}$ |
| $t_{cd} \approx \frac{\beta^2 R_d^2}{\hat{m}_{ev}^2 D_{10}} = \frac{\beta^2 t_{diff}}{\hat{m}_{ev}^2}$ | $t_{cd}^H \approx \frac{\beta^2 t_{diff}^H}{\left(\frac{\hat{m}_{ev}}{Le}\right)^2}$ |

$$\left(\frac{\beta}{\hat{\Gamma}(\beta)}\right)^2 \frac{R_0^2}{\ln^2(1+B_M) D_{10}} \max[1, Le]$$
$$= \max\left[t_{cd}, t_{cd}^H\right] > \max\left[t_{conv}, t_{conv}^H\right] > \max\left[t_{diff}, t_{diff}^H\right] \tag{12.22}$$

The time scale of the evaporation process, $t_{ev}$, could be one of the three above mentioned, but the condition (12.21) is certainly satisfied when $\max\left[t_{cd}, t_{cd}^H\right] \ll t_{osc}$.

In [12] this analysis was quantitatively applied to define a region on a $T_\infty - T_s$ graph where the assumption (12.21) holds for any value of the equivalent drop radius $R_d$.

The graphs of Fig. 12.4 report the range of drop and gas temperatures, relative to each drop size, where the above discussed condition may hold, for different drop species (water, methanol, ethanol, acetone, benzene, n-hexane, n-octane, n-dodecane and n-hexadecane). For any given curve corresponding to a drop size, the region where the condition holds is that on the right, i.e. for any drop temperature (on the graph abscissa), the gas temperature must be larger than that defined by the mentioned curve. The results of this analysis show that the inequality (12.21) may become acceptable for small drops (few tenths of micron) in hot gaseous environment and for high volatility fluids, conditions that are reached in many applicative cases.

## 12.2  Unsteady Heat and Mass Transfer in a Shrinking Spherical Drop

In many commonly used approaches to model heating and evaporation of a single-component drop into a gaseous environment, the drop is treated as a lump mass with no internal structure, assuming implicitly that the temperature distribution inside the drop is uniform. To relieve this assumption, the heat transfer inside the drop must be modelled, accounting for the time variation of the temperature field and for the variation of the drop size due to evaporation. To date, the most advanced analytical treatment of this problem is that of Sazhin and co-workers, reported in [14, 15] and explained in more details in [16]. It will be briefly described here, on the basis of the cited publications. Spherical symmetry is assumed and, in a spherical coordinate system, the problem can be reduced to the solution of the equation

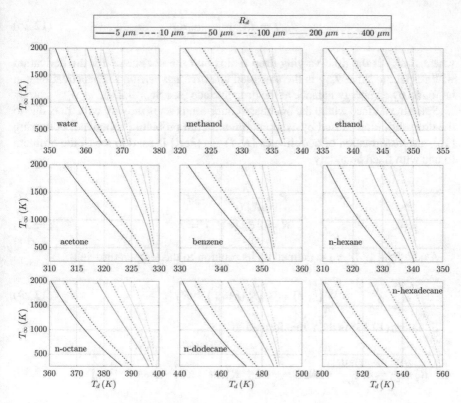

**Fig. 12.4** Range of drop and gas temperatures, as function of drop size and composition, where the quasi-steady condition may hold

$$\frac{\partial T}{\partial t} = \alpha \left( \frac{\partial^2 T}{\partial R^2} + \frac{2}{R} \frac{\partial T}{\partial R} \right) + Q\,(R) \tag{12.23}$$

where $Q(R)$ is a source term that may be used to take into account the effect of radiative heating of the semi-transparent droplet, as in [17]. To solve Eq. (12.23), initial conditions must be set

$$T\,(R, 0) = T_0\,(R) \tag{12.24}$$

as well as boundary conditions on the drop centre and on the drop surface

$$\left( \frac{\partial T}{\partial R} \right)_{R=0} = 0; \quad \left( \frac{\partial T}{\partial R} \right)_{R=R_d(t)} + \frac{h_c}{k_L} T = \frac{h_c}{k_L} T_{eff}\,(t) \tag{12.25}$$

The function $T_{eff}\,(t)$ on the RHS of the boundary condition at the drop surface is used to account for the effect of evaporation on the energy balance of the drop by setting

$$T_{eff} = T_\infty + \frac{\rho_L}{h_c} h_{lv} \dot{R}_d \qquad (12.26)$$

where $R_d(t)$ is the time-varying drop radius and the dot stands for the derivation with respect to time. $T_{eff}$ is the so-called *effective temperature*. This equation can be found by an energy balance at the drop surface (see Sect. 9.2.4).

Since the surface where the boundary condition is set is moving with time, due to the drop shrinking caused by evaporation, a Stefan problem is obtained. Following [16], a transformation of coordinates $(R, t) \to (\hat{r}, t')$ is used to recast the problem to one with fixed boundary

$$\hat{r} = \frac{R}{R_d(t)}; \ \tau = t \qquad (12.27)$$

$$R = \hat{r} R_d(\tau); \ t = \tau \qquad (12.28)$$

A new field $W(\hat{r}, \tau)$ is defined by its relation to the temperature field

$$W(\hat{r}, \tau) = T\left(R(\hat{r}), \tau\right) \hat{r} \, R_d^{3/2} e^{\frac{R_d \dot{R}_d}{4\alpha} \hat{r}^2} \qquad (12.29)$$

Equation (12.23) is then transformed to

$$R_d^2 \frac{\partial W}{\partial \tau} = \alpha \frac{\partial^2 W}{\partial r^2} + \ddot{R}_d R_d^3 \frac{r^2}{4\alpha} W + R_d^2 RP(R) \left(\frac{1}{R_d^{1/2}} e^{-\frac{R_d \dot{R}_d}{4\alpha} \hat{r}^2}\right)^{-1} \qquad (12.30)$$

and the boundary condition on the drop surface becomes

$$\left(\frac{\partial W}{\partial \hat{r}}\right)_{\hat{r}=1} + H_0 W(1, t) = \mu_0(t) \qquad (12.31)$$

where $R_{d,0} = R_d(0)$, $H_0(\tau) = \left(\frac{h_c R_d}{k_L} - 1 - \frac{R_d \dot{R}_d}{2\alpha}\right)$ and $\mu_0(t) = R_d^{5/2} e^{\frac{R_d \dot{R}_d}{4\alpha}} \frac{h_c}{k_L} T_{eff}$; in the new set of coordinates this condition is set on the fixed boundary $\hat{r} = 1$. The initial conditions is correspondingly transformed to

$$W(\hat{r}, 0) = \hat{r} T_0\left(\hat{r} R_{d,0}\right) R_{d,0}^{3/2} e^{\frac{R_{d,0} \dot{R}_d(0)}{4\alpha} \hat{r}^2} = W_0(\hat{r}) \qquad (12.32)$$

An explicit analytical solution of Eq. (12.30) was found in [14, 15] for some special cases. The effect of radiation was assumed negligible ($Q(R) = 0$), which is justified for many engineering applications, the function $H_0$ was assumed constant, i.e. $H_0(\tau) = H_0 = const.$, which is justified for conditions like, for example, those found in sprays for internal combustion engines (see [14] for details). The relaxation of this last condition was analysed in [16], Appendix D, and leads to a more complex integral representation of the solution. The analytical explicit representation of the solution that will be reported below was found for $\ddot{R}_d = 0$ [15] i.e. when the radius can be assumed to vary linearly with time

$$R_d(t) = R_{d,0}(1 + at) \tag{12.33}$$

an assumption that may be justified when a relatively short time interval is analysed. A more involved expression of the solution for a general time-dependence of the drop radius is described in [16], pp. 123–126, to which the interested reader should refer for a more detailed analysis. Under the above mentioned assumptions, Eq. (12.30) simplifies to

$$R_d^2 \frac{\partial W}{\partial \tau} = \alpha \frac{\partial^2 W}{\partial \hat{r}^2} \tag{12.34}$$

and the inhomogeneous boundary condition on $\hat{r} = 1$ is transformed into an homogenous one by the following transformation:

$$W = Z + \frac{\mu_0(t)}{1 + H_0}\hat{r} \tag{12.35}$$

after which the Eq. (12.34) transforms to

$$R_d^2 \frac{\partial Z}{\partial t} = \alpha \frac{\partial^2 Z}{\partial \hat{r}^2} - R_d^2 \frac{\dot{\mu}_0(t)}{1 + H_0}\hat{r} \tag{12.36}$$

and the boundary condition on $\hat{r} = 1$ to

$$\left(\frac{\partial Z}{\partial \hat{r}}\right)_{\hat{r}=1} + H_0 Z(1, \tau) = 0 \tag{12.37}$$

The solution is then sought under the form

$$Z(\hat{r}, \tau) = \sum_{n=1}^{\infty} c_n(\tau) z_n(\hat{r}) \tag{12.38}$$

where the functions $z_n(\hat{r})$ are the solution of the eigenvalue problem

$$\frac{d^2 z_n}{d\hat{r}^2} + \lambda^2 z_n = 0 \tag{12.39}$$

$$z_n(0) = 0; \quad \left(\frac{\partial z_n}{\partial \hat{r}}\right)_{\hat{r}=1} + H_0 z_n(1) = 0 \tag{12.40}$$

i.e.

$$z_n(\hat{r}) = \sin(\lambda_n \hat{r}) \tag{12.41}$$

with eigenvalues given by the solution of the transcendental equation

$$\lambda_n \cos(\lambda_n) + H_0 \sin \lambda_n = 0 \tag{12.42}$$

These functions are orthogonal for the inner product $\langle f, g \rangle = \int_0^1 f\, g\, d\hat{r}$, as it can be easily verified by direct integration, and they can be used to expand the function $-\frac{\hat{r}}{1+H_0}$ in Eq. (12.36) as

$$-\frac{\hat{r}}{1+H_0} = \sum_{n=1}^{\infty} f_n z_n(\hat{r}) \tag{12.43}$$

where

$$f_n = -\frac{\sin(\lambda_n^2)}{\|z_n\|^2 \lambda_n^2} \tag{12.44}$$

$$\|z_n\|^2 = \langle z_n, z_n \rangle = \frac{1}{2}\left(1 - \frac{\sin 2\lambda_n}{2\lambda_n}\right) = \frac{1}{2}\left(1 + \frac{H_0}{\left[H_0^2 + \lambda_n^2\right]}\right) \tag{12.45}$$

Substituting Eqs. (12.38), (12.43) and (12.44) into Eq. (12.36) yields

$$\sum_{n=1}^{\infty}\left[R_d^2 \frac{\partial c_n}{\partial t} + \alpha \lambda_n^2 c_n - R_d^2 \dot{\mu}_0(t)\, f_n\right] z_n = 0 \tag{12.46}$$

where Eq. (12.39) has been used. From the properties of the complete set of orthogonal functions $z_n$ it stems that

$$R_d^2 \frac{\partial c_n}{\partial t} + \alpha \lambda_n^2 c_n - R_d^2 \dot{\mu}_0(t)\, f_n = 0 \tag{12.47}$$

Since the initial conditions for Eq. (12.36) are

$$Z(\hat{r}, 0) = W_0(\hat{r}) - \frac{\mu_0(0)}{1+H_0}\hat{r} \tag{12.48}$$

and the function $W_0(\hat{r})$ can be expanded in terms of $z_n$ as

$$W_0(\hat{r}) = \sum w_n z_n \tag{12.49}$$

then the initial conditions for Eqs. (12.47) are

$$c_n(0) = w_n + \mu_0(0)\, f_n \tag{12.50}$$

The general solution of Eq. (12.47) is given in [16] as

$$c_n(t) = c_n(0) \, e^{\frac{\alpha\lambda_n^2}{aR_{d,0}^2}\left(\frac{1}{1+at}-1\right)} + f_n \int_0^t \frac{d\mu_0(\tau)}{d\tau} e^{\frac{\alpha\lambda_n^2}{aR_{d,0}^2}\left(\frac{1}{1+at}-\frac{1}{1+a\tau}\right)} d\tau \qquad (12.51)$$

Finally the solution for the temperature field inside the drop can then be written as

$$T(R,t) = e^{-\frac{R_{d,0}a}{4\alpha R_d(t)}R^2} \left[ \sum_{n=1}^{\infty} c_n(t) \frac{1}{R_d^{1/2}} \frac{\sin\left(\frac{\lambda_n R}{R_d(t)}\right)}{R} + \frac{\mu_0(t)}{1+H_0} \frac{1}{R_d^{3/2}} \right] \qquad (12.52)$$

The boundary condition on $R = 0$ is satisfied, thanks to the properties of the functions $\frac{sin(x)}{x}$ (i.e. $\lim_{x \to 0} \frac{sin(x)}{x} = 0$) appearing in the series.

The application of this analytical modelling to the study of the evaporation of fuel drops showed that the effect of taking into account the internal temperature distribution and drop shrinking may become not negligible for many engineering applications, in particular with reference to the drop lifetime. As an example, in [16] it was shown that for a $5\,\mu m$ n-dodecane drop the estimated lifetime increases by more than 5%.

## 12.3   Unsteady Heat and Mass Transfer Outside a Shrinking Drop

In this section we analyse a way to model the unsteady heat and mass transfer through the gas phase when a drop is suddenly injected into a hot gas. We will consider the case of a single component spherical drop evaporating in a gaseous environment and we will assume a constant reference value for gaseous mixture properties (density, diffusion coefficient, conductivity and heat capacity). The species conservation equations for the transient problem, in the region outside the drop, are

$$\rho_G \frac{\partial \chi^{(p)}}{\partial t} = -\nabla_j n_j^{(p)}; \quad p = (0, 1) \qquad (12.53)$$

where the summation convention has been used, $\rho_G$ is the mass density of the gas mixture; the species fluxes, in a spherical coordinate system, can be written as

$$n_R^{(p)} = \rho_G v_R \chi^{(p)} - D_{10}\rho_G \frac{\partial \chi^{(p)}}{\partial R} \qquad (12.54)$$

where $v_R$ is the mixture radial velocity (mass averaged). The continuity equation can be derived summing Eqs. (12.53) over $p$ and integrating over the drop surface, yielding the following general solution

$$\rho_G v_R = \frac{m_T(t)}{4\pi R^2} \tag{12.55}$$

where $m_T(t) = m^{(0)}(t) + m^{(1)}(t)$ is the total mass flow rate, $m^{(0)}$ and $m^{(1)}$ are the mass flow rates of the ambient gas ($p = 0$) and of the evaporating species ($p = 1$), respectively. The substitution of Eqs. (12.54) and (12.55) into Eq. (12.53) yields, for $p = 1$

$$\rho_G \frac{\partial \chi^{(1)}}{\partial t} = -\frac{m_T(t)}{4\pi R^2} \frac{\partial \chi^{(1)}}{\partial R} + D_{10}\rho_G \left( \frac{\partial^2 \chi^{(1)}}{\partial R^2} + \frac{2}{R} \frac{\partial \chi^{(1)}}{\partial R} \right) \tag{12.56}$$

while the equation for $p = 0$ can be dismissed since it is linearly dependent on Eqs. (12.56) and (12.55). The energy equation for the gas mixture is Eq. (9.40), observing that for the single component drop evaporation the gaseous mixture is binary, then

$$\frac{\partial \rho_G \hat{h}}{\partial t} + \nabla_k \left( n_k^{(0)} H^{(0)} + n_k^{(1)} H^{(1)} \right) = k_G \nabla^2 T \tag{12.57}$$

and since $n_k^{(T)} = \rho_G v_k = n_k^{(0)} + n_k^{(1)}$ and $n_k^{(1)} = \rho_G v_k - n_k^{(0)}$ then

$$n_k^{(0)} H^{(0)} + n_k^{(1)} H^{(1)} = \rho_G v_k H^{(1)} + n_k^{(0)} \left( H^{(0)} - H^{(1)} \right) \tag{12.58}$$

and the energy equation (12.57) becomes

$$\frac{\partial \rho_G \hat{h}}{\partial t} + \nabla_k \left( \rho_G v_k H^{(1)} \right) + \nabla_k \left[ n_k^{(0)} \left( H^{(0)} - H^{(1)} \right) \right] = k_G \nabla^2 T \tag{12.59}$$

We have seen that the flux $n_k^{(0)}$ is nil for a non-shrinking drop, since the diffusion of component $p = 0$ into the liquid drop is assumed negligible and the drop surface is still. This is not true in the present case; however the assumption that the vapour flux $n_k^{(1)}$ (due to evaporation) is much larger than the gas flux $n_k^{(0)}$ (which it only due to drop shrinking since gas diffusion inside the liquid is neglected) allows to neglect the third term on the LHS, and, after assuming ideal gas behaviour of both components of the gas mixture, the energy equation, in a spherical coordinate system, becomes

$$\rho_G \hat{c}_{pG} \frac{\partial T}{\partial t} = -\frac{m_T(t)}{4\pi R^2} \hat{c}_p^{(1)} \frac{\partial T}{\partial R} + k_G \left( \frac{\partial^2 T}{\partial R^2} + \frac{2}{R} \frac{\partial T}{\partial R} \right) \tag{12.60}$$

where Eq. (12.55) has been used, $\hat{c}_{pG}$ is the specific heat of the mixture and $\hat{c}_p^{(1)}$ that of the pure species $p = 1$. Since the problem is one-dimensional, i.e. only radial vector components are non-nil, in the following the velocity and the fluxes will be written without the subscript that indicate the vector component (i.e. $v$ and $n^{(p)}$ instead of $v_R$ and $n_R^{(p)}$).

## 12.3.1  Initial and Boundary Conditions

The evaporating drop is assumed to be injected into a hot environment, the initial conditions for the temperature and vapour mass fraction distribution through the gas phase will be therefore considered uniform and equal to the value at infinity

$$\chi^{(1)}(R, 0) = \chi_\infty^{(1)} \text{ for } R > R_d \tag{12.61a}$$

$$T(R, 0) = T_\infty \text{ for } R > R_d \tag{12.61b}$$

The boundary conditions are assumed to be generally time dependent, but uniform over the drop surface and of Dirichlet type

$$\chi^{(1)}(R = R_d(t)) = \chi_s^{(1)}(t); \quad \chi^{(1)}(R = \infty) = \chi_\infty^{(1)} \tag{12.62a}$$

$$T(R = R_d(t)) = T_s(t); \quad T(R = \infty) = T_\infty \tag{12.62b}$$

It is worth to notice that the function $m_T(t)$ appearing in Eqs. (12.55), (12.67) and (12.68) is fully determined by the drop radius shrinking. In fact, let consider the species jump balance at the drop surface (see Chap. 6)

$$\rho_L^{(p)}\left[v_L^{(p)}(R_d) - V_n\right] - \rho_G^{(p)}\left[v_G^{(p)}(R_d) - V_n\right] = 0 \tag{12.63}$$

where $V_n$ is the normal component of the surface velocity, the subscripts $G$ and $L$ are used for the values inside the gas and the liquid phase respectively. We have already seen that absorption of the species $p = 0$ is negligible and the velocity of the liquid phase is nil, then

$$n^{(0)} = \rho_G^{(0)} v_G^{(0)} = \rho_G^{(0)} \dot{R}_d \tag{12.64}$$

$$n^{(1)} = \rho_G^{(1)} v_G^{(1)} = \left(\rho_G^{(1)} - \rho_L\right) \dot{R}_d \tag{12.65}$$

where $V_n = \dot{R}_d$, i.e. interface velocity $V_n$ is equal to the time derivative of the drop radius and $m_T(t)$ can be obtained as

$$m_T(t) = 4\pi R_d^2 \rho_G v_G = 4\pi R_d^2 (\rho_G - \rho_L) \dot{R}_d \tag{12.66}$$

To simplify the problem it is useful to introduce the non-dimensional variables $X = \frac{\chi^{(1)} - \chi_\infty^{(1)}}{\chi_0^{(1)} - \chi_\infty^{(1)}}$; $\hat{T} = \frac{T - T_\infty}{T_0 - T_\infty}$ where $T_0 = T_s(0)$, $\chi_0^{(1)} = \chi_s^{(1)}(0)$; then the conservation equations (12.56) and (12.60) and the initial and boundary conditions (12.61a), (12.62a) yield the following problem

$$\rho_G \frac{\partial X}{\partial t} = -\frac{m_T(t)}{4\pi R^2} \frac{\partial X}{\partial R} + D_{10}\rho_G \nabla^2 X \tag{12.67}$$

$$\rho_G \frac{\partial \hat{T}}{\partial t} = -\frac{m_T(t)}{4\pi R^2} \frac{\hat{c}_p^{(1)}}{\hat{c}_{pG}} \frac{\partial \hat{T}}{\partial R} + \frac{k_G}{\hat{c}_{pG}} \nabla^2 \hat{T} \tag{12.68}$$

$$X(R_d(t)) = X_s(t); \quad X(R = \infty) = 0; \quad \hat{T}(R_d(t)) = T_s(t); \quad \hat{T}(R = \infty) = 0 \tag{12.69}$$

where $X_s(t) = \frac{X_s^{(1)} - X_\infty^{(1)}}{X_0^{(1)} - X_\infty^{(1)}}$ and $\hat{T}_s(t) = \frac{T_s(t) - T_\infty}{T_0 - T_\infty}$. This differential problem, where the interface between liquid and gas moves with time due to drop radius shrinking during the evaporation process, is again a Stefan problem since the boundary conditions are set on a moving boundary.

Again the problem can be transformed to a fixed boundary one by a change of variables; to this end consider the following coordinates transformation

$$\zeta = \zeta(r, t) = \frac{R_d(t)}{R} = \frac{R_{d,i}\, \varphi(t)}{R} \tag{12.70a}$$

$$\tau = \tau(r, t) = \frac{t D_v}{R_{d,i}^2} \tag{12.70b}$$

where $R_{d,i} = R_d(0)$ and $R_d(t) = \varphi(t) R_{d,i}$.

In the following, to distinguish from the derivative with respect to time $t$, the symbol $\overset{\circ}{F}$ is used for the derivative of the function $F$ with respect to the non-dimensional time $\tau$, i.e. $\overset{\circ}{F} = \frac{dF}{d\tau}$ or $\overset{\circ}{F} = \left(\frac{\partial F}{\partial \tau}\right)_\zeta$. Since $\dot{R}_d = \overset{\circ}{\varphi}\frac{D_{10}}{R_{d,i}}$, where $\overset{\circ}{\varphi} = \frac{d\varphi}{d\tau}$, the function $m_T(t)$ (Eq. (12.66)) can now be written as

$$m_T(\tau) = 4\pi(\rho_G - \rho_L) D_{10} R_{d,i} \varphi^2(\tau)\, \overset{\circ}{\varphi}(\tau) \tag{12.71}$$

and the energy and species conservation equations become

$$\overset{\circ}{X} = \left[\frac{\widehat{m}_T(\tau)\zeta^4}{\varphi^3(\tau)} - \frac{\overset{\circ}{\varphi}\zeta}{\varphi(\tau)}\right] X' + \frac{\zeta^4}{\varphi^2(\tau)} X'' \tag{12.72a}$$

$$\overset{\circ}{\hat{T}} = \left[c_0 \frac{\widehat{m}_T(\tau)\zeta^4}{\varphi^3(\tau)} - \frac{\overset{\circ}{\varphi}\zeta}{\varphi(\tau)}\right] \hat{T}' + Le \frac{\zeta^4}{\varphi^2(\tau)} \hat{T}'' \tag{12.72b}$$

where $c_0 = \frac{\hat{c}_p^{(1)}}{\hat{c}_{p\,G}}$, $Le = \frac{k_G}{\rho_G \hat{c}_{pG} D_{10}}$ is the Lewis number, the prime means derivation with respect to $\zeta$ ($X' = \left(\frac{\partial X}{\partial \zeta}\right)_\tau$), and $\widehat{m}_T = \frac{m_T}{4\pi \rho_G D_{10} R_{d,i}} = \left(1 - \frac{\rho_L}{\rho_G}\right)\varphi^2(\tau)\,\overset{\circ}{\varphi}(\tau)$.

The boundary conditions now become

$$X\left(\zeta=1,\tau\right)=X_s\left(\tau\right); \ X\left(\zeta=0,\tau\right)=0 \tag{12.73}$$

$$\hat{T}\left(\zeta=1,\tau\right)=\hat{T}_s\left(\tau\right); \ \hat{T}\left(\zeta=0,\tau\right)=0 \tag{12.74}$$

and, in this new coordinate system, this is now a fixed boundary problem. We can now notice that the solution of this problem depends on three time-dependent functions $X_s\left(\tau\right)$, $\hat{T}_s\left(\tau\right)$ and $\varphi\left(\tau\right)$.

Assuming equilibrium conditions at the drop surface, the first one, $X_s\left(\tau\right)$, can be written as a function of the second one, $\hat{T}_s\left(\tau\right)$; in fact, in such a case, $\chi_s^{(1)}$ is a unique function of the surface temperature

$$\chi_s^{(1)} = \frac{\rho_{G,s}^{(1)}}{\rho_G} = \frac{P_{v,s}\left(T_s\right)Mm^{(1)}}{T_s\bar{R}\rho_G} \tag{12.75}$$

and then $X_s\left(\tau\right)=X_s\left(T_s\left(\tau\right)\right)$; $P_{v,s}\left(T_s\right)$ is the saturation vapour pressure at temperature $T_s$ and $\bar{R}$ is the universal gas constant. The two remaining functions, $\hat{T}_s\left(\tau\right)$ and $\varphi\left(\tau\right)$, can be defined by solving the mass and energy balance equations over the liquid drop.

In heat and mass transfer, the problem set by the simultaneous solution of conservation equation in two connected phases is called *conjugate problem* (see [18]), and it is usually a quite formidable task to solve it analytically and even numerically. Here we will simplify this task by performing integral mass and energy balances over the liquid drop.

## 12.3.2  Energy and Mass Balances over the Liquid Drop

The mass balance over the single component liquid drop is simplified by the assumption of nil liquid velocity (no recirculation), and it reduces to a simple balance of the flux of component $p=1$ at the interface. The flux $n^{(1)}$ at drop surface can be evaluated from Eq. (12.65)

$$n^{(1)} = \left(\rho_G^{(1)} - \rho_L\right)\dot{R}_d \tag{12.76}$$

but it can also be found by using Eq. (12.54) as

$$n^{(1)} = \rho_G v_G \chi_s^{(1)} - \rho_G D_{10}\left(\frac{\partial\chi^{(1)}}{\partial R}\right)_{R=R_d} \tag{12.77}$$

Transforming Eqs. (12.76) and (12.77) into the new coordinate system (Eqs. (12.70)), eliminating $n^{(1)}$ and using the definition of $X$ yields

$$\mathring{\varphi}\varphi = -\hat{\rho}B_0\frac{W\left(\tau\right)}{1+B_0\left[1-X_s\left(\tau\right)\right]} \tag{12.78}$$

where $\hat{\rho} = \frac{\rho_G}{\rho_L}$, $B_0 = \frac{\chi_0^{(1)} - \chi_\infty^{(1)}}{1 - \chi_0^{(1)}}$ and $W(\tau) = \left(\frac{\partial X}{\partial \zeta}\right)_{\zeta=1}$ is a function of the non-dimensional time $\tau$.

Equation (12.78) can be integrated with the condition $\varphi(0) = 1$ to yield the following analytical expression

$$\varphi(\tau) = \left\{1 - 2\hat{\rho}B_0 \int_0^\tau \frac{W(s)}{1 + B_0[1 - X_s(s)]}ds\right\}^{1/2} \tag{12.79}$$

and it is clear that the evaluation of the radius shrinking needs the evaluation of the function $W(\tau) = \left(\frac{\partial X}{\partial \zeta}\right)_{\zeta=1}$ by the solution of the problem (12.72).

The integral energy equation over the drop can be written as (see Sect. 9.2.3)

$$M_d \hat{c}_{pL} \frac{d\overline{T}_d(t)}{dt} = \dot{Q} - m_{ev}h_{LV} \tag{12.80}$$

where $\overline{T}_d$ is the instantaneous average drop temperature $\overline{T}_d = \frac{\int_{V_d} T\,dV}{V_d}$, $V_d = \frac{4}{3}\pi R_d^3$ is the drop volume and $h_{LV}$ is the latent heat of vaporisation of species $p = 1$. In this equation, $m_{ev}$ is obtained from the mass balance

$$m_{ev} = -\frac{dM_d}{dt} = -4\pi R_d^2 \rho_L \dot{R}_d = -4\pi R_{d,i} \rho^{(L)} D_{10} \varphi^2 \overset{\circ}{\varphi} \tag{12.81}$$

and the heat rate can be defined as

$$\dot{Q} = -4\pi R_d^2 k_G \left(\frac{\partial T}{\partial r}\right)_{r=R_d} = -4\pi R_{d,i} k_G \varphi^2(t) \left(\frac{\partial \hat{T}}{\partial \zeta}\right)_{\zeta=1}(T_0 - T_\infty) \tag{12.82}$$

The closure of the problem is obtained by assuming that the drop surface temperature is equal to the average drop temperature, $T_s(t) = T_d(t)$, which is equivalent to assume the infinite conductivity approximation [16]. Substitution of Eq. (12.82) into Eq. (12.80) yields

$$\frac{1}{3}\varphi \overset{\circ}{\hat{T}}_s = -\gamma \hat{\rho} Le Z(\tau) + \overset{\circ}{\varphi}\frac{1}{Ja} \tag{12.83}$$

where $Z = \left(\frac{\partial \hat{T}}{\partial \zeta}\right)_{\zeta=1}$, $Ja = \frac{\hat{c}_{pL}(T_0 - T_\infty)}{h_{LV}}$ is the Jakob number (named after the German physicist Max Jakob, 1879–1955) and $\gamma = \frac{\hat{c}_{pG}}{\hat{c}_{pL}}$. The integration of Eq. (12.83) yields the following analytical expression

$$\hat{T}_s(\tau) = \int_0^\tau \left[\frac{3}{Ja}\frac{\overset{\circ}{\varphi}}{\varphi(s)} - 3\hat{\rho}Le\frac{Z(s)}{\varphi(s)}\right]ds + \hat{T}_s(0) \tag{12.84}$$

The connection with the problem in the gas phase is given by the two functions $W(\tau) = \left(\frac{\partial X}{\partial \zeta}\right)_{\zeta=1}$ and $Z = \left(\frac{\partial \hat{T}}{\partial \zeta}\right)_{\zeta=1}$.

### 12.3.3 A Comparison with a Quasi-steady Solution

As already pointed out, the problem set by Eqs. (12.72) and (12.79), (12.84) is a quite formidable one and, at present, no analytical solution has been obtained, despite of the simplification introduced by the integral approach to the conservation equations for the liquid phase. In [19] this problem was solved numerically implementing an implicit finite difference scheme. The number of grid points in the range $\zeta \in [0, 1]$ was set equal to 100, and a grid independence study assured that with a number of grid points larger than 50 the accuracy in evaluating the temperature field was better than 0.02 K, and analogously for the mass fraction field. The initial conditions set by the problem have a discontinuity in $\zeta = 1$; to avoid the instability caused by this discontinuity at the initial time $\tau = 0$, the initial conditions were implemented using smoother functions of the form

$$X(\zeta, 0) = e^{-\frac{1}{\delta_X}\left(\frac{1-\zeta}{\zeta}\right)} \tag{12.85}$$

$$\hat{T}(\zeta, 0) = e^{-\frac{1}{\delta_T}\left(\frac{1-\zeta}{\zeta}\right)} \tag{12.86}$$

where $\delta_X$ and $\delta_T$ were chosen sufficiently small (equal to $10^{-3}$) to approximate the real initial condition to an acceptable extent; a parametric analysis was performed to assure independence of the numerical results of the value of $\delta_X$ and $\delta_T$, (refer to [19] for further details). Samples of the results reported in [19] are shown in Fig. 12.5; time evolution of non-dimensional temperature and mass fraction distributions are reported for a n-octane drop injected at 300 K in air at 773 K and 1 bar.

To notice that, due to the chosen definition of non-dimensional temperature, since the drop temperature increases, the non-dimensional temperature at the drop surface ($\zeta = 1$) decreases from the initial value, while the non-dimensional vapor mass fraction increases.

The values of the gradient of the non-dimensional temperature and mass fraction distributions at the drop surface (functions $W(\tau) = \left(\frac{\partial X}{\partial \zeta}\right)_{\zeta=1}$ and $Z(\tau) = \left(\frac{\partial \hat{T}}{\partial \zeta}\right)_{\zeta=1}$) predicted by this model are compared, in Fig. 12.6, to those predicted by the quasi-steady model, i.e. considering a temperature and vapour distribution through the gas corresponding, at each time, to the steady-state solution of the energy and the species equations. Large differences, particularly at the initial stage of the process, are observed and as a consequence, heat and evaporation rates at the beginning of the process are much larger than those predicted by a quasi-steady approximation. Heat and evaporation rates decrease rapidly to reach values during the main evaporation period that are lower than those predicted by quasi-steady approximation.

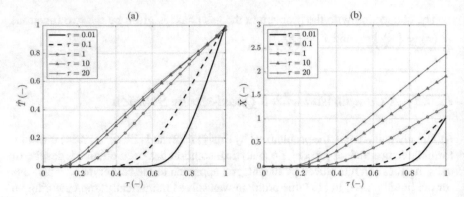

**Fig. 12.5** Time evolution of non-dimensional temperature and mass fraction distributions for an n-octane drop injected at 300 K in air at 773 K and 1 bar

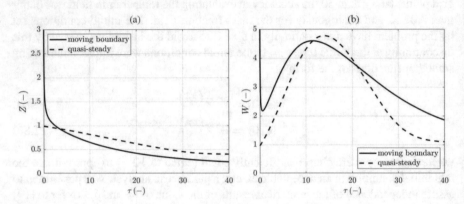

**Fig. 12.6** Gradients of temperature and mass fraction distributions at the drop surface predicted by assuming moving or quasi-steady boundaries

The results reported in this and the previous sections show that taking into account the time-dependence of the phenomena that characterise drop evaporation, namely unsteady heat and mass transfer and drop shrinking, yields non neglectful differences from those obtained under a quasi-steady approximation.

## 12.4   Transient Heat Conduction in Ellipsoids of Revolution

The solution of the energy equation in a spherical drop, which was discussed in Sect. 12.2, allowed us to take into account the effect of a non-uniform distribution of temperature over the entire drop, thus eliminating the rather unphysical assumption of infinite conductivity, which is often used when drop evaporation is analytically modelled. An extension to a spheroidal drop is the natural continuation of this line of development, however it immediately faces quite tough mathematical problems. In this section we will analyse the transient heat conduction in spheroids, assuming that the shape does not change with time. For an evaporating drop, the assumption of a steady non-spherical shape may appear too limiting, since a non-spherical drop is expected to deform under the effect of surface tension at least, thus changing the shape with time. It can be observed that a drop floating in a moderate convective environment may attain a relatively stable oblate shape [20–22], which, to some extent, may justify the interest in heat transfer in a spheroidal body of fixed shape. It is also worth to notice that, at present, analytical solutions for the heat transfer in a spheroid deforming with time are not available.

The first attempt to find a solution for the unsteady heat transfer in a spheroid dates back to 1873, when Mathieu (Émile Léonard Mathieu, a French mathematician, 1835–1890) reported an approximate solution (see [23], p. 267) of this problem. The exact solution was later given by Niven (Charles Niven, a Scottish mathematician and physicist, 1845–1923) in his paper [24], for stationary Dirichlet and Robin conditions at the surface and this was the first time that the so-called *spheroidal wave functions* were described and used.

### 12.4.1 The Energy Equation in Spheroidal Coordinates

The energy conservation equation, when the fluid velocity is nil and the thermo-physical properties are assumed constant is given by (cfr Sect. 9.2.1)

$$\frac{\partial T}{\partial t} - \alpha \nabla^2 T = 0 \tag{12.87}$$

Assuming that the body shape is a spheroid (either oblate or prolate) the most convenient coordinate systems are the spheroidal (either oblate or prolate) coordinates, which are defined by the equations (see Chap. 4 for all the details):

|  | Prolate | Oblate |  |

$$
\begin{aligned}
&\text{Prolate} && \text{Oblate} \\
x &= a\sqrt{(\zeta^2 - 1)(1 - \eta^2)}\cos\varphi & x &= a\sqrt{(\zeta^2 + 1)(1 - \eta^2)}\cos\varphi \\
y &= a\sqrt{(\zeta^2 - 1)(1 - \eta^2)}\sin\varphi & y &= a\sqrt{(\zeta^2 + 1)(1 - \eta^2)}\sin\varphi \\
z &= a\zeta\eta & z &= a\zeta\eta
\end{aligned}
\tag{12.88}
$$

In these coordinate systems, Eq. (12.87) transforms to

$$\frac{\partial T}{\partial t} - \alpha \frac{1}{g^{1/2}} \left\{ \frac{\partial}{\partial \zeta}\left[ W^{(\zeta)}\frac{\partial T}{\partial \zeta}\right] + \frac{\partial}{\partial \eta}\left[ W^{(\eta)}\frac{\partial T}{\partial \eta}\right] + \frac{\partial}{\partial \varphi}\left[ W^{(\varphi)}\frac{\partial T}{\partial \varphi}\right] \right\} = 0 \tag{12.89}$$

(the explicit form of the functions $W^{(k)}$ and $g^{1/2}$ are reported in Chap. 4).

In the present analysis we will assume that the boundary conditions are uniform at the outer surface of the spheroid, and that the initial conditions are independent of the $\varphi$-coordinate, then we can search for a solution that is independent of $\varphi$. Using the method of separation of variables, such solution can then be searched under the form

$$\Psi = e^{-\alpha\omega^2 t}\Phi\left(\zeta, \eta\right) \tag{12.90}$$

where the function $\Phi\left(\zeta, \eta\right)$ satisfies the Helmholtz equation

$$\nabla^2\Phi\left(\zeta, \eta\right) + \omega^2\Phi\left(\zeta, \eta\right) = 0 \tag{12.91}$$

as it can be verified by substitution.

The problem is then reduced to the solution of the Helmholtz equation, that in these coordinate systems has the form

Prolate $\dfrac{\partial}{\partial\zeta}\left[\left(\zeta^2 - 1\right)\dfrac{\partial\Phi}{\partial\zeta}\right] + \dfrac{\partial}{\partial\eta}\left[\left(1 - \eta^2\right)\dfrac{\partial\Phi}{\partial\eta}\right] + a^2\omega^2\Phi\left(\zeta^2 - \eta^2\right) = 0$

$$\tag{12.92}$$

Oblate $\dfrac{\partial}{\partial\zeta}\left[\left(\zeta^2 + 1\right)\dfrac{\partial\Phi}{\partial\zeta}\right] + \dfrac{\partial}{\partial\eta}\left[\left(1 - \eta^2\right)\dfrac{\partial\Phi}{\partial\eta}\right] + a^2\omega^2\Phi\left(\zeta^2 + \eta^2\right) = 0$

$$\tag{12.93}$$

and it is separable (see Chap. 4). The separation $\Phi = X\left(\zeta\right) N\left(\eta\right)$ yields, for the prolate case

$$\frac{\partial}{\partial\zeta}\left[\left(\zeta^2 - 1\right)\frac{\partial X}{\partial\zeta}\right] - \left[\kappa_p + a^2\omega^2\left(1 - \zeta^2\right)\right] X = 0 \tag{12.94a}$$

$$\frac{\partial}{\partial\eta}\left[\left(\eta^2 - 1\right)\frac{\partial N}{\partial\eta}\right] - \left[\kappa_p + a^2\omega^2\left(1 - \eta^2\right)\right] N = 0 \tag{12.94b}$$

and for the oblate one

$$\frac{\partial}{\partial\zeta}\left[\left(\zeta^2 + 1\right)\frac{\partial X}{\partial\zeta}\right] - \left[\kappa_o - a^2\omega^2\left(\zeta^2 + 1\right)\right] X = 0 \tag{12.95a}$$

$$\frac{\partial}{\partial\eta}\left[\left(\eta^2 - 1\right)\frac{\partial N}{\partial\eta}\right] - \left[\kappa_o - a^2\omega^2\left(1 - \eta^2\right)\right] N = 0 \tag{12.95b}$$

where $\kappa_o$ and $\kappa_p$ are the separation constants for the oblate and prolate case, respectively. To notice that for the prolate case the equations for $X\left(\zeta\right)$ and for $N\left(\eta\right)$ are identical. Equations (12.94), (12.95) are different forms of the so-called *spheroidal wave equation* (see Appendix)

$$\frac{\partial}{\partial z}\left[\left(z^2 - 1\right)\frac{\partial Z}{\partial z}\right] - \left[\lambda + \gamma^2\left(1 - z^2\right) - \frac{\mu^2}{1 - z^2}\right] Z = 0 \tag{12.96}$$

which has the general solution, for $\mu = m = 0, 1, 2, \ldots,$

$$Z(z) = A \, Ps_n^m (z, \gamma^2) + B \, Qs_n^m (z, \gamma^2) \tag{12.97}$$

where $Ps_n^m (z, \gamma)$ and $Qs_n^m (z, \gamma)$ (for $n = m, m + 1, \ldots$) are the *spheroidal wave functions* of the first and the second kind, respectively. Consider now the case $\mu = 0$ then

$$\frac{\partial}{\partial z} \left[ (z^2 - 1) \frac{\partial Z}{\partial z} \right] - [\lambda + \gamma^2 (1 - z^2)] Z = 0 \tag{12.98}$$

Setting $\lambda = \kappa_p$ this equation transforms to Eqs. (12.94a) and (12.94b), and thus the solution of these equations are

$$X^{(p)} = Ps_n^0 (\zeta, a^2 \omega^2) ; \quad N^{(p)} = Ps_n^0 (\eta, a^2 \omega^2) \tag{12.99}$$

where the separation constant $\kappa_p$ takes the values $\lambda_n (\gamma^2)$ and $\gamma^2 = a^2 \omega^2$.
For the oblate spheroid case, setting $\gamma^2 = -a^2 \omega^2$ transforms (12.98) into (12.95b):

$$\frac{\partial}{\partial \eta} \left[ (\eta^2 - 1) \frac{\partial N^{(o)}}{\partial \eta} \right] - [\kappa_o - a^2 \omega^2 (1 - \eta^2)] N^{(o)} = 0 \tag{12.100}$$

Choosing $\kappa_0 = \lambda$ yields the solution $N^{(o)} = Ps_n^0 (\eta, -a^2 \omega^2)$.
Choosing again $\kappa_0 = \lambda$, $\gamma^2 = -a^2 \omega^2$ and $z = i\zeta$ Eq. (12.98) transforms to Eq. (12.95a):

$$\frac{\partial}{\partial \zeta} \left[ (\zeta^2 + 1) \frac{\partial X^{(o)}}{\partial \zeta} \right] - [\kappa_o - a^2 \omega^2 (\zeta^2 + 1)] X^{(o)} = 0 \tag{12.101}$$

and then the solution is $X^{(o)} = Ps_n^0 (i\zeta, -a^2 \omega^2)$.
Finally, solutions of the Helmholtz equation $\nabla^2 \Phi (\zeta, \eta) + \omega^2 \Phi (\zeta, \eta) = 0$ are of the form

$$
\begin{array}{cc}
\text{Prolate} & \text{Oblate} \\
\Phi_n = Ps_n^0 (\zeta, a^2 \omega^2) \, Ps_n^0 (\eta, a^2 \omega^2) & \Phi_n = Ps_n^0 (i\zeta, -a^2 \omega^2) \, Ps_n^0 (\eta, -a^2 \omega^2)
\end{array}
\tag{12.102}
$$

## 12.4.2 The Niven Solution for the Energy Equation

The most general solution of the energy equation (12.87) can be found as follows [24]. First notice that the functions

$$\Psi_n = e^{-\alpha \omega^2 t} \Phi_n (\zeta, \eta; \omega^2) \tag{12.103}$$

are solution of the energy equation for any $\omega$ and $n = 0, 1, \ldots$, as above shown. Then consider boundary conditions of the general type

$$\Psi_n\left(\zeta_0, \eta; \omega^2\right) + c\frac{\partial}{\partial\zeta}\Psi_n\left(\zeta_0, \eta; \omega^2\right) = 0 \tag{12.104}$$

(where $c$ is a constant) which, since they must hold for any time, yields

$$\Phi_n\left(\zeta_0, \eta; \omega^2\right) + c\frac{\partial}{\partial\zeta}\Phi_n\left(\zeta_0, \eta; \omega^2\right) = 0 \tag{12.105}$$

For each value of $n$, Eqs. (12.105) yield a countable infinity of values of $\omega$; since each countable infinity may be different for different values of $n$, the *spectral constants* are defined by two subscripts, i.e. $\omega_{nk}^2$, where $n, k = 1, 2, \ldots$. As an example, consider the prolate case (but the same reasoning applies to the oblate one), using Eq. (12.102), the conditions (12.105) become

$$Ps_n^0\left(\zeta_0, a^2\omega^2\right) + c\frac{\partial}{\partial\zeta}Ps_n^0\left(\zeta_0, a^2\omega^2\right) = 0 \tag{12.106}$$

and this equation has solutions for an infinite series of the spectral values $a^2\omega_{nk}^2$, i.e. for any $n$ there exist a series of spectral values of $\omega_{nk}$, and the corresponding functions $T_{nk} = \Phi_n\left(\zeta, \eta; \omega_{nk}^2\right)$ are solution of the energy equation (12.87) and satisfy the homogeneous boundary condition (12.105). It can be shown (see Sect. 12.4.3) that $T_{nk} = \Phi_n\left(\zeta, \eta; \omega_{nk}^2\right)$ are orthogonal, i.e. that the following equation holds

$$\text{if } \left(\omega_{pl}^2 - \omega_{nk}^2\right) \neq 0 \Rightarrow \int \Phi_p\left(\zeta, \eta; \omega_{pl}^2\right)\Phi_n\left(\zeta, \eta; \omega_{nk}^2\right) dV = 0 \tag{12.107}$$

We are now ready to build up the most general solution of the energy equation, which satisfies the given boundary condition, by a linear combination of all the solutions that we have found, then

$$T = \sum_{nk} c_{nk}e^{-\alpha\omega_{nk}^2 t}T_{nk}\left(\zeta, \eta; a^2\omega_{n,k}^2\right) \tag{12.108}$$

where

$$T_{nk}\left(\zeta, \eta; a^2\omega_{nk}^2\right) = Ps_n^0\left(\zeta, a^2\omega_{nk}^2\right)Ps_n^0\left(\eta, a^2\omega_{nk}^2\right) \text{ prolate} \tag{12.109}$$

$$T_{nk}\left(\zeta, \eta, a^2\omega_{nk}^2\right) = Ps_n^0\left(i\zeta, a^2\omega_{nk}^2\right)Ps_n^0\left(\eta, -a^2\omega_{nk}^2\right) \text{ oblate} \tag{12.110}$$

and $Ps_n^0\left(\zeta, a^2\omega_{nk}^2\right)$ are spheroidal wave function of the first kind.

The coefficients $c_{nk}$ are unknown and they can be found by satisfying a given initial condition. Since we are dealing with a rotational symmetric problem, the initial condition must be independent of the azimuthal angle $\varphi$ and then it can be set as

$$T(0, \zeta, \eta, \varphi) = f_0(\zeta, \eta) \tag{12.111}$$

and from Eq. (12.108):

$$f_0(\zeta, \eta) = \sum_{nk} c_{nk} T_{nk}(\zeta, \eta; a^2 \omega_{n,k}^2) \tag{12.112}$$

We can now use the orthogonality of $T_{nk}$ to find the constants $c_{nk}$, in fact

$$\int_V f_0(\zeta, \eta) T_{pl} dV = \sum_{nk} c_{nk} \int_V T_{nk} T_{pl} dV = c_{pl} \int_V T_{pl} T_{pl} dV \tag{12.113}$$

where the last equality comes from Eq. (12.107). Then the general form of the coefficients $c_{nk}$ is

$$c_{nk} = \frac{\int_V f_0(\zeta, \eta) T_{nk} dV}{\int_V T_{nk} T_{nk} dV} \tag{12.114}$$

### 12.4.3 The Orthogonality of $T_{nk}$

Considering again the Helmholtz equation (12.91), the functions $T_{nk} = \Phi_n (\zeta, \eta; \omega_{nk}^2)$ are solutions of that equation corresponding to the specific values $\omega = \omega_{nk}$ of the separation constant, i.e.

$$\nabla^2 T_{nk}(\zeta, \eta) + \omega_{nk}^2 T_{nk}(\zeta, \eta) = 0 \tag{12.115}$$

and the boundary conditions, which determine $\omega_{nk}^2$, are

$$T_{nk}(\zeta_0, \eta; a^2 \omega_{nk}^2) + c \nabla_\zeta T_{nk}(\zeta_0, \eta; a^2 \omega_{nk}^2) = 0 \tag{12.116}$$

Let us now consider two equations like Eq. (12.115) corresponding to two constants $\omega_{nk}$ and $\omega_{pl}$; writing Eq. (12.115) once for $\omega_{nk}$ and the other for $\omega_{pl}$, multiplying the first equation by $T_{pl}$ and the second by $T_{nk}$ and subtracting one from the other results in

$$T_{pl} \nabla^2 T_{nk} - T_{nk} \nabla^2 T_{pl} = (\omega_{pl}^2 - \omega_{nk}^2) T_{nk} T_{pl} \tag{12.117}$$

Let now integrate over the ellipsoid volume both sides of Eq. (12.117)

$$\int_V (T_{pl} \nabla^2 T_{nk} - T_{nk} \nabla^2 T_{pl}) dV = (\omega_{pl}^2 - \omega_{nk}^2) \int_V T_{nk} T_{pl} dV \tag{12.118}$$

or in curvilinear coordinates

$$2\pi \int_{\zeta_0}^{\infty} \int_{-1}^{+1} \left(T_{pl}\nabla^2 T_{nk} - T_{nk}\nabla^2 T_{pl}\right) g^{1/2} d\eta d\zeta$$

$$= 2\pi \left(\omega_{pl}^2 - \omega_{nk}^2\right) \int_{\zeta_0}^{\infty} \int_{-1}^{+1} T_{nk} T_{pl} g^{1/2} d\eta d\zeta \qquad (12.119)$$

Considering the following identities (we suspend the summation convention in this section)

$$T_{pl}\nabla^2 T_{nk} = \sum_{j=1}^{3} \nabla_j \left(T_{pl}\nabla_j T_{nk}\right) - \sum_{j=1}^{3} \nabla_j T_{pl} \cdot \nabla_j T_{nk} \qquad (12.120a)$$

$$T_{nk}\nabla^2 T_{pl} = \sum_{j=1}^{3} \nabla_j \left(T_{nk}\nabla_j T_{pl}\right) - \sum_{j=1}^{3} \nabla_j T_{nk} \cdot \nabla_j T_{pl} \qquad (12.120b)$$

and then subtracting Eq. (12.120b) from (12.120a), yields

$$\left(T_{pl}\nabla^2 T_{nk} - T_{nk}\nabla^2 T_{pl}\right) = \sum_{j=1}^{3} \nabla_j \left(T_{pl}\nabla_j T_{nk}\right) - \sum_{j=1}^{3} \nabla_j \left(T_{nk}\nabla_j T_{pl}\right)$$

$$= \sum_{j=1}^{3} \nabla_j \left(T_{pl}\nabla_j T_{nk} - T_{nk}\nabla_j T_{pl}\right) \qquad (12.121)$$

Taking the integral of both sides over the ellipsoid volume yields

$$\int_V \left(T_{pl}\nabla^2 T_{nk} - T_{nk}\nabla^2 T_{pl}\right) dV = \int_V \sum_{j=1}^{3} \nabla_j \left(T_{pl}\nabla_j T_{nk} - T_{nk}\nabla_j T_{pl}\right) dV =$$

$$= \int_A \left[T_{pl}\nabla_\zeta T_{nk} - T_{nk}\nabla_\zeta T_{pl}\right]_{\zeta=\zeta_0} dA \qquad (12.122)$$

where the symbol $[f]_{\zeta=\zeta_0}$ is used to indicate that $f$ is evaluated over the ellipsoid surface: $\zeta = \zeta_0$. The last equality is obtained applying the divergence theorem (see Chap. 6) and observing that the component of the gradient of $T_{ab}$ (i.e. the vector $\nabla T_{ab}$) normal to the spheroid surface is just $\nabla_\zeta T_{ab}$, since the surface is defined by $\zeta = \zeta_0$. The term inside the surface integral can be further reduced, using the boundary condition (12.116)

$$T_{nk}\left(\zeta_0, \eta; a^2\omega_{nk}^2\right) = -c\nabla_\zeta T_{nk}\left(\zeta_0, \eta; a^2\omega_{nk}^2\right) \qquad (12.123)$$

$$T_{pl}\left(\zeta_0, \eta; a^2\omega_{pl}^2\right) = -c\nabla_\zeta T_{pl}\left(\zeta_0, \eta; a^2\omega_{pl}^2\right) \qquad (12.124)$$

then

$$\left[T_{pl}\nabla_{\zeta}T_{nk} - T_{nk}\nabla_{\zeta}T_{pl}\right]_{\zeta=\zeta_0} = -c \left[\begin{array}{c} \nabla_{\zeta}T_{pl}\left(\zeta_0, \eta; a^2\omega_{pl}^2\right)\nabla_{\zeta}T_{nk}\left(\zeta_0, \eta; a^2\omega_{nk}^2\right) + \\ -\nabla_{\zeta}T_{nk}\left(\zeta_0, \eta; a^2\omega_{nk}^2\right)\nabla_{\zeta}T_{pl}\left(\zeta_0, \eta; a^2\omega_{pl}^2\right) \end{array}\right]_{\zeta=\zeta_0} = 0$$

(12.125)

which holds also for $\frac{1}{c} \to 0$, i.e. for boundary conditions of the second kind (von Neumann).

This proves that $\int_V \left(T_{pl}\nabla^2 T_{nk} - T_{nk}\nabla^2 T_{pl}\right) dV$ is nil and from Eq. (12.118) we obtain

$$\left(\omega_{pl}^2 - \omega_{nk}^2\right)\int_V T_{nk}T_{pl}dV = 0 \qquad (12.126)$$

which shows that :

$$\text{if } \left(\omega_{pl}^2 - \omega_{nk}^2\right) \neq 0 \Rightarrow \int_V T_{nk}T_{pl}dV = 0 \qquad (12.127)$$

yielding the orthogonality of the set $T_{nk}\left(\zeta, \eta; a^2\omega_{pl}^2\right)$ when $\omega_{pl}^2$ are given by Eq. (12.116).

As a final comment, it is fair to say that the implementation of the solution given by Eqs. (12.108) and (12.114) is not straightforward, since the analytical evaluation of the integrals in Eq. (12.114) requires series expansion of the spheroidal functions and a huge analytical effort, which may not be completely justified in view of available numerical solutions of this problem.

# References

1. Rayleigh, L.: On the capillary phenomena of jets. Proc. R. Soc. Lond. A **29**, 71–97 (1879)
2. Chandrasekhar, S.: Hydrodynamic and Hydromagnetic Stability. Dover, New York (1981)
3. Lamb, H.: Hydrodynamics, 6th edn. Cambridge University Press, Cambridge (1932)
4. Chandrasekhar, S.: The oscillations of a viscous liquid globe. Proc. Lond. Math. Soc. **9**, 141–149 (1959)
5. Quan, S., Lou, J., Schmidt, D.P.: Modeling merging and breakup in the moving mesh interface tracking method for multiphase flow simulations. J. Comput. Phys. **228**(7), 2660–2675 (2009)
6. Takaki, R., Yoshiyasu, N., Arai, Y., Adachi, K.: Dynamic pattern formation of an evaporating drop. In: Proceedings of the First International Symposium for Science on Form, KTK Scientific Publishers, Tokyo (1986)
7. Haywood, R.J., Renksizbulut, M., Raithby, G.D.: Transient deformation and evaporation of droplets at intermediate Reynolds numbers. Int. J. Heat Mass Transf. **37**(9), 1401–1409 (1994)
8. Kowalewski, T.A.: Transient evaporation of oscillating droplet. In: Advanced Modelling and Simulation in Engineering, Pultusk (1994)
9. Mashayek, F.: Dynamics of evaporating drops. Part I: formulation and evaporation model. Int. J. Heat Mass Transf. **44**(8), 1517–1526 (2001)
10. Deng, Z.T., Litchford, R.J., Jeng, S.M.: Two-dimensional simulation of droplet evaporation at high pressure. In: AIAA Paper, pp. 29-3122 (1992)
11. Mashayek, F.: Dynamics of evaporating drops. Part II: free oscillations. Int. J. Heat Mass Transf. **44**(8), 1527–1541 (2001)
12. Tonini, S., Cossali, G.E.: An evaporation model for oscillating spheroidal drops. Int. Commun. Heat Mass Transf. **51**, 18–24 (2014)

13. Basaran, O.A.: Nonlinear oscillations of viscous liquid drops. J. Fluid Mech. **241**, 169–198 (1992)
14. Sazhin, S.S., Krutitskii, P.A., Gusev, I.G., Heikal, M.R.: Transient heating of an evaporating droplet. Int. J. Heat Mass Transf. **53**, 2826–2836 (2010)
15. Sazhin, S.S., Krutitskii, P.A., Gusev, I.G., Heikal, M.R.: Transient heating of an evaporating droplet with presumed time evolution of its radius. Int. J. Heat Mass Transf. **54**, 1278–1288 (2011)
16. Sazhin, S.: Droplet and Sprays. Springer, Berlin (2014)
17. Sazhin, S.S., Krutitskii, P.A., Martynov, S.B., Mason, D., Heikal, M.R., Sazhina, E.M.: Transient heating of semitransparent spherical body. Int. J. Therm. Sci. **46**(5), 444–457 (2007)
18. Dorfman, A.S.: Conjugate Problems in Convective Heat Transfer. CRC Press, Boca Raton (2009)
19. Tonini, S., Cossali, G.E.: Modeling of liquid drop heating and evaporation: the effect of drop shrinking. Comput. Therm. Sci. **10**(3), 273–283 (2018)
20. Bakhshi, A., Ganji, D.D., Gorji, M.: Deformation and breakup of an axisymmetric falling drop under constant body force. Int. J. Partial Differ. Equ. Appl. **3**(1), 1–6 (2015)
21. Pruppacher, J.A., Bear, K.V.: A wind tunnel investigation of the internal circulation and shape of water drops falling at terminal velocity in air. Q. J. R. Meteorol. Soc. **96**, 47–256 (1970)
22. Pruppacher, H.R., Pitter, R.L.: A semi-empirical determination of the shape of cloud and rain drops. J. Atmos. Sci. **28**, 86–94 (1971)
23. Mathieu, E.: Course de Physique Mathematique. Gauthier-Villars, Paris (1873)
24. Niven, C.: On the conduction of heat in ellipsoids of revolution. Proc. R. Soc. Lond. **171**, 117–151 (1880)

# Chapter 13
# Multi-component Drop Evaporation

In many engineering applications multi-component liquids are used in spray systems, just think of fuel sprays in internal combustion engines where the fuels (Diesel, gasoline or even biofuels) are blend of a large number of different hydrocarbons [1]. In those cases the species may have quite different volatilities and diffusivities as well as other thermo-physical characteristics and this may strongly influence the rates of heating and vaporisation of the spray droplets.

The approaches to the modelling of heating and evaporation of multi-component drops can be grouped into two families: those that assume that a multi-component fuel is made of a finite number of components, each one with its own different properties (the *discrete component* models, refer to [2–6], for example) and those based on a probabilistic approach (e.g. *distillation curve* model [7] or the *continuous thermodynamics* approach [8–10]). The first approach is applicable when a relatively small number of components needs to be taken into account, whereas the second one is used when considering a large number of components to reduce the number of variables compared to the discrete component approach. As above mentioned, in real fuels used in some applications (like internal combustion engines) the number of different components is very large and this has motivated the development of particular methods based on the so-called "quasi-components" [11] to simplify the analytical treatment of these cases for applications to Diesel, gasoline and biofuel droplets [12–16] with remarkable results. It is out of the scope of this book to review the above described approaches and the interested reader is invited to refer to [1, 17] for more information.

## 13.1  A Multi-component Model for Spherical Drops

In this section we will analyse the heat and mass transfer from a spherical multi-component drop, with radius $R_d$, floating in a gaseous environment. The analysis reported here is based on [18]. Quasi-steadiness is assumed, as well as constant

© Springer Nature Switzerland AG 2021
G. E. Cossali and S. Tonini, *Drop Heating and Evaporation: Analytical Solutions in Curvilinear Coordinate Systems*, Mathematical Engineering,
https://doi.org/10.1007/978-3-030-49274-8_13

thermo-physical properties of the gas phase. Under these simplifying conditions, the species conservation equations in spherical coordinates are obtained from (Eq. 8.7) as

$$0 = \frac{dn_R^{(p)}}{dR} + \frac{2}{R}n_R^{(p)}$$
(13.1)

where $n_R^{(p)}$ are the radial physical components of the mass flux of species $p$, and

$$n_R^{(p)} = \rho v_R \chi^{(p)} + j_R^{(p)}$$
(13.2)

where $j_R^{(p)}$ are the radial physical components of the diffusional fluxes. In this section we will assume that Fick's law can be used to model diffusional fluxes (see Sect. 7.6), i.e.

$$j_R^{(p)} = -\rho D_p^{(m)} \nabla_R \chi^{(p)} = -\rho D_p^{(m)} \frac{d\chi^{(p)}}{dR}$$
(13.3)

Substitution of Eqs. (13.3) and (13.2) into Eq. (13.1) yields

$$\frac{d}{dR}\left(R^2 \rho v_R \chi^{(p)} - R^2 \rho D_p^{(m)} \frac{d\chi^{(p)}}{dR}\right) = 0$$
(13.4)

where the index $p = 1, \ldots, n$ represent the $p$th evaporating species, while $p = 0$ is used for the non-evaporating species (gas). In Eq. (13.4) $D_p^{(m)}$ is the mass diffusion coefficient of species $p$ in the gaseous mixture, and it can be evaluated using one of the existing rules (see again Sect. 7.6), like Blanc's law or Wilke's law

$$D_p^{(m)} = \left(\sum_{\substack{k=0 \\ k\neq p}}^{n} \frac{y_{ref}^{(k)}}{D_{pk}}\right)^{-1}$$
(13.5a)

$$D_p^{(m)} = \left(1 - y_{ref}^{(p)}\right)\left(\sum_{\substack{j=0 \\ j\neq p}}^{n} \frac{y_{ref}^{(j)}}{D_{pj}}\right)^{-1}$$
(13.5b)

In Eqs. (13.5), $D_{pk}$ are the binary diffusion coefficients of species $p$ in species $k$, $y_{ref}^{(k)}$ are the molar fraction of species $k$ in the gas phase, calculated at a reference condition (see Chap. 10). It is assumed that no gas absorption takes place and then the flux of the component $p = 0$ is nil at the surface and the flux $\mathbf{n}^{(0)}$ in the gas mixture is nil everywhere. As already pointed out, only $n$ out of $n + 1$ Eqs. (13.1) are independent since the sum over $p$ yields the mass conservation equation

$$0 = \frac{d\left(R^2 n_R^{(T)}\right)}{dR}$$
(13.6)

where $n_R^{(T)} = \sum_{p=1}^{n} n_R^{(p)} = \rho v_R$ is the total mass flux, then only the Eqs. (13.4) for $p = 1, \ldots n$ need to be considered. Equation (13.6) can be readily integrated to yield

$$v_R = \frac{m_{ev}^T}{4\pi R^2 \rho} \qquad (13.7)$$

where $m_{ev}^T = 4\pi R^2 n_R^T = \sum_{p=1}^{n} m_{ev}^{(p)}$ denotes the total evaporation rate, i.e. the sum of the evaporation rates of each component $m_{ev}^{(p)}$.

Substitution of Eq. (13.7) into Eq. (13.4) and the change of variable $\zeta = \frac{R_d}{R}$ yields a set of constant coefficients linear ODEs

$$\frac{m_{ev}^T}{4\pi} \frac{d}{d\zeta} \chi^{(p)} + R_d \rho D_p^{(m)} \frac{d^2 \chi^{(p)}}{d\zeta^2} = 0 \qquad (13.8)$$

Equation (13.8) has the solution

$$\chi^{(p)} = \lambda^{(p)} e^{-\frac{m_{ev}^{(T)}}{4\pi \rho R_d D_p^{(m)}} \zeta} + \nu^{(p)} ; \quad p = 1, \ldots n \qquad (13.9)$$

and $\chi^{(0)}$ is simply calculated from the unity constrain $\chi^{(0)} = 1 - \sum_{k=1}^{n} \chi^{(k)}$.

The constants $\nu^{(p)}$ have a special physical meaning, in fact substituting Eq. (13.9) into Eqs. (13.3) and (13.2) and setting $\zeta = 1$, equivalent to $R = R_d$, yields

$$n_R^{(p)} (R_d) = \nu^{(p)} \frac{m_{ev}^T}{4\pi R_d^2} \qquad (13.10)$$

or

$$\nu^{(p)} = \frac{4\pi R_d^2 n_R^{(p)} (R_d)}{m_{ev}^T} = \frac{m_{ev}^{(p)}}{\sum_{k=1}^{n} m_{ev}^{(k)}} \qquad (13.11)$$

i.e. $\nu^{(p)}$ is the ratio between the evaporation rate of species $p$ and the total evaporation rate.

The closure of the problem, which consists in calculating the $2n$ constants $\lambda^{(p)}$ and $\nu^{(p)}$, is obtained imposing the boundary conditions at the drop surface, $\zeta = 1$, and at infinity, $\zeta = 0$

$$\chi^{(k)} (\zeta = 1) = \chi_s^{(k)} \qquad (13.12a)$$
$$\chi^{(k)} (\zeta = 0) = \chi_\infty^{(k)} \qquad (13.12b)$$

yielding, in particular

$$\nu^{(p)} = \frac{\chi_s^{(p)} - \chi_\infty^{(p)} e^{-\frac{m_{ev}^{(T)}}{4\pi \rho R_d D_p^{(m)}}}}{1 - e^{-\frac{m_{ev}^{(T)}}{4\pi \rho R_d D_p^{(m)}}}} \qquad (13.13)$$

The values of $\chi_s^{(p)}$ can be calculated by the equation

$$\chi_s^{(p)} = \frac{\rho_{G,s}^{(p)}}{\rho_G} = \frac{P_{v,s}^{(p)}\,(T_s)\,Mm^{(p)}}{T_s \bar{R}\rho_G} \tag{13.14}$$

where the partial pressure of each component, $P_{v,s}^{(p)}$, can be evaluated by Raoult's law (named after the French chemist François-Marie Raoult, 1830–1901)

$$P_{v,s}^{(p)} = y_L^{(p)}\,\check{P}_{v,s}^{(p)} \tag{13.15}$$

where $\check{P}_{v,s}^{(p)}$ is the equilibrium vapor pressure of the pure component $p$, and $y_L^{(p)}$ is the mole fraction of the component $p$ in the liquid phase. Here it is assumed that the liquid drop composition is uniform over the entire volume. It can be observed that even if the initial composition of the liquid drop is uniform, the differential evaporation of the components produces a variation of the relative concentration over the surface, then gradients of the species concentration appear inside the drop and a differential migration of components takes place. This problem has been considered in [12] and an analytical solution for the transport of species inside the drop was proposed (see also [17]).

In Eqs. (13.13) the only unknown is the total evaporation rate $m_{ev}^{(T)}$, which can be implicitly calculated from the identity $\sum_{k=1}^{n} \nu^{(p)} = 1$, which yields the trascendental equation

$$\sum_{p=1}^{n} \frac{\chi_\infty^{(p)} - \chi_s^{(p)}}{\left(e^{-\frac{m_{ev}^{(T)}}{4\pi\rho R_d D_p^{(m)}}} - 1\right)} = 1 - \sum_{k=1}^{n} \chi_\infty^{(p)} \tag{13.16}$$

The evaluation of the evaporation rate of each species is then reduced to the solution (necessarily approximate) of the single trascendental Eq. (13.16) in the single unknown and the subsequent use of (13.13) to calculate the parameters $\nu^{(p)}$ and finally $m_{ev}^{(p)} = \nu^{(p)} m_{ev}^{(T)}$.

The solution of the trascendental equation (13.16) for $m_{ev}^{(T)}$ requires a numerical algorithm with a rather simple implementation. A similar approach was also reported in [19], where instead a numerical solutions of a system of $n + 1$ trascendental equations was required for calculation of $m_{ev}^{(T)}$ and $\nu^{(p)}$. This model was used in [18] to predict the evaporation of drops made by a mixture of n-tetradecane and n-hexadecane with different initial composition and compared with the experimental results reported in [20, 21], where drops of $50\,\mu m$ initial diameter were levitated in air at the same temperature of $304\,K$ (see [20, 21] for the details of the experiments). Figure 13.1 shows the comparison.

The two extreme cases of pure n-tetradecane and pure n-hexadecane are represented by the two straight lines, since under such conditions the $D^2$-law holds. The other cases show the typical behaviour of a bi-component evaporating drop: the curve slope changes from an initial value close to that corresponding to the more

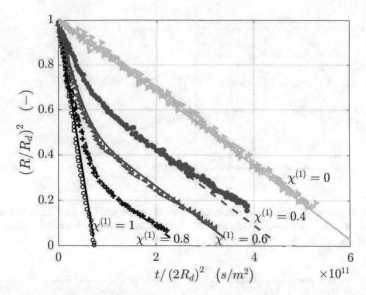

**Fig. 13.1** Comparison between model predictions (lines) and experimental data of [20] (symbols) of non-dimensional drop size temporal evolution for five n-tetradecane/n-hexadecane drops with different initial composition; $R_{d,0} = 25\,\mathrm{mm}$, $T_\infty = 304\,\mathrm{K}$, $\mathrm{Re} = 0$

volatile pure component (n-tetradecane) to reach, when the more volatile species is almost totally evaporated, the slope of the curve corresponding to the evaporation of the pure heaviest component (n-hexadecane). Figure 13.2 shows the evolution of the drop size together with the drop composition for a case with an initial composition containing 80% in mass of n-tetradecane, pointing out that the change of the slope in the curve representing the drop size (precisely the ratio $R_d^2\,(t)\,/\,R_d^2\,(0)$) corresponds to the time when the more volatile species is almost disappeared. Figure 13.2b shows also the instantaneous evaporation rates of the two alkanes. The lighter component has a much larger evaporation rate at the beginning, which decreases with time due to the decrease of its concentration in the liquid that causes a continuous decrease of $\chi^{(n-C_{14}H_{30})}$ at the drop surface. The evaporation rate of hexadecane (the heaviest component) initially is comparatively very low, then it increases while its concentration in the liquid drop increases and it reaches a maximum at the time when tetradecane has almost totally disappeared.

Investigation with three-component drops is reported in Fig. 13.3. Figure 13.3a shows the comparison between the experimental measurements of [20] and the model predictions of the non-dimensional drop size evolution for two cases, one initially composed by 33% n-octane, 33% n-decane and 34% n-dodecane and the other one initially composed by 33% n-octane, 33% n-dodecane and 34% n-hexadecane. Figure 13.3b plots the transient profiles of the liquid composition for the second test case, showing that the changes of slope in the drop size profile correspond to the time when the compositions of the more volatile species becomes almost nil.

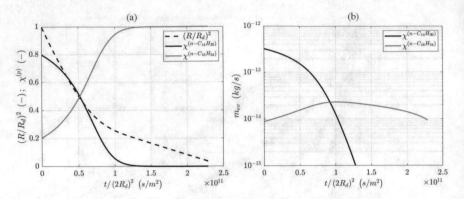

**Fig. 13.2** Temporal evolution of **a** non-dimensional drop size and species volume fraction concentration and **b** instantaneous evaporation rates for an n-tetradecane/n-hexadecane drop; $R_{d,0} = 25\,\text{mm}$, $T_\infty = 304\,\text{K}$, $\text{Re} = 0$, $\chi_{l,0}^{(n-C_{14}H_{30})} = 0.4$

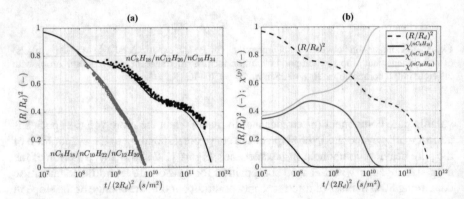

**Fig. 13.3** **a** Comparison between model predictions (lines) and experimental measurements of [20] (symbols) of non-dimensional drop size temporal evolution for two alkane drops with different initial composition; $R_{d,0} = 25\,\text{mm}$, $T_\infty = 299.8\,\text{K}$ and $T_\infty = 296.2\,\text{K}$, $\text{Re} = 0$; **b** temporal evolution of non-dimensional drop size and species volume fraction concentration; $R_{d,0} = 25\,\text{mm}$, $T_\infty = 296.2\,\text{K}$, $\text{Re} = 0$

In [18] this model was extended to the case of a drop evaporating under convective conditions (i.e. $\text{Re} > 0$), by applying a *film theory* approach (see Chap. 10), through the definition of different Sherwood numbers (and then different film thicknesses) for each species.

## 13.2 Multi-component Drop Evaporation Solving the Stefan–Maxwell Equations

The analysis of multi-component drop evaporation reported in the previously section was based on the Fick's law to evaluate the diffusional fluxes, but we have seen in Chap. 7 that a more accurate way to treat multi-component diffusion is through the use of the Stefan–Maxwell equations, which for a mixture of $n + 1$ species, neglecting Soret effect and diffusion due to pressure gradients and to external forces, can be written as

$$\nabla y^{(p)} = \sum_{k=0}^{n} \frac{1}{cD_{pk}} \left( y^{(p)} \mathbf{N}^{(k)} - y^{(k)} \mathbf{N}^{(p)} \right) \tag{13.17}$$

In this equation $\mathbf{N}^{(p)}$ is the molar flux of the $p$-component, which is related to the mass flux by $\mathbf{n}^{(p)} = \mathbf{N}^{(p)} Mm^{(p)}$, where $Mm^{(p)}$ is the molar mass of the $p$-component.

We consider again a multi-component spherical drop and we assume spherical symmetry, then only the radial components of the species molar fluxes, $N_R^{(k)}$, are non-nil. As in the previous section, quasi-steadiness is assumed and the absence of absorption of species $p = 0$ (air) into the liquid yield a nil value of $N_R^{(0)}$ on the surface and everywhere. Under these hypotheses

$$N_R^{(k)} = \frac{m_{ev}^{(k)}}{4\pi R^2 Mm^{(k)}} \quad \text{for } k = 1 \ldots n \tag{13.18}$$

$$N_R^{(0)} = 0 \tag{13.19}$$

where $m_{ev}^{(k)}$ is the mass evaporation rate of the $k$th–component. Using the same change of variable as in the previous section, $\zeta = \frac{R_0}{R}$, and introducing the non-dimensional quantities $\hat{m}_{ev}^{(k)} = \frac{m_{ev}^{(k)}}{4\pi R_d Mm^{(k)} c D_{ref}}$, where $D_{ref}$ is an arbitrary reference value for the diffusion coefficients, Eqs. (13.17) yield the ODE system

$$\frac{dy^{(p)}}{d\zeta} = -\sum_{k=0}^{n} \varphi^{pk} \left( y^{(p)} \hat{m}_{ev}^{(k)} - y^{(k)} \hat{m}_{ev}^{(k)} \right) \tag{13.20}$$

where $\varphi^{pk} = \frac{D_{ref}}{D_{pk}}$. It is worth to notice that the arbitrary value $D_{ref}$, which has been introduced to conveniently non-dimensionalise the evaporation rates, disappears from Eq. (13.20) when performing the products $\hat{m}_{ev}^{(k)} \varphi^{pk}$. The ODE systems (13.20) can be written in a more compact form; observing that $\hat{m}_{ev}^{(0)} = 0$, $\sum_{k=0}^{n} y^{(k)} = 1$ and setting $\varphi^{pp} \equiv \varphi^{0p} \equiv \varphi^{p0}$, Eqs. (13.20) can be written in a matrix form

$$\frac{d}{d\zeta} \Psi = \mathcal{A} \Psi + \mathbf{B} \tag{13.21}$$

where

$$
\mathcal{A} = \begin{bmatrix}
-\sum_{k=1}^{n} \varphi^{1k} \hat{m}_{ev}^{(k)} & \hat{m}_{ev}^{(1)} \left( \varphi^{12} - \varphi^{10} \right) & \cdots & \hat{m}_{ev}^{(1)} \left( \varphi^{1n} - \varphi^{10} \right) \\
\hat{m}_{ev}^{(2)} \left( \varphi^{21} - \varphi^{20} \right) & -\sum_{k=1}^{n} \varphi^{2k} \hat{m}_{ev}^{(k)} & \cdots & \hat{m}_{ev}^{(2)} \left( \varphi^{2n} - \varphi^{20} \right) \\
\cdots & \cdots & \cdots & \cdots \\
\hat{m}_{ev}^{(n)} \left( \varphi^{n1} - \varphi^{n0} \right) & \hat{m}_{ev}^{(n)} \left( \varphi^{n2} - \varphi^{n0} \right) & \cdots & -\sum_{k=1}^{n} \varphi^{nk} \hat{m}_{ev}^{(k)}
\end{bmatrix}
\tag{13.22}
$$

$$
\Psi = \left[ y^{(1)} \cdots y^{(n)} \right]^{T}
\tag{13.23}
$$

$$
\mathbf{B} = \left[ \hat{m}_{ev}^{(1)} \varphi^{10}, \hat{m}_{ev}^{(2)} \varphi^{20}, \ldots, \hat{m}_{ev}^{(n)} \varphi^{n0} \right]^{T}
\tag{13.24}
$$

To find a simple solution of Eq. (13.21) let us make the assumption of a constant value of the molar density $c$; we have seen that this kind of assumption is common in most of the drop evaporation models, but it may became influential for evaporation in high temperature environments [22].

In this case the general solution of Eq. 13.21 in matrix form is

$$
\Psi = e^{\mathcal{A}\zeta} \mathbf{C}_0 - \mathcal{A}^{-1} \mathbf{B}
\tag{13.25}
$$

as it can be proven by direct substitution. In this equation we make use of the *matrix exponential* $e^{\mathcal{A}\zeta}$ that can be defined as

$$
e^{\mathcal{A}\zeta} = \mathcal{I} + \sum_{k=1}^{\infty} \frac{1}{k!} \mathcal{A}^{k}
\tag{13.26}
$$

where $\mathcal{I}$ is the identity matrix. The vector $\mathbf{C}_0$, containing unknown constants and appearing in Eq. (13.25), has to be found imposing the boundary conditions

$$
y^{(k)} \left( \zeta = 1 \right) = y_{s}^{(k)}
\tag{13.27a}
$$

$$
y^{(k)} \left( \zeta = 0 \right) = y_{\infty}^{(k)}
\tag{13.27b}
$$

which can be written as

$$
\Psi \left( 1 \right) = \left[ y_{s}^{(1)}, \ldots, y_{s}^{(n)} \right] = \Psi_{s}
\tag{13.28}
$$

$$
\Psi \left( 0 \right) = \left[ y_{\infty}^{(1)}, \ldots, y_{\infty}^{(n)} \right] = \Psi_{\infty}
\tag{13.29}
$$

Substituting Eqs. (13.28) and (13.29) into Eq. (13.25) yields

$$
\Psi_{s} = e^{\mathcal{A}} \cdot \mathbf{C}_0 - \mathcal{A}^{-1} \mathbf{B}
\tag{13.30}
$$

$$
\Psi_{\infty} = \mathbf{C}_0 - \mathcal{A}^{-1} \mathbf{B}
\tag{13.31}
$$

and now $\mathbf{C}_0$ can be eliminated since from the second equation $\mathbf{C}_0 = \Psi_{\infty} + \mathcal{A}^{-1} \mathbf{B}$. After substituting into the Eq. (13.30) one obtains

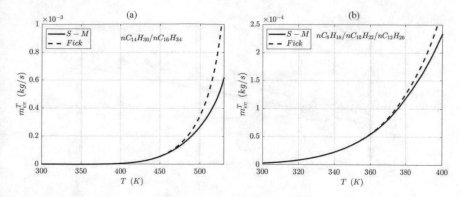

**Fig. 13.4** Effect of drop/gas temperature on total mass evaporation rate, as predicted by the Stefan–Maxwell model $(S - M)$ and the Fick's law based model, for **a** 50% n-tetradecane/ 50% n-hexadecane drop and **b** 33% n-octane/33% n-decane/34% n-dodecane drop

$$\Psi_s = e^{\mathcal{A}} \cdot \left\{ \Psi_\infty + \mathcal{A}^{-1}\mathbf{B} \right\} - \mathcal{A}^{-1}\mathbf{B} \tag{13.32}$$

This matrix equation can be written in a more convenient form with some manipulations; first left-multiply both side by $\mathcal{A}$ to get

$$\mathcal{A}\Psi_s + \mathbf{B} = \mathcal{A}e^{\mathcal{A}}\Psi_\infty + \mathcal{A}e^{\mathcal{A}}\mathcal{A}^{-1}\mathbf{B} \tag{13.33}$$

then making use of the identity

$$\mathcal{A}e^{\mathcal{A}} = e^{\mathcal{A}}\mathcal{A}$$

which can easily be derived from the definition of $e^{\mathcal{A}}$, Eq.(13.26), the following matrix equation is obtained

$$e^{\mathcal{A}} \left( \mathcal{A}\Psi_\infty + \mathbf{B} \right) - \left( \mathcal{A}\Psi_s + \mathbf{B} \right) = 0 \tag{13.34}$$

which represents a system of $n$ trascendental equations for the $n$ unknowns $\left( \hat{m}_{ev}^{(1)}, \ldots, \hat{m}_{ev}^{(n)} \right)$.

The solution of Eq.(13.34), which requires a numerical algorithm of relatively simple implementation, yields the values of $\hat{m}_{ev}^{(k)}$, then all the evaporation rates $m_{ev}^{(k)}$ are found.

In [23] this approach was compared to that described in the previous section, where the Fick's law approach was used, and it was found that the difference between the two approaches become more important at higher temperatures of gas and drops. Figure 13.4 shows a comparison of the predicted total evaporation rate $m_{ev}^{(T)} = \sum_{k=1}^{n} m_{ev}^{(k)}$ for binary (50%tetradecane-50%hexadecane) and ternary (33%n-octane-33%n-decane-34%n-dodecane) liquid mixtures for drop evaporating in an isothermal environment ($T_s = T_\infty$), while Fig. 13.5 shows the effect of gas temperature on the predicted evaporation rate for a binary solution (50%ethanol-50%acetone).

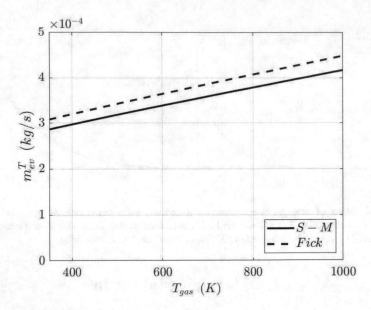

**Fig. 13.5** Effect of gas temperature on total mass evaporation rate, as predicted by the Stefan–Maxwell model ($S - M$) and the Fick's law based model, for a 50% ethanol/50% acetone drop ($T_s = 325$ K)

Varying the drop composition, it was observed that the largest deviation of the evaporation rates calculated by the two approaches is found when none of the species mass fraction prevails on the other.

## 13.3   Drop Evaporation from Multi-component Ellipsoidal Drops

In this last section we will consider an extension of the model for multi-component *spherical* drops described in Sect. 13.1 to the case of *deformed* (ellipsoidal) multi-component drops (more information can be found in [24]). The way to treat this particular case relies on the general results described in Chap. 10 for the 1-D cases. Assuming again quasi-steadiness and constant properties of the gas phase, the species conservation equations can be written in curvilinear coordinates $u^k$, for the *physical* components of the fluxes $\hat{n}_k^{(p)}$, as

$$\sum_{k=1}^{3} \frac{\partial \left( \frac{\sqrt{g}}{h_k} \hat{n}_k^{(p)} \right)}{\partial u^k} = 0 \tag{13.35}$$

When uniform boundary conditions on the drop surface and at infinity are assumed, the vapour diffusion and convection is one-dimensional, and this allows to retain only the derivatives with respect to the coordinate orthogonal to the surface. When dealing with spheroidal drops, the *natural* coordinate system is the spheroidal (oblate or prolate) one (see Chap. 4), and Eq. (13.35) becomes

$$\frac{\partial}{\partial \zeta}\left(h_\eta h_\varphi \hat{n}_\zeta^{(p)}\right) = 0 \tag{13.36}$$

where the scale factors can be found in Chap. 4.

In these systems the drop surface is defined by the simple equation $\zeta = \zeta_0$. The $\zeta$−component of the mass fluxes are

$$n_\zeta^{(p)} = \rho v_\zeta \chi^{(p)} + j_\zeta^{(p)} = \rho v_\zeta \chi^{(p)} - \rho D_p^{(m)} \frac{1}{h_\zeta} \frac{d\chi^{(p)}}{d\zeta} \tag{13.37}$$

where diffusional fluxes are modelled using Fick's law. Equation (13.36) then yields the $n$ equations ($p = 1, \ldots n$)

$$\frac{\partial}{\partial \zeta}\left(h_\eta h_\varphi \rho v_\zeta \chi^{(p)} - \rho D_p^{(m)} \frac{h_\eta h_\varphi}{h_\zeta} \frac{d\chi^{(p)}}{d\zeta}\right) = 0 \tag{13.38}$$

Again the flux of the component $p = 0$ (gas) is nil everywhere, $n_\zeta^{(0)} = 0$. The mass conservation equation is analogously written as

$$\frac{\partial}{\partial \zeta}\left(h_\eta h_\varphi \rho v_\zeta\right) = 0 \tag{13.39}$$

and its integration yields

$$\rho v_\zeta = \frac{1}{h_\eta h_\varphi} \cdot C\,(\eta, \varphi) \tag{13.40}$$

where $C\,(\eta, \varphi)$ is an unknown function of $\eta$ and $\varphi$, and substituting into (13.38) yields

$$\frac{\partial}{\partial \zeta}\left(C\,(\eta, \varphi)\,\chi^{(p)} - \rho D_p^{(m)} \frac{h_\eta h_\varphi}{h_\zeta} \frac{d\chi^{(p)}}{d\zeta}\right) = 0 \tag{13.41}$$

Since in spheroidal coordinates the group $\frac{h_\zeta}{h_\eta h_\varphi}$ is separable, i.e.

$$\frac{h_\zeta}{h_\eta h_\varphi} = B\,(\zeta)\,H\,(\eta, \varphi) \tag{13.42}$$

where

$$
B\left(\zeta\right) = \begin{cases} \frac{1}{a\left(\zeta^2-1\right)} & \text{prolate} \\ \frac{1}{a\left(\zeta^2+1\right)} & \text{oblate} \end{cases} \;\; ; \;\; H\left(\eta, \varphi\right) = 1 \tag{13.43}
$$

equation (13.41) becomes

$$
H\left(\eta, \varphi\right) C\left(\eta, \varphi\right) \frac{d\chi^{(p)}}{d\zeta} = \rho D_p^{(m)} \frac{d}{d\zeta}\left(\frac{1}{B\left(\zeta\right)} \frac{d\chi^{(p)}}{d\zeta}\right) \tag{13.44}
$$

Since $\chi^{(p)}$ depends only on $\zeta$, the RHS depends only on $\zeta$ and this must hold also for the LHS, then necessarily

$$
H\left(\eta, \varphi\right) C\left(\eta, \varphi\right) = M_1 = \text{const.} \tag{13.45}
$$

Introducing the function

$$
F\left(\zeta\right) = \int_{\zeta_0}^{\zeta} B\left(s\right) ds \tag{13.46}
$$

the solutions of Eqs. (13.44) can be written as

$$
\chi^{(p)} = \lambda^{(p)} e^{\frac{M_1}{\rho D_p^{(m)}} F(\zeta)} + \nu^{(p)} \tag{13.47}
$$

where $\lambda^{(p)}$ and $\nu^{(p)}$ are arbitrary constants, which can be found by applying the Dirichlet boundary conditions yielding the explicit formulas

$$
\lambda^{(p)} = \frac{\chi_s^{(p)} - \chi_\infty^{(p)}}{\left(1 - e^{Z^{(p)}}\right)} \tag{13.48}
$$

$$
\nu^{(p)} = \frac{\chi_\infty^{(p)} - \chi_s^{(p)} e^{Z^{(p)}}}{\left(1 - e^{Z^{(p)}}\right)} \tag{13.49}
$$

where

$$
Z^{(p)} = \frac{M_1}{\rho D_p^{(m)}} F\left(\infty\right) \tag{13.50}
$$

The local vapour fluxes of each species on the drop surface ($\zeta = \zeta_0$) can be calculated from Eqs. (13.37) using equations (13.47), (13.40), (13.42) and (13.45), yielding

$$
n_\zeta^{(p)} = \frac{M_1}{h_\eta h_\varphi H\left(\eta, \varphi\right)} \nu^{(p)} \tag{13.51}
$$

The evaporation rate for each species are found by integrating the species flux over the drop surface

$$m_{ev}^{(p)} = \int_{A_d} \hat{n}_\zeta^{(p)} dA = \int_0^{2\pi} \int_{-1}^1 \hat{n}_\zeta^{(p)} h_\eta h_\varphi d\eta d\varphi = M_1 \nu^{(\alpha)} K \qquad (13.52)$$

where $K = \int_0^{2\pi} \int_{-1}^1 \frac{1}{H(u,v)} d\eta d\varphi$ is a constant depending only on the geometry of the problem that for the spheroidal and ellipsoidal cases is equal to $4\pi$ [25].

The total evaporation rate is

$$m_{ev}^{(T)} = \sum_{p=1}^n m_{ev}^{(p)} = M_1 K \sum_{p=1}^n \nu^{(\alpha)} \qquad (13.53)$$

but it can also be found observing that $n^{(T)} = \sum_{k=1}^n n_\zeta^{(k)} = \rho v_\zeta$ (the condition $n_\zeta^{(0)} = 0$ has been used) and then

$$m_{ev}^{(T)} = \int_{A_d} \hat{n}_\zeta^{(T)} dA = \int_0^{2\pi} \int_{-1}^1 \rho v_\zeta h_\eta h_\varphi d\eta d\varphi = \int_0^{2\pi} \int_{-1}^1 C(\eta, \varphi) d\eta d\varphi = M_1 K \qquad (13.54)$$

where the third equality comes from (13.40) and the last one from (13.45).

Comparison of Eqs. (13.52), (13.53) and (13.54) yields

$$\nu^{(p)} = \frac{m_{ev}^{(p)}}{m_{ev}^{(T)}} \qquad (13.55)$$

$$\sum_{\alpha=1}^n \nu^{(\alpha)} = 1 \qquad (13.56)$$

Since $\nu^{(p)}$ are functions of $M_1$ (see Eqs. 13.48 and 13.50), Eq. (13.56) is a trascendental equation for the unknown $M_1$, or $m_{ev}^{(T)}$ through equation (13.54). Then the solution of (13.56), which can be obtained by standard numerical methods, yields the total mass flow rate and, through (13.55), the partial mass flow rates.

The model is then an extension of what discussed in Sect. 13.1 to spheroidal multi-component drops; for more information on the model and an extension to convection condition see [24].

# References

1. Sazhin, S.: Modelling of fuel droplet heating and evaporation: recent results and unsolved problems. Fuel **196**, 69–101 (2017)
2. Faeth, G.M.: Evaporation and combustion of sprays. Prog. Energy Combust. Sci. **9**, 1–76 (1983)
3. Tong, A.Y., Sirignano, W.A.: Multicomponent transient droplet vaporization with internal circulation: integral equation formulation. Numer. Heat Transf. **10**, 253–278 (1986)
4. Lage, P.L.C., Hackenberg, C.M., Rangel, R.H.: Nonideal vaporization of dilating binary droplets with radiation absorption. Combust. Flame **101**, 36–44 (1995)

5. Torres, D.J., O'Rourke, P.J., Amsden, A.A.: Efficient multi-component fuel algorithm. Combust. Theory Model. **7**, 67–86 (2003)
6. Maqua, C., Castanet, G., Lemoine, F.: Bi-component droplets evaporation: temperature measurements and modelling. Fuel **87**, 2932–2942 (2008)
7. Burger, M., Schmehl, R., Prommersberger, K., Schäfer, O., Koch, R., Wittig, S.: Droplet evaporation modelling by the distillation curve model: accounting for kerosene fuel and elevated pressures. Int. J. Heat Mass Transf. **46**, 4403–4412 (2003)
8. Tamim, J., Hallett, W.L.H.: Continuous thermodynamics model for multicomponent vaporization. Chem. Eng. Sci. **50**, 2933–2942 (1995)
9. Lippert, A.M., Reitz, R.D.: Modelling of multicomponent fuels using continuous distributions with application to droplet evaporation and sprays. SAE Technical Paper 972882 (1997)
10. Zhu, G.-S, Reitz, R.D.: A model for high-pressure vaporization of droplets of complex liquid mixture using continuous thermodynamics. Int. J. Heat Mass Transf. **45**, 495–507 (2002)
11. Sazhin, S.S., Elwardany, A., Sazhina, E.M., Heikal, M.R.: A quasi-discrete model for heating and evaporation of complex multicomponent hydrocarbon fuel droplets. Int. J. Heat Mass Transf. **54**, 4325–4332 (2011)
12. Sazhin, S.S., Elwardany, A.E., Krutitskii, P.A., Depredurand, V., Castanet, G., Lemoine, F., Sazhina, E.M., Heikal, M.R.: Multi-component droplet heating and evaporation: numerical simulation versus experimental data. Int. J. Therm. Sci. **50**(7), 1164–1180 (2011)
13. Elwardany, A.E., Sazhin, S.S.: A quasi-discrete model for droplet heating and evaporation: application to Diesel and gasoline fuels. Fuel **97**, 685–694 (2012)
14. Sazhin, S.S., Shishkova, I.N., Al Qubeissi, M.: Heating and evaporation of a two-component droplet: hydrodynamic and kinetic models. Int. J. Heat Mass Transf. **79**, 704–712 (2014)
15. Sazhin, S.S., Al Qubeissi, M., Nasiri, R., Gun'ko, V.M., Elwardany, A.E., Lemoine, F., Grisch, F., Heikal, M.R.: A multi-dimensional quasi-discrete model for the analysis of Diesel fuel droplet heating and evaporation. Fuel **129**, 238–266 (2014)
16. Al Qubeissi, M., Sazhin, S.S., Elwardany, A.E.: Modelling of blended Diesel and biodiesel fuel droplet heating and evaporation. Fuel **187**, 349–355 (2017)
17. Sazhin, S.: Droplet and Sprays. Springer, Berlin (2014)
18. Tonini, S., Cossali, G.E.: A novel formulation of multi-component drop evaporation models for spray applications. Int. J. Therm. Sci. **89**, 245–253 (2015)
19. Zhang, L., Kong, S.-C.: Multicomponent vaporization modeling of bio-oil and its mixtures with other fuels. Fuel **95**, 471–480 (2012)
20. Wilms, J.: Evaporation of multicomponent droplets. PhD thesis, Universität Stuttgart (2005)
21. Wilms, J., Weigand, B.: Composition measurements of binary mixture droplets by rainbow refractometry. J. Appl. Opt. **46**, 2109–2118 (2007)
22. Tonini, S., Cossali, G.E.: A novel vaporisation model for a single-component drop in high temperature air streams. Int. J. Therm. Sci. **75**, 194–203 (2014)
23. Tonini, S., Cossali, G.E.: A multi-component drop evaporation model based on analytical solution of Stefan-Maxwell equations. Int. J. Heat Mass Transf. **92**, 184–189 (2016)
24. Tonini, S., Cossali, G.E.: An analytical approach to model heating and evaporation of multi-component ellipsoidal drops. Heat Mass Transf. **55**(5), 1257–1269 (2019)
25. Tonini, S., Cossali, G.E.: One-dimensional analytical approach to modelling evaporation and heating of deformed drops. Int. J. Heat Mass Transf. **9**, 301–307 (2016)

# Appendix
# Special Functions

Special functions are found to be of particular importance in mathematical analysis or mathematical physics or in other applications. This definition is possibly vague, since there is no consensus on a general definition of special functions, but there exists a common agreement on a large amount of them, like those reported in this appendix. Many special functions, and all those reported here, appear either as integrals of some elementary functions or as solutions of differential equations.

The following description of some kinds of special functions is limited to those used in this book, without any claim of completeness, and the interested reader is invited to refer to specialised books like [1–3] and other that are specifically mentioned in the text, for more extensive treatments.

## A.1    The Gamma, Digamma and Beta Functions and Elliptic Integrals

Although practically all special functions admit an integral representation, i.e. they can be written as the result of a definite integral, those reported in this section have a quite direct relationship with a definite integral. Beside the *elliptic integrals*, the *Gamma* function was for some time called *Euler integral of second kind* (named after the celebrated Swiss mathematician Leonhard Euler, 1707–1783); the *Digamma* function is the logarithmic derivative of the *Gamma* function while the *Beta* function was also called the *Euler integral of first kind*.

### A.1.1    The Gamma Function

The Gamma function is a quite ubiquitous special function, since it appears in the most disparate areas of physics and mathematics, and often in mathematical relations among other special functions. The treatment of the Gamma function given in this

© Springer Nature Switzerland AG 2021

G. E. Cossali and S. Tonini, *Drop Heating and Evaporation: Analytical Solutions in Curvilinear Coordinate Systems*, Mathematical Engineering, https://doi.org/10.1007/978-3-030-49274-8

appendix is deliberately brief and the interested reader can refer, for a more complete treatment, to specialised books like [4, 5].

The first representation of the Gamma function was given by Daniel Bernoulli (a Swiss mathematician and physicist, 1700–1782) and Leonhard Euler in two letters written independently to Goldbach (Christian Goldbach, a German mathematician, 1690–1764) in October 1729 [6]; at that time Daniel Bernoulli and Leonhard Euler were working together in St. Petersburg.

The Gamma function can be seen as an extension of the factorial to complex and real number arguments, and it was the research on this problem that led Euler to suggest the first integral representation

$$\Gamma(z) = \int_0^\infty e^{-t} t^{z-1} dt; \qquad \text{Re}\{z\} > 0 \qquad (A.1)$$

The notation $\Gamma(x)$ is due to Legendre [7] (see also [8]), and the relation with the factorial is

$$\Gamma(n+1) = n! \qquad (A.2)$$

Some special values of $\Gamma(z)$ are the following

$$\Gamma(1) = 1; \quad \Gamma\left(\frac{1}{2}\right) = \pi^{1/2}; \quad \Gamma'(1) = \gamma \qquad (A.3)$$

where

$$\gamma = \lim_{n\to\infty} \left(\sum_{k=1}^n \frac{1}{k} - \ln(n)\right) = 0.5772156649.... \qquad (A.4)$$

is the Euler's constant. The Gamma function can be defined for any complex number $z$ (see [3] for a detailed analysis) with the exception of negative integer numbers. The Gamma function has no zeros in the complex plane [3] and Fig. A.1 shows $\Gamma(z)$ for real values of $z$.

The Gamma function satisfies an important recurrence relation

$$\Gamma(z+1) = z\Gamma(z) \qquad (A.5)$$

that can be found directly from the integral representation, in fact

$$\Gamma(z+1) = \int_0^\infty e^{-t} t^z dt = -\left[e^{-t} t^z\right]_0^\infty + z \int_0^\infty e^{-t} t^{z-1} dt \qquad (A.6)$$

and since the first term in the most RHS is nil, Eq. (A.5) is proven. Equation (A.5) is useful to find values of the function without performing other integrations, for example

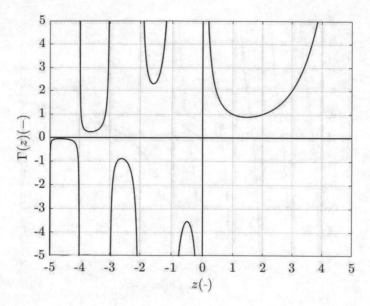

**Fig. A.1** Gamma function $\Gamma(z)$ for real values of $z$

$$\Gamma(3/2) = \frac{3}{2}\Gamma(1/2) = \frac{3}{2}\pi^{1/2}$$

$$\Gamma(5/2) = \frac{5}{2}\Gamma(3/2) = \frac{15}{4}\pi^{1/2} \tag{A.7}$$

$$\ldots$$

There are other two important relations satisfied by $\Gamma(z)$

$$\Gamma(z)\,\Gamma(z-1) = \frac{\pi}{\sin(z\pi)}; \quad z \neq 0, \pm 1, \pm 2, \ldots \tag{A.8}$$

$$\Gamma(2z) = \frac{2^{2z-1}}{\pi^{1/2}}\Gamma(z)\,\Gamma(z+1/2); \quad 2z \neq 0, -1, -2, \ldots \tag{A.9}$$

which can again be obtained directly from Eq. (A.1), see [3] for a derivation.

## A.1.2 The Digamma Function

Another important function related to the Gamma function is its *logarithmic derivative*, or

$$\psi(z) = \frac{d\ln\Gamma(z)}{dz} = \frac{\Gamma'(z)}{\Gamma(z)} \tag{A.10}$$

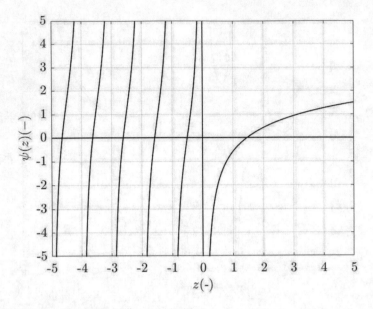

**Fig. A.2** Digamma function $\psi(z)$ for real values of $z$

called *Digamma* function or *Psi* function. Figure A.2 shows its graph for real values of $z$.

The analogous of the relations (A.5), (A.8) and (A.9) are

$$\psi(z+1) - \psi(z) = \frac{1}{z} \tag{A.11}$$

$$\psi(1-z) - \psi(z) = \frac{\pi}{\tan(z\pi)}; \quad z \neq 0, \pm 1, \ldots \tag{A.12}$$

$$\psi(2z) - \frac{1}{2}[\psi(z) + \psi(z+1/2)] = \ln 2; \quad 2z \neq 0, -1, -2, \ldots \tag{A.13}$$

and some special values can be found applying the definition or these relations, for example

$$\psi(1) = \frac{\Gamma'(1)}{\Gamma(1)} = -\gamma \tag{A.14}$$

and from (A.12)

$$\psi(1/2) = \psi(1) - 2\ln 2 = -\gamma - 2\ln 2 \tag{A.15}$$

Values of $\psi(z)$ for negative argument can be calculated from those with positive argument by (A.12)

$$\psi(-z) = \psi(1+z) + \frac{\pi}{\tan(z\pi)} \tag{A.16}$$

For many different integral representations of $\psi(z)$ see [1].

## A.1.3 The Beta Function

A third function that appears relatively often in the theory of special functions and in many applications is the so-called *Beta* function, a name that was given to this function, also called *Euler integral of first kind*, by Binet (a French mathematician, 1786–1856) on 1839 [9] in a paper (see also [8]), where he proposed the symbol $B$, which is the Greek capital letter *beta*.

The integral representation of the Beta function is

$$B(a, b) = \int_0^1 t^{a-1} (1-t)^{b-1} \, dt \tag{A.17}$$

and it bears a simple relation to the Gamma function

$$B(a, b) = \frac{\Gamma(a)\,\Gamma(b)}{\Gamma(a+b)} \tag{A.18}$$

It satisfy various identities (see [1]), the following

$$B(a, 1-a) = \frac{\pi}{\sin(\pi u)} \tag{A.19}$$

setting $a = 1/2$, in conjunction with Eq. (A.18) yields the result

$$\pi = \frac{\pi}{\sin(\pi/2)} = B(1/2, 1/2) = \frac{\Gamma^2(1/2)}{\Gamma(1)} = \Gamma^2(1/2) \tag{A.20}$$

i.e. $\Gamma(1/2) = \pi^{1/2}$. Equation (A.18) suggest also a relation with binomial coefficient for $a$, $b$ integers

$$\binom{m}{n} = \frac{m!}{n!\,(m-n)!} = \frac{1}{m+1} \frac{(m+1)!}{n!\,(m-n)!} = \frac{1}{m+1} \frac{\Gamma(m+2)}{\Gamma(n+1)\,\Gamma(m-n+1)} \tag{A.21}$$

$$= \frac{1}{(m+1)\,B(n+1, m-n+1)}$$

Due to Eq. (A.18) all the properties of this function can be derived from those of $\Gamma(z)$.

## A.1.4    The Elliptic Integrals

An integral of the form

$$I = \int F(t, w)\, dt \tag{A.22}$$

where $F$ is a rational function of $w$ and $t$ (i.e. it can be written as the ratio between two polynomials in $w$ and $t$) and $w^2$ is a polynomial of third or fourth degree in $t$ (i.e. $w^2 = \sum_{k=0}^{4} a_k t^k$ where $a_4 \neq 0$ or $a_3 \neq 0$ or both) is said to be an *elliptic integral*. The name is due to the fact that these integrals were found when trying to calculate the length of an arc of ellipse. The first to use elliptic integral was probably Fagnano (Giulio Carlo Fagnano, an Italian mathematician, 1682–1766) while studying the rectification of the lemniscate [10], while the foundation of the elliptic integrals and elliptic functions theory is due to Euler.

The most important elliptic integrals are those called *incomplete (Legendre)* elliptic integral of first, $F(\phi, t)$, second, $E(\phi, t)$ and third, $\Pi(\phi, \alpha^2, t)$ kind, defined as

$$F(\phi, t) = \int_0^\phi \frac{dx}{\sqrt{1 - k^2 \sin^2(x)}} = \int_0^{\sin(\phi)} \frac{dt}{\sqrt{1 - t^2}\sqrt{1 - k^2 t^2}} \tag{A.23}$$

$$E(\phi, t) = \int_0^\phi \sqrt{1 - k^2 \sin^2(x)}\, dx = \int_0^{\sin(\phi)} \frac{\sqrt{1 - k^2 t^2}}{\sqrt{1 - t^2}}\, dt \tag{A.24}$$

$$\Pi(\phi, \alpha^2, t) = \int_0^\phi \frac{1}{\sqrt{1 - k^2 \sin^2(x)}\left(1 - \alpha^2 \sin^2(x)\right)}\, dx$$
$$= \int_0^{\sin(\phi)} \frac{dt}{\sqrt{1 - t^2}\sqrt{1 - k^2 t^2}\left(1 - \alpha^2 t^2\right)} \tag{A.25}$$

where $\phi$ is the *amplitude* (or argument), $k$ is the *elliptic modulus* (or eccentricity) while $\alpha^2$ in the third one is called *characteristic*. Sometimes the parameter $m = k^2$ is used instead of the modulus, or also the *complementary* modulus $k'$ defined by $k^2 + k'^2 = 1$. The *complete* elliptic integrals are instead

$$K(k) = F\left(\frac{\pi}{2}, t\right) \quad \text{first kind} \tag{A.26}$$

$$E(k) = E\left(\frac{\pi}{2}, t\right) \quad \text{second kind} \tag{A.27}$$

$$\Pi(\alpha^2, t) = \Pi\left(\frac{\pi}{2}, \alpha^2, t\right) \quad \text{third kind} \tag{A.28}$$

Some special values are the following

$$F\left(0,k\right)=0;\ \ F\left(\phi,0\right)=\phi;\ \ F\left(\frac{\pi}{2},1\right)=\infty;\ \ \lim_{\phi\to0}\frac{F\left(\phi,k\right)}{\phi}=1 \quad (A.29)$$

$$E\left(0,k\right)=0;\ \ E\left(\phi,0\right)=\phi;\ \ E\left(\frac{\pi}{2},1\right)=1;\ \ \lim_{\phi\to0}\frac{E\left(\phi,k\right)}{\phi}=1 \quad (A.30)$$

$$\Pi\left(\phi,0,t\right)=F\left(\phi,t\right) \quad (A.31)$$

$$K\left(0\right)=E\left(0\right)=\frac{\pi}{2};\ \ K\left(1\right)=\infty;\ \ E\left(1\right)=1 \quad (A.32)$$

The complete integrals satisfy some connection formulas and identities like

$$\frac{\pi}{2}=E\left(k'\right)K\left(k\right)-K\left(k\right)K'\left(k\right)+E\left(k\right)K\left(k'\right) \quad (A.33)$$

$$K\left(\frac{1}{k}\right)=\begin{cases}kK\left(k\right)-ikK\left(k'\right)\ \ \text{if}\ \text{Im}\left(k^2\right)>0\\kK\left(k\right)+ikK\left(k'\right)\ \ \text{if}\ \text{Im}\left(k^2\right)<0\end{cases} \quad (A.34)$$

$$E\left(\frac{1}{k}\right)=\begin{cases}\frac{1}{k}\left[E\left(k\right)+iE\left(k'\right)-k'^2K\left(k\right)-ik^2K\left(k'\right)\right]\ \ \text{if}\ \text{Im}\left(k^2\right)>0\\\frac{1}{k}\left[E\left(k\right)-iE\left(k'\right)-k'^2K\left(k\right)+ik^2K\left(k'\right)\right]\ \ \text{if}\ \text{Im}\left(k^2\right)<0\end{cases} \quad (A.35)$$

and many others (see [1]). The incomplete elliptic integral can be useful for the evaluation of definite integrals like, for example

$$\int_1^x\frac{dt}{\sqrt{t^3-1}}=\frac{1}{3^{1/4}}F\left(\arccos\left(\frac{\sqrt{3}+1-x}{\sqrt{3}-1+x}\right),\sqrt{\frac{2-\sqrt{3}}{4}}\right)\ ;x>1 \quad (A.36)$$

$$\int_x^1\frac{dt}{\sqrt{1-t^3}}=\frac{1}{3^{1/4}}F\left(\arccos\left(\frac{\sqrt{3}-1+x}{\sqrt{3}+1-x}\right),\sqrt{\frac{2+\sqrt{3}}{4}}\right)\ ;x<1 \quad (A.37)$$

$$\int_0^x\frac{dt}{\sqrt{1+t^4}}=\text{sign}\left(x\right)\frac{1}{2}F\left(\arccos\left(\frac{1-x^2}{1+x^2}\right),\sqrt{\frac{1}{2}}\right) \quad (A.38)$$

Finally $K\left(k\right)$ and $E\left(k\right)$ can be expressed in term of the *hypergeometric* function (see the last section of the appendix)

$$K\left(k\right)=\frac{\pi}{2}\,_2F_1\left(\frac{1}{2};\frac{1}{2};1;k^2\right) \quad (A.39)$$

$$E\left(k\right)=\frac{\pi}{2}\,_2F_1\left(-\frac{1}{2};\frac{1}{2};1;k^2\right) \quad (A.40)$$

## A.2 The Bessel Functions

The solutions of the Bessel's equation

$$z^2\frac{d^2W}{dz^2}+z\frac{dW}{dz}+\left(z^2-\nu^2\right)W=0 \quad (A.41)$$

for an arbitrary complex number $\nu$, are called *Bessel functions of order* $\nu$. The first appearance of these functions (precisely the Bessel function of order $\nu = 0$) is in a memoir of Daniel Bernoulli, written in 1732–1733, but published in 1738 [11], where it was used to describe the oscillating modes of an hanging chain. This equation appears often when solving problems in cylindrical coordinates, that is why Bessel functions are also called *cylindrical* functions. In this appendix we will summarise some of the properties of this class of functions and the interested reader can refer to [12] for a complete treatment.

When $\nu = n$ is an integer (in this section $n$ always indicates a positive integer), the most general solution of Eq. (A.41) is

$$W = A J_n (z) + B Y_n (z) \tag{A.42}$$

where $J_n (z)$ are the Bessel functions of *first kind* of order $n$, and $Y_n (z)$ are the Bessel functions of *second kind* (of order $n$), sometime called *Weber* functions (named after the German mathematician Heinrich Martin Weber, 1842–1913 [13]). It should be noticed that for non integer order the two independent solutions are $J_\nu (z)$ and $J_{-\nu} (z)$ while for integer orders these two functions are related by the identity

$$J_n (z) = (-1)^n J_n (z) \tag{A.43}$$

### A.2.1   Bessel Function of First Kind

The Bessel function of first kind can be defined by the series:

$$J_\nu (z) = \left(\frac{1}{2}z\right)^n \sum_{k=0}^{\infty} (-1)^k \frac{\left(\frac{1}{4}z^2\right)^k}{k\Gamma (\nu + k + 1)} \tag{A.44}$$

where $\Gamma (x)$ is the Gamma function or, for integer order, by the integral forms

$$J_n (z) = \frac{1}{\pi} \int_0^{\pi} \cos (z \sin (\theta) - n\theta) \, d\theta \tag{A.45}$$

$$J_n (z) = \frac{i^{-1}}{\pi} \int_0^{\pi} e^{iz \cos(\theta)} \cos (n\theta) \, d\theta \tag{A.46}$$

but many others integral forms, equivalent to those above reported, can be used (see [3], pp. 114–115). A sketch of the first four Bessel functions are reported in Fig. A.3. The functions $J_n$ satisfy some important recurrence relations. The first one is

$$J_{n-1} (z) + J_{n+1} (z) = \frac{2n}{z} J_n (z) \tag{A.47}$$

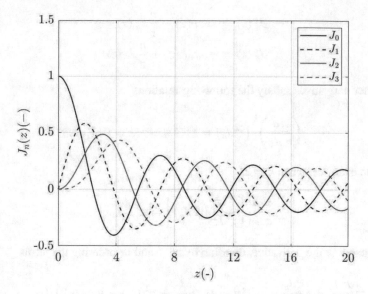

**Fig. A.3** Bessel functions of first kind $J_n(z)$ for real values of $z$

which shows how all the $J_n$ of arbitrary order can be expressed in terms of $J_0$ and $J_1$; in fact repeatedly applying (A.47) yields the sequence:

$$J_2(z) = \frac{2}{z} J_1(z) - J_0(z) \tag{A.48}$$

$$J_3(z) = \frac{4}{z} J_2(z) - J_1(z) = \left(\frac{8}{z^2} - 1\right) J_1(z) - \frac{4}{z} J_0(z) \tag{A.49}$$

$$J_4(z) = \frac{6}{z} J_3(z) - J_2(z) = \left(\frac{48}{z^3} - \frac{4}{z}\right) J_1(z) - \left(\frac{24}{z^2} + 1\right) J_0(z) \tag{A.50}$$

$$\cdots$$

Another recurrence relation involves the first derivative of $J_n$

$$J_{n-1}(z) - J_{n+1}(z) = 2J_n'(z) \tag{A.51}$$

where $J_n'(z) = \frac{dJ_n}{dz}$. An important special case is

$$\frac{dJ_0}{dz} = -J_1(z) \tag{A.52}$$

which is obtained from (A.51), making use of the identity (A.43). A combination of (A.47) and (A.51) can be used to yield other recurrence relations like [1]

$$J_n'(z) = J_{n-1}(z) - \frac{n}{z} J_n(z) \tag{A.53}$$

$$J_n'(z) = -J_{n+1}(z) + \frac{n}{z} J_n(z) \tag{A.54}$$

Higher derivatives satisfy the following relations

$$\left(\frac{1}{z}\frac{d}{dz}\right)^k (z^n J_n) = z^{n-k} J_{n-k}; \quad k = 0, 1, \ldots \tag{A.55}$$

As an example, when $k = 2$

$$\frac{1}{z}\frac{d}{dz}\left[\frac{1}{z}\frac{d}{dz}(z^n J_n)\right] = z^{n-2} J_{n-2} \tag{A.56}$$

then expanding the derivative, dividing by $z^{n-4}$ and re-ordering the terms

$$z^2\frac{d^2 J_n}{dz^2} + z\frac{d J_n}{dz} = -(2n-2)z\frac{d J_n}{dz} + z^2 J_{n-2} - n(n-2) J_n \tag{A.57}$$

Using (A.47) and (A.51) yields

$$z^2 J_{n-2}(z) = 2(n-1)z J_{n-1}(z) - z^2 J_n(z) \tag{A.58}$$

$$(2n-2)z J_n'(z) = (2n-2)z J_{n-1}(z) - (2n-2)n J_n(z) \tag{A.59}$$

and substituting into the RHS of (A.57) yields

$$z^2\frac{d^2 J_n}{dz^2} + z\frac{d J_n}{dz} = (n^2 - z^2) J_n(z) \tag{A.60}$$

which is again the Bessel equation.

Integral relations among different $J_n(z)$ can be deduced by the recurrence relations (A.51), for example

$$\int z^{n+1} J_n(z)\, dz = z^{n+1} J_{n+1}(z) \tag{A.61}$$

can be obtained from (A.55) with $k = 1$

$$\frac{d}{dz}(z^n J_n) = z^n J_{n-1} \tag{A.62}$$

integrating both side and changing $n$ with $n + 1$.

Many other integrals involving Bessel functions can be found in [1], pp. 240–246.

As it can be inferred from Fig. A.3, the functions $J_n(z)$ and $J_n'(z)$ have an infinite numbers of positive real simple zeros and, with the exception of $z = 0$, the functions $J_n(z)$ of different order have no common zeroes. An important property of Bessel functions, related to the zeroes, is the orthogonality. Defining with $z_{n,k}$ the $k$th-zero of the function $J_n(z)$ the following orthogonality relation holds

$$\int_0^1 t J_n\left(z_{n,k}\right) J_n\left(z_{n,j}\right) dt = \frac{1}{2}\delta_{jk}\left[J_n'\left(z_{n,k}\right)\right]^2 \tag{A.63}$$

where $\delta_{jk}$ is the Kronecker's symbol.

Consider now the positive solutions, say $z = s_k$, of the equation

$$A J_n(z) + B J_n'(z) = 0 \tag{A.64}$$

with $A$, $B$ real and $B \neq 0$, the following orthogonality relation holds

$$\int_0^1 t J_n\left(s_k\right) J_n\left(s_j\right) dt = \delta_{jk}\left(\frac{\frac{A^2}{B^2} - n^2 + s_k^2}{2s_k^2}\right) \left[J_n\left(s_k\right)\right]^2 \tag{A.65}$$

## A.2.2 Bessel Functions of Second Kind

Bessel functions of second kind $Y_n(z)$ are the second independent solution of the Bessel equation (A.41) when $\nu$ is an integer and it can be defined, for non integer $\nu$, as

$$Y_\nu(z) = \frac{J_\nu(z) \cos(\nu\pi) - J_{-\nu}(z)}{\sin(\nu\pi)} \tag{A.66}$$

while for $\nu = n$ positive integer

$$Y_n(z) = \lim_{\nu \to n} Y_\nu(z) \tag{A.67}$$

and a series definition is

$$Y_n(z) = \frac{\left(\frac{z}{2}\right)^{-n}}{\pi} \sum_{k=0}^{n-1} \frac{(n-1-k)!}{k!} \left(\frac{z}{2}\right)^{2k} + 2\pi \ln\left(\frac{z}{2}\right) J_n(z) + \tag{A.68}$$

$$-\frac{\left(\frac{z}{2}\right)^n}{\pi} \sum_{k=0}^{\infty} \left[\psi(k+1) + (-1)^k \psi(n+k+1) \frac{\left(\frac{z}{2}\right)^{2k}}{k!(n+k)!}\right]$$

where $\psi(x) = \frac{\Gamma'(x)}{\Gamma(x)} = \frac{d \ln \Gamma(x)}{dx}$ is the *Digamma* function. Figure A.4 shows the first four Bessel functions of second kind.

The identity

$$Y_{-n}(x) = (-1)^n Y_n(x) \tag{A.69}$$

can be used to evaluate the function for negative orders.

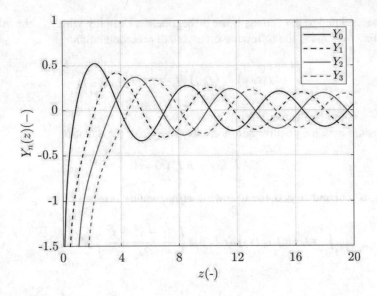

**Fig. A.4** Bessel functions of second kind $Y_n(z)$ for real values of $z$

Recurrence relations (A.47), (A.51), (A.53) and (A.54) hold substituting $J_n$ with $Y_n$, and the same holds for the high order derivatives (Eq. (A.55)) and the integral relation (A.61). An integral representation is (*Schläfli's integral*)

$$Y_n(z) = \frac{1}{\pi} \int_0^\pi \sin(z \sin(\theta) - n\theta)\, d\theta - \frac{1}{\pi} \int_0^\infty \left(e^{nt} + e^{-nt} \cos(n\pi)\right) e^{-z \sinh(t)}\, dt \tag{A.70}$$

and many others can be found in [1], p. 224.

Although Bessel functions of second kind are a second independent solution of the Bessel equation, many relations exist between the two kind of functions. For example [1] if $z_k$ are the zeros of the linear combination

$$F_n(z) = J_n(z) \cos(\pi t) + Y_n \sin(\pi t) \tag{A.71}$$

where $t$ is a parameter, i.e. $F_n(z_k) = 0$, then

$$F_n'(z_k) = F_{n-1}(z_k) = -F_{n+1}(z_k) \tag{A.72}$$

while if $s_k$ is a zero of $F_n'(z)$, i.e. $F_n'(s_k) = 0$ then

$$F_n(s_k) = \frac{s_k}{n} F_{n-1}(s_k) = \frac{s_k}{n} F_{n+1}(s_k) \tag{A.73}$$

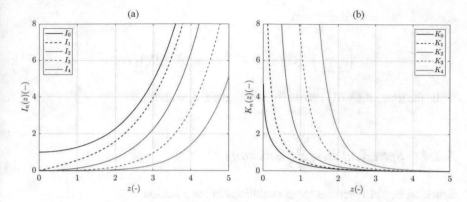

**Fig. A.5** Modified Bessel functions of **a** first kind $I_n(z)$ and **b** second kind $K_n(z)$ for $z \in [0, 5]$

## A.2.3 Modified Bessel Functions

Modified Bessel functions are the solution of the modified Bessel equation

$$z^2 \frac{d^2 W}{dz^2} + z \frac{dW}{dz} - \left(z^2 + \nu^2\right) W = 0 \tag{A.74}$$

and are represented by $I_\nu(z)$ (modified Bessel function of first kind) and $K_\nu(z)$ (modified Bessel function of second kind). Figure A.5 shows the first five modified Bessel functions of first and second kind.

Since Eq. (A.74) can be obtained from (A.41) by the substitution $z \to \pm i z$, there exists obvious relationships among the modified and non modified Bessel functions

$$I_\nu(z) = i^{-\nu} J_\nu(iz) \tag{A.75}$$

$$K_\nu(z) = \frac{\pi}{2} \frac{I_{-\nu}(z) - I_\nu(z)}{\sin(\nu \pi)} \tag{A.76}$$

and the limiting value, for $\nu \to n$, must be taken in the second equation when $n$ is an integer. For integer orders, using the symbol $Z_n(z)$ to indicate either $I_n(z)$ or $K_n(z)$, the following recurrence relations hold

$$Z_{n-1}(z) - Z_{n+1}(z) = \frac{2n}{z} Z_n(z) \tag{A.77}$$

$$Z_{n-1}(z) + Z_{n+1}(z) = 2 Z_n'(z) \tag{A.78}$$

$$Z_n'(z) = Z_{n-1}(z) - \frac{n}{z} Z_n(z) \tag{A.79}$$

$$Z_n'(z) = Z_{n+1}(z) + \frac{n}{z} Z_{n+1}(z) \tag{A.80}$$

and for higher order derivatives

$$\left(\frac{1}{z}\frac{d}{dz}\right)^k \left(z^n Z_n\right) = z^{n-k} Z_{n-k}; \quad k = 0, 1, \ldots \tag{A.81}$$

while the integral (A.61) holds also substituting $J_n$ with $Z_n$.

## A.2.4 Spherical Bessel Functions

Spherical Bessel functions are the solutions of the equation

$$z^2\frac{d^2w}{dz^2} + 2z\frac{dw}{dz} + \left[z^2 - n(n+1)\right]w = 0 \tag{A.82}$$

The name came from the fact that Eq. (A.82) is found when solving the Helmholtz equation in spherical coordinate by the separation of variable method, and that this equation can be transformed to the Bessel equation setting

$$w = z^{-1/2}W \tag{A.83}$$

in fact, substituting $w$ into Eq. (A.82) yields the Bessel equation

$$z^2\frac{d^2W}{dz^2} + z\frac{dW}{dz} + \left(z^2 - \left(n + \frac{1}{2}\right)^2\right)W = 0 \tag{A.84}$$

and the two independent solutions of this Eq. (A.82) are $J_{n+\frac{1}{2}}(z)$ and $J_{-\left(n+\frac{1}{2}\right)}(z)$ (since $\nu = n + \frac{1}{2}$ is not an integer). The two independent solutions of Eq. (A.82) can be written as

$$j_n(z) = \sqrt{\frac{\pi}{2z}}J_{n+\frac{1}{2}}(z) = (-1)^n\sqrt{\frac{\pi}{2z}}Y_{-n-\frac{1}{2}}(z) \tag{A.85}$$

$$y_n(z) = (-1)^{n+1}\sqrt{\frac{\pi}{2z}}J_{-n-\frac{1}{2}}(z) = \sqrt{\frac{\pi}{2z}}Y_{n+\frac{1}{2}}(z) \tag{A.86}$$

where the last equalities come from Eq. (A.66) with $\nu = n + \frac{1}{2}$, and $j_n(z)$, $y_n(z)$ are the *spherical Bessel functions of first and second kind* respectively.

Analogously to the Bessel functions, the solution of the equation

$$z^2\frac{d^2w}{dz^2} + 2z\frac{dw}{dz} - \left[z^2 - n(n+1)\right]w = 0 \tag{A.87}$$

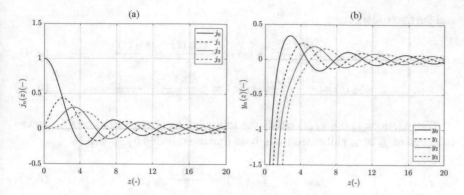

**Fig. A.6** Spherical Bessel functions of the **a** first kind $j_n(z)$ and **b** second kind $y_n(z)$ for $z \in [0, 20]$

are the *modified* spherical Bessel functions, defined in terms of the modified Bessel functions $I_\nu(z)$ and $K_\nu(z)$ as

$$i_n(z) = \sqrt{\frac{\pi}{2z}} I_{n+\frac{1}{2}}(z) \tag{A.88}$$

$$k_n(z) = \sqrt{\frac{\pi}{2z}} K_{n+\frac{1}{2}}(z) \tag{A.89}$$

Figure A.6 shows the graphs for some spherical Bessel functions. Explicit formulas for $j_n(z)$ and $y_n(z)$ are [1]

$$
\begin{aligned}
j_n(z) = \sin\left(z - \frac{n}{2}\pi\right) \sum_{k=0}^{\lfloor n/2 \rfloor} (-1)^k \frac{a_{2k}\left(n + \frac{1}{2}\right)}{z^{2k+1}} \\
+ \cos\left(z - \frac{n}{2}\pi\right) \sum_{k=0}^{\lfloor (n-1)/2 \rfloor} (-1)^k \frac{a_{2k+1}\left(n + \frac{1}{2}\right)}{z^{2k+2}}
\end{aligned} \tag{A.90}
$$

$$
\begin{aligned}
y_n(z) = -\cos\left(z - \frac{n}{2}\pi\right) \sum_{k=0}^{\lfloor n/2 \rfloor} (-1)^k \frac{a_{2k}\left(n + \frac{1}{2}\right)}{z^{2k+1}} \\
+ \sin\left(z - \frac{n}{2}\pi\right) \sum_{k=0}^{\lfloor (n-1)/2 \rfloor} (-1)^k \frac{a_{2k+1}\left(n + \frac{1}{2}\right)}{z^{2k+2}}
\end{aligned} \tag{A.91}
$$

where

$$a_k(n + 1/2) = \begin{cases} \frac{(n+k)!}{2^k k!(n-k)!} & k = 0, \dots, n \\ 0 & k = n+1, \dots \end{cases} \tag{A.92}$$

and the first orders are

$$j_0(z) = \frac{\sin(z)}{z}; \quad j_1(z) = -\frac{\cos(z)}{z} + \frac{\sin(z)}{z^2} \tag{A.93}$$

$$y_0(z) = -\frac{\cos(z)}{z}; \quad y_1(z) = -\frac{\sin(z)}{z} - \frac{\cos(z)}{z^2} \tag{A.94}$$

The functions $j_n(z)$, $y_n(z)$ satisfy the following recurrence relations ($f_n$ is taken to represent $j_n$ or $y_n$) inherited from those that are satisfied by $J_\nu$ and $Y_\nu$

$$f_{n-1}(z) + f_{n+1}(z) = \frac{2n+1}{z} f_n(z) \tag{A.95}$$

$$-f_{n+1}(z) + \frac{n}{z} f_n(z) = f_n'(z) \tag{A.96}$$

and the first Eq. (A.95) together with Eq. (A.93) or (A.94) allow to write the explicit form of any $j_n$ and $y_n$.

## A.3   The Legendre Functions

The Legendre functions are named after Legendre, who introduced them in a memoir on the gravitational attraction of homogeneous spheroids, published in 1785 [14] but approved for publication in 1783–1784 [15], and their name was first proposed in 1875 [16].

The Legendre differential equation

$$\left(1 - x^2\right) \frac{d^2 W}{dx^2} - 2x \frac{dW}{dx} + \nu(\nu+1) W = 0 \tag{A.97}$$

appears when solving the Laplace equation in spherical coordinates assuming rotational symmetry. The two independent solutions of this equation are the Legendre functions of first kind, $P_\nu(x)$, and second kind, $Q_\nu(x)$; the parameter $\nu$ is called *degree* of the functions. When the parameter $\nu$ is and integer, i.e. $\nu = n$, the functions $P_n(x)$ are called Legendre *polynomials*, while $Q_n(x)$ are sometimes improperly called Legendre polynomials of *second kind* although only $P_n(x)$ are polynomials.

Legendre polynomials form a complete set of orthogonal functions on the interval $[-1, 1]$ with unitary weight since

$$\int_{-1}^{1} P_n(x) P_m(x) \, dx = \frac{n}{2n+1} \delta_{nm} \tag{A.98}$$

These functions can be derived also through the Rodrigues' formula

**Table A.1** First five Legendre polynomials

| $n$ | $P_n(x)$ |
|---|---|
| 0 | 1 |
| 1 | $x$ |
| 2 | $\frac{3}{2}x^2 - \frac{1}{2}$ |
| 3 | $\frac{5}{2}x^3 - \frac{3}{2}x$ |
| 4 | $\frac{35}{8}x^4 - \frac{30}{8}x^2 + 3$ |

$$P_n(x) = \frac{1}{2^n n!} \frac{d^n}{dx^n}\left[\left(x^2 - 1\right)^n\right] \tag{A.99}$$

and they can be explicitly written as

$$P_n(x) = \frac{1}{2^n} \sum_{k=0}^{\lfloor n/2 \rfloor} (-1)^k \binom{n}{k}\binom{2(n-k)}{n} x^{n-2k} = 2^n \sum_{k=0}^{n} \binom{n}{k}\binom{\frac{n+k-1}{2}}{n} x^k \tag{A.100}$$

where in the most RHS the generalised binomial coefficient $\binom{a}{n} = \frac{a(a-1)\dots(a-k+1)}{k!}$ has been used. It is also easy to see that $P_{-(n+1)}(x) = P_n(x)$, as it stems from Eq. (A.97) by the substitution $\nu = -n - 1$; then it can be assumed $n \geq 0$. The first Legendre polynomials are reported in Table A.1, and Fig. A.7 shows the graphs of the first five $P_n(x)$ and it is evident that

$$P_n(x) = (-1)^n P_n(-x) \tag{A.101}$$

The polynomials are standardised in such a way that $P_n(1) = 1$. These polynomials satisfy the recurrence relation

$$P_{n+1}(x) = \frac{(2n+1)}{(n+1)} x P_n(x) - \frac{n}{(n+1)} P_{n-1}(x) \tag{A.102}$$

which can be used to evaluate any $P_n$ starting from $P_0(x) = 1$ and $P_1(x) = x$. The first derivative of $P_n$ can be expressed through $P_n(x)$ and $P_{n-1}(x)$ as

$$\left(x^2 - 1\right)\frac{dP_n}{dx} = nx P_n(x) - n P_{n-1}(x) \tag{A.103}$$

Another useful expression can be obtained evaluating $\frac{dP_{n+1}}{dx}$ and $\frac{dP_{n-1}}{dx}$ from (A.103), subtracting the results and using twice (A.102) to eliminate on the RHS $P_{n-2}(x)$ and $P_{n+1}(x)$ yielding

$$\frac{dP_{n+1}}{dx} - \frac{dP_{n-1}}{dx} = (2n+1) P_n(x) \tag{A.104}$$

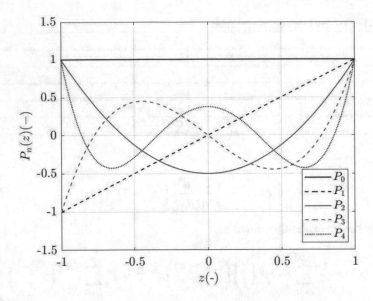

**Fig. A.7** First five Legendre functions of the first kind $P_n(z)$ for $z \in [-1, 1]$

which is very useful to evaluate indefinite integrals of $P_n(x)$

$$\int P_n(x) \, dx = \frac{1}{2n+1} \int \frac{d}{dx} (P_{n+1} - P_{n-1}) \, dx = \frac{P_{n+1}(x) - P_{n-1}(x)}{2n+1}$$

(A.105)

Special values of these polynomials are

$$P_n(-1) = (-1)^n$$

(A.106)

$$P_n(0) = \begin{cases} \frac{(-1)^k}{2^{2k}} \binom{2k}{k} & n = 2k \\ 0 & n = 2k + 1 \end{cases}$$

(A.107)

$$P_n'(1) = \frac{n(n+1)}{2}; \quad P_n'(-1) = (-1)^{n+1} \frac{n(n+1)}{2}$$

(A.108)

For $\nu = n$ integer the functions $Q_n(x)$ can be defined by the relation

$$Q_n(x) = \frac{1}{2} P_n \ln \left( \frac{1+x}{1-x} \right) - W_{n-1}(x)$$

(A.109)

where

$$W_{-1}(x) = 0$$

(A.110)

$$W_{n-1}(x) = \sum_{k=1}^{n} \frac{P_{k-1}(x) P_{n-k}(x)}{k}; \quad n > 0$$

(A.111)

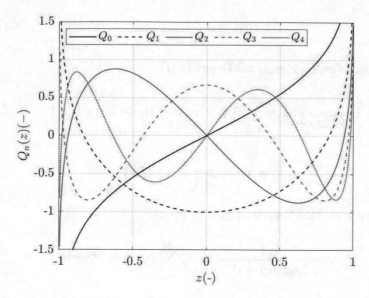

**Fig. A.8** First five Legendre functions of the second kind $Q_n(z)$ for $z \in [-1, 1]$

Figure A.8 shows the first five $Q_n(x)$.

The functions $Q_n$ satisfies the recurrence relations

$$(n+2) Q_{n+2} = (2n+3) x Q_{n+1} - (n+1) Q_n = 0 \qquad \text{(A.112)}$$

$$\left(1 - x^2\right) \frac{d Q_n}{dx} = (n+1) x Q_n - n Q_{n-1} \qquad \text{(A.113)}$$

which are identical to (A.102) and (A.103).

An integral representation of the Legendre polynomials [1] is

$$P_n(\cos \theta) = \frac{2^{1/2}}{\pi} \int_0^\theta \frac{\cos \left(\left(n + \frac{1}{2}\right) t\right)}{\sqrt{\cos(t) - \cos(\theta)}} dt \qquad \text{(A.114)}$$

for $\theta \in (0, \pi)$, which is known also as Mehler–Dirichlet formula [3]; see [1, 2] for other integral representations.

## A.3.1  Generating Functions

Generating functions are often used to define in a compact way a sequence of functions by defining them as the coefficients of a power series. The generating function of the Legendre polynomial is

$$G(z, y) = \frac{1}{\sqrt{1 - 2zy + z^2}} \tag{A.115}$$

in fact the following identity can be proven [1]

$$\frac{1}{\sqrt{1 - 2zy + z^2}} = \sum_{n=0}^{\infty} P_n(y)z^n; \quad |z < 1|, \quad -1 < y < -1 \tag{A.116}$$

and

$$P_n(y) = \frac{1}{n!} \left( \frac{\partial^n G(z, y)}{\partial z^n} \right)_{z=0} \tag{A.117}$$

Equation (A.116) yields the following identity

$$\frac{1}{[\cosh(\eta) - y]^{1/2}} = \sqrt{2} \sum_{n=0}^{\infty} P_n(y) e^{-(n+1/2)|\eta|} \tag{A.118}$$

in fact, Eq. (A.116) can be written as

$$\frac{1}{\sqrt{\frac{1+z^2}{2z} - s}} = \sqrt{2} \sum_{n=0}^{\infty} P_n(y)z^{n+1/2} \tag{A.119}$$

and the substitution $\frac{1+z^2}{2z} = \cosh\eta$ transform the LHS of (A.119) into the LHS of (A.118); observing that equation $z^2 - \cosh(\eta)\, 2z + 1 = 0$ has two solutions: $z = e^{\pm|\eta|}$ and only that lower than one can be retained (see the conditions in (A.116)), the RHS of (A.119) transform into that of (A.118).

Multiplying both sides of (A.118) by $P_k(y)$ and integrating over the interval $\eta \in -1, 1$, yields another useful identity

$$\int_{-1}^{1} \frac{P_k(y)}{[\cosh(\eta) - y]^{1/2}} dy = \sqrt{2} \frac{2}{2k + 1} e^{-(k+1/2)|\eta|} \tag{A.120}$$

where the RHS is obtained using Eq. (A.98). Finally, derivating both side of (A.120) with respect to $\eta$ yields the following identity

$$\int_{-1}^{1} \frac{P_k(y)}{[\cosh(\eta) - y]^{3/2}} dy = \frac{2\sqrt{2}}{\sinh(|\eta|)} e^{-(k+1/2)|\eta|} \tag{A.121}$$

Equations (A.118)–(A.120) are used in Chap. 11.

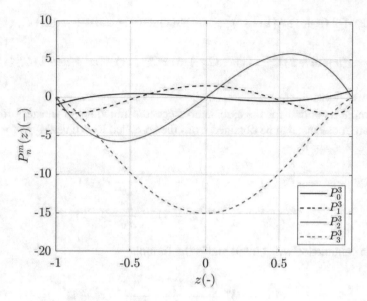

**Fig. A.9** Associated Legendre functions of the first kind $P_n^m(z)$ for $m = 3$ and $z \in [-1, 1]$

## A.3.2 Associated Legendre Functions

The Eq. (A.97) can be seen as a special case of the *generalised* or *associated* Legendre equation:

$$(1 - x^2) \frac{d^2 W}{dx^2} - 2x \frac{dW}{dx} + \left[ \nu(\nu + 1) - \frac{\mu}{1 - x^2} \right] W = 0 \qquad (A.122)$$

The two independent solutions of Eq. (A.122) are $P_\nu^\mu(x)$ and $Q_\nu^\mu(x)$, the *associated Legendre functions* of first and second kind respectively, where $\mu$ is the *order* and $\nu$ the *degree*. When restricted to the interval $x \in [-1, 1]$, $P_\nu^\mu(x)$ and $Q_\nu^\mu(x)$ are also called *Ferrers functions* of first and second kind (named after the British mathematician Norman Macleod Ferrers, 1829–1903) [1]. Clearly $P_\nu^0(x) = P_\nu(x)$, $Q_\nu^0(x) = Q_\nu(x)$.

Figure A.9 shows the graphs of four associated Legendre function of first kind.

Also these functions satisfy some important recurrence relations, and writing $C_\nu^\mu(x)$ for both $P_\nu^\mu(x)$ and $Q_\nu^\mu(x)$ the following are fundamental (see [1] for a full list of them)

$$C_\nu^{\mu+1}(x) + 2(\mu+1)x\left(1-x^2\right)_\nu^{-1/2} C_\nu^{\mu+1}(x) + (\nu-\mu)(\nu+\mu+1)C_\nu^\mu(x) = 0 \tag{A.123}$$

$$(\nu-\mu+2)C_{\nu+2}^\mu(x) - (2\nu+3)xC_{\nu+1}^\mu(x) + (\nu+\mu+1)C_\nu^\mu(x) = 0 \tag{A.124}$$

A simple definition for the associated Legendre functions of integer order and degree are those that can be obtained from the so-called Rodriguez' type formulas, i.e.

$$P_n^m(x) = (-1)^m \left(x^2-1\right)^{m/2} \frac{d^m}{dx^m} P_n(x) \tag{A.125}$$

$$Q_n^m(x) = (-1)^m \left(x^2-1\right)^{m/2} \frac{d^m}{dx^m} Q_n(x) \tag{A.126}$$

which in conjunction with (A.99) yields the formula

$$P_n^m(x) = (-1)^m \frac{\left(x^2-1\right)^{m/2}}{2^n n!} \frac{d^{m+n}}{dx^{m+n}} \left(x^2-1\right)^n \tag{A.127}$$

Again for integer order and degree the following condition holds

$$P_n^m(x) = 0 \text{ for } m > n \tag{A.128}$$

as it can be easily derived from Eq. (A.125), remembering that $P_n(x)$ is a polynomial of degree $n$.

Sometimes the function

$$\mathbf{Q}_\nu^\mu(x) = e^{-i\mu\pi} \frac{Q_\nu^\mu}{\Gamma(\mu+\nu+1)} \tag{A.129}$$

is used instead of $Q_\nu^\mu(x)$ as the second independent solution of Eq. (A.122) since, differently from $Q_\nu^\mu(x)$, it is real valued for $x \in (1, \infty)$ for all real values of $\nu$ and $\mu$ (see [1, 17]).

### A.3.3  Toroidal, Conical, Ring and Mehler Functions

The functions $P_{\lambda-1/2}^\mu(x)$, $Q_{\lambda-1/2}^\mu(x)$ are just special cases of the associate Legendre functions, but they have a special place among them since they are often found when solving problems in *spheroidal* and in *toroidal* coordinates (see Chap. 4). When $\lambda$ is an integer, $\lambda = n$, the functions $P_{n-1/2}^\mu(x)$, $Q_{n-1/2}^\mu(x)$ are also called *toroidal* functions [3] (or even *ring* functions [1]). Most of their properties are shared with the most general associate Legendre functions.

When $\lambda$ is an imaginary value, $\lambda = i\tau$, the functions $P_{i\tau-1/2}^{\mu}(x)$, $Q_{i\tau-1/2}^{\mu}(x)$ are called *Mehler* function (or even *conical* functions) [1]. In the interval $x \in -1, +1$, the functions $P_{i\tau-1/2}^{\mu}(x)$ are real valued, while in general $Q_{i\tau-1/2}^{\mu}(x)$ are complex valued.

When rotational symmetry is imposed to the solutions of the Laplace equation in toroidal coordinates, only the functions $P_{\lambda-1/2}(x)$, $Q_{\lambda-1/2}(x)$ (i.e. with $\mu = 0$) are found. The use of the Mehler functions for analytical solution is highly facilitated by the Mehler–Fock transform.

### A.3.3.1 The Mehler–Fock Transform

The Mehler–Fock transform is named after the German mathematician Gustav Ferdinand Mehler, 1835–1895, who first used for electrodynamic problems [18], and the Russian mathematician Vladimir Aleksandrovich Fock, 1898–1974, that later developed it [19]. The Mehler–Fock transform of a function $f(x)$ is defined as

$$F(\tau) = \int_1^{\infty} f(x) \, P_{i\tau-1/2}(x) \, dx \tag{A.130}$$

where $\tau \geq 0$ and the function $f(x)$ is piecewise continuous in the interval $(1, \infty)$ (for a detailed treatment of this integral transform refer for example to [20]). The inversion formula is given by

$$f(x) = \int_0^{\infty} \tau \tanh(\pi\tau) \, P_{i\tau-1/2}(x) \, F(\tau) \, d\tau \tag{A.131}$$

Effective calculation of the integral (A.130) is usually obtained through the various integral representation of the Mehler functions [20], like

$$P_{i\tau-1/2}(\cosh(\xi)) = \frac{2}{\pi} \int_0^{\xi} \frac{\cosh(\tau s)}{\sqrt{2(\cosh(\xi) - \cosh(s))}} \, ds \tag{A.132}$$

and then

$$
\begin{aligned}
F(\tau) &= \int_0^{\infty} f(\cosh(\xi)) \, P_{i\tau-1/2}(\cosh\xi) \sinh(\xi) \, d\xi = \\
&= \int_0^{\infty} f(\cosh\xi) \frac{2}{\pi} \left[ \int_0^{\xi} \frac{\cosh(\tau s)}{\sqrt{2(\cosh(\xi) - \cosh(s))}} \, ds \right] \sinh(\xi) \, d\xi = \\
&= \int_0^{\infty} \frac{2}{\pi} \left[ \int_s^{\infty} \frac{f(\cosh\xi) \sinh(\xi)}{\sqrt{2(\cosh(\xi) - \cosh(s))}} \, d\xi \right] \cosh(\tau s) \, ds = \\
&= \int_0^{\infty} \frac{2}{\pi} \left[ \int_{\cosh(s)}^{\infty} \frac{f(x)}{\sqrt{2(x - \cosh(s))}} \, dx \right] \cosh(\tau s) \, ds
\end{aligned} \tag{A.133}
$$

The effective calculation of integrals like those appearing in the last equation can be found, for example, in [21], pp. 218–226, for some functions $f(x)$.

## A.4 Spheroidal Wave Functions

Spheroidal wave functions are found when solving the Helmholtz equation in spheroidal coordinates by separation of variables. They are called *oblate* or *prolate* spheroidal wave function depending on which coordinate system the Helmholtz equation is written. In this section we will report only few information about this class of special functions, more information can be found in Chap. 30 of [1] and a complete treatment can be found in [22].

The *spheroidal wave functions* $Ps_n^m(x, \gamma^2)$ are the solution of the *spheroidal wave* differential equation

$$\frac{d}{dx}\left[(1 - x^2)\frac{dF}{dx}\right] + \left(\lambda + \gamma^2(1 - x^2) - \frac{\mu^2}{1 - x^2}\right)F = 0 \qquad (A.134)$$

with $\mu = m = 0, 1, 2, \ldots$. This equation contains three real parameters: $\lambda$, $\gamma^2$ and $\mu$, where $\gamma^2$ is a positive real number for the *prolate* spheroidal wave functions and negative for the *oblate* spheroidal wave functions.

The solution exists bounded in $x = \pm 1$ for numerable specified values of the constant $\lambda$, which are the so-called eigenvalues $\lambda_n^m(\gamma^2)$ with $n = m, m + 1, m + 2, \ldots$. A second independent solution of Eq. (A.134) is the spheroidal wave functions of 2nd kind $Qs_n^m(z, \gamma^2)$, and the general solution is

$$A\, Ps_n^m(z, \gamma^2) + B\, Qs_n^m(z, \gamma^2) \qquad (A.135)$$

To notice that for $\gamma = 0$: $Ps_n^m(x, 0) = P_n^m(x)$, $Qs_n^m(x, 0) = Q_n^m(x)$, i.e. they become the associated Legendre functions.

The eigenvalues $\lambda_n^m(\gamma^2)$ are analytic functions of $\gamma^2$, and $\lambda_m^m(\gamma^2) < \lambda_{m+1}^m(\gamma^2) < \lambda_{m+2}^m(\gamma^2)$. To notice that to each function $Ps_n^m(x, \gamma^2)$ is associated the eigenvalue (function of $\gamma^2$) $\lambda_n^m(\gamma^2)$.

### A.4.1 Angular and Radial Functions

The functions corresponding to each eigenvalue $\lambda_n^m(\gamma^2)$, i.e. $Ps_n^m(x, \gamma^2)$ are normalised such that

$$\int_{-1}^{1}\left[Ps_n^m(x, \gamma^2)\right]^2 dx = \frac{2}{2n + 1}\frac{(n + m)!}{(n - m)!} \qquad (A.136)$$

When $\gamma^2 > 0$, the functions $S_n^m (\eta, \gamma^2) = Ps_n^m (\eta, \gamma^2)$, defined on $\eta \in [-1, +1]$, are the so-called *prolate angular* spheroidal wave functions (while they are called *oblate* when $\gamma^2 < 0$); when $\gamma = 0$: $Ps_n^m (x, 0) = P_n^m (x)$.

The angular functions satisfy the following orthogonality condition

$$\int_{-1}^{1} Ps_n^m (x, \gamma^2) \, Ps_k^m (x, \gamma^2) \, dx = \frac{2}{2n + 1} \frac{(n + m)!}{(n - m)!} \delta_{nk} \qquad (A.137)$$

(to notice that $\gamma^2$ is the same in both functions).

The angular functions $S_n^m (\eta, \gamma^2) = Ps_n^m (\eta, \gamma^2)$ can be expanded in series of Legendre functions

$$S_n^m (\eta, \gamma^2) = Ps_n^m (\eta, \gamma^2) = \sum_{k=-R}^{\infty} (-1)^k a_{n,k}^m (\gamma^2) P_{n+2k}^m (\eta) \qquad (A.138)$$

$$R = \left\lfloor \frac{n - m}{2} \right\rfloor \qquad (A.139)$$

The coefficients $a_{n,k}^m (\gamma^2)$ are the solution $f_k = a_{n,k}^m$ of the equation [1]

$$A_k (n, m, \gamma^2) \, f_{k-1} + \left( B_k (n, m, \gamma^2) - \lambda_n^m (\gamma^2) \right) f_k + C_k (n, m, \gamma^2) \, f_{k+1} = 0 \qquad (A.140)$$

where the coefficients $A_k (n, m, \gamma^2)$, $B_k (n, m, \gamma^2)$, $C_k (n, m, \gamma^2)$, are given as (see again [1], Eq. 30.8.3)

$$A_k = -\gamma^2 \frac{(n - m + 2k - 1)(n - m + 2k)}{(2n + 4k - 3)(2n + 4k - 1)} \qquad (A.141a)$$

$$B_k = (n + 2k)(n + 2k + 1) - 2\gamma^2 \frac{(n + 2k)(n + 2k + 1) - 1 + m^2}{(2n + 4k - 1)(2n + 4k + 3)} \qquad (A.141b)$$

$$C_k = -\gamma^2 \frac{(n + m + 2k - 1)(n + m + 2k + 2)}{(2n + 4k + 3)(2n + 4k + 5)} \qquad (A.141c)$$

To notice that the function $\lambda_n^m (\gamma^2)$ (eigenvalue) associated to the spheroidal wave function appears linearly in Eq. (A.140).

For the spheroidal wave function of second kind, $Q_n^m (x, \gamma^2)$, the procedure is similar with some added peculiarities (see [1], Sect. 30.8).

The radial functions $R_n^m (\zeta, \gamma)$ satisfy the same Eq. (A.134) but they are defined on the interval $x \in [1, \infty)$. The radial functions can be expanded in series of spherical bessel functions

$$R_n^m(z, \gamma) = \frac{(1 - z^{-2})^{m/2}}{A_n^{-m}(\gamma^2)} \sum_{2k > m-n} a_{n,k}^{-m}(\gamma^2) \, j_{n+2k}(\gamma z) \tag{A.142}$$

$$a_{n,k}^{-m}(\gamma^2) = \frac{(n-m)! \, (n+m+2k)!}{(n+m)! \, (n-m+2k)!} a_{n,k}^m(\gamma^2) \tag{A.143}$$

$$A_n^{\pm m}(\gamma^2) = \sum_{2k \geq \mp m-n} (-1)^k \, a_{n,k}^{\pm m}(\gamma^2) \tag{A.144}$$

(for the functions of second kind, the spherical Bessel functions of second type $y_{n+2k}(\gamma z)$ are used). The connection between $R_n^m(z, \gamma^2)$ and $Ps_n^m(z, \gamma^2)$ is given by the following relation (see [1], Eq. 30.11.8)

$$R_n^m(z, \gamma) = K_n^m(\gamma) \, Ps_n^m(\zeta, \gamma^2) \tag{A.145}$$

$$K_n^m(\gamma) = \frac{\sqrt{\pi}}{2} \frac{(-1)^m}{A_n^{-m}(\gamma^2)} \left(\frac{\gamma}{2}\right)^m \begin{cases} \dfrac{a_{n,(m-n)/2}^{-m}(\gamma^2)}{\Gamma(m+3/2) Ps_n^m(0,\gamma^2)} & \text{for } (n-m) \text{ even} \\[2ex] \dfrac{\gamma}{2} \dfrac{a_{n,(m-n+1)/2}^{-m}(\gamma^2)}{\Gamma(m+5/2)\left(\frac{d Ps_n^m(z,\gamma^2)}{dz}\right)_{z=0}} & \text{for } (n-m) \text{ odd} \end{cases} \tag{A.146}$$

When Eq. (A.134) is obtained from a separation of variable approach to the solution of Helmholtz equation in spheroidal coordinate, assuming rotational symmetry, the parameter $\mu$ is nil, $\mu = 0$, and the angular functions $S_n^0(\eta, \gamma^2) = Ps_n^0(\eta, \gamma^2)$ can be expanded in terms of Legendre polynomials

$$S_n^0(\eta, \gamma^2) = Ps_n^0(\eta, \gamma^2) = \sum_{k=-R}^{\infty} (-1)^k \, a_{n,k}^0(\gamma^2) \, P_{n+2k}(\eta) \tag{A.147}$$

With the substitution $\zeta = \gamma x$, Eq. (A.134) can also be written as

$$(\zeta^2 - \gamma^2) \frac{d^2 F}{d\zeta^2} + 2\zeta \frac{dF}{d\zeta} + \left(\zeta^2 - \lambda - \gamma^2 - \frac{\gamma^2 \mu^2}{\zeta^2 - \gamma^2}\right) F = 0 \tag{A.148}$$

$$\frac{d}{d\zeta}\left[(\zeta^2 - \gamma^2)\frac{dF}{d\zeta}\right] + \left(\zeta^2 - \lambda - \gamma^2 - \frac{\gamma^2 \mu^2}{\zeta^2 - \gamma^2}\right) F = 0 \tag{A.149}$$

## A.5    The Hypergeometric Function

The term *hypergeometric series* has been used with different meanings; the first use of this term seems to be due to Wallis (John Wallis, an English mathematician, 1616–1703) in 1656 [23] to refer to a sort of generalization of the geometric series. The first systematic study is due to Gauss [24] and that is why this special function is sometimes called *Gauss hypergeometric function*, and a characterisation in terms of solution of a differential equation is due to Riemann [25].

The hypergeometric function $_2F_1 (a, b, c, z)$ is a function of four parameters, and it is symmetric with respect to the first two, as it stems from its definition by the series

$$_2F_1 (a, b, c, z) = \sum_{n=0}^{\infty} \frac{(a)_n (b)_n}{(c)_n n!} z^n \tag{A.150}$$

and its analytical continuation (see [1]); $(a)_n$ is the Pochhammer symbol (named after the Prussian mathematician Leo August Pochhammer, 1841–1920) defined as

$$(a)_n = a (a + 1) (a + 2) \dots (a + n - 1) = \frac{\Gamma (a + n)}{\Gamma (a)} \tag{A.151}$$

for $n$ integer $((a)_0 = 1)$. The function is undefined for $c = 0, -1, -2, \dots$

A peculiarity of this functions, which was pointed out by Gauss [24] (see also [26]) is that practically any transcendental function that appears in analysis may be obtained as a special case of the hypergeometric series, and in fact the following identities can be proven

$$_2F_1 \left( \frac{1}{2}, 1, \frac{3}{2}, z^2 \right) = \frac{\ln \left( \frac{1+z}{1-z} \right)}{2z} \tag{A.152}$$

$$_2F_1 \left( \frac{1}{2}, 1, \frac{3}{2}, -z^2 \right) = \frac{\arctan (z)}{z} \tag{A.153}$$

$$_2F_1 \left( \frac{1}{2}, \frac{1}{2}, \frac{3}{2}, z^2 \right) = \frac{\arcsin (z)}{z} \tag{A.154}$$

$$_2F_1 \left( \frac{1}{2}, \frac{1}{2}, \frac{3}{2}, -z^2 \right) = \frac{\ln \left( z + \sqrt{1 + z^2} \right)}{z} \tag{A.155}$$

$$_2F_1 (a, b, b, z) = (1 - z)^{-a} \tag{A.156}$$

$$_2F_1 \left( -a, a, \frac{1}{2}, \sin^2 (z) \right) = \cos (2az) \tag{A.157}$$

$$_2F_1 (a, 1 - b, 1 + a, z) = a z^{-a} B (z, a, b) \tag{A.158}$$

$$_2F_1 \left( 1 + \nu, -\nu, 1, \frac{1-z}{2} \right) = P_\nu (z) \tag{A.159}$$

as well as many others (see [1] and also [21]).

The hypergeometric function is a solution of the hypergeometric differential equation

$$z (1 - z) \frac{d^2 W}{dz^2} + [c - (a + b + 1) z] \frac{dW}{dz} - abW = 0 \tag{A.160}$$

Precisely, there are different pairs of solutions that are so-called *numerically satisfactory*, i.e. a pair such that in a neighborhood of a singularity (i.e. points where the coefficients of the linear differential equation have poles) one member becomes dominant and the other recessive, so to avoid numerical cancellation of terms in the linear combination of the two solutions $AW_1 + BW_2$ (see [1], Sect. 2.7). For the hypergeometric equation those points are $z = 0, 1, \infty$ and the three pair of solutions are

$$
\begin{aligned}
W_1 &= {}_2F_1\,(a, b, c, z) \\
W_2 &= z^{1-c}\,{}_2F_1\,(a - c + 1, b - c + 1, 2 - c, z)
\end{aligned}
\qquad \text{around } z = 0
$$

$$
\begin{aligned}
W_1 &= {}_2F_1\,(a, b, a + b + 1 - c, 1 - z) \\
W_2 &= (1 - z)^{c-a-b}\,{}_2F_1\,(c - a, c - b, c - a - b + 1, 1 - z)
\end{aligned}
\qquad \text{around } z = 1
$$

$$
\begin{aligned}
W_1 &= z^{-a}\,{}_2F_1\left(a, a - c + 1, a - b + 1, \tfrac{1}{z}\right) \\
W_2 &= z^{-b}\,{}_2F_1\left(b, b - c + 1, b - a + 1, \tfrac{1}{z}\right)
\end{aligned}
\qquad \text{around } z = \infty
$$

$$\text{(A.161)}$$

There exists other pair of independent solution, up to a total of twenty-four, known as Kummer's solutions (named after the German mathematician Ernst Eduard Kummer, 1810–1893) [1]). For further properties and details the interested reader should refer to [1, 3].

# References

1. Olver, F.W., Lozier, D.W., Boisvert, R.F., Clark, C.W. (eds.): NIST Handbook of Mathematical Functions. Cambridge University Press, Cambridge (2010)
2. Abramowitz, M., Stegun, I.A.: Handbook of Mathematical Functions with Formulas, Graphs, and Mathematical Tables, 10th edn. Dover Publication Inc., New York (1972)
3. Lebedev, N.N.: Special Functions and Their Applications. Dover, New York (1972)
4. Artin, E.: The Gamma Function. Holt, Rinehart and Winston, New York (1964)
5. Bonnar, J.: The Gamma Function. Createspace Independent Publishers, Scotts Valley (2014)
6. Fuss, P.H.: Correspondance mathématique et physique de quelques célèbres géomètres du XVIIIeme siècle St. Pétersbourg (in French), vols. 1 and 2 (1843)
7. Legendre, A.M.: Mémoires de la classe des sciences mathématiques et physiques de l'Institut de France, Paris (in French) (1809)
8. Cajori, F.: A History of Mathematical Notations. Dover, New York (1993)
9. Jacques, P.M.: Binet mémoire sur les intégrales définies Eulériennes et sur leur application à la théorie des suites; ainsi qu'à l'évaluation des fonctions des grands nombres (in French). Journal de l'École Polytechnique **16**, 123–343 (1839)
10. Fagnano, G.C.: Produzioni Matematiche (in Italian), Stamperia Gavelliana-Pesaro (1750)
11. Bernoulli, D.: Theoremata de oscillationibus corporum filo flexili connexorum et catenae verticaliter suspensae (in Latin). Comm. Acad. Sci. Imp. Petrop. (**1732–1733**), 108–122 (1738)
12. Watson, G.N.: Treatise on the Theory of Bessel Functions. Cambridge University Press, Cambridge (1962)
13. Weber, H.M.: Ueber eine Darstellung willkürlicher Functionen durch Bessel'sche Functionen (in German). Math. Ann. **6**(2), 146–161 (1873)
14. Legendre, A.M.: Reserches sur l'attraction des sphéroïdes homogènes. Mémoires de mathématique et de physique Académie Royale des Sciences (in French) **10**, 411–434 (1785)
15. Laden, H.N.: An historical and critical development of the theory of Legendre polynomials before 1900, MA thesis University of Maryland (1938)
16. Todhunter, I.: An Elementary Treatise on Laplace's, Lamé's, and Bessel's Functions (in French). MacMillan, London (1875)

17. Olver, F.W.J.: Connection formulas for second order differential equations having an arbitrary number of turning points of arbitrary multiplicities. SIAM J. Math. Anal. **8**(4), 673–700 (1977)
18. Mehler, F.G.: Ueber eine mit den Kugel- und Cylinderfunctionen verwandte Function und ihre Anwendung in der Theorie der Electricitätsvertheilung (in German). Math. Ann. **18**, 161–194 (1881)
19. Fock, V.A.: On the representation of an arbitrary function by an integral involving Legendre functions with complex index (In Russian). Dokl. Akad. Nauk SSSR **39**, 253–256 (1943)
20. Ditkin, V.A., Prudnikov, A.P.: Integral Transforms and Operational Calculus. Pergamon Press, Oxford (1965)
21. Prudnikov, A.P., Brychkov, Y.A., Marichev, O.I.: Integrals and Series, vol. 3. Gordon & Breach Science Publishers, New York (1990)
22. Flammer, C.: Spheroidal Wave Functions. Stanford University Press, Stanford (1957)
23. Dutka, J.: The early history of the hypergeometric function. Arch. Hist. Exact Sci. **31**, 15–34 (1984)
24. Gauss, C.F.: Disquisitiones generales circa seriem $1 + \frac{\alpha\beta}{1.\gamma}x + \frac{\alpha(\alpha+1)\beta(\beta+1)}{1.2\gamma(\gamma+1)}x^2 + \frac{\alpha(\alpha+1)(\alpha+2)\beta(\beta+1)+\beta+2}{1.2.3\gamma(\gamma+1)(\gamma+2)}x^3 +$ . . . etc. pars prior. Commentationes Societatis Regiae Scientiarum Gottingensis (in Latin), Göttingen (1812)
25. Riemann, B.: Beiträge zur Theorie der durch die Gauss'sche Reihe F($a$, $\beta$, $\gamma$, $x$) darstellbaren Functionen. Abhandlungen der Mathematischen Classe der Königlichen Gesellschaft der Wissenschaften zu Göttingen (in German), vol. 7, pp. 3–22. Verlag der Dieterichschen Buchhandlung, Göttingen (1857)
26. Papadopoulos, A.: Looking backward: from Euler to Riemann. In: Ji, L., Papadopoulos, A., Yamada, S. (eds.) From Riemann to Differential Geometry and Relativity. Springer, Cham (2017)

Printed in the United States
by Baker & Taylor Publisher Services